Agricultural Economic Transformation and Sustainable Development

Agricultural Economic Transformation and Sustainable Development

Guest Editors

Fotios Chatzitheodoridis

Efstratios Loizou

Achilleas Kontogeorgos

Basel • Beijing • Wuhan • Barcelona • Belgrade • Novi Sad • Cluj • Manchester

Guest Editors

Fotios Chatzitheodoridis	Efstratios Loizou	Achilleas Kontogeorgos
Department of Management Science and Technology	Department of Management Science and Technology	Department of Agriculture
University of Western Macedonia	University of Western Macedonia	International Hellenic University
Kozani	Kozani	Thessaloniki
Greece	Greece	Greece

Editorial Office
MDPI AG
Grosspeteranlage 5
4052 Basel, Switzerland

This is a reprint of the Special Issue, published open access by the journal *Sustainability* (ISSN 2071-1050), freely accessible at: https://www.mdpi.com/journal/sustainability/special_issues/Q121BPS99H.

For citation purposes, cite each article independently as indicated on the article page online and as indicated below:

Lastname, A.A.; Lastname, B.B. Article Title. *Journal Name* **Year**, *Volume Number*, Page Range.

ISBN 978-3-7258-3515-7 (Hbk)
ISBN 978-3-7258-3516-4 (PDF)
https://doi.org/10.3390/books978-3-7258-3516-4

Contents

About the Editors

Fotios Chatzitheodoridis

Prof. Fotios Chatzitheodoridis is a Full Professor of Regional Economics in Rural Development and Policy in the Department of Management Science and Technology, University of Western Macedonia (Greece). He holds a Ph.D. (1998) in Rural Sustainable Development from Aegean University and a BSc in Economics (with a major in Development and Programming) (1991) from the Aristotle University of Thessaloniki. His main research interests include sustainable and integrated development, regional planning, environmental economics, and policy analysis. His technical experience has specialized in developing and implementing EU-funded projects at local and national levels. He has been involved in several European and national studies and co-financed projects. In the past, he worked in the Greek Ministry of Rural Development and Foods, at the Aristotle University of Thessaloniki, at the University of Central Greece, and in several local development companies advancing diverse research.

Efstratios Loizou

Efstratios Loizou has been a Professor of Applied Economics in the Department of Management Science and Technology at the University of Western Macedonia, Greece, since 2023. He was previously involved in the Departments of Regional Development and Cross-Border Studies (2019–2023) and the Division of Agricultural Economics, the Department of Agriculture, which he left in 2019. In 2018, on sabbatical, he served for 1 year as a senior researcher in the ERA chair at the Department of Bioeconomy and Systems Analysis, Institute of Soil Science and Plant Cultivation (IUNG), in Poland. He is a graduate of the Department of Economics at the University of Macedonia, Greece (1992), and holds an MSc. in Agricultural Economics from the Mediterranean Agronomic Institute of Chania (1995). In 2001, he received a PhD in Agricultural Economics from the Aristotle University of Thessaloniki, the Department of Agricultural Economics, the School of Agriculture. He is active in the fields of agricultural, rural, and regional economics with expertise in input–output analysis, regional–rural modeling, inter-industry linkages and impact analysis, rural and regional development, common agricultural policy, the agri-food sector and food industries, precision agriculture, and innovations in agriculture, energy, and the bioeconomy.

He has served, since 1997, as a member of several national and EU projects (Horizon, ERASMUS+, FP, INTERREG, TEMPUS, and DG AGRI TENDERS), mainly specializing in CAP program evaluation, impact analysis at the regional level, the bioeconomy, etc. He has published manuscripts in several referenced journals, conference volumes, collective volumes, research reports, and conferences.

Achilleas Kontogeorgos

Dr. Achilleas Kontogeorgos is a professor, specializing in "Quality Management in Agriculture", in the Department of Agriculture at the International University of Greece. He holds a PhD in Agricultural Economics from the School of Agriculture at the Aristotle University of Thessaloniki, where he studied Economics and Quality Management at the postgraduate level. He has taught undergraduate and postgraduate courses on agricultural economics and policy, the cooperative economy, and quality management, in addition to courses with a focus on entrepreneurship in the primary sector. He has published more than 80 research papers in international scientific journals and scientific conferences on the management of agricultural enterprises. He has also participated in several research projects focusing on advancing the quality of agricultural products.

Preface

Sustainability has emerged as one of the most critical issues of our generation, requiring coordinated efforts across multiple disciplines to address the complex interplay between economic growth, environmental health, and societal equity. This Special Issue brings together contributions from leading experts, researchers, and practitioners, providing an in-depth inquiry into innovative solutions and strategies for sustainable development.

The manuscripts presented in this reprint cover a wide range of topics, including advancements in renewable energy, sustainable urban planning, climate resilience, and the circular economy. Through attentive analysis and thoughtful discussion, these studies highlight the significance of interdisciplinary collaboration and innovation in tackling such global challenges.

We extend our gratitude to all authors, reviewers, and editorial staff who have contributed to the success of this Special Issue. Your expertise and dedication have ensured the excellent quality and relevance of the articles presented herein. We also acknowledge the invaluable support of the academic community and funding agencies, whose contributions facilitated the research featured in this reprint.

This Special Issue is intended for scholars, practitioners, and policymakers who are dedicated to fostering sustainable development. We hope that the insights and innovations presented within this reprint inspire meaningful action and drive progress toward a more sustainable future.

Fotios Chatzitheodoridis, Efstratios Loizou, and Achilleas Kontogeorgos
Guest Editors

 *sustainability*

Article

Integration of Water Resources Management Strategies in Land Use Planning towards Environmental Conservation

Stavros Kalogiannidis [1,*], Dimitrios Kalfas [2,*], Grigoris Giannarakis [1] and Maria Paschalidou [3]

1. Department of Business Administration, University of Western Macedonia, 51100 Grevena, Greece; ggiannarakis@uowm.gr
2. Department of Agriculture, Faculty of Agricultural Sciences, University of Western Macedonia, 53100 Florina, Greece
3. Department of Regional and Cross-Border Development, University of Western Macedonia, 50100 Kozani, Greece; drdcbs00008@uowm.gr
* Correspondence: aff00056@uowm.gr (S.K.); kalfdimi@otenet.gr (D.K.)

Abstract: Water resources management is a critical component of environmental conservation and sustainable development. This study examines the integration of water resources management strategies into land use planning and its impact on environmental conservation, with a focus on the case of Greece. This study employed a quantitative research methodology using a cross-sectional survey research design. The target population consisted of environmental experts in Greece, and a sample of 278 participants was selected based on the Krejcie and Morgan table for sample size determination. Data were collected through an online survey questionnaire, and the statistical analysis was conducted using SPSS version 23. The relationships between the study variables were examined through regression analysis. The findings support the hypotheses, demonstrating the importance of integrating water resources management strategies into land use planning to achieve both sustainable development and environmental conservation. This paper discusses various strategies and approaches that can be adopted to effectively manage water resources while considering the impacts of land use decisions on the environment. Better public awareness and better enforcement of water conservation rules result from this integration, which makes it possible for land use authorities and water management agencies to collaborate more effectively. This study acknowledges the need for strategic planning and cooperation between water management and land use authorities to address the growing challenges of water resources management and environmental protection. Emphasizing stakeholder participation, adaptive management, and continuous monitoring can lead to successful outcomes and a more resilient and sustainable future.

Keywords: sustainable development; water resources management; land use; nature conservation; mitigation of climate change; Greece

Citation: Kalogiannidis, S.; Kalfas, D.; Giannarakis, G.; Paschalidou, M. Integration of Water Resources Management Strategies in Land Use Planning towards Environmental Conservation. *Sustainability* **2023**, *15*, 15242. https://doi.org/10.3390/su152115242

Academic Editor: Teodor Rusu

Received: 16 September 2023
Revised: 22 October 2023
Accepted: 24 October 2023
Published: 25 October 2023

1. Introduction

1.1. Water Resources Management and Land Use Planning

Water is a finite resource essential for various human activities and the health of ecosystems [1,2]. The ever-increasing demands for water due to population growth, urbanization, and industrialization have raised concerns about water scarcity and environmental degradation. Integrating water resources management (WRM) strategies into land use planning has emerged as a crucial approach to address these challenges [3–5]. Water resources are vital for ecosystem function and the well-being of human societies. However, growing demands of urbanization, agriculture, and industrialization pose significant challenges to the sustainable management of water resources [6,7]. Land use planning, on the other hand, plays a pivotal role in shaping the physical development of regions and can greatly influence the quality and availability of water resources. Integrating water resources management strategies into land use planning processes can foster more holistic and effective approaches to environmental conservation [8].

The increasing demands for water resources from various sectors, such as agriculture, industry, and urban development, often lead to over-extraction and depletion of water sources [3,9–11]. Additionally, poor land use decisions, such as deforestation and urban sprawl, can exacerbate the degradation of water bodies, leading to reduced water quality and availability [1]. Climate change further intensifies these challenges, causing irregular precipitation patterns and exacerbating droughts and floods. The fragmented nature of water resources management and land use planning can hinder effective conservation efforts [12–15].

Mengistu et al. (2023) noted that when water agencies take into account land management strategies, they may better manage the land use origins of their water problems, such as stormwater concerns from impermeable surfaces or groundwater over pumping [7,16]. When water management organizations work with land use planners, they have access to additional compliance options [17–19]. Research, for instance, shows that imposed irrigation restrictions during droughts may significantly reduce total water use, but voluntary measures are not nearly as successful [20,21]. It is difficult for a water management group operating alone to enforce such restrictions. Working with a land use authority may help a water agency become much more effective since it can codify a policy, help with enforcement, and raise public awareness [22–24]. A water management agency may also consider land management strategies, such as landscape-scale conservation strategies and site-scale green stormwater infrastructure, to address difficulties with source water protection, flood control, and water quality [25–27].

The relationship between water resources management and land use planning is inherently interconnected [28]. Land use decisions directly influence the availability and quality of water resources, while water availability and quality significantly impact land use options [7,29]. The integration of water resources management strategies in land planning is to achieve a stable state both to meet human needs and conserve natural ecosystems. This paper emphasizes the importance of adopting an integrated approach to ensure the sustainable use and conservation of water resources.

1.2. Purpose of the Study

The study focused on assessing the integration of water resources management strategies in land use planning towards environmental conservation concerning a case study in Greece.

1.3. Objectives of the Study

1. To examine the effect of land use planning on environmental conservation.
2. To evaluate the relationship between water resources management strategies and level of environmental conservation.
3. To examine the benefits of integration of water resources management strategies in land use planning on environmental conservation.

1.4. Research Hypotheses

Hypothesis 1 (H1). *Aspects of land use planning have an effect on environmental conservation.*

Hypothesis 2 (H2). *There is a relationship between water resources management strategies and level of environmental conservation.*

Hypothesis 3 (H3). *Benefits of integration of water resources management strategies in land use planning have an impact on environmental conservation.*

1.5. Significance of the Study

The findings of this study hold considerable significance due to its potential to address crucial challenges related to sustainable development, environmental protection, and water resources management.

2. Literature Review

2.1. Water Resources Management

The conventional sectoral approach to managing water resources has made it difficult to properly address issues with supply and demand for diverse water users or the cumulative impact of various land use activities on water quality [30]. Over the past 30 years, the integrated water resources management (IWRM) method has been pushed as a superior and more successful alternative to the conventional sectoral and top-down strategy to halt the deterioration of water resources [1,8,31]. Despite thousands of articles on the subject of IWRM, the validity of evidence for effective IWRM techniques is still being debated. The definition of IWRM contains some of the responses to this query [32]. International water resources management (IWRM) is defined as "the integrated and coordinated management of both land and water resources as a way of balancing resource conservation with addressing social, ecological, and economic development demands" [6].

New issues are developing that can impact availability and demand, as well as the quality of the water, in addition to the present needs for water and the effects of pollution [8]. These include increased climatic variability, intensifying land use (in both urban and agricultural settings), new water uses like those connected to fracking, increased desalinization, water for making snow, atmospheric changes, and a resurgence of interest in hydropower generation as a substitute for electricity generation that emits greenhouse gases [20,33]. Only lately have we begun to address the issue of how to sustain water for environmental services, as most water management has traditionally been human-centric [34]. The way we utilize, and safeguard water resources will alter significantly as a result of these trade-offs. Meanwhile, we need to pay attention to the importance of the green water cycle and more specifically to the effect of rainwater falling on land and plants before the phenomenon of its evapotranspiration to the atmosphere [28,35]. The blue water cycle, or the fraction of rainfall that ends up in rivers, lakes, and groundwater, is the current focus of most human operations. Up to 65% of all precipitation went via the green water cycle, whereas only 33% went through the blue water cycle [2].

Pacheco-Vega (2020) noted that due to the fact that various jurisdictions have varied legislation, especially in international trans-boundary watersheds, applying integrated water resources management (IWRM) at large river basin sizes is particularly challenging [36,37]. In addition to making data integration challenging, this also raises ethical and moral issues [38,39]. Processes also switch to unique scales, which is more significant [40,41]. By contrast, lowland systems generally re-mobilize and carry accumulated sediments that have been deposited in the riverbed over time. For instance, soil erosion in head water systems immediately joins streams [42]. A headwater basin problem's downstream effects frequently take a while to become apparent, and the gap between the damage and the solution is frequently too wide to draw any conclusions [43]. The situation of hydroelectric dams that accumulate sediments above the dam is an excellent illustration of this [7]. The majority of the phosphorus that is accessible is absorbed by the sediments, depriving the downstream area of this crucial nutrient for maintaining aquatic biota [22,44,45].

Policy and management practices of this process require long-term commitment on the part of researchers so as to withdraw conclusions about the level of efficiency and success of such a process. The accounting and regulation of non-point sources (NPS) of pollution is particularly challenging [46]. The creation of a fair legal foundation for NPS is extremely difficult. Because it is believed that best management practices (BMP) are solving the issue, long-term financing commitments from government agencies are needed to enable the monitoring of the efficacy of management [1].

The integrated water resources planning (IWRP) process is shown graphically in Figure 1. It includes public input as a crucial step to make sure that different viewpoints are taken into account and addressed. It also considers future demands for water, current and future supply options, and economic approaches to meet water objectives. The IWRP process for a community should incorporate land use components, much like a community comprehensive master plan should do. This can happen at the demand forecasting stage, the water supply planning stage, or even the public process stage. When assessing water management plan choices, the IWRP's comprehensive and adaptable approach makes it possible to take into account a wide range of variables [31].

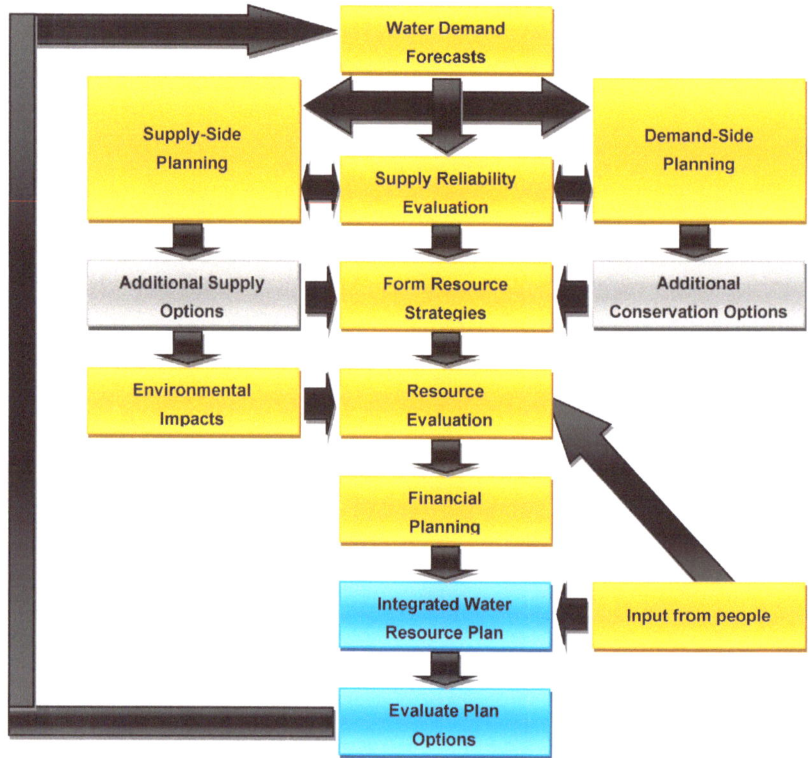

Figure 1. Integrated water resources planning/management.

Sang and Ode Sang (2015) noted that the provision of safe drinking water, wastewater services, and/or stormwater management are some of the most important services performed by water management organizations [47]. Making plans is essential to their success. Water management companies are often reluctant to impose rules on their clients since they are service organizations [48]. Water management organizations may provide water conservation initiatives; however, they are often optional or paid for rather than mandated. Making contact with land use planners may provide access to knowledge on public involvement and engagement, as well as policies to better water management [31]. A water agency might decide to voluntarily work with a land use authority to draft its plan even though a state does not have such a legislative mandate. In reality, any water management organization may benefit from and adopt the strategies adopted by governments with legislative obligations [8,10]. Additionally, they may devise unique strategies for integrating land use into their water management plans. The goal of proving the link points in legislation is to prove that land use issues are already, as legally required, integrated into water management plans [19,29].

2.2. Land Use Planning

Although the land use planning system offers certain regulatory procedures for altering land use, these are currently not frequently employed [28,33]. Because the planning system is not designed for substantial natural resource decision making, there is a huge gap between theoretically possible sustainable land use regulation and actual implementation [49]. According to Kalfas et al. (2023), making minor adjustments to the planning process will not result in significant changes to land use [31]. The planning system, according to them, only ostensibly acknowledges the need for change, and the instruments for land use planning that are offered cannot provide the desired results [50–52]. To effectively ensure that inappropriate land use practices are avoided, changes must be made to the land use planning process. To make natural resources management decision making a priority, these changes necessitate reforms to the institutional, regulatory, and policy frameworks [2,53].

The formation of suitable zones and overlays, as well as the creation of strategic statements outlining the intended use throughout a local government area, are ways in which the land use planning system may limit changes to land use [43]. Outside of the Melbourne metropolitan region, there is a desire for small country lifestyle lots, and zones and overlays can help to regulate their growth [6]. For instance, the zoning regulations might regulate how rural residential subdivision lots are created, trying to reduce conflicts where these mostly residential lots meet agricultural property. A covenant imposing catchment objectives, such as plant cover, may be included in any new lots that are produced, depending on how many of these lots the local government decides to build [54,55].

The multidisciplinary method of sustainable land use development is a well-organized set of linked procedures that may be altered depending on the nature and scope of the research area [56]. The technique focuses on conflict based decision making processes and subsequent proposals for harmony in the supply of landscape features as natural capital and complex natural resources and the demands and impacts of human activities (Figure 2).

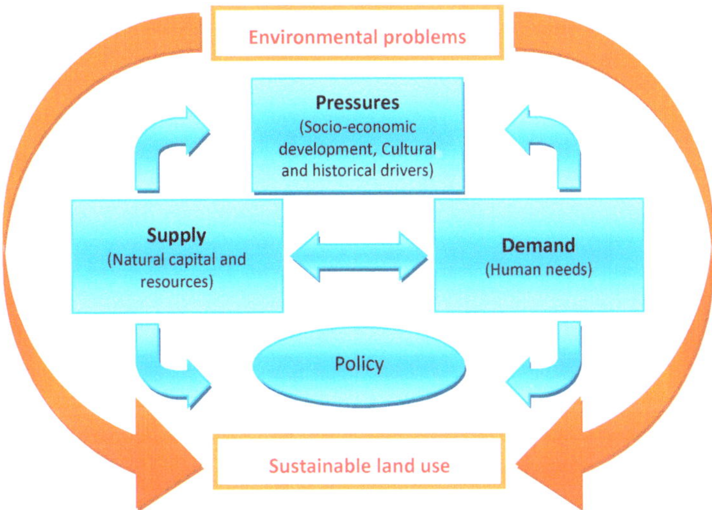

Figure 2. Integrated approach to sustainable land use management.

2.3. Integrated Land Use and Water Management Planning

Water resources management involves the sustainable use, allocation, and protection of water sources. Land use planning, on the other hand, involves the allocation and regulation of land for various purposes, such as residential, commercial, agricultural, and conservation areas. The integration of these two fields aims to ensure that land use decisions consider the impact on water availability, quality, and ecosystem health [57]. Owing to varying state mandates or owing to local discretion, water planning and land use authorities may function on distinct timetables. Therefore, local agencies perform best when they develop durable policies that guarantee continued cooperation. Setting up local processes that self-perpetuate and can endure shifting government should be one objective of integrated planning [13,42].

Communities practicing integrated planning, for instance, have to include coordinated growth reviews between planners and employees from water management organizations [48,58]. Without coordination, local communities that engage in integrated planning run the risk of being disrupted if staff turnover, changes in local government, or competing local agendas make integrated planning less urgent. State and federal water management programs may benefit from integrated planning between a water management agency and a land use authority [28]. The Urban Water Management Planning Act of California, the Aquifer Protection Program of Connecticut, and Western guaranteed water supply legislation have all been made more easily implementable because of coordinated planning [8,30,59].

Piemontese (2020) noted that communities must plan for integrated land use and water management in order to address the interconnected issues of climate change, population expansion, and diminishing water resources [2]. In order to make it easier to grasp both national trends and potential areas for growth, Mengistu et al. (2023) analyzed the possibilities for integrated planning with a special emphasis on comprehensive planning and water management planning [7]. The relationship between land use and water management may be strengthened by action at all levels, including local implementation, regional cooperation, and statewide legislation [60–62].

The cooperation of the involved stakeholders, which in this case include water management organizations and local planning agencies, is considered essential for the implementation of integrated land and water use planning. Councils and governing bodies from other localities may provide the leadership and assistance that is essential for success [2,46]. The public, developers, corporations, and nonprofit groups are important stakeholders that planners should include in their planning processes. In order to close the gap between drinking water, wastewater, and stormwater services, local water management organizations may also take on the additional task of holistically coordinating water management [1,20]. Similarly, it may be necessary for land use planners to collaborate with several local water management organizations. Long-term success requires developing structured procedures for integrating land use planning and water management [1,8,63].

In terms of improper management of water resources, wetlands are often turned into agricultural land by draining and the construction of embankments, which results in a significant loss of ecosystem goods and services that help to clean and regulate surface and subsurface waterways [17,64]. In addition to reducing biological diversity, conversion—often to irrigated agriculture and in deltas to significant crops like rice—hinders groundwater recharge, taints downstream and subsurface drinking water sources, and worsens the quality of coastal waterways. Even shoddily built irrigation canals and ditches lose water, resulting in waterlogged soils and decreased yield. Consumptive irrigation practices, as well as irrigation-related dams and barrages, waste valuable water, destroy coastal ecosystems downstream, pollute estuaries, and alter salt levels that harm coastal residents and their fisheries [36,37].

2.4. Environmental Conservation

The term "environment" refers to the whole of a living organism's surroundings, including natural forces and other living entities, which provide opportunities for development and growth as well as risks and harm [20,65]. A physical or social component might make up the environment. The constructed environment, natural environment, weather, water, land, atmosphere, etc. are all considered to be a part of the physical environment [62,66]. The environment is dynamic and flexible. In other words, the term "environment" may be used to refer to all interactions between humans and the land, water, and air. It encompasses every aspect of the natural and biological environment, as well as how they interact [8]. The environment and the organism are in constant communication. For instance, the earth's atmosphere, which contains the gases oxygen and carbon dioxide, is crucial to the ecosystem and to life as we know it. These gases exist as a result of living things acting upon them and are a prerequisite for life. Another example is the connection between dirt and vegetarians [67]. The relationship between humans and their surroundings is a particularly perplexing issue. The cultural environment must be taken into account in the case of man in addition to the biological and physical environment [30].

The practice of protecting the environment entails the wise management of our natural resources. This implies that we can utilize the resources, but only properly and intelligently [47,68]. Examples include recycling, conserving trees, cutting down on trash, and utilizing renewable resources instead of depleting our natural resources. Additionally, it mandated that all natural resources be owned collectively [8]. The management of all the earth's territory should ensure the long-term existence of humans as an ecosystem component. Since there is only one planet, we must take great care to avoid destroying the potential of the natural world. This tends to lead to the creation and adoption of a global communication strategy by many nations [1].

The idea of environmental management is one that is always evolving and changing. It mostly has to do with managing the environment that surrounds a company or activity. In general, it depicts the organizational layout, responsibility hierarchies, procedures, and prerequisites for putting environmental business policy into practice [20]. Setting goals and evaluating progress, managing information and communications, and assisting in decision making are the main responsibilities of effective environmental management. Internal and external audits of different projects and their execution are also included in environmental management [1].

In the European Community, environmental protection and sustainable development policies are a crucial part of the long- and medium-term strategy that is the foundation of the region's long-term growth [30,69]. Given this environmental program, it is possible to conclude that the European Union is advancing environmental goals outside of the borders of its member states. This fact alone makes the community's policies more effectively disseminated in order to achieve sustainable development [70]. The primary goals of EU policies include the preservation of the environment via the use of economic and legal tools, as well as the deployment of suitable countermeasures to pollution. Based on technical and scientific evidence and taking into account the actual environmental situations in the various E.U. areas, the European Community is developing and promoting its environmental policies [7]. The European Parliament has fiercely established itself as a co-legislator with expanded authority in the domain of environmental protection after the implementation of the Treaty of Lisbon, exercising democratic oversight over all European institutions. In terms of international cooperation and legislation that take on a worldwide or cross-border character, environmental protection against the major issue of global warming is presently a top priority [20].

There is no one strategy to address the challenge of environmental protection, but combining existing possibilities and increasing efficiency across all social and economic sectors of states would help to address the issue of resources and distribution [32,71,72]. Today, international legislation and cooperation are focused mostly on environmental protection, which has a global or cross-border scope. Global efforts for prevention are required because of the environmental issues' permanence, purpose, and intertemporal character [20].

3. Methodology
3.1. Research Design

The study used a quantitative research methodology and a cross-sectional survey research design. The cross-sectional research strategy relies on a thorough investigation of a group or event in order to unearth the roots of numerous fundamental approaches related to the research topic or study subject. Because of the cross-sectional research design, it was simple to focus on integrating water resources management techniques in land use planning for environmental protection.

3.2. Target Population

The study focused on a variety of environmental specialists in Greece. This community provided the most appropriate sample for the study to learn how to integrate water resources management methods into land use planning for environmental protection.

3.3. Sample Size and Sampling Technique

Using the table from Krejcie and Morgan (1970), the optimal sample size from the population will be calculated. Krejcie and Morgan (1970) developed a table for calculating sample size for a certain population as presented in Table 1. Based on the target population of 1000 participants, a corresponding sample size of 278 participants as per Krejcie and Morgan (1970) was used for this study [73]. A purposive sampling technique was used to select the representative sample for the study.

Table 1. Table for determining sample size from a given or known population.

N	n	N	n	N	n
10	10	220	140	1200	291
15	14	230	144	1300	297
20	19	240	148	1400	302
25	24	250	152	1500	306
30	28	260	155	1600	310
35	32	270	159	1700	313
40	36	280	162	1800	317
45	40	290	165	1900	320
50	44	300	169	2000	322
55	48	320	175	2800	338
60	52	340	181	3000	341
65	56	360	186	3500	346
70	59	380	191	4000	351
75	63	400	196	4500	354
80	66	420	201	5000	357
85	70	440	205	6000	361
90	73	460	210	7000	364
95	76	480	214	8000	367
100	80	500	217	9000	368
110	86	550	226	10,000	370
120	92	600	234	15,000	375
130	97	650	242	20,000	377
140	103	700	248	30,000	379
150	108	750	254	40,000	380
160	113	800	260	50,000	381
170	118	850	265	75,000	382
180	123	900	269	1,000,000	384

Equation (1) shows the equation of Krejcie and Morgan.

$$n = \frac{\chi^2 NP(1-P)}{d^2(N-1) + \chi^2 P(1-P)} \tag{1}$$

where:

    n = Sample size;
    N = Population size (75,000);
    X^2 = Chi-square for specified confidence level at 1 degree of freedom (3.841);
    d = Desired Margin of Error (expressed as a portion = 0.05);
    P = Population portion (0.05 in this table).

3.4. Data Collection

Data from the chosen environmental experts in Greece were gathered via an online survey form. Data collection started only when participants gave informed consent, and it was confirmed that they were willing to participate in the study. The information gathered will be useful in establishing relationships between the study variables and in addressing the research questions. The questionnaire had a variety of investigative questions about the incorporation of water resources management methods into land use planning for environmental conservation (see Appendix A).

3.5. Data Analysis

SPSS was used to code and analyze the quantitative data. The findings were tabulated, and frequencies and percentages were used to analyze them. The cumulative predictive ability of the multiple independent factors on the study's dependent variable was calculated using regression analysis [31,74]. A multiple regression model is required in this situation to determine various predicted values (Equation (2)).

$$Y = \beta_0 + \beta_1 X_1 + \beta_2 X_2 + \varepsilon \ldots \ldots 1 \tag{2}$$

where:

    Y = Environmental conservation;
    β_0 = constant (coefficient of intercept);
    X_1 = Aspects of land use planning;
    X_2 = Water resources management strategies;
    X_3 = Benefits of integration;
    ε = Represents the error term in the multiple regression model.

The hypotheses of the study were tested at the 5% (0.05) level of significance throughout the study.

3.6. Ethical Considerations

To confirm participants' desire to engage in the study, the researcher made sure informed consent was obtained. Additionally, privacy and confidentiality were maintained while handling the responses' data. Finally, respondents were given the option to reply to questions based on how well they understood the different opinion questions. This helped to increase the number of replies to certain inquiries.

4. Results

This section presents the different results obtained after analysis using SPSS.

4.1. Demographic Characteristics

Table 2 shows that the most respondents (60.8%) were male, and the females were only 39.2%. Most participants (38.8%) had a bachelor's degree, 31.3% had master's degrees, and only 2.9% had PhDs. Most participants (48.2%) had an experience of above 10 years in the environmental sector followed by 33.1% had an experience of 5–10 years, and only 18.7% had an experience of less than 5 years as an environmental expert.

Table 2. Showing demographic data of study respondents.

Characteristic	Frequency	Percentage (%)
Gender		
Male	169	60.8
Female	109	39.2
Education Qualification		
Certificate	17	6.1
Diploma	58	20.9
Bachelors	108	38.8
Masters	87	31.3
PhD	8	2.9
Experience in the environment sector		
Below 5 years	52	18.7
5–10 years	92	33.1
Above 10 years	134	48.2
Total	278	100

Source: Authors' own work (2023).

4.2. Descriptive Statistics

The study examined the different aspects of land use planning, and the results are presented in Figure 3.

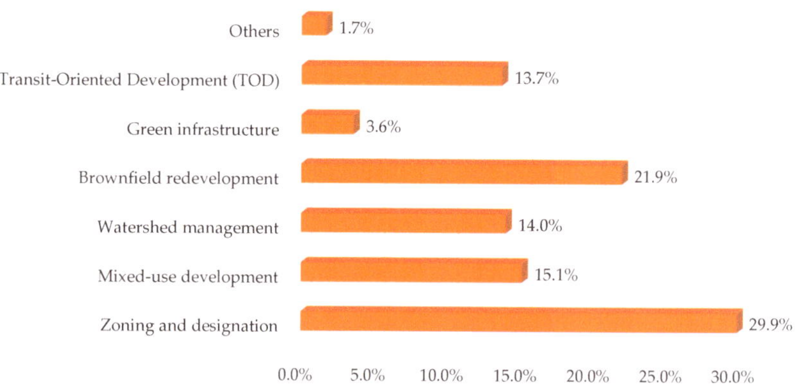

Figure 3. Aspects of land use planning. Source: Authors' own work (2023).

In regard to the aspects of land use planning in Figure 3, the majority of respondents (29.9%) identified "Zoning and designation" as the most significant aspect of land use planning. Zoning and designation involve categorizing land for specific purposes, such as residential, commercial, industrial, or agricultural, and are fundamental in urban and regional planning. "Brownfield redevelopment" was the second most mentioned aspect, with 21.9% of respondents recognizing its importance. Brownfield redevelopment refers to the revitalization and repurposing of previously developed and possibly contaminated land. "Mixed-use development" was noted by 15.1% of participants, indicating the significance of incorporating a mix of residential, commercial, and other land uses in urban planning to create vibrant and sustainable communities. "Transit-Oriented Development (TOD)" was mentioned by 13.7% of the respondents, suggesting that many consider the integration of transportation and land use planning as crucial. Only 1.7% of participants mentioned "other aspects," including the systematic assessment of land and water potential. This implies that this aspect is relatively less emphasized in the context of land use planning according to the respondents.

The study also explored the different water resources management strategies, and the results are presented in Table 3.

Table 3. Water resources management strategies.

	Frequency	Percentage (%)
Watershed Management	48	17.3
Integrated Water Resources Management (IWRM)	83	29.9
Ecosystem-based approaches	59	21.2
Stakeholder engagement and public participation	17	6.1
Water-Smart Landscaping and Irrigation	32	11.5
Water Quality Monitoring	36	12.9
Others	3	1.1
Total	278	100

Source: Authors' own work (2023).

The results in Table 3 show that integrated water resources management (IWRM) was identified by the highest percentage of participants (29.9%) as the key strategy. IWRM is a comprehensive and holistic approach to managing water resources that considers the interconnectedness of water systems and involves multiple stakeholders in decision making. Ecosystem-based approaches" were recognized by 21.2% of the respondents, highlighting the importance of considering ecological systems and their services in water resources management. Watershed management strategies" were chosen by 17.3% of participants. Watershed management focuses on the protection and sustainable use of entire watershed areas to ensure a consistent supply of clean water. Water quality monitoring" was noted by 12.9% of respondents, indicating the significance of regularly assessing and ensuring the quality of available water resources. A small proportion (1.1%) of participants mentioned "other strategies" like utilizing riparian zones, emphasizing the importance of maintaining natural areas along rivers and water bodies to support biodiversity.

The study also examined the different benefits of integration of water resources management strategies in land use planning on environmental conservation and the results are in Table 4.

Table 4. Benefits of Integration of water resources management strategies in land use planning.

	Frequency	Percentage (%)
Sustainable water supply	43	15.5
Ecosystem protection	71	25.5
Flood and drought mitigation	96	34.5
Water quality improvement	62	22.3
Others	6	2.2
Total	278	100

Source: Authors' own work (2023).

The results in Table 4 show that most participants (34.5%) identified flood and drought mitigation. This result implies that many individuals are concerned about reducing the impact of extreme weather events on their communities. This was followed by ecosystem protection (25.5%) meaning that some participants considered the preservation of natural habitats and biodiversity as a crucial outcome of such integration. Furthermore, 22.3% of participants acknowledged that water quality improvement is a significant advantage. This indicates that a sizable portion of the respondents valued the enhancement of the overall quality of water resources. Also, 15.5% of the respondents identified sustainable water supply as a significant benefit of integrating water resources management strategies in land use planning. This suggests that a portion of participants recognized the importance of ensuring a consistent and reliable water supply. Finally, the lowest portion (2.2%) mentioned other benefits of integration of water resources management strategies in land use planning.

The study also established the different aspects of environmental conservation, and the results are presented in Figure 4.

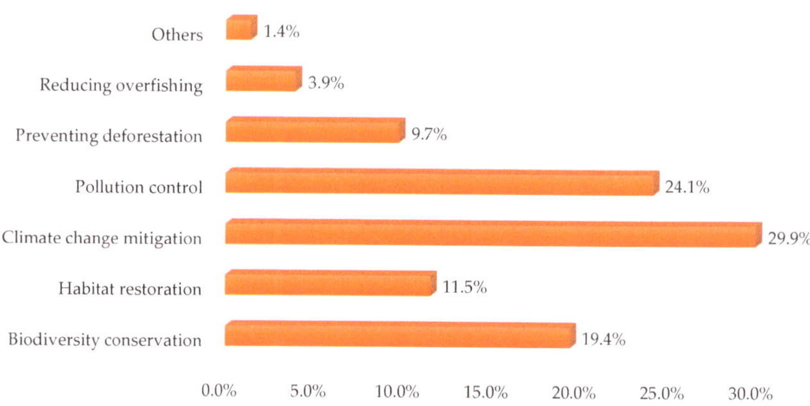

Figure 4. Aspects of environmental conservation. Source: Authors' own work (2023).

In regard to aspects of environmental conservation (Figure 4), the highest percentage of respondents (29.9%) recognized climate change mitigation as a major aspect of environmental conservation. This suggests that many participants believe that addressing climate change is crucial for environmental preservation. This was followed by 24.1% of respondents who identified pollution control as a significant aspect of environmental conservation. This highlights the importance of reducing pollution to protect the environment. A sizable portion of participants (19.4%) acknowledged the importance of conserving biodiversity. This result indicates that protecting the variety of species and ecosystems on Earth is a key concern for some respondents. Furthermore, 11.5% of participants considered habitat restoration as a significant aspect of environmental conservation. This suggests that rehabilitating natural habitats is an important goal for a portion of the respondents. A small percentage (1.4%) mentioned other aspects of environmental conservation, including afforestation and reducing deforestation, which were not explicitly listed in the survey.

4.3. Regression Analysis

The effect of integration of water resources management strategies in land use planning towards environmental conservation was established using regression analysis as presented in Tables 5–7.

Table 5. Model Summary.

Model	R	R Square	Adjusted R Square	Std. Error of the Estimate
	0.642 [a]	0.627	0.601	0.1257

[a] Predictors (constant): social costs of fires, economic costs of fires, environmental costs of fires, aspects of land use planning, water resources management strategies, and benefits of integration.

Table 6. ANOVA analysis.

Model	Sum of Squares	df	Mean Square	F	Sig.
Regression	76.204	3	28.031	73.261	0.014
Residual	71.051	380	0.413		
Total	147.255	382			

Dependent environmental conservation, predictors (constant): aspects of land use planning, water resources management strategies, benefits of integration.

Table 7. Regression coefficients.

Model	Unstandardized Coefficients		Standardized Coefficients	t	Sig.
	B	Std. Error	Beta		
(Constant)	0.588	0.126		1.941	0.027
Aspects of land use planning	0.168	0.054	0.371	1.124	0.024
Water resources management strategies	0.424	0.072	0.062	0.817	0.001
Benefits of Integration	0.126	0.041	0.052	0.817	0.012

Dependent variable: environmental conservation.

Environmental conservation was the dependent variable. The dependent variable and independent variable are regressed, yielding an R^2 value of 0.627. This shows that the independent factors account for 62.7% of the variance in the dependent variable. Additionally, the regression findings show that none of the study's independent variables had any impact on 37.3% of the changes.

The F-statistic of 73.261 at prob. (Sig) = 0.014 conducted at 5% level of significance means that there is a significant linear relationship that exists between the independent variables (aspects of land use planning, water resources management strategies, and benefits of integration) and the dependent variable (environmental conservation) as a whole.

The results in the table above confirm environmental conservation was measured in terms of aspects of land use planning, water resources management strategies, and benefits of integration since $p < 0.05$.

Since the significance level of 0.024 is less than 0.05%, we confirm that aspects of land use planning such as zoning and designation, mixed-use development, and green infrastructure, have an effect on environmental conservation. Therefore, we accept hypothesis H1 that aspects of land use planning have an effect on environmental conservation.

Also, there is a relationship between water resources management strategies and environmental conservation since the significance level of 0.001 is less than 0.05%. This is an indication that water resources management strategies such as integrated water resources management (IWRM), ecosystem-based approaches, and water quality monitoring greatly influence environmental conservation. We therefore accept H2 that there is a relationship between water resources management strategies and level of environmental conservation.

Since the significance level of 0.01224 is less than 0.05%, we confirm that benefits of integration of water resources management strategies in land use planning have an effect on environmental conservation. Therefore, we accept Hypothesis H3 that benefits of integration of water resources management strategies in land use planning have an impact on environmental conservation. This suggests that when water management agencies collaborate with land use planners, they can better address water-related challenges and enhance environmental conservation efforts.

5. Discussion

This study examined the efficacy of integration of water resources management strategies in land use planning in regard to environmental conservation. The study showed that there is an association between water resources management strategies and level of environmental conservation. By ensuring that human activities are balanced with the preservation and protection of natural resources and ecosystems, land use planning plays a critical role in environmental conservation [40,41].

The regression analysis reveals a significant positive relationship between aspects of land use planning and environmental conservation. This result supports the idea that land use decisions, including zoning and designation, mixed-use development, and green infrastructure, play a crucial role in shaping the environment. This finding aligns with Mengistu et al. (2023), who emphasize the importance of land use planning in protecting natural ecosystems and water resources [7]. Zoning regulations, for instance, can help control the growth of residential areas and mitigate conflicts between urban development and agricultural land. This is in line with the idea that responsible land use decisions can help maintain the quality and availability of water resources [12]. Some important elements of land use planning that support environmental protection were presented by the research [43,51]. Zoning in land use is the process of separating property into several zones or regions with designated allowable land uses [12]. Certain locations may be designated as conservation zones, protected areas, or green spaces by land use planners. By doing this, urban development is reduced, and natural areas, important ecosystems, and biodiversity hotspots are preserved [13,14]. Implementing urban development boundaries aids in limiting the growth of urban areas in environmentally vulnerable areas. These limits prevent development from going any farther, safeguarding forests, farmland, and other priceless natural resources [29,38,75,76].

Instead of transforming virgin land, land use planning might concentrate on redeveloping degraded or abandoned industrial areas (brownfields) [77,78]. The natural limits of watersheds may be taken into consideration when planning land use. Maintaining healthy aquatic ecosystems, preventing erosion, and protecting water quality are all benefits of regulating land uses appropriately within these borders. Conservation easements, which limit some property uses to conserve natural features, agricultural land, or animal habitats, may be established by land use planners working with landowners. These contracts guarantee long-term preservation while permitting a few specific land uses [19,47]. Urban areas that include green spaces, parks, and natural corridors benefit from increased biodiversity, recreational possibilities, and stormwater runoff management, which relieves pressure on existing infrastructure. Participating in land use planning procedures with the community encourages a feeling of ownership and accountability for environmental preservation [20,65]. Making better judgments may start with educating the public on the significance of sustainable land use practices [17,33,64].

The analysis also shows a significant positive relationship between water resources management strategies and environmental conservation. Water management strategies such as integrated water resources management (IWRM), ecosystem-based approaches, and water quality monitoring have a substantial impact on environmental conservation [22,79]. This result is consistent with the literature, which highlights the importance of sustainable water resources management for ecosystem health. IWRM, for example, is an approach that considers the entire water cycle, including the needs of human communities and natural ecosystems [20,54]. Such strategies help protect water bodies, maintain water quality, and support biodiversity, contributing to environmental conservation [2,20,32]. Artificial intelligence or a spiking neural network based architecture could help in this direction, as is done in the energy sector [80]. In order to maintain the sustainable use of water resources, it aims to balance opposing needs. IWRM can provide a framework for decision making in the context of land use planning that takes into consideration the water demands of various sectors, ecological requirements, and the general well-being of communities [22,54,79].

Planning based on watersheds is important in water management towards environment conservation. Watersheds act as the natural organizational units for controlling the usage of land and water. Utilizing watershed-based planning enables a comprehensive strategy to address human needs while conserving and restoring ecosystems [12]. Improved water infiltration, decreased stormwater runoff, and improved water quality may all be achieved by including green infrastructure, such as green areas, wetlands, and permeable surfaces, in land use planning [61,81]. Another method of managing water resources is via land use zoning, in which areas are designated for conservation, sustainable agriculture, and urban growth in accordance with their capacity to hold water [6]. Additionally, encouraging water-saving habits, putting in effective irrigation systems, and promoting water reuse may all help to lower water demand and encourage sustainable land use techniques [1,34,61].

The findings indicate that the benefits of integrating water resources management strategies into land use planning also have a positive impact on environmental conservation. Yelling (2007) supports this idea by emphasizing the advantages of integrated planning. Collaboration between water management and land use planning authorities can lead to more effective policies, better enforcement of regulations, and increased public awareness [6]. This integrated approach is vital for addressing complex issues like water source protection, flood control, and water quality management [46,82]. For successful environmental protection and sustainable development, the management of water resources must be included in land use planning [83]. Communities may develop resilient, environmentally balanced, and socially thriving settings by taking water-related issues into account when making choices about where to utilize their property [7]. Although there are difficulties, the advantages of integration are obvious, resulting in better water quality, increased ecosystem health, and a more sustainable future. To address challenges and adopt integrated strategies for the welfare of current and future generations, policymakers, planners, and stakeholders must collaborate [13,70].

6. Conclusions

The study shows that the integration of water resources management strategies in land use planning is paramount for achieving environmental conservation and sustainable development. The study's goals were achieved, and the research hypotheses were tested, confirming the relationship between the degree of environmental conservation, water resources management, and land use planning. The research verified that several facets of land use planning, such as mixed-use development, green infrastructure, and zoning and designation, significantly influence environmental preservation. This emphasizes the necessity for urban and regional planners to prioritize the preservation of natural ecosystems in their practices and to take the ecological repercussions of their decisions into account. Sustainable land use practices can help lessen the detrimental effects of industrialization and urbanization on the availability and quality of water. The research also confirmed the premise that the degree of environmental conservation and water resources management techniques, like ecosystem-based approaches and integrated water resources management (IWRM), are significantly correlated.

These tactics are essential for guaranteeing the prudent use of water resources and the preservation of natural ecosystems. They offer a comprehensive method of managing water resources that takes into account the requirements of both environmental preservation and human society. The results of the study showed how including strategies for managing water resources in land use planning has a favorable effect on environmental protection. Better public awareness and better enforcement of water conservation rules result from this integration, which makes it possible for land use authorities and water management agencies to collaborate more effectively. Together, these organizations can create policies that safeguard source waters, manage flooding, and improve water quality, all of which improve the ecosystem as a whole. The demand for water resources is rising in a world where urbanization and population growth are both ongoing trends, posing a number of difficulties for sustainable water management.

Findings also emphasize the need for strategic planning and cooperation between water management and land use authorities to address the growing challenges of water resources management and environmental protection. Emphasizing stakeholder participation, adaptive management, and continuous monitoring can also lead to successful outcomes and a more resilient and sustainable future. This study can guide policymakers, urban planners, researchers, and communities in making informed decisions that benefit both people and the planet by adopting an integrated approach that considers the interconnections between land and water, policymakers and planners can ensure the optimal use of water resources while safeguarding the environment for future generations. The integration of water resources management strategies in land use planning is crucial for achieving environmental conservation, sustainable development, and improved resilience to environmental challenges. By considering the interconnectedness of water systems and land use, societies can ensure long-term water availability, ecosystem health, and overall environmental well-being. The challenges are significant, but the benefits are equally promising, making this integration a critical endeavor for a better future. Emphasizing stakeholder participation, adaptive management, and continuous monitoring will lead to successful outcomes and a more resilient and sustainable future. This study can guide policymakers, urban planners, researchers, and communities in making informed decisions that benefit both people and the planet.

6.1. Recommendations

Despite the benefits of integrating water resources management strategies into land use planning, several challenges may arise. These challenges may include conflicting interests among stakeholders, limited financial resources, and inadequate data. To address these issues, collaboration among various sectors, governments, and international organizations is essential. Additionally, investing in research and data collection can help make informed decisions.

6.2. Areas for Future Research

The future requires more adaptive and resilient approaches as climate change impacts intensify. Incorporating climate change projections into integrated planning and utilizing advanced technologies like remote sensing and geographic information systems (GIS), artificial intelligence or a spiking neural network based architecture can enhance decision making.

Author Contributions: Conceptualization, S.K. and G.G.; methodology, S.K. and D.K.; software, M.P. and G.G.; validation, S.K., G.G. and D.K.; formal analysis, D.K.; investigation, S.K. and G.G.; resources, M.P. and G.G.; data curation, D.K. and M.P; writing—original draft preparation, S.K.; writing—review and editing, M.P.; visualization, D.K. and M.P.; supervision, G.G. and S.K.; project administration, S.K. and D.K.; funding acquisition, G.G. All authors have read and agreed to the published version of the manuscript.

Funding: This research received no external funding.

Institutional Review Board Statement: Not applicable.

Informed Consent Statement: Informed consent was obtained from all subjects involved in the study.

Data Availability Statement: Data available upon request.

Acknowledgments: This research did not receive any specific grant from funding agencies in the public, commercial, or not-for-profit sectors. The authors thank the editor and the anonymous reviewers for the feedback and their insightful comments on the original submission. All errors and omissions remain the responsibility of the authors.

Conflicts of Interest: The authors declare no conflict of interest.

Appendix A

Questionnaire
Section A: Demographic Information:
Qtn 1: Gender
1. Male
2. Female

Qtn 2: Education Qualification
1. Certificate
2. Diploma
3. Bachelor's
4. Master's
5. PhD

Qtn 3: Experience in the Environment Sector
1. Below 5 years
2. 5–10 years
3. Above 10 years

Section B: Objective related questions

Qtn 1: What common aspect of land use planning do you know? Choose one below.
1. Zoning and designation
2. Brownfield redevelopment
3. Mixed-use development
4. Transit-oriented development (tod)
5. Watershed management
6. Green infrastructure
7. Others . . . . Specify

Qtn 2: What water resources' management strategies do you know, choose one below:
1. Integrated Water Resources Management (IWRM)
2. Ecosystem-based approaches
3. Stakeholder engagement and public participation
4. Water-Smart Landscaping and Irrigation
5. Water Quality Monitoring
6. Others . . . . Specify

Qtn 4: Are you conversant with the benefits of integration of water resources' management strategies in land use planning on environmental conservation.
1. Yes
2. No

Qtn 5: What common benefit of integration of water resources' management strategies in land use planning on environmental conservation do you know? Choose one
1. Sustainable water supply
2. Ecosystem protection
3. Flood and drought mitigation
4. Water quality improvement
5. Others . . . . Specify

References

1. Bithas, K.; Kollimenakis, A.; Maroulis, G.; Stylianidou, Z. The Water Framework Directive in Greece. Estimating the Environmental and Resource Cost in the Water Districts of Western and Central Macedonia: Methods, Results and Proposals for Water Pricing. *Procedia Econ. Financ.* **2014**, *8*, 73–82. [CrossRef]
2. Piemontese, L. Sustainable Land and Water Management for a Greener Future Large-Scale Insights in Support of Agroecological Intensification. Ph.D. Thesis, Stockholm University, Stockholm, Sweden, 2020.
3. Duda, A.M. *Co-Managing Land and Water for Sustainable Development*; United Nations Convention to Combat Desertification (UNCCD): New York, NY, USA, 2017.
4. Duda, A.M. Integrated management of land and water resources based on a collective approach to fragmented international conventions. *Philos. Trans. R. Soc. Lond. Ser. B Biol. Sci.* **2003**, *358*, 2051–2062. [CrossRef] [PubMed]
5. Aylward, B.; Barbier, E.B. Valuing environmental functions in developing countries. *Biodivers. Conserv.* **1992**, *1*, 34–50. [CrossRef]
6. Yelling, J.A. Compensation and land use. In *Slums and Slum Clearance in Victorian London*; Routledge—Taylor & Francis Group: London, UK, 2007; pp. 75–90; ISBN 9780203717066.
7. Mengistu, A.G.; Woldesenbet, T.A.; Dile, Y.T.; Bayabil, H.K.; Tefera, G.W. Modeling impacts of projected land use and climate changes on the water balance in the Baro basin, Ethiopia. *Heliyon* **2023**, *9*, e13965. [CrossRef] [PubMed]
8. Sofios, S.; Arabatzis, G.; Baltas, E. Policy for management of water resources in Greece. *Environmentalist* **2008**, *28*, 185–194. [CrossRef]

9.   Moltz, H.L.N.; Wallace, C.W.; Sharifi, E.; Bencala, K. Integrating Sustainable Water Resource Management and Land Use Decision-Making. *Water* **2020**, *12*, 2282. [CrossRef]
10. Ekstedt, K. Local Water Resource Assessment in Messinia, Greece. Master's Thesis, Stockholm University, Stockholm, Sweden, 2013.
11. Costanza, R.; d'Arge, R.; de Groot, R.; Farber, S.; Grasso, M.; Hannon, B.; Limburg, K.; Naeem, S.; O'Neill, R.V.; Paruelo, J.; et al. The value of the world's ecosystem services and natural capital. *Ecol. Econ.* **1998**, *25*, 3–15. [CrossRef]
12. Penning de Vries, F.W.T.; Acquay, H.; Molden, D.J.; Scherr, S.; Valentin, C.; Cofie, O. *Integrated Land and Water Management for Food and Environmental Security*; IWMI: Colombo, Sri Lanka, 2003.
13. Schismenos, S.; Emmanouloudis, D.; Stevens, G.J.; Katopodes, N.D.; Melesse, A.M. Soil governance in Greece: A snapshot. *Soil Secur.* **2022**, *6*, 100035. [CrossRef]
14. Karamesouti, M.; Detsis, V.; Kounalaki, A.; Vasiliou, P.; Salvati, L.; Kosmas, C. Land-use and land degradation processes affecting soil resources: Evidence from a traditional Mediterranean cropland (Greece). *CATENA* **2015**, *132*, 45–55. [CrossRef]
15. Fynn, O.F.; Dzikunoo, E.A.; Chegbeleh, L.P.; Yidana, S.M. Enhancing adaptation to climate change through groundwater-based irrigation. *Sustain. Water Resour. Manag.* **2023**, *9*, 36. [CrossRef]
16. Nkosi, M.; Mathivha, F.I.; Odiyo, J.O. Impact of Land Management on Water Resources, a South African Context. *Sustainability* **2021**, *13*, 701. [CrossRef]
17. Kourgialas, N.N. A critical review of water resources in Greece: The key role of agricultural adaptation to climate-water effects. *Sci. Total Environ.* **2021**, *775*, 145857. [CrossRef] [PubMed]
18. Qin, T.; Feng, J.; Li, C.; Zhang, X.; Yan, D.; Liu, S.; Wang, J.; Lv, X.; Abebe, S.A. Risk assessment and configuration of water and land resources system network in the Huang-Huai-Hai watershed. *Ecol. Indic.* **2023**, *154*, 110712. [CrossRef]
19. Tang, X.; Adesina, J.A. Integrated Watershed Management Framework and Groundwater Resources in Africa—A Review of West Africa Sub-Region. *Water* **2022**, *14*, 288. [CrossRef]
20. Zatsepina, E.; Sorokina, O.; Kosinsky, V.; Petrova, L.; Fomkin, I. Environmental Aspects of Land Management. In Proceedings of the IV International Scientific and Practical Conference "Anthropogenic Transformation of Geospace: Nature, Economy, Society" (ATG 2019), Volgograd, Russia, 1–4 October 2019; Atlantis Press: Volgograd, Russia, 2020; pp. 321–325.
21. Dias, F.T.; Mazon, G.; Cembranel, P.; Birch, R.; de Andrade Guerra, J.B.S.O. Land Use and Global Environmental Change: An Analytical Proposal Based on A Systematic Review. *Land* **2023**, *12*, 115. [CrossRef]
22. Kokkoris, I.P.; Drakou, E.G.; Maes, J.; Dimopoulos, P. Ecosystem services supply in protected mountains of Greece: Setting the baseline for conservation management. *Int. J. Biodivers. Sci. Ecosyst. Serv. Manag.* **2017**, *14*, 45–59. [CrossRef]
23. Kalfas, D.; Chatzitheodoridis, F.; Loizou, E.; Melfou, K. Willingness to Pay for Urban and Suburban Green. *Sustainability* **2022**, *14*, 2332. [CrossRef]
24. Tzanopoulos, J.; Kallimanis, A.S.; Bella, I.; Labrianidis, L.; Sgardelis, S.; Pantis, J.D. Agricultural decline and sustainable development on mountain areas in Greece: Sustainability assessment of future scenarios. *Land Use Policy* **2011**, *28*, 585–593. [CrossRef]
25. Li, C.; Peng, C.; Chiang, P.-C.; Cai, Y.; Wang, X.; Yang, Z. Mechanisms and applications of green infrastructure practices for stormwater control: A review. *J. Hydrol.* **2019**, *568*, 626–637. [CrossRef]
26. Pereira, B.; David, L.M.; Galvão, A. Green Infrastructures in Stormwater Control and Treatment Strategies. *Proceedings* **2020**, *48*, 7.
27. Barbosa, A.E.; Fernandes, J.N.; David, L.M. Key issues for sustainable urban stormwater management. *Water Res.* **2012**, *46*, 6787–6798. [CrossRef] [PubMed]
28. Zdruli, P.; Ziadat, F.; Nerilli, E.; D'Agostino, D.; Lahmer, F.; Bunning, S. Sustainable development of land resources. In *Zero Waste in the Mediterranean: Natural Resources, Food and Knowledge*; International Centre for Advanced Mediterranean Agronomic Studies (CIHEAM), Ed.; Presses de Sciences Po: Paris, France, 2016; pp. 91–113; ISBN 978-2-7246-1921-8.
29. Mulder, J.; Kimball, K.R.; Larson, D.T.; Douglas, L.J.; Rovig, L.R.; Labau, D.; Sizemore, K. *Integrating Water Resources and Land Use Planning*; Utah Water Research Laboratory: Logan, UT, USA, 1979.
30. UN—OECD. *Integrated Water Resources Management in Eastern Europe, the Caucasus and Central Asia*; UN—OECD: New York, NY, USA, 2014.
31. Kalfas, D.; Kalogiannidis, S.; Chatzitheodoridis, F.; Toska, E. Urbanization and Land Use Planning for Achieving the Sustainable Development Goals (SDGs): A Case Study of Greece. *Urban Sci.* **2023**, *7*, 43. [CrossRef]
32. Koulelis, P.; Solomou, A.; Fassouli, V. Sustainability Constraints in Greece. Focusing on Forest Management and Biodiversity. In Proceedings of the 9th International Conference on Information and Communication Technologies in Agriculture, Food and Environment (HAICTA 2020), Thessaloniki, Greece, 24–27 September 2020; pp. 592–603.
33. Izakovičová, Z.; Špulerová, J.; Petrovič, F. Integrated Approach to Sustainable Land Use Management. *Environments* **2018**, *5*, 37. [CrossRef]
34. Angelo, M.J. Integrating Water Management and Land Use Planning: Uncovering the Missing Link in the Protection of Florida's Water Resources. *Univ. Florida J. Law Public Policy* **2001**, *12*, 223–249.
35. Rugland, E. *Integrating Land Use and Water Management. Planning and Practice*, 1st ed.; Lincoln Institute of Land Policy: Cambridge, MA, USA, 2022; ISBN 978-1-55844-439-3.
36. Pacheco-Vega, R. Governing Urban Water Conflict through Watershed Councils—A Public Policy Analysis Approach and Critique. *Water* **2020**, *12*, 1849. [CrossRef]

37. Cotler, H.; Cuevas, M.L.; Landa, R.; Frausto, J.M. Environmental Governance in Urban Watersheds: The Role of Civil Society Organizations in Mexico. *Sustainability* **2022**, *14*, 988. [CrossRef]
38. Bitterman, P.; Koliba, C.; Singer, A. A network perspective on multi-scale water governance in the Lake Champlain Basin, Vermont. *Ecol. Soc.* **2023**, *28*, 44. [CrossRef]
39. Dimadama, Z.; Zikos, D. Social Networks as Trojan Horses to Challenge the Dominance of Existing Hierarchies: Knowledge and Learning in theWater Governance of Volos, Greece. *Water Resour. Manag.* **2010**, *24*, 3853–3870. [CrossRef]
40. Young, M.D.; Ros, G.H.; de Vries, W. A decision support framework assessing management impacts on crop yield, soil carbon changes and nitrogen losses to the environment. *Eur. J. Soil Sci.* **2021**, *72*, 1590–1606. [CrossRef]
41. Manos, B.D.; Papathanasiou, J.; Bournaris, T.; Voudouris, K. A DSS for sustainable development and environmental protection of agricultural regions. *Environ. Monit. Assess.* **2010**, *164*, 43–52. [CrossRef]
42. IAEA. *Integrated Assessment of Climate, Land, Energy and Water*; Non-Serial Publications; International Atomic Energy Agency (IAEA): Vienna, Austria, 2020; ISBN 978-92-0-113720-3.
43. Papadopoulou, C.-A.; Papadopoulou, M.P.; Laspidou, C. Implementing Water-Energy-Land-Food-Climate Nexus Approach to Achieve the Sustainable Development Goals in Greece: Indicators and Policy Recommendations. *Sustainability* **2022**, *14*, 4100. [CrossRef]
44. Ruckelshaus, M.; McKenzie, E.; Tallis, H.; Guerry, A.; Daily, G.; Kareiva, P.; Polasky, S.; Ricketts, T.; Bhagabati, N.; Wood, S.A.; et al. Notes from the field: Lessons learned from using ecosystem service approaches to inform real-world decisions. *Ecol. Econ.* **2015**, *115*, 11–21. [CrossRef]
45. Chan, K.M.A.; Balvanera, P.; Benessaiah, K.; Chapman, M.; Díaz, S.; Gómez-Baggethun, E.; Gould, R.; Hannahs, N.; Jax, K.; Klain, S.; et al. Why protect nature? Rethinking values and the environment. *Proc. Natl. Acad. Sci. USA* **2016**, *113*, 1462–1465. [CrossRef] [PubMed]
46. Dogaru, L. The Importance of Environmental Protection and Sustainable Development. *Procedia-Soc. Behav. Sci.* **2013**, *93*, 1344–1348. [CrossRef]
47. Sang, N.; Ode Sang, Å. *A Review on the State of the Art in Scenario Modelling for Environmental Management: Potential for Application in Achieving the Swedish Environmental Objectives*; Rapport; Swedish Environmental Protection Agency: Stockholm, Sweden, 2015.
48. Tzanakakis, V.A.; Angelakis, A.N.; Paranychianakis, N.V.; Dialynas, Y.G.; Tchobanoglous, G. Challenges and Opportunities for Sustainable Management of Water Resources in the Island of Crete, Greece. *Water* **2020**, *12*, 1538, Correction in *Water* **2022**, *14*, 1024. [CrossRef]
49. Rodríguez, M.I.; Grindlay, A.L.; Cuevas, M.M.; Zamorano, M. Integrating land use planning and water resource management: Threshold scenarios—A tool to reach sustainability. *WIT Trans. Ecol. Environ.* **2015**, *192*, 231–242. [CrossRef]
50. Singh, S.; Biswas, R. Analysis of Land Use Change Effects/Impacts on Surface Water Resources in Delhi. *Urban Sci.* **2022**, *6*, 92. [CrossRef]
51. Kyriakopoulos, G.L. Land Use Planning and Green Environment Services: The Contribution of Trail Paths to Sustainable Development. *Land* **2023**, *12*, 1041. [CrossRef]
52. Ackerschott, A.; Kohlhase, E.; Vollmer, A.; Hörisch, J.; von Wehrden, H. Steering of land use in the context of sustainable development: A systematic review of economic instruments. *Land Use Policy* **2023**, *129*, 106620. [CrossRef]
53. Martino, S.; Kenter, J.O. Economic valuation of wildlife conservation. *Eur. J. Wildl. Res.* **2023**, *69*, 32. [CrossRef]
54. Gugerell, K.; Endl, A.; Gottenhuber, S.L.; Ammerer, G.; Berger, G.; Tost, M. Regional implementation of a novel policy approach: The role of minerals safeguarding in land-use planning policy in Austria. *Extr. Ind. Soc.* **2020**, *7*, 87–96. [CrossRef]
55. Kalogiannidis, S.; Kalfas, D.; Chatzitheodoridis, F.; Papaevangelou, O. Role of Crop-Protection Technologies in Sustainable Agricultural Productivity and Management. *Land* **2022**, *11*, 1680. [CrossRef]
56. Nolon Blanchard, J.C. *Integrating Water Efficiency into Land Use Planning in the Interior West: A Guidebook for Local Planners*; Elliott, D.L., Nolon, J.R., Eds.; Land Use Law Center for Western Resource Advocates: New York, NY, USA, 2018.
57. Duan, Y.; Tang, J.; Li, Z.; Yang, Y.; Dai, C.; Qu, Y.; Lv, H. Optimal Planning and Management of Land Use in River Source Region: A Case Study of Songhua River Basin, China. *Int. J. Environ. Res. Public Health* **2022**, *19*, 6610. [CrossRef] [PubMed]
58. Sgroi, M.; Vagliasindi, F.G.A.; Roccaro, P. Feasibility, sustainability and circular economy concepts in water reuse. *Curr. Opin. Environ. Sci. Health* **2018**, *2*, 20–25. [CrossRef]
59. Kalogiannidis, S.; Papadopoulou, C.-I.; Loizou, E.; Chatzitheodoridis, F. Risk, Vulnerability, and Resilience in Agriculture and Their Impact on Sustainable Rural Economy Development: A Case Study of Greece. *Agriculture* **2023**, *13*, 1222. [CrossRef]
60. Ma, P.; Lyu, S.; Diao, Z.; Zheng, Z.; He, J.; Su, D.; Zhang, J. How Does the Water Conservation Function of Hulunbuir Forest & ndash; Steppe Ecotone Respond to Climate Change and Land Use Change? *Forests* **2022**, *13*, 2039.
61. Thoidou, E.; Foutakis, D. Spatial Planning in view of new challenges: Land take and some evidence from Greece. In Proceedings of the Planning for Transition, AESOP 2019 Annual Congress, Venice, Italy, 9–13 July 2019.
62. Thoidou, E. Spatial Planning and Climate Adaptation: Challenges of Land Protection in a Peri-Urban Area of the Mediterranean City of Thessaloniki. *Sustainability* **2021**, *13*, 4456. [CrossRef]
63. Kalogiannidis, S.; Chatzitheodoridis, F.; Kalfas, D.; Patitsa, C.; Papagrigoriou, A. Socio-Psychological, Economic and Environmental Effects of Forest Fires. *Fire* **2023**, *6*, 280. [CrossRef]

64. Kourgialas, N.N.; Psarras, G.; Morianou, G.; Pisinaras, V.; Koubouris, G.; Digalaki, N.; Malliaraki, S.; Aggelaki, K.; Motakis, G.; Arampatzis, G. Good Agricultural Practices Related to Water and Soil as a Means of Adaptation of Mediterranean Olive Growing to Extreme Climate-Water Conditions. *Sustainability* **2022**, *14*, 13673. [CrossRef]
65. Elhag, M.; Boteva, S. Quantitative Analysis of Different Environmental Factor Impacts on Land Cover in Nisos Elafonisos, Crete, Greece. *Int. J. Environ. Res. Public Health* **2020**, *17*, 6437. [CrossRef]
66. Kyvelou, S.S.; Gourgiotis, A. Landscape as Connecting Link of Nature and Culture: Spatial Planning Policy Implications in Greece. *Urban Sci.* **2019**, *3*, 81. [CrossRef]
67. Randolph, J. *Environmental Land Use Planning and Management*, 2nd ed.; Island Press: Washington, DC, USA, 2011; ISBN 9781597267304.
68. Sobhani, P.; Esmaeilzadeh, H.; Sadeghi, S.M.M.; Wolf, I.D.; Deljouei, A. Prioritizing Water Resources for Conservation in a Land of Water Crisis: The Case of Protected Areas of Iran. *Water* **2022**, *14*, 4121. [CrossRef]
69. Kalogiannidis, S.; Loizou, E.; Kalfas, D.; Chatzitheodoridis, F. Local and Regional Management Approaches for the Redesign of Local Development: A Case Study of Greece. *Adm. Sci.* **2022**, *12*, 69. [CrossRef]
70. Papanastasis, V.P.; Arianoutsou, M.; Papanastasis, K. Environmental conservation in classical Greece. *J. Biol. Res.* **2010**, *14*, 123–135.
71. Atay, I. Water Resources Management in Greece: Perceptions about Water Problems in the Nafplion Area. Master's Thesis, Stockholm University, Stockholm, Sweden, 2012.
72. Koulelis, P.P.; Proutsos, N.; Solomou, A.D.; Avramidou, E.V.; Malliarou, E.; Athanasiou, M.; Xanthopoulos, G.; Petrakis, P.V. Effects of Climate Change on Greek Forests: A Review. *Atmosphere* **2023**, *14*, 1155. [CrossRef]
73. Krejcie, R.V.; Morgan, D.W. Determining Sample Size for Research Activities. *Educ. Psychol. Meas.* **1970**, *30*, 607–610. [CrossRef]
74. Kalogiannidis, S.; Kalfas, D.; Chatzitheodoridis, F.; Lekkas, E. Role of Governance in Developing Disaster Resiliency and Its Impact on Economic Sustainability. *J. Risk Financ. Manag.* **2023**, *16*, 151. [CrossRef]
75. Mitchell, D.P. Land administration trends and their implications for Australian natural resource management. *J. Spat. Sci.* **2009**, *54*, 35–47. [CrossRef]
76. Mitchell, D.; Grenfell, R.; Bell, K. Assessing the Suitability of Land Administration to Support Natural Resource Management at the Parcel Level. In Proceedings of the International Federation of Surveyors (FIG) Working Week, Paris, France, 13–17 April 2003; International Federation of Surveyors (FIG), Ed.; International Federation of Surveyors (FIG): Athens, Greece, 2003; pp. 1–15.
77. Elimam, H. *Environmental Conservation and Sustainable Development*; LAP Lambert Academic Publishing: London, UK, 2014; ISBN 978-3659609282.
78. McDonach, K.; Yaneske, P.P. Environmental management systems and sustainable development. *Environmentalist* **2002**, *22*, 217–226. [CrossRef]
79. Vlek, P.L.G.; Khamzina, A.; Azadi, H.; Bhaduri, A.; Bharati, L.; Braimoh, A.; Martius, C.; Sunderland, T.; Taheri, F. Trade-Offs in Multi-Purpose Land Use under Land Degradation. *Sustainability* **2017**, *9*, 2196. [CrossRef]
80. Brusca, S.; Capizzi, G.; Lo Sciuto, G.; Susi, G. A new design methodology to predict wind farm energy production by means of a spiking neural network–based system. *Int. J. Numer. Model. Electron. Netw. Devices Fields* **2019**, *32*, e2267. [CrossRef]
81. Chatzitheodoridis, F.; Michailidis, A.; Theodosiou, G.; Loizou, E. Local cooperation: A dynamic force for endogenous rural development. In *Balkan and Eastern European Countries in the Midst of the Global Economic Crisis*; Contributions to Economics; Physica: Heidelberg, Germany, 2013; pp. 121–132. [CrossRef]
82. Papadopoulou, C.I.; Loizou, E.; Melfou, K.; Chatzitheodoridis, F. The knowledge based agricultural bioeconomy: A bibliometric network analysis. *Energies* **2021**, *14*, 6823. [CrossRef]
83. Mishra, P.K. Environmental Conservation and Management. In *Environmental Geography*; IGNOU—Indira Gandhi National Open University: New Delhi, India, 2021; pp. 185–195.

 *sustainability*

Article

Technological Integration and Obstacles in China's Agricultural Extension Systems: A Study on Disembeddedness and Adaptation

Xinran Hu [1,†], Bin Xiao [2,*,†] and Zhihui Tong [1,3,*]

[1] School of Agricultural Economics and Rural Development, Renmin University of China, Beijing 100872, China; huxr1199@ruc.edu.cn
[2] School of Humanities and Law, Northeastern University, Shenyang 110819, China
[3] Rural Governance Research Center of Renmin University of China, Beijing 100872, China
* Correspondence: 2010009@stu.neu.edu.cn (B.X.); tmggyx@126.com (Z.T.)
† These authors contributed equally to this work.

Abstract: In light of China's evolving agricultural technology extension system, this study investigates a critical issue known as "technological disembeddedness". This phenomenon, observed in the context of the country's push towards administrative and market-oriented extension, reflects a significant disconnect between the formalized methods of technology extension, such as classroom instruction, and the practical needs of farmers. As a consequence, the envisioned improvements in agricultural production efficiency have not materialized as expected. The analysis, based on fieldwork conducted in Shandong Province from 2019 to 2020, identifies that different stakeholder interests have further exacerbated the situation. Agricultural technology extension, driven by diverse agendas, has been utilized as a tool for profit, resulting in a stark disparity in farmers' access to technology and the emergence of multiple, formalized extension models. This marginalized small-scale farmers and undermined the initial objectives of the extension system. The study proposes a fundamental shift in approach. It advocates for a social-centric perspective on technology extension, suggesting that the solution lies in harnessing local community dynamics to gradually build a technology extension system that aligns with the practical realities of farmers' production and daily lives. In summary, the study identifies "technological disembeddedness" as a primary challenge within China's agricultural technology extension system. It underscores the need to reorient the approach towards a more socially connected model, with a focus on the local community's role in creating a technology extension system that genuinely serves the needs of farmers.

Keywords: agricultural technology extension; technological disentanglement; social relationships

Citation: Hu, X.; Xiao, B.; Tong, Z. Technological Integration and Obstacles in China's Agricultural Extension Systems: A Study on Disembeddedness and Adaptation. *Sustainability* **2024**, *16*, 859. https://doi.org/10.3390/su16020859

Academic Editors: Fotios Chatzitheodoridis, Efstratios Loizou and Achilleas Kontogeorgos

Received: 25 December 2023
Revised: 12 January 2024
Accepted: 15 January 2024
Published: 19 January 2024

1. Research Background and Problem Statement

Agricultural technology plays a crucial role in improving agricultural production efficiency and driving the modernization of agriculture [1]. In 2023, China's central "Document No. 1" emphasized the need to strengthen support for agricultural science, technology, and equipment to accelerate the development of a strong agricultural sector (*source: "Opinions of the Central Committee of the Communist Party of China and the State Council on Key Tasks for Comprehensive Advancement of Rural Revitalization in 2023"*). Since the beginning of the reform and opening-up policy, China has undergone reforms in its grassroots agricultural technology extension system, adapting to the processes of marketization and the management of grassroots agricultural technology personnel. These reforms built upon the existing organizational structure for agricultural technology developed since the founding of the People's Republic. The reforms involved the transfer and delegation of certain powers (*referred to as "two releases" in 1989–1990 and 2001–2003, related to the delegation of management rights over personnel, finances, and resources of township-level agricultural technology stations from counties to townships*) and the centralization of other powers (*referred to as "two receptions" in 1991–2000 and 2004–2008, related to the centralization of certain rights*) [2]. These changes have gradually shaped the agricultural technology extension system, which is primarily led by government administrations and market entities, with the involvement of research institutions. The aim has been to improve the efficiency of agricultural technology extension through diverse participation.

However, practical observations have shown that this system's effectiveness is not ideal in practice. It is characterized by a trend towards a singular mode of extension, where many training activities are conducted for the sake of completing tasks, and formalism in agricultural technology extension is prevalent. A significant challenge is the coexistence of low agricultural technology adoption rates and low acceptance levels of agricultural technology extension. The gap between the agricultural technology extension system and its intended goals remains a central issue in contemporary agricultural technology extension. Existing research has primarily focused on the following aspects:

First, examining the effectiveness of the agricultural technology extension system: Although researchers acknowledge that training-based agricultural technology extension effectively increases grain production [3,4], raises agricultural income [5,6], and reduces the use of agricultural chemicals [7,8], there is a widespread consensus that China's agricultural technology extension system still operates at a low level of performance [9,10]. This situation is attributed to insufficient government-led funding for agricultural technology extension [11], management problems at grassroots extension organizations [12], the scarcity of agricultural technology extension resources leading to only a small portion of farmers accessing public extension services [13,14], and the participation of diverse extension entities that has not completely resolved the issues [15]. The extension of agricultural technology based on kinship and consanguinity remains the primary channel for farmers to access new technologies [16].

Second, an attempt is made to explain the reasons behind the poor performance of the current agricultural technology extension system. The government, as a provider of public services, plays a primary role in the provision of public welfare-oriented extension services [17]. At the micro-level, the low educational levels and generally low income of agricultural technology extension personnel significantly affect the performance of technology extension [2,18]. From a meso-level perspective, a continuous process of interaction and bargaining between the state, grassroots agricultural technology extension agencies, and farmers has resulted in each forming its own logic of action and path dependency, leading to institutional dilemmas in the current agricultural technology extension system [19]. These contradictions have further exacerbated issues related to macro-level funding insufficiency and institutional mechanisms not functioning smoothly [20]. As administrative power is decentralized in rural society, it is continually eroded [21].

With China's market-oriented reforms, agricultural technology extension increasingly incorporates market-driven forces, including agribusiness, into the extension system [22,23]. While this diversifies the forms of technology extension, it also makes agricultural technology extension vulnerable to market dynamics, supply–demand relationships, and price fluctuations [24]. Additionally, market entities, in pursuit of maximizing organizational profits, are compelled to engage in technological innovation, promotion, and commercialized services [25], leading to the promotion of proprietary agricultural technologies. The agricultural technologies offered by these market entities are often highly profitable advanced technologies [26]. In this process, agricultural enterprises are burdened with dual goals, and as a secondary objective, agricultural technology extension can experience a certain degree of deviation due to coexistence with the primary goal of profit maximization. From the perspective of public economics, as agricultural technology is a quasi-public good, if government subsidies are insufficient to offset the income spillovers generated by the private sector supply of agricultural technology services, private sector entities are likely to withdraw from or reduce their supply [27]. Moreover, agricultural colleges and research institutions, as extension entities, continue to be centered around government-led extension efforts. They face internal capacity constraints and a limited reserve workforce, making it challenging to take on a significant amount of agricultural technology extension work beyond their core responsibilities [28].

In summary, existing research generally acknowledges that the current effectiveness of agricultural technology extension is not promising. It recognizes that promoting entities, represented by the government and the market, are constrained by their own characteristics and cannot maximize the efficiency of technology extension. However, much of the existing research tends to view technology extension as a top-down process of knowledge transfer, often overlooking the fact that farmers, as recipients of technology, are also disseminators of technology. While existing research reveals the negative impact of government and market systems on technology extension, it often fails to connect the social context in which farmers are situated and the behaviors of farmers within their social context, thus not offering a comprehensive explanation for the suboptimal extension outcomes.

Agricultural technology is an integral part of farmers' production and daily lives, deeply embedded within local communities. It cannot exist independently of the local social and institutional environment. Some scholars have recognized that social networks influence technology adoption [29,30] and have attempted to explore their relationship through quantitative models. However, these attempts often lack corresponding theoretical analyses and fail to explain why the technology extension system constructed by the state is not the primary choice for technology adoption in rural communities.

This paper aims to introduce the concept of "technology disembeddedness", situating agricultural technology within the local social context. It dissects the underlying logic of technology extension in this domain and attempts to explain why the current agricultural technology extension system in China is suboptimal in practice within rural communities. Furthermore, it seeks to reveal the deep-seated logic of various stakeholders in technology extension and proposes potential pathways for escaping this predicament.

2. Theoretical Foundation and Analytical Framework

When examining the process of technology extension from an economic perspective, promoters tend to choose the most efficient pathways for extension, while recipients are inclined to accept agricultural technologies that maximize their individual benefits. However, rural life is deeply rooted in its local context [31], and agricultural technology is inseparable from this rural social fabric. It relies on the rural social environment for extension and is simultaneously "embedded" within the networks of village communities.

In fact, technology extension follows its own spontaneous order, and this order is built upon the foundation of rural communities. The existing literature has also discussed this viewpoint. Some scholars emphasize the embedded characteristics of social networks and their impact on technology extension [32], suggesting that the key to effectively connecting technology with smallholder farmers lies in the fact that technology is embedded within the social structure of rural communities. Additionally, individual farmers' interactive learning through social networks is a driving force for the extension of new technologies [33]. Farmers tend to rely on "acquaintance relationships" as their primary source of technology acquisition. It can be said that the process of technology extension is also a collective assimilation process [34].

Regarding technology adoption, traditional agricultural technology extension relies on social interactions among farmers, permeating rural communities and giving rise to farmers' spontaneous information exchange relationships [35]. However, to promote agricultural and rural modernization and enhance the rate of scientific and technological conversion, the government has established a system for technology extension. Nevertheless, some scholars have observed that this system's performance is suboptimal in practice, sometimes even conflicting with the needs of farmers [36]. They attribute this phenomenon to factors such as the lag in grassroots technology extension system development and the aging of knowledge among agricultural technicians [37]. However, they have not analyzed the relationship between rural communities and agricultural technology or technology extension.

As Granovetter elaborates in his work, "The economic actions of actors are not only individualistic but also embedded in social relationships. Apart from the influence of individual rationality and personal preferences, individuals are 'embedded' within social networks, constantly exchanging information with others in their social environment. They are influenced by these interactions, which may change their preferences and, ultimately, be reflected in their decision-making" [38]. When applied to the context of technology extension in agriculture, the pathways through which farmers transmit and receive agricultural technology information closely align with their social networks. Therefore, the technology extension methods that best suit farmers' preferences should be embedded within rural communities. However, with the reform of the technology extension system, the previously deep-reaching connections into rural communities have gradually disappeared. Simultaneously, the growing influence of market forces has caused the technology extension to be gradually disembedded from rural communities and farmers' lives.

While the term "technological disembeddedness" has not been introduced into research on technology extension, with the rise of new economic sociology, there is an increasing focus on the influence of social relationships on economic issues. "embeddedness" and "disembeddedness" have increasingly entered the field of rural development. In contrast to "embeddedness", the term "disembeddedness" refers to a state in which things detach from their original social structures, networks of relationships, cultural concepts, and other elements of social systems [39,40]. Polanyi's concept of "disembeddedness" pertains to the state in which an economic system detaches from the operation of a social system due to the development of self-regulating markets driven by the multiple commodifications of land, labor, and nature [41,42]. While Polanyi emphasized that self-regulating markets never fully emerged, this tendency has become increasingly pronounced, causing significant disruptions to the functioning of social systems [43]. Taking agricultural disembeddedness as an example, the influx of capital into rural areas and the market-oriented transformation have caused agricultural operations to disentangle from rural social networks, becoming disconnected from the production and daily lives of farmers. This has led to the separation of the multifunctionality of agriculture [44]. Some scholars also point out that the disembedded development of rural areas has resulted in contemporary environmental issues in rural regions [45].

Contemporary technology extension, to some extent, has been disembedded from social structures and relationship networks embedded in the daily production and lives of farmers, resulting in a series of adverse consequences. This paper introduces the concept of "technological disembeddedness", where "technology" specifically refers to agricultural technology, and defines "technological disembeddedness" as the promotion of technology detaching from the actual production and lives of farmers, as well as from their social relationship networks and village orders, operating outside the realms of village production and social life. Subsequently, the paper will elucidate the reasons behind the unsatisfactory outcomes of agricultural technology extension by focusing on technological disembeddedness through the analysis of social relationship networks from an embeddedness theory perspective. It will explore the specific manifestations of technological disembeddedness, investigate the negative consequences it leads to, and delve into the underlying reasons for technological disembeddedness and the suboptimal effects of technology extension. Based on these considerations, the paper constructs the following analytical framework (see Figure 1).

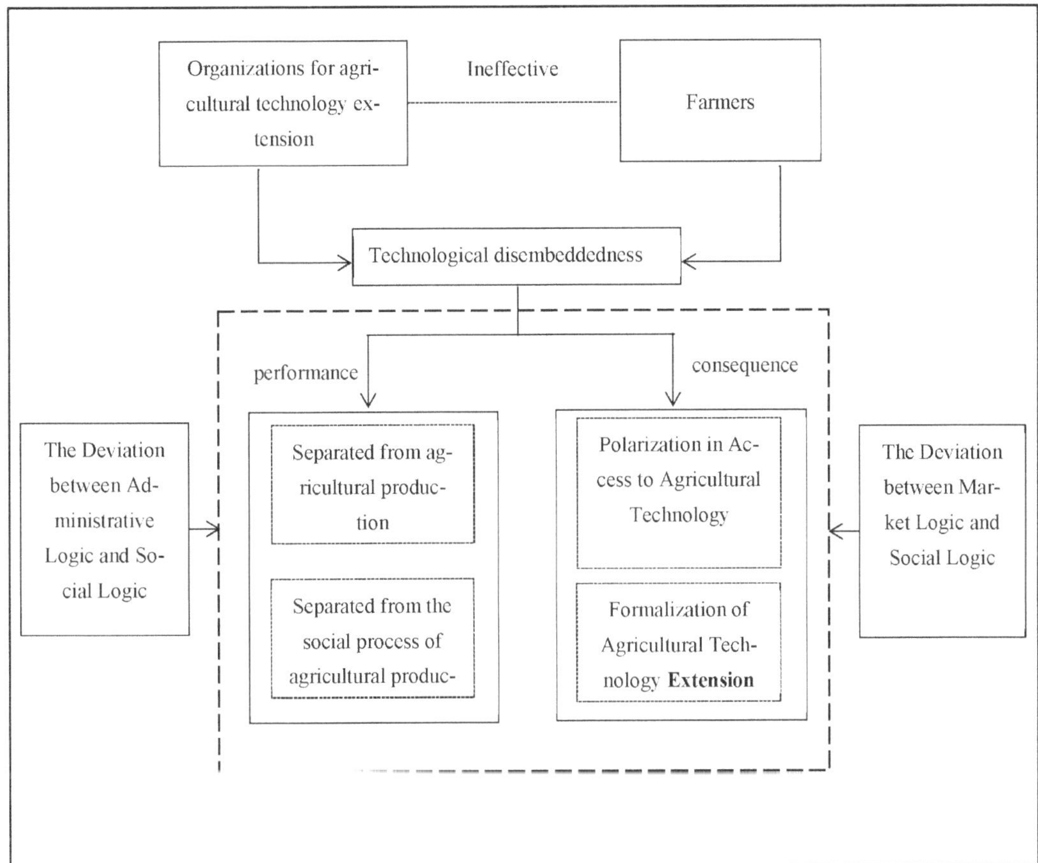

Figure 1. Analysis framework diagram.

3. Research Design

3.1. Method Selection

This study employs an exploratory, single-case research method. The reason for selecting this research method is because the core research question addressed in this paper is why China's current agricultural technology extension system is performing suboptimally in practice. Case study research is particularly well-suited for addressing questions that are rich in explanatory elements and seek to understand "how" and "why" [46]. Furthermore, by extracting key elements from a single case, refining, mining, and articulating them clearly, this method plays a crucial role in uncovering the inherent associations behind phenomena, formulating testable theoretical hypotheses, and advancing research [47]. The angle of inquiry in this study pertains to a portion that has not been deeply explored in the existing literature, making it appropriate for an exploratory single-case research approach.

Regarding data collection, the author's research team conducted four field surveys in Ma Lianzhuang Town, Laixi City, Shandong Province, from 2019 to 2020, accumulating nearly two months of research time. Substantial data were collected through the following means: First, the research team conducted in-depth interviews with a wide range of individuals, including government employees, village cadres, farmers, and various actors in the agricultural supply chain. The interviews primarily focused on topics such as agricultural industry development, agricultural technology extension, and the primary channels for accessing agricultural technology. Through this method, a total of 109 interview materials were collected, covering all seven communities (new villages) in Ma Lianzhuang Town. (*Here, "communities (new villages)" refers to the new village units created after the amalgamation of villages. Ma Lianzhuang Town in Laixi City completed the amalgamation of the previous 77 administrative villages into 7 communities (new villages) by the end of 2019. The specific villages mentioned in the case below are all the administrative villages before the amalgamation.*) In total, 217 h of interview recordings were obtained, which were then transcribed and organized into approximately 870,000 words of interview data (as shown in Table 1).

Table 1. Interviewees and main interview topics.

Interviewee	Number of Interviewees	Interview Duration	Interview Topics
Director and Staff of Malianzhuang Town Agricultural Service Center	3	405 min	1. Agricultural industry development status 2. Agricultural technology extension work 3. Challenges in agricultural technology extension work
Farmers (various categories, including field crops, sweet melon, and grape growers)	45	5446 min	1. Personal history and current status of industry development 2. Use of technology in the cultivation process 3. Channels for accessing agricultural technology 4. Specific steps for cultivating crops and encountered issues 5. Purchase of agricultural production materials
Village Cadres from All 7 Communities (new villages) in the Town	52	6133 min	1. Village industry situation 2. Role of villages in the agricultural technology extension process 3. Farming activities of village cadres themselves
Agricultural Technicians	4	482 min	1. Personal experiences 2. How they conduct agricultural technology extension and where they acquire agricultural knowledge 3. Roles played in current agricultural technology extension
Owners of Agricultural Supply Chain (e.g., agricultural supply stores and upstream and downstream businesses)	5	558 min	1. Extension of agricultural technology information in the supply chain 2. Coverage of business operations 3. Personal experiences of the business owners

Secondly, the research team collected case data through a combination of direct observation and participant observation. Observational evidence often supplements research content [46]. During our research in Ma Lianzhuang Town, we established a favorable interactive relationship with the local government, which allowed us to attend certain meetings and gain access to government office premises. By observing the working style of the local government and interactions with village cadres who came for administrative matters, we obtained nearly 50,000 words of observational records, scene notes, and research reflections. These observational data, along with the information obtained from in-depth interviews, contributed to triangulation for validation [46].

Finally, the research team collected internal documents from the local government, including county annals, agricultural statistical data, and relevant policy documents.

3.2. Case Selection

The decision to focus on Malianzhuang Town in Laixi City as the subject for this case study is grounded in several key considerations, with the addition of a map providing visual context for a clearer understanding of the region:

Representativeness: Malianzhuang Town serves as a representative microcosm of China's agricultural technology extension system, showcasing the challenges faced in its practical implementation. The region's developmental logic shares commonalities with other areas, and its diverse cultivation structure enhances the general applicability of the study. (See Figure 2 for the geographical location of Malianzhuang Town within Shandong Province).

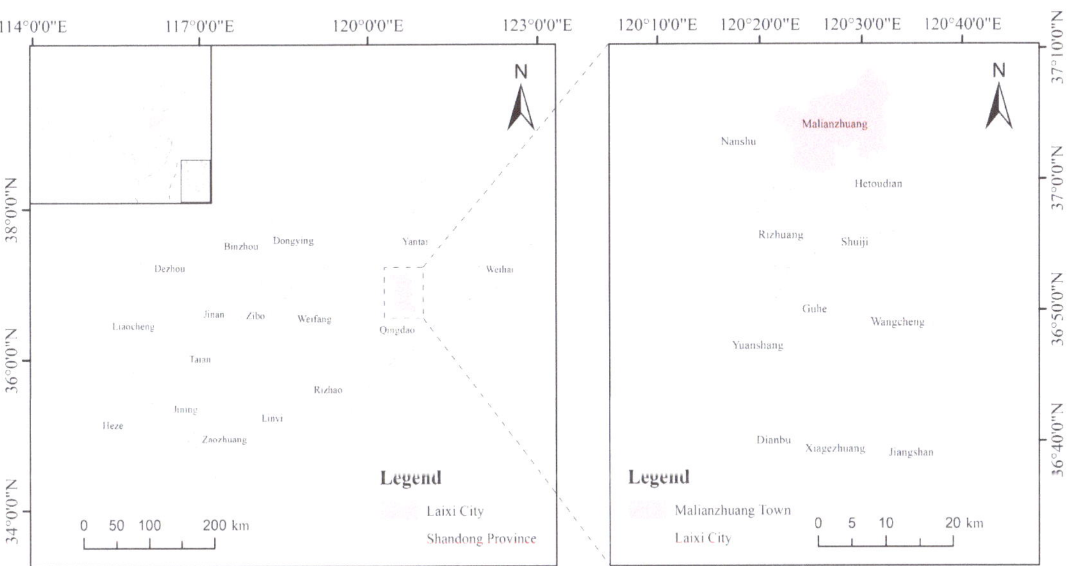

Figure 2. Geographical location map of Malianzhuang, Shandong Province, China.

Typicality: Positioned as a typical agricultural township, Malianzhuang experiences continuous evolution in agricultural technology parallel to its industrial development. Situated upstream of the Daguchuan River and adhering to environmental protection requirements, the absence of industrial activities has resulted in a unique cultivation structure dominated by field crops and high-yield crops, particularly fruits and melons. The role of agricultural technology in the growth of these crops is visually depicted in Figure 2, underscoring its significance as valuable research material for investigating the development and dissemination of agricultural technology.

Exploratory Value: Malianzhuang commenced agricultural industry transformation in the 1990s, marking a period of substantial changes in agricultural technology dissemination methods. These transformative shifts provide the basis for longitudinal comparative analysis, facilitating the exploration of deeper reasons behind the suboptimal performance of the current agricultural technology extension system. (Refer to Figure 2 for a visual representation of the spatial layout of Malianzhuang Town and its surrounding areas.)

3.3. Case Background Introduction

Malianzhuang Town is situated in the northernmost part of Qingdao City, bordered by Laiyang City to the east and adjacent to Zhaoyuan City to the north. It is located upstream of the Daguchuan River and is renowned for the abundant production of staple food crops such as peanuts, wheat, and maize. In recent years, Malianzhuang Town has vigorously encouraged the local populace to engage in the cultivation of fruits and melons. Currently, it has established a year-round fruit cultivation pattern, encompassing sweet melons in spring, grapes in summer, apples (pears) in autumn, and strawberries in winter. According to statistical data from 2019, the town collectively cultivates approximately 10,071 hectares of sweet melons, 1386 hectares of strawberries, 2271 hectares of pear trees, 6704 hectares of apple trees, and 5114 hectares of grapevines. It has earned the reputation of being the "Land of Fruits" and the "Sweet Melon Town".

The cultivation of high-yield crops in the area can be traced back to the early 1990s, when local farmers spontaneously initiated the planting of watermelons, sweet melons, apples, and other fruits. In 1996, the Malianzhuang Town government commenced guiding farmers in sweet melon cultivation. In 2000, Laiyang City, considering the local topography and the spontaneously emerging cultivation pattern, introduced an agricultural industry development plan termed "Southern Vegetables and Northern Fruits". This plan elevated the status of fruit and melon cultivation to a pivotal component of the county-level overall layout. In 2005, Malianzhuang Town constructed a sweet melon market to facilitate sweet melon sales. Over the subsequent decade, through the establishment of brands and trademark registration, the local sweet melon cultivation area rapidly expanded to encompass approximately 1340 hectares. From 2014 to the present, with the diversification of planting varieties, the area has established a relatively stable development pattern.

The historical development of fruit and melon cultivation in Malianzhuang Town is illustrated in Figure 3.

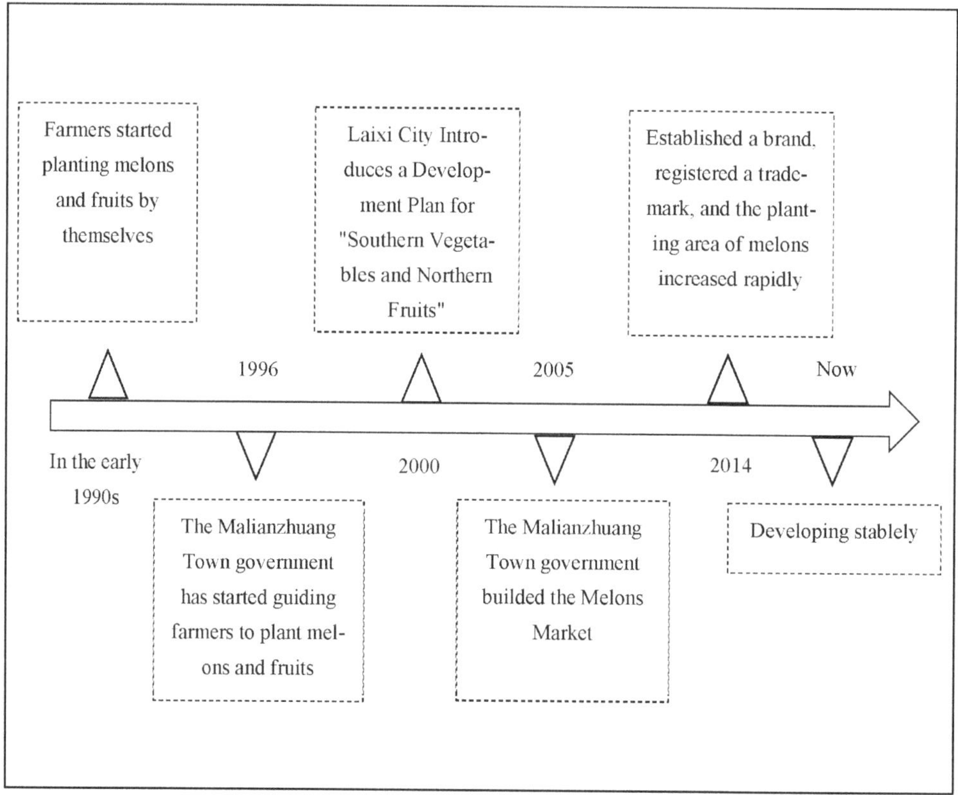

Figure 3. The development history of melon and fruit planting in Malianzhuang Town

4. Case Analysis Results

In examining the case analysis, three key aspects have surfaced, each revealing profound insights into the repercussions of the evolving landscape of agricultural technology extension. Two primary research conclusions emerge prominently. Firstly, the detachment of agricultural technology extension from the actual processes of agricultural production is evident. Secondly, there is a discernible disconnection between agricultural technology extension and the broader social fabric of agriculture. Additionally, the examination exposes a consequential negative outcome stemming from this technological disembedding. The overarching pattern that emerges is one of technological disembeddedness, as illustrated in Figure 4.

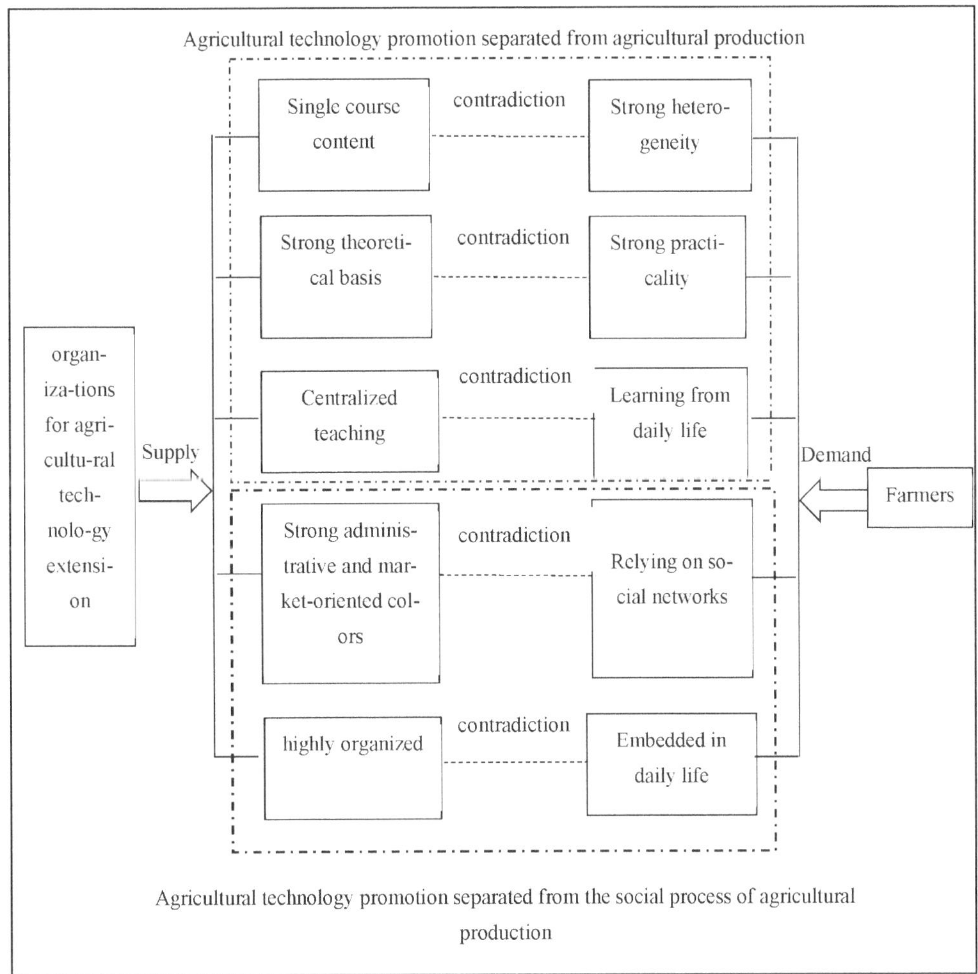

Figure 4. Performance of technical disembeddedness.

4.1. Agricultural Technology Disembeddedness from Agricultural Production Practice

(1) Mismatch between Technology Supply and Farmer Production Demands

The misalignment between agricultural technology supply and farmer production demands is one of the manifestations of technological disembeddedness. In the current agricultural technology dissemination system in China, training is the primary form of interaction between external entities and farmers [48]. Regardless of whether the disseminating entities are local agricultural extension agencies, market-based agricultural input suppliers like agribusinesses, or educational and research institutions such as universities, the attempt is made to disseminate technology through "centralized teaching" or "centralized teaching with field visits". However, this approach becomes disconnected from the actual practices of farmers.

Firstly, the singularity of training content conflicts with the complexity of agricultural production. Agricultural production varies depending on the specific geographic location, the methods of cultivation, and various other factors. Different soil conditions, planting methods, and other factors significantly impact crop growth. Moreover, even different fields within the same farm may face substantially different challenges. For local governments, although they understand the vital role of training in enhancing farmers' technical capabilities, they are often burdened by the heavy load of daily administrative work, coupled with the pressure of meeting superior-level assessment goals. This has led to a certain degree of formality in training, with the training's primary goal being the fulfillment of superiors' objectives. As a result, training sometimes takes on a superficial nature, wherein similar or identical course content is presented in many training sessions, even though the participating members exhibit high levels of heterogeneity. This has led to a situation where farmers across the entire county (or district) receive homogeneous technical content, which cannot cater to the diverse technological needs of individual farmers' production processes.

Secondly, the theoretical nature of the training content conflicts with the practical aspects of agricultural production. On one hand, agricultural production is highly practical, with farmers acquiring their production experience through day-to-day farming activities. The abstract nature of theoretical explanations makes it difficult for farmers to comprehend and, more importantly, fails to address their production issues. On the other hand, disseminating entities are constrained by factors such as time and space, making it challenging to effectively integrate theoretical training with specific practical farming contexts. This disconnect between training content and the actual demands of agricultural production contradicts the nature of agricultural production [24]. As a result, theoretical-oriented agricultural training has little impact on the farmers' actual production. Over time, agricultural training has become increasingly theoretical and formalized, becoming disengaged from practical production. Some of the farmers in Malianzhuang Town have participated in training organized by the township's agricultural technology department and higher-level departments. "The county offers technical training every year, and I attended it twice. *They are all oral lectures with no technical guidance, which did not have much impact" (interview with ZJS, 6 August 2020)*. This illustrates the real situation of technology disembeddedness and reflects the dissatisfaction of farmers with the current formal agricultural technology dissemination content. This dissatisfaction is due to their inability to access agricultural knowledge that meets their specific needs.

In the long run, the misalignment between the supply of agricultural technology knowledge and the demands of agricultural production directly results in low receptiveness among farmers and relatively passive participation in agricultural technology training. Even when farmers participate in training sessions, due to factors like the authority of village cadres, social obligations, and social prestige, the effectiveness of agricultural technology extension is limited. In comparison to practical application, farmers have weaker comprehension and assimilation capabilities of theoretical knowledge, and homogenized training content cannot address the specific challenges that farmers face in their practical production processes. This has resulted in the disengagement of agricultural technology extension from actual agricultural production practices in China.

(2) Contradictions Between the Organization of Agricultural Technology Dissemination and Farmers' Production Habits

The disembeddedness of agricultural technology extension from agricultural production practice is further evident in the impact of its organizational methods on the behavioral logic of farmers. For farmers, their knowledge systems and behavioral logic are developed through daily life practices and village social interactions. This determines the routine nature of their agricultural production knowledge acquisition, with their production skills gained through actual agricultural activities and interactions with other farmers. However, in the context of current agricultural technology training, local governments often tend to offer centralized training sessions, primarily due to considerations of time, organizational costs, and performance assessment metrics. These training sessions typically have fixed schedules, which do not align with the daily routines of farmers, especially during peak agricultural seasons when they have limited time for training.

For instance, in Malianzhuang Town, where farmers primarily cultivate sweet melons, one cycle of sweet melon production takes about six months, from planting in December to harvesting in May of the following year. Throughout this period, farmers are engaged in continuous, fragmented agricultural activities. They have to attend to their sweet melon fields nearly every other day. Some melon farmers even cultivate two cycles of sweet melons, leaving them with limited time and energy for formal training. (***Data derived from interviews with several farmers in Malianzhuang Town***).

The process of agricultural technology extension carries multiple objectives as it attempts to penetrate villages from the outside, and it inherently involves a long and challenging journey. Throughout this process, farmers gradually realize the conflicts between the form and content of agricultural technology extension and their own agricultural production practices. This realization may lead them to reject or resist the new forms of agricultural technology extension. They avoid the possibility of "classroom-style" extension impacting the logical systems they have constructed in their own behavior. They continue to acquire new technology and implement technical improvements through their traditional technology extension channels.

It is observed that administrative-driven agricultural technology dissemination often results in deviations from policy goals, leading to poor or even counterproductive dissemination outcomes. This contributes to one of the manifestations of technological disembeddedness.

In conclusion, administrative-driven agricultural technology extension primarily relies on classroom-based training. However, this form and content of agricultural technology extension do not align well with the processes of agricultural production. For farmers, agricultural technology is acquired and disseminated through their actual agricultural practices. While government efforts, such as training, have produced some results, the fixed format and excessive theoretical content limit its effectiveness. Consequently, government-led dissemination diverges from the specific production practices of farmers and their habitual knowledge acquisition patterns, failing to meet the actual technological needs of farmers, thus resulting in lower farmer engagement. Moreover, combined with the presence of performance assessment metrics in different regions, local governments often view this as an administrative task, leading to a high degree of formalization in training. Consequently, it often falls short of policy goals, resulting in poor training outcomes and even alienation, ultimately leading to the disembeddedness of agricultural technology dissemination from agricultural production.

4.2. Disembeddedness of Agricultural Technology Extension from Agricultural Social Processes

(1) Disconnection from Farmers' Social Networks

For agricultural production, farmers' adoption of technology is influenced not only by economic incentives but also by their social relationships. Given that farmers within the same region often share planting experiences and encounter similar issues, technology and expertise are frequently passed on through informal channels. The degree of farmers' adoption of a particular technology also depends on the intimacy of their social relationships. Thus, social relationships play a significant role in the dissemination of agricultural technology, rendering it a social process [49]. Agricultural technology dissemination involves not only the exchange of technical information but also social interactions among various stakeholders.

As GYL, a farmer from Lugezhuang Village, who was among the first to cultivate sweet melons locally, explained, "Back then, everyone in the village was growing watermelons using greenhouses to cultivate the seedlings. However, one year, a heavy rainfall destroyed all the watermelon seedling greenhouses. There wouldn't be enough time to rebuild the greenhouses and then grow watermelons. So, that's why we switched to growing sweet melons. When I first started growing them, there were no techniques for roasting melons. At that time, we were growing honeydew melons. In the first year, we had no fruit, and we didn't use seed grafting. Grafting cucumber seeds onto sweet melons just didn't work because the flowers didn't pollinate. None of it worked. It was only in the second year that we got some fruit. I went to the Xinhua bookstore to buy books to learn, but none of them were related to sweet melons. Later, I heard that there was someone in Zhuangtou who was growing melons on a small piece of land (the cultivation methods for small-scale melon and sweet melon are the same). So, I contacted a relative, and he introduced me to learn there, and that's how I started roasting melons. But even after roasting them, the market didn't recognize these melons, so we had to 'create a market' ourselves. It took us three years just to sell honeydew melons, three years at Lai Xi market, and another three years at Zhaoyuan market. Eventually, we gradually opened up the market, and traders started coming. It was only seven or eight years later that people in the village began growing sweet melons. They learned the techniques from me. Sometimes they would come to my (melon) greenhouse (to learn), and other times they would ask me to visit their greenhouses. The village never organized any training sessions" **(interview with GYL, 4 August 2020)**.

From the interview data, it becomes evident that in the early stages of the sweet melon industry in Lugezhuang Village, a lack of technical knowledge about growing sweet melons posed challenges in producing market-quality agricultural products. To overcome this predicament, GYL, as the recipient of agricultural technology, took the initiative to expand his social network beyond the village, engaging with agricultural technology providers from other areas. This expansion allowed him to access agricultural technology information. It demonstrates that in this initial phase, informal agricultural technology was transferred across regions, and over time, GYL transitioned from being a technology recipient to becoming a technology owner. Meanwhile, agricultural technology began to disseminate within the village. This transition marks the onset of the second phase. It was initiated by GYL's gradual success in entering the market and earning the first substantial profits. When the village had its "first mover", who achieved significant financial gains by cultivating sweet melons, other villagers started emulating and learning from him. During the process of mutual interactions, their behavior and decision-making patterns evolved. In this context, GYL, as the owner of agricultural technology, took on the role of a technology extension in this phase. He taught fellow villagers the techniques for growing sweet melons. Due to the frequency and complexity of interactions in rural communities, a multifaceted pattern of interaction emerged between the disseminators and recipients, enabling the smooth operation of agricultural technology dissemination and adding implicit credibility to the process.

In summary, farmers' technology needs are predominantly acquired through their social relationships, and social networks serve as important conduits for agricultural technology dissemination. With the commercialization reform of agricultural technology dissemination, the dissemination methods have become more diversified, with government and market-based training gradually becoming the primary means of dissemination. Paradoxically, these varied dissemination methods have adopted a standardized training approach, which has removed agricultural technology from its embeddedness within social networks. This detached mode of agricultural technology dissemination fails to cater to farmers' actual needs and is met with their resistance.

It is evident that under government and market-driven standardization of agricultural technology dissemination, the unique logic of agricultural technology dissemination is gradually eroded. It not only results in a lack of trust in the disseminating bodies but also reduces farmers' acceptance of new agricultural technology. Furthermore, it fails to meet the practical needs of farmers for agricultural technology.

(2) Disconnection from Farmers' Daily Life

Agricultural production itself constitutes an integral part of farmers' daily lives, and it unfolds as a part of daily life, interwoven with social interactions resulting from production. In fact, the initial agricultural technology dissemination system was founded based on this characteristic. During the collectivization period, the government established "seven stations and eight institutions" in towns, with personnel from these subsidiary organizations providing guidance in the fields. This significantly enhanced the efficiency of agricultural technology dissemination, and it was in alignment with the practicality of agriculture, integrated into the daily production and lives of farmers. However, this segmented institutional arrangement placed a considerable burden on local township governments. With the market-driven reform of agricultural technology dissemination, the original agricultural technology dissemination system disintegrated, and the dissemination approach gradually shifted toward classroom-style training. While this approach became more organized, it gradually disconnected from the production and lives of farmers as it was primarily driven by administrative tasks and market efficiency.

Lugezhuang Village is one of the major apple-growing villages in Malianzhuang Town. The village party secretary, YYX, explained, *"Now, the government organizes technical training through the village and community about 2–3 times a year. The content covers daily orchard management and learning about production processes. During regular days, some experienced orchardists give lectures to everyone. If anyone has questions, they'll ask them"* **(interview with YYX, 15 August 2020)**.

The primary characteristic of this grassroots agricultural technology dissemination, where farmer technicians serve as the main disseminators, is its flexibility. It is not constrained by geographic limitations embedded in daily life and interpersonal relationships. Simultaneously, the implicit reputation network constructed within village life also serves as a reciprocal incentive for farmer technicians to engage in agricultural technology dissemination.

However, the current training approach, characterized by classroom-style instruction, relocates farmers from the fields to the classroom, where agricultural technology knowledge is imparted. This removes them from the actual production and life settings. Farmers not only find it challenging to understand this way, but it also detaches them from practical production. Even if they comprehend the material during the training, implementing it upon return becomes problematic. Over time, farmers lose interest in attending these training sessions. As described by YYX, *"The agricultural technology department organizes training once or twice in the spring and autumn, and for the rest of the time, they mainly focus on administrative work. They don't have time to participate in farmers' daily lives"* **(interview with YYX, 15 August 2020)**.

Although these training sessions provide farmers with new technology and knowledge, the results achieved by the nationally driven agricultural technology dissemination system, as well as market-driven agricultural technology dissemination, still fall short of expectations. On the other hand, market-driven agricultural technology dissemination also leads to the increasing detachment of technology from farmers' daily lives. Driven by the pursuit of profit and market share, it naturally assumes a state of technological disembeddedness.

4.3. Adverse Consequences of Agricultural Technology Disembeddedness

(1) Polarization in Access to Agricultural Technology

In practical agricultural technology dissemination, whether driven by the government or the market, new agricultural entities and large-scale growers are often prioritized as the primary training targets [50]. This is partly because organizing such training involves lower costs and is relatively straightforward in terms of coordinating participants. Additionally, larger-scale operators tend to exhibit better understanding and acceptance of the technologies being disseminated. This dynamic results in new agricultural entities and large-scale growers having a higher likelihood of receiving agricultural technology services. A staff member from the Agricultural Service Center in Malianzhuang Town revealed the logic behind their training approach, stating, *"One is the overarching trend of national policy, which encourages the development of large-scale farming operations, and we are committed to complying with this policy. Furthermore, large-scale growers, after training, tend to yield better results compared to smallholders. We prefer engaging with large-scale growers; they are relatively more open-minded"* **(interview with YZR, 12 August 2020)**.

From the perspective of agricultural technology promoters, regardless of the scale of cultivation or individual qualifications, new agricultural entities are considered superior to smallholders. Not only can they assist in achieving administrative objectives, but they also demonstrate better results in terms of demonstrating and disseminating technology. They are even expected to drive the development of smallholders. However, in practice, owing to the socio-economic status disparities among farmers, the demonstration effect of large-scale growers on smallholders is not particularly pronounced [13]. This situation results in the creation of "technological barriers" or even exacerbates the polarization in technology acquisition between large-scale growers and smallholders. It is noteworthy that the operational scale of large-scale farms in China varies based on the type of agricultural activity. For those engaged in the cultivation of grains such as rice, wheat, and corn, the land management scale should be at least 10 hectares. For those involved in the cultivation of vegetables, fruits, horticultural crops, or other crops, the land management scale should be at least 8 hectares. In the case of aquaculture, the land management scale should be at least 5 hectares.

WXS, the village secretary of Xiawazi Village in Malianzhuang Town, began cultivating pear trees in 2014, with an orchard area exceeding 600 mu (approximately 40 hectares). His operations span multiple villages, including Xiawazi Village, Sunjia Village, Jijia Village, and Beishankou Village. He is renowned as a large-scale pear grower, and the government supports his endeavors through multiple subsidies and development projects, including comprehensive land development, water, electricity, and road projects. He stated, *"My farm has benefited from national subsidies. The irrigation channels were dug by the water conservancy department, electricity was provided by the power company, and the roads were constructed by a construction company. We also receive training sessions organized by the government. They invite experts from Yantai and Laiyang to give lectures"*. However, when asked about how to develop and uplift the two-thirds of growers who are smallholders, he explained, *"There is some assistance, but at most, it means helping them get an extra one yuan when selling the produce. Managing together is not feasible, and each household does not have the necessary facilities. Some production materials, such as plant growth regulators, can be shared among the community, but for fertilizers and pesticides, we use high-quality ones which are more expensive, something the average folks cannot afford"* **(interview with WXS, 3 June 2020)**.

Clearly, "supporting the strong and the large" has become the core logic of organizing agricultural technology training. Although it allows agricultural technology extension departments to fulfill their training tasks at the lowest organization and coordination cost, it results in polarization between large-scale growers and smallholders in terms of technology training and adoption. Large-scale growers receive various economic and technical subsidies, while smallholders are effectively excluded and may not be included in the scope of agricultural technology training. This logic considerably restricts the space for smallholders to access agricultural technology. In other words, approximately 90% of China's total agricultural operators have not received effective agricultural technology training, leading not to improvements in their production techniques but rather to the wastage of significant resources *(data from the Third National Agricultural Census in 2016 show that China had a total of 314.22 million agricultural operators. Large-scale agricultural operators accounted for only 4.1% of the overall agricultural operator population. This demonstrates that smallholder farmers remain a vast and important segment of the agricultural landscape, both currently and in the foreseeable future)*.

(2) Formalization of Agricultural Technology Extension

In addition to indirectly excluding smallholders from agricultural technology training, technology disembeddedness leads to the gradual formalization of agricultural technology extension. In practical agricultural technology extension, multiple promotional entities become ensnared in conflicting objectives and gradually shift towards standardizing and task-oriented agricultural technology extension, ultimately reducing it to mere formality in order to fulfill their mandates. Faced with formal and often less practical agricultural technology training, farmers frequently opt not to participate. Even when some farmers partake in these training programs due to various administrative pressures and profit incentives, it is often seen as a perfunctory obligation. Importantly, the formalization of agricultural technology extension not only fails to enhance farmers' technical proficiency but also engenders resentment among them, leading to explicit or implicit resistance in practice.

ZQS, a pear orchard owner in Zhanjia Village, is among the younger generation in the village. He commented, *"I attended the training sessions led by new agricultural entities, and basically, it was the people working in the village who participated. They have no land of their own and are not very receptive. The money spent on these sessions is wasted. The organizers are aware of this, but why do they still conduct the training? It's because they all have vested* interests; all that money needs to be spent" **(interview with ZQS, 31 May 2020)**.

A local agricultural supply store owner in Malianzhuang Town expressed, *"To promote fertilizers and agricultural materials, we usually hire lecturers. However, regular folks are quite pragmatic. Without freebies, they won't attend, and they won't even come for food"* **(interview with LRB, 5 August 2020)**.

It is evident that both government-driven and market-led agricultural technology extensions have strayed from their original purpose of enhancing farmers' technical abilities, becoming tools to fulfill administrative tasks and gain economic profits instead. This has led farmers to resist these efforts. Agricultural technology extension, influenced by both governmental and market forces, has increasingly drifted away from what rural communities genuinely require in terms of agricultural technology dissemination.

5. The Logical Paradox in Agricultural Technology Extension System Construction

As analyzed in the preceding sections, the current agricultural technology extension methods have become detached from agricultural production and social processes. This detachment has resulted in the formalization of agricultural technology extension and the exclusion of smallholder farmers, among other issues. The fundamental reason for these challenges lies in the inherent conflict between the supply logic of agricultural technology and the actual logic of its dissemination. The conflict between these two types of logic has led to suboptimal outcomes in agricultural technology extension.

5.1. Discrepancy between Administrative Logic and Social Logic

Generally, the core department responsible for agricultural technology extension is the local agricultural technology extension department, a direct subsidiary of government functional departments with administrative characteristics. While it does consider practical work outcomes, it is primarily oriented towards achieving upper-level tasks due to factors such as organizational costs and superior assessments. In contrast, the actual extension of agricultural technology is aimed at increasing efficiency, considering the localized nature of agricultural production. Agricultural technology extension is embedded in farmers' daily production, life, and social interactions and operates according to social logic. These two types of logic, administrative logic and social logic, are inherently contradictory in practical agricultural technology extension.

In the early years of the People's Republic of China, the government established agricultural technology stations at the township level, and their staff provided field-based agricultural technology guidance, which was in line with the reality of agricultural production, as previously discussed. However, with the advent of market-oriented reforms and concurrent adjustments to grassroots administrative structures, many townships abolished specialized technical extension stations and established new agricultural offices or comprehensive service stations, incorporating agricultural technology extension within them. As a result, agricultural technology extension took on a more administrative coloration and often organized training sessions. Although this practice facilitated effective management of agricultural technology extension and improved administrative efficiency, the results were not as effective compared to the previous method, where agricultural technicians directly promoted technologies. This is because including agricultural technology extensions in new departments resulted in standardized assessments from upper authorities. Once agricultural technology extension was integrated into new departments, agricultural technicians fell under the dual jurisdiction of both the higher-level agricultural technology extension institutions and local governments. They had to fulfill the task of agricultural technology extension while also facing administrative assessments. However, the effectiveness of agricultural technology extension was often difficult to measure in the short term, leading to course-based training becoming the primary mode of technology extension. The agricultural department could use quantifiable indicators such as the number of training participants and the frequency of classes to highlight its workload, while whether the training actually improved the technical abilities of farmers became less important. This gradual shift led agricultural technology extensions to detach from the needs of farmers' production and lives, resulting in formalization.

In summary, the standardized assessments and administrative operations of agricultural technology departments emphasize the formal extension of technology, such as classroom lectures, and neglect the practicality of technology extension for farmers. This detachment from the social logic of agricultural technology extension based on social relationships ultimately leads to a suboptimal outcome in technology extension.

5.2. Discrepancy between Market Logic and Social Logic

After the implementation of economic reforms and opening-up policies, agricultural production resources in China began to gradually transition from state control to private capital operation [51]. This shift was accompanied by the rise of a new wave of agricultural input production companies. Over the years, these companies formed a supply chain for agricultural inputs, and the extension of agricultural technology became an integral part of the current agricultural technology extension system. In practice, market-driven agricultural technology extension becomes a vital component of the agricultural technology extension system. Despite being embedded in local communities through rural agricultural input shops, the primary objective of market-driven agricultural technology extension remains the sale of agricultural inputs. Agricultural technology services are provided to farmers only when they purchase these inputs. Thus, agricultural technology services have become tools for promoting agricultural inputs and maximizing profits for the input providers. The core logic governing this approach is driven by profit maximization. Under this logic, market entities possess a strong incentive to provide agricultural technology services, yet they tend to promote only the information that benefits their own financial interests. This may even lead to the irregular practice mentioned earlier, where market entities financially incentivize farmers to attend training sessions to compete in the sales market. Over time, the various costs borne by market entities are subtly transferred to the farmers, further increasing their production costs. Furthermore, some market entities forcefully implement technology upgrades to improve product profitability, leading to an excessive emphasis on agricultural technology extension.

The market-driven extension of agricultural technology runs contrary to the logic of its promotion in local communities. In fact, the spillover and quasi-public nature of agricultural technology imply that its promotion within rural communities should have a "shared" attribute. As mentioned earlier, agricultural technology is integrated into the production and daily lives of farmers, disseminated through social relationship networks, and its acquisition and promotion take place within the farmers' daily interactions. It involves frequent reciprocal acts of mutual benefit, guided by the principles of "receiving favors" and "returning favors" among farmers. When the market logic of agricultural technology extension enters villages, agricultural input sales personnel naturally promote technologies that yield higher profits while excluding cost-effective, practical technologies. This not only increases farmers' costs but also hinders the sustainable development of agriculture.

In summary, government-led and market-driven agricultural technology extensions each have their own inherent logic. Government-led extension is primarily focused on meeting upper-level assessment targets, while market-driven extension places profit maximization at its core. Both of these logics deviate from the reality of agricultural production and diverge from the social logic of agricultural technology extension, which relies on social relationships. This ultimately results in the formalization and instrumentalization of agricultural technology extension, detaching it from rural communities.

6. Conclusions and Policy Recommendations

Agricultural technology promotion has been a pivotal force in reshaping agricultural production methods and expediting rural modernization in China. Despite state-led efforts to promote market-oriented reforms in this domain, aiming for a multi-participant system involving the government, market entities, and other stakeholders, the practical outcomes have fallen short. This paper introduces the concept of "technological disembedding" to elucidate the root cause of underperformance, attributing it to the conflict between the government and market-driven logic in agricultural technology promotion and the actual logic of agricultural technology dissemination.

In China, traditional agricultural practices have been deeply embedded in farmers' daily lives, promoted through interpersonal networks and social logic. However, the current promotion system, guided by the government and market, diverges from this approach, adopting formal and profit-centric characteristics. This deviation results in a disconnect from practical agricultural needs, leading to poor outcomes and farmer opposition. To address this, the paper proposes a shift towards a multi-dimensional agricultural technology promotion system, considering farmers' characteristics, production realities, and social dynamics. This involves increased investment in training for farmer technicians and informal promoters, leveraging local influencers to disseminate advanced technology widely.

Comparison with the European Union: A Holistic Approach

In contrast, the European Union (EU) has embraced a more holistic approach to agricultural technology promotion, aligning policies with the socio-economic and environmental dimensions of farming. The EU emphasizes sustainability, agroecology, and farmer empowerment. While both regions grapple with the challenge of balancing administrative and market-driven logic, the EU has seen success in fostering collaboration between government, research institutions, and farmers through agri-environmental schemes. Unlike the formalized approach in China, the EU's promotion models encourage diversified channels and informal networks.

Lessons from the EU: A Nuanced Strategy

Drawing lessons from the EU, China should pivot towards a comprehensive strategy that harmonizes administrative, market-driven, and informal promotion methods. Strengthening the role of informal agricultural technology promoters as the "last mile" ensures effective technology dissemination. This comparative analysis underscores the need for a nuanced and context-specific approach, promoting a balance between formalized training and diverse channels of technology promotion to avoid the pitfalls of technological disembedding.

Comparison with the United States: Market Orientation and Farmer Engagement

Simultaneously, a comparison with the United States reveals two distinct models of agricultural technology promotion. In the U.S., technology dissemination places greater emphasis on market orientation and farmer engagement, adopting more flexible and diversified approaches. Collaboration between government, universities, and agricultural enterprises fosters technological innovation, and farmers actively participate in training and adopt new technologies. This open and flexible model brings U.S. agricultural technology promotion closer to practical needs, enhancing the acceptance of technology.

Lessons from the U.S.: A Flexible, Market-oriented Strategy

Drawing lessons from the U.S., China can adopt a more flexible, market-oriented strategy, promoting collaboration among the government, universities, and enterprises to stimulate farmer engagement.

Author Contributions: X.H.: Conceptualized data analysis and conducted the literature review to help draft the article. Z.T.: Supervised dissertations, provided revision suggestions, and organized research projects. B.X.: Provided thesis ideas, prepared the essay, and revised the paper. All authors have read and agreed to the published version of the manuscript.

Funding: There was no foundation support for this project.

Institutional Review Board Statement: The study was conducted according to the guidelines of the Declaration of Helsinki. The project was approved by the Ph.D. Program Committee of Renmin University of China and Northeastern University and by the Institutional Ethics Committee of the Renmin University of China and Northeastern University (Project Identification Code: 10/11/2021).

Informed Consent Statement: Informed consent was obtained from all subjects involved in the study.

Data Availability Statement: The raw data supporting the conclusions of this article will be made available by the authors without undue reservation.

Conflicts of Interest: The authors declare no conflicts of interest.

References

1. Chavas, J.P.; Nauges, C. Uncertainty, Learning, and Technology Adoption in Agriculture. *Appl. Econ. Perspect. Policy* **2020**, *42*, 42–53. [CrossRef]
2. Huang, J.K.; Hu, R.F.; Zhi, H.Y. 30-year Development and Reform of Grassroots Agricultural Technology Extension System: Policy Evaluation and Suggestions. *J. Agrotech. Econ.* **2009**, *1*, 4–11. Available online: https://d.wanfangdata.com.cn/periodical/ChlQZXJpb2RpY2FsQ0hJTmV3UzIwMjMxMjI2Eg9ueWpzamoyMDA5MDEwMDEaCGxkcjVoZzhz (accessed on 14 January 2024).
3. Kuehne, G.; Llewellyn, R.; Pannell, D.J.; Wilkinson, R.; Dolling, P.; Ouzman, J.; Ewing, M. Predicting farmer uptake of new agricultural practices: A tool for research, extension and policy. *Agric. Syst.* **2017**, *156*, 115–125. [CrossRef]
4. Spielman, D.J.; Ekboir, J.; Davis, K. The art and science of innovation systems inquiry: Applications to Sub-Saharan African agriculture. *Technol. Soc.* **2009**, *31*, 399–405. [CrossRef]
5. Zheng, Y.Y.; Zhu, T.H.; Jia, W. Does Internet use promote the adoption of agricultural technology? Evidence from 1 449 farm households in 14 Chinese provinces. *J. Integr. Agric.* **2022**, *21*, 282–292. [CrossRef]
6. Norton, G.W.; Alwang, J. Changes in Agricultural Extension and Implications for Farmer Adoption of New Practices. *Appl. Econ. Perspect. Policy* **2020**, *42*, 8–20. [CrossRef]
7. Ogutu, S.O.; Fongar, A.; Gödecke, T.; Jäckering, L.; Mwololo, H.; Njuguna, M.; Wollni, M.; Qaim, M. How to make farming and agricultural extension more nutrition-sensitive: Evidence from a randomised controlled trial in Kenya. *Eur. Rev. Agric. Econ.* **2020**, *47*, 95–118. [CrossRef]
8. Gao, T.Z.; Feng, H.; Lu, Q. Can Digital Agricultural Extension Services Promote Farmers' Green Production Technology Choices?—Based on Micro-survey Data from Three Provinces in the Yellow River Basin. *J. Agrotech. Econ.* **2023**, *9*, 23–38. [CrossRef]
9. Cai, J.Y.; Hu, R.F.; Hong, Y. Impact of farmer field schools on agricultural technology extension-evidence from greenhouse vegetable farms in China. *Appl. Econ.* **2022**, *54*, 2727–2736. [CrossRef]
10. Ruzzante, S.; Labarta, R.; Bilton, A. Adoption of agricultural technology in the developing world: A meta-analysis of the empirical literature. *World Dev.* **2021**, *146*, 105599. [CrossRef]
11. Wang, J.Y.; Chen, S.B. Analysis of the Effectiveness of the Reform of the Agricultural Technology Promotion System of "Supporting Business with Money"—Based on Field Research in Jiangxia, Xiangyang, and Zengdu, Hubei Province. *Issues Agric. Econ.* **2013**, *34*, 97–103. [CrossRef]
12. Gao, Y.; Wang, Q.N.; Chen, C.; Wang, L.Q.; Niu, Z.H.; Yao, X.; Yang, H.R.; Kang, J.L. Promotion methods, social learning and environmentally friendly agricultural technology diffusion: A dynamic perspective. *Ecol. Indic.* **2023**, *154*, 11. [CrossRef]

13. Tong, D.J.; Huang, W. Differences in Socio-economic Status, Access to Extension Service and Agricultural Technology Diffusion. *Chin. Rural. Econ.* **2018**, *11*, 128–143. Available online: https://d.wanfangdata.com.cn/periodical/zgncjj201811009 (accessed on 1 January 2024).

14. Hu, R.F.; Cai, Y.Q.; Chen, K.Z.; Huang, J.K. Effects of inclusive public agricultural extension service: Results from a policy reform experiment in western China. *China Econ. Rev.* **2012**, *23*, 962–974. [CrossRef]

15. Wang, X.Y.; Wang, W.L.; Zhuang, X.C. Comparative Analysis of the Diversified Development of Agricultural Technology Extension Organizations—Based on a Survey in Sanming City, Fujian Province. *Jiangsu Agric. Sci.* **2017**, *45*, 281–286. [CrossRef]

16. Dhehibi, B.; Rudiger, U.; Moyo, H.P.; Dhraief, M.Z. Agricultural Technology Transfer Preferences of Smallholder Farmers in Tunisia's Arid Regions. *Sustainability* **2020**, *12*, 421. [CrossRef]

17. Birke, F.M.; Lemma, M.; Knierim, A. Perceptions towards information communication technologies and their use in agricultural extension: Case study from South Wollo, Ethiopia. *J. Agric. Educ. Ext.* **2019**, *25*, 47–62. [CrossRef]

18. Gatdet, C. The Ethiopian agricultural extension services: A mixed perspective. *Cogent Food Agric.* **2022**, *8*, 11. [CrossRef]

19. Cook, B.R.; Satizábal, P.; Curnow, J. Humanising agricultural extension: A review. *World Dev.* **2021**, *140*, 19. [CrossRef]

20. Hu, R.F.; Sun, Y.D. The Get-rid-of and the Reply of Agricultural Technology Extension System. *Reform* **2018**, *2*, 89–99. Available online: https://d.wanfangdata.com.cn/periodical/ChlQZXJpb2RpY2FsQ0hJTmV3UzIwMjMxMjI2Eg1nYWlnMjAxODAyMDA5Ggg4YjVxMXVrNg== (accessed on 4 January 2024).

21. Darnhofer, I.; Lamine, C.; Strauss, A.; Navarrete, M. The resilience of family farms: Towards a relational approach. *J. Rural Stud.* **2016**, *44*, 111–122. [CrossRef]

22. Xia, J.Y. Developing diversified agricultural technology extension services for modern agricultural construction. *China Agric. Technol. Ext.* **2010**, *26*, 4–7. [CrossRef]

23. Arslan, C.; Wollni, M.; Oduol, J.; Hughes, K. Who communicates the information matters for technology adoption. *World Dev.* **2022**, *158*, 106015. [CrossRef]

24. Liao, C.D.; Ross, H.; Jones, N.; Palaniappan, G. System lacks systems thinking: Top-down organization and actor agency in China's agricultural extension system. *Syst. Res. Behav. Sci.* **2023**, *41*, 82–99. [CrossRef]

25. Chi, Z.X.; Guo, Y.Y.; Zhang, X.Y.; Zhang, Y.Y. Discussion on Several Issues of Agricultural Intermediary Organizations. *Issues Agric. Econ.* **2004**, 50–52+55.

26. Li, Z.J.; Hu, R.F.; Zhang, C.; Xiong, Y.K.; Chen, K. Governmental regulation induced pesticide retailers to provide more accurate advice on pesticide use to farmers in China. *Pest Manag. Sci.* **2022**, *78*, 184–192. [CrossRef] [PubMed]

27. Chen, X.Y.; Li, T.S. Diffusion of Agricultural Technology Innovation: Research Progress of Innovation Diffusion in Chinese Agricultural Science and Technology Parks. *Sustainability* **2022**, *14*, 15008. [CrossRef]

28. Dou, P.H.; Chen, S.B. A Study on the Mechanism of the Role of Rural Cooperative Economic Organizations in the Diversified Agricultural Technology Extension System: Based on Field Investigation of 7 Farmers' Professional Cooperatives in Zaozhuang, Shandong Province. *Theory Mon.* **2011**, *8*, 172–176. [CrossRef]

29. Genius, M.; Koundouri, P.; Nauges, C.; Tzouvelekas, V. Information Transmission in Irrigation Technology Adoption and Diffusion: Social Learning, Extension Services, and Spatial Effects. *Am. J. Agr. Econ.* **2014**, *96*, 328–344. [CrossRef]

30. Benyishay, A.; Mobarak, A.M. Social Learning and Incentives for Experimentation and Communication. *Rev. Econ. Stud.* **2019**, *86*, 976–1009. [CrossRef]

31. Fei, X.T. *Rural China's Fertility System*; Peking University Press: Beijing, China, 1998.

32. Xu, J.B.; Cui, Z.D.; Wang, T.Y.; Wang, J.J.; Yu, Z.G.; Li, C.X. Influence of Agricultural Technology Extension and Social Networks on Chinese Farmers' Adoption of Conservation Tillage Technology. *Land* **2023**, *12*, 1215. [CrossRef]

33. Nakano, Y.; Tsusaka, T.W.; Aida, T.; Pede, V.O. Is farmer-to-farmer extension effective? The impact of training on technology adoption and rice farming productivity in Tanzania. *World Dev.* **2018**, *105*, 336–351. [CrossRef]

34. Sah, U.; Singh, S.K.; Pal, J.K. Farmer-To-Farmer Extension (F2FE) approach for speedier dissemination of agricultural technologies: A review. *Indian J. Agric. Sci.* **2021**, *91*, 1419–1425. [CrossRef]

35. Takahashi, K.; Muraoka, R.; Otsuka, K. Technology adoption, impact, and extension in developing countries' agriculture: A review of the recent literature. *Agric. Econ.* **2020**, *51*, 31–45. [CrossRef]

36. Gao, Y.; Zhao, D.Y.; Yu, L.L.; Yang, H.R. Influence of a new agricultural technology extension mode on farmers' technology adoption behavior in China. *J. Rural Stud.* **2020**, *76*, 173–183. [CrossRef]

37. Wang, J.M.; Zhou, N.; Zhang, L. Innovative Research on the Evaluation Mechanism of Agricultural Technology Extension Behavior in China: Based on a Survey of 759 Agricultural Technicians in 16 Counties. *Issues Agric. Econ.* **2011**, *32*, 28–32+110. [CrossRef]

38. Granovetter, M. Economic-Action and Social-Structure—The Problem of Embeddedness. *Am. J. Sociol.* **1985**, *91*, 481–510. [CrossRef]

39. Fu, P. Embeddedness: Divergence and controversy. *Sociol. Stud.* **2009**, *24*, 141–164+245. [CrossRef]

40. D'Cruz, P.; Noronha, E.; Banday, M.U.L.; Chakraborty, S. Place Matters: (Dis)embeddedness and Child Labourers' Experiences of Depersonalized Bullying in Indian Bt Cottonseed Global Production Networks. *J. Bus. Ethics* **2022**, *176*, 241–263. [CrossRef]

41. Polanyi, K. *The Economy as an Instituted Process*; Polanyi, K., Arensburg, C.M., Pearson, H.W., Eds.; Westview Press: Boulder, CO, USA, 1957.

42. Polanyi, K. The Great Transformation. In *Readings in Economic Sociology*; John Wiley and Sons Ltd.: Hoboken, NJ, USA, 2008.

43. Monteiro, C.; Lima, R. Embeddedness and Disembeddedness in Economic Sociology in Three Time Periods. *Sociol. Antropol.* **2021**, *11*, 43–67. [CrossRef]
44. Huang, Z.F. Dis-embeddedness and Re-embeddedness: Agricultural Operation and Governance in Village Order. *China Rural. Surv.* **2018**, *3*, 51–64. Available online: https://d.wanfangdata.com.cn/periodical/zgncgc201803004 (accessed on 5 January 2024).
45. Geng, Y.H. Dis-Embedding Development: An Explanation Frame of Rural Environmental Problems. *J. Nanjing Agric. Univ. Soc. Sci. Ed.* **2017**, *17*, 21–30+155–156.
46. Yin, R.K. The Case-Study Crisis—Some Answers. *Adm. Sci. Q.* **1981**, *26*, 58–65. [CrossRef]
47. Zhang, J. The Goal of Case Analysis: From Story to Knowledge. *Soc. Sci. China* **2018**, *8*, 126–142+207. Available online: https://d.wanfangdata.com.cn/periodical/zgshkx201808009 (accessed on 1 January 2024).
48. Kong, X.Z. Comparison among Different Countries' Agricultural Technology Extension Systems, Current Situations and China's Countermeasures. *Reform* **2012**, *1*, 12–23. Available online: https://d.wanfangdata.com.cn/periodical/ChlQZXJpb2RpY2FsSQ0hJTmV3UzIwMjMxMjl2Eg1nYWlnMjAxMjAxMDAzGggxNXJvd2M0ZA== (accessed on 18 January 2024).
49. Thornton, P.H.; Lounsbury, W.O.M. The Institional Logics Perspective: A New Approach to Culture, Structure and Process. In *The Institional Logics Perspective: A New Approach to Culture, Structure and Process*; OUP: Oxford, UK, 2012.
50. Sun, X.H. Study on Tilting Massive Production Operators in Grass-roots Agricultural Technology Service under Background of Scale Operation. *J. Northwest A F Univ. Soc. Sci. Ed.* **2017**, *17*, 80–86. [CrossRef]
51. Chen, Y.Y. The Marketization of China's Agricultural Inputs and the Hidden Dynamics of Agrarian Capitalization. *Open Times* **2018**, *3*, 95–111+119–110. Available online: https://d.wanfangdata.com.cn/periodical/ChlQZXJpb2RpY2FsSQ0hJTmV3UzIwMjMxMjl2Eg1rZnNkMjAxODAzMDExGgg5a2dkZjg2bw== (accessed on 3 January 2024).

sustainability

Article

The Regional Heterogeneity of the Impact of Agricultural Market Integration on Regional Economic Development: An Analysis of Pre-COVID-19 Data in China

Xinru Miao [1,*], Shaopeng Wang [2,*], Jiqin Han [1], Zhaoyi Ren [3], Teng Ma [3] and Henglang Xie [3]

[1] College of Economics and Management, Nanjing Agricultural University, Nanjing 210095, China; jhan@njau.edu.cn

[2] Faculty of Agriculture, Universiti Putra Malaysia, Serdang 43400, Malaysia

[3] School of Business and Economics, Universiti Putra Malaysia, Serdang 43400, Malaysia

* Correspondence: 2020206006@stu.njau.edu.cn (X.M.); gs67018@student.upm.edu.my (S.W.)

Abstract: The abrupt onset of the COVID-19 pandemic in late 2019 significantly disrupted China's domestic agricultural production and supply chain stability. Local governments, responding to urgent circumstances, implemented various trade restrictions that profoundly affected regional economic development. This study, covering data from 2010 to 2019 across 31 provinces, investigates agricultural market integration and regional economic development. Employing a dynamic spatial panel Durbin model, it systematically analyzes the complex relationship between these variables. International trade variables related to agricultural products are then introduced to examine their "substitution effect" in promoting regional economic development through agricultural market integration. The research findings are summarized as follows: (1) disregarding international agricultural trade, a one-unit increase in the agricultural market integration index corresponds to a 0.156% rise in regional economic development. (2) In an open economy, the substitution coefficients for agricultural imports, exports, and total trade concerning market integration are -0.00097, -0.0012, and -0.0038, respectively. (3) The strength of the substitution effect from the international agricultural market to the domestic market varies regionally, with coefficients of -0.00099 and -0.00217 for the eastern and western regions, respectively.

Keywords: agricultural market integration; agricultural foreign trade; substitution effect; regional economic development

Citation: Miao, X.; Wang, S.; Han, J.; Ren, Z.; Ma, T.; Xie, H. The Regional Heterogeneity of the Impact of Agricultural Market Integration on Regional Economic Development: An Analysis of Pre-COVID-19 Data in China. *Sustainability* **2024**, *16*, 1734. https://doi.org/10.3390/su16051734

Academic Editors: Fotios Chatzitheodoridis, Efstratios Loizou and Achilleas Kontogeorgos

Received: 7 December 2023
Revised: 17 January 2024
Accepted: 7 February 2024
Published: 20 February 2024

1. Introduction

The abrupt onset of the COVID-19 pandemic in late 2019 significantly disrupted the stability of China's domestic agricultural production and supply chain. Due to lockdown measures, various stages of agricultural activities, including planting, harvesting, processing, and transportation, were affected [1,2]. This led to a reduction in agricultural production and supply shortages, resulting in price fluctuations in the agricultural market. Consumers faced food shortages and rising prices, while farmers encountered difficulties in selling their agricultural products [3,4]. Simultaneously, the Chinese government implemented a series of measures to strengthen quarantine and health requirements for imported agricultural products to prevent the spread of the pandemic. It restricted agricultural product exports to ensure domestic market supply. These measures created uncertainty and difficulties in international agricultural trade. In the post-COVID-19 era, overcoming the downward pressure on China's domestic economic development caused by the severe global economic recession, accelerating the construction of agricultural market integration, and implementing a series of policies to support farmers and agricultural producers to alleviate the negative impact of the pandemic on them and promote regional economic development will be key issues [5,6].

The total volume of China's agricultural import and export trade increased from USD 121.96 billion in 2010 to USD 228.427 billion in 2019, with an average annual growth rate of 6.48%. In 2020, the total volume of agricultural imports and exports amounted to USD 246.83 billion, representing a year-on-year increase of 8.0%. The trade deficit reached USD 94.77 billion, reflecting a growth of 32.9%. China's domestic agricultural market is closely connected with the international market and is significantly influenced by the international agricultural market. From the perspective of the domestic market, the sales revenue of agricultural products increased from CNY 267.848 billion in 2010 to CNY 1858.08 billion in 2019, with an average annual growth rate of 21.37%. In 2020, the sales revenue of agricultural products amounted to CNY 2220.53 billion, representing a 19.5% increase. However, the growth rate declined. Against the backdrop of a significant increase in the agricultural trade deficit and a decrease in the sales scale of the domestic agricultural market, this situation is not conducive to the improvement of per capita disposable income for residents. Some scholars argue that significant fluctuations in agricultural product prices can lead to chaotic transactions in the agricultural market and a reduction in overall social welfare [7,8]. They emphasize that market integration, including the agricultural market, is a necessary condition for achieving cross-regional price stability, ensuring agricultural supply security and thereby improving the well-being of the population [9].

Regional economic development demands stable market prices [10] and seeks to avoid market segmentation. Agricultural markets are closely tied to residents' welfare and exert a substantial impact on economic development [11]. From 2012 to 2020, there was significant fluctuation in the retail prices of agricultural products nationwide, with even more pronounced volatility in fresh agricultural product prices. Taking the average wholesale and retail prices of 15 vegetables (potatoes, tomatoes, cucumbers, bell peppers, eggplants, carrots, Chinese cabbage, cabbage, green beans, hot peppers, leeks, radishes, celery, garlic sprouts, and rapeseed) as an example, the price difference increased from CNY 1.92/kg to CNY 2.20/kg. Considering the average wholesale and retail prices of four fruits (Fuji apples, bananas, citrus, and watermelons) as another example, the price difference rose from CNY 1.88/kg in 2012 to CNY 3.71/kg in 2020. Historical experience suggests that the cause of this phenomenon is the excessively high circulation costs of agricultural products between regions [12]. Statistical data also confirm this point. In 2016, logistics costs in China accounted for 14.9% of GDP, and by 2020, this proportion had only decreased by 0.2%. In contrast, logistics costs in the United States have consistently been kept below 10% of GDP. The high circulation costs reflect a low level of agricultural market integration among regions and a severe state of market segmentation in China. This ultimately poses challenges to regional economic development [13].

The segmentation of agricultural markets is the result of the combined influence of factors such as local government policies, information technology levels, and geographical location [14–16]. It manifests in three specific categories: the first category involves the motivation of local governments to protect their local agricultural markets and ensure the market share of local enterprises [17]. This protective policy leads to market segmentation [18]. The second category is the lag in the development of information technology. It prevents timely and effective communication of production costs and sales prices between agricultural producers and consumers in different regions [19]. This communication barrier results in the invisible isolation of two markets. Producers cannot adjust production strategies promptly based on consumer demand information, and consumers cannot purchase satisfactory agricultural products. This leads to efficiency losses in agricultural markets [20,21]. The third category arises from the geographical differences between regions, leading to high transportation costs for the sale of agricultural products between different areas [22]. Breaking this segmentation and establishing an integrated agricultural market depend on transportation infrastructure. This is evident in two aspects: first, the development of transportation shortens transit time, enhances the circulation efficiency of agricultural products, and reduces the likelihood of price fluctuations in agricultural markets due to geographical distances [23]. Second, convenient transportation helps lower agricultural transportation costs, strengthen economic connections between regions, and enhance the level of agricultural market integration [24,25]. In conclusion, even advanced economies experience segmentation in agricultural markets, but the trend towards market integration is inevitable [26].

China, with its vast land area and large population, boasts a massive consumer market for agricultural products. The spatial geographic conditions and climatic environments vary significantly among different regions [27]. This diversity provides an excellent research context to explore the regional heterogeneity of agricultural market integration and its impact on regional economic development under complex natural and socio-economic conditions. The impact of the COVID-19 pandemic has revealed the vulnerability of the existing agricultural production and market supply chains [28]. Simultaneously, it has created a demand to identify new "points of equilibrium" for agricultural market integration and regional economic development in the post-pandemic era [29,30]. Based on this, the analysis focuses on the impact of agricultural market integration on regional economic development from a regional heterogeneity perspective. At the same time, the research incorporates international trade variables to explore their "substitution effects" in promoting regional economic development through agricultural market integration, while considering variations across different regions. In comparison to existing research, our marginal contributions are primarily evident in three aspects. Firstly, we provided a precise estimation of the agricultural market integration index and analyzed its temporal trends. Secondly, we constructed a dynamic spatial Durbin model to investigate the spatial effects of agricultural market integration on regional economic development and regional heterogeneity. Finally, we introduced variables related to international trade of agricultural products to examine their "substitution effects" in promoting regional economic development through agricultural market integration, with a specific focus on differences between the eastern and western regions. Furthermore, to ensure the model accurately describes the relationships between variables, we addressed estimation biases caused by endogeneity issues within the model. Additionally, we utilized officially published statistical data from 31 provinces, municipalities, and autonomous regions in China from 2010 to 2019 to ensure the scientific validity and accuracy of the results.

The structure of this paper is organized as follows: the first section comprises the introduction, the second section presents the conceptual framework, the third section outlines the research hypotheses, the fourth section details the research design, the fifth section reports the results, the sixth section delves into the extended analysis (substitution relationships and regional heterogeneity), the seventh section engages in discussions, and the final section provides the conclusion.

2. Conceptual Framework

2.1. Meaning and Key Features

2.1.1. Agricultural Market Integration

Agricultural market integration refers to the process of consolidating dispersed agricultural markets from different regions into a unified and coordinated market system. This process aims to eliminate barriers caused by local government policies, information technology levels, geographical locations, and other factors. The goal is to enable the free and efficient circulation and trade of agricultural products across diverse regions [31]. The core objective of agricultural market integration is to establish a unified domestic market for agricultural products, encompassing the production, distribution, and sales stages. This is aimed at enhancing the operational efficiency of the agricultural sector and the overall economy [32]. Key features of agricultural market integration include market consolidation, information sharing, smooth supply chains, and consistent government policies [33,34]. In summary, agricultural market integration is expected to facilitate easier market access for agricultural products, reduce transaction costs, enhance the competitiveness of agricultural products, improve food supply, and ultimately increase the welfare of farmers and other market participants [35].

In existing research, three main methods measure the level of agricultural market integration. The first method is the price index approach, which compares prices of identical commodities in different regional markets, analyzes the trend of price changes, and assesses the degree of market integration [36]. This method typically takes into account transaction costs, taxes, and other factors affecting prices to more accurately evaluate the state of market integration. Studies indicate that prices for similar goods in different regions of developed countries tend to remain stable over long periods. For example, in the United States, the time required for price convergence is 5 to 10 years [37]. Conversely, in developing countries such as China and India, the time for price convergence is shorter [38,39]. The second method is the production approach, measuring the level of market integration by evaluating factors such as output structure and production efficiency in different regions [40,41]. Large differences in economic and industrial structures between regions imply a high degree of regional specialization in production, requiring close collaboration between regions and leading to an increase in the degree of market integration.

2.1.2. Regional Economic Development

Regional economic development refers to the process, within a specific geographical area, of achieving economic growth, enhancing the standard of living for the population, and improving social welfare through effective resource allocation and economic policies. It is typically measured at the level of cities, states, provinces, counties, or urban agglomerations. Regional economic development differs from regional economic growth [42], wherein economic growth focuses on increasing the regional gross domestic product (GDP) and creating more employment opportunities, fostering the development of import and export trade, including agricultural products [43]. It is evident that regional economic growth is only a component of regional economic development and cannot fully replace indicators of regional economic development in research. It is essential to identify indicators that encompass both economic growth and the development of social welfare [44].

Regional economic development involves the effective allocation of resources to ensure that the region's resources are fully optimized and utilized [45]. The process and purpose of optimizing various resources aim to improve the quality of life for local residents [46], including enhancements in education, healthcare, housing, and food security. It also typically considers the sustainability of development to ensure that economic growth does not adversely affect the environment, society, and future generations [47]. It is evident that regional economic development emphasizes a broader, comprehensive set of goals, encompassing factors related to the economy, urban and rural areas, and the environment [48,49]. In this study, residents' disposable income is chosen as an alternative indicator for economic development because it considers not only economic growth but also the potential to enhance overall societal and individual well-being, improve the quality of life, and promote environmental sustainability.

2.2. Positive Effects of Agricultural Market Integration on Regional Economic Development

Summarizing existing research findings, the positive effects of agricultural market integration on regional economic development mainly include the following four aspects.

2.2.1. Facilitating Price Discovery

Agricultural market integration, achieved through the amalgamation of market information and the interconnection of markets across diverse regions, facilitates a more precise and equitable determination or discovery of market prices for agricultural products. This process enhances the efficiency of agricultural markets, assisting farmers and other market participants in making informed decisions. Simultaneously, it contributes to the equilibrium of supply and demand, thereby maintaining the stability of the agricultural product market. These outcomes are primarily achieved through four key approaches and methods: firstly, providing transparent market information to assist stakeholders in gaining a better understanding of market conditions [50]; secondly, balancing supply and demand to prevent excessive fluctuations in various agricultural product prices [51]; thirdly, mitigating information asymmetry and enhance market transparency [52]; lastly, implementing additional positive measures, such as supporting contract pricing and exploring international markets, thereby fostering international trade in agricultural products [53].

2.2.2. Reducing Transaction Costs

Agricultural market integration, overcoming geographical barriers, enhances market efficiency, transparency, and competitiveness, thereby contributing to the reduction in various transaction-related costs. This not only benefits farmers, agricultural producers, wholesalers, retailers, and consumers but also fosters the growth of agricultural markets and the development of agriculture. The impact of market integration on reducing costs related to agricultural products is evident in three key aspects: firstly, it reduces storage costs, decreasing both product loss and quality deterioration [54]. Secondly, it minimizes intermediary trading links, improving the operational efficiency of agricultural markets [55]. Thirdly, it enhances payment convenience, providing more flexible payment options that can shorten transaction cycles [56]. Lastly, it supports e-commerce and online transactions, further lowering transaction costs for agricultural products and enhancing market efficiency [57].

2.2.3. Increasing Farmers' Income

Agricultural market integration can enhance farmers' income through various means, such as providing more sales opportunities, improving supply chain management, and boosting both yield and quality. These factors collectively contribute to enhancing farmers' economic conditions and rural community livelihoods, fostering sustainable agricultural development. The role of agricultural market integration in increasing farmers' income is predominantly evident in the following three aspects: firstly, it improves agricultural output and quality. Increased production is fueled by the expanding market demand [58], and the enhancement of agricultural product quality is linked to the adoption of food safety standards, which are often required in integrated agricultural markets. This fosters the branding of agricultural products, leading to a significant boost in farmers' income [59]. Secondly, market diversification plays a crucial role. By enabling farmers to sell their products to more distant regions, this strategy diversifies production and business risks, augments the sales volume of agricultural products, and stabilizes farmers' income [60]. Thirdly, supply chain optimization is a vital factor. This enhances the turnover rate of goods and reduces the backlog of agricultural product inventory [61].

2.2.4. Improving Food Supply

Agricultural market integration positively influences the food supply by improving the sustainability, diversity, and traceability of agricultural products. This is accomplished through elevating food safety standards, ensuring a higher-quality food supply for consumers. It plays a crucial role in addressing the continually increasing demand for food, proving essential for nutrition and health and contributing significantly to the overall well-being of individuals. Its specific effects are evident in the following: firstly, the diversification of supply channels ensures a varied source of agricultural products for the public, reducing the risk of food shortages [62]. Secondly, the integrated agricultural market facilitates the prompt handling of food safety incidents, aiding in the swift identification and resolution of food safety issues, thereby reducing health problems resulting from food quality concerns [63]. Thirdly, it promotes sustainable agricultural production by encouraging farmers to adopt sustainable agricultural practices, minimizing the adverse environmental impact of agriculture, and ensuring a long-term food supply [64].

In summary, agricultural market integration stands as a pivotal driver for regional economic development. It optimizes the allocation of resources in the agricultural industry, boosts the scale of domestic agricultural trade, and ensures regional food security. This is accomplished through initiatives such as promoting price discovery, reducing transaction costs, increasing farmers' income, and enhancing food supply. Collectively, these efforts play a vital role in fostering robust regional economic development.

2.3. Theoretical Analysis

The theoretical foundation of how agricultural market integration promotes regional economic development includes the following aspects.

2.3.1. Allocation Efficiency Theory

Agricultural market integration can unite diverse regions into a cohesive whole, strategically allocating resources across different areas to maximize overall benefits, guided by principles of efficiency [65,66]. This approach is designed to cater to the optimal requirements of various agricultural producers for production factors. When resources, including land, water, and labor, are fully optimized, farmers and other producers can enhance the efficiency of agricultural output [67]. In essence, agricultural market integration facilitates the optimal utilization of agricultural resources throughout the entire industry chain—from production and transportation to consumption—preventing the waste of limited agricultural resources. Agricultural products from different regions can be cultivated in the most efficient locations, thereby reducing unnecessary redundancy and enhancing the operational efficiency of different regions and the entire domestic economy.

2.3.2. Specialization and Comparative Advantage Theory

Different regions exhibit diverse geographical conditions, resources, climates, and production technologies [68]. According to the theories of specialization and comparative advantage, the key to promoting regional economic development through agricultural market integration lies in each region focusing on the production of agricultural products where it holds a relative advantage. This approach aims to achieve more efficient output and enhance regional economic benefits. The comparative advantage theory suggests that each region should concentrate on producing agricultural products in which it has endowment advantages [69], meaning products with lower production costs and higher quality in that specific region. Consequently, each region can enhance its production efficiency. The theory of specialization in production holds that agricultural market integration encourages different regions to meet the specific demands of certain markets through specialized production [70]. This implies that each region leverages industry chain advantages developed over time, concentrating on producing specific types of agricultural products for sale. With producers possessing in-depth knowledge of the products they cultivate, they can employ the most suitable production equipment and techniques, thus reducing costs and increasing their output. Furthermore, specialized production within the context of agricultural market integration promotes the standardization of agricultural product quality standards, as well as the rapid dissemination and progress of production technologies, fundamentally driving regional economic development [71].

2.3.3. Economies of Scale Theory

The most direct outcome of agricultural market integration is the creation of a vast market scale, capable of encompassing a larger consumer base. This, in turn, generates increased demand for agricultural products and encourages all producers to expand their production scale to gain more profits [72]. Engaging in large-scale agricultural production allows individual farmers or producers to reduce unit production costs and enhance economic efficiency [73]. Large-scale production often provides greater capacity to invest in new technology research and development as well as the adoption of innovative methods to improve production efficiency and the quality of agricultural products [74]. Since the fixed production costs of individual producers can be spread across a greater number of products, this directly reduces unit production costs, ultimately lowering the prices of agricultural products and making them more competitive in the market. Importantly, participants in the entire agricultural market have a profit motive to reduce production costs and product prices, ultimately benefiting consumers and fostering regional economic development.

2.3.4. Risk Theory

From the perspective of farmers and other producers, agricultural market integration introduces diversification into demand markets, effectively averting sales crises in individual regions caused by preferences or other changes in demand [75]. On the side of agricultural market supply, integration encompasses a multitude of producers from different regions, helping to alleviate the supply impact on the entire agricultural market in the face of natural disasters or seasonal production fluctuations in a specific region [76]. For example, if a particular region experiences a natural disaster, other regions can promptly reallocate agricultural products to ensure a stable supply in the affected area. This shared risk helps reduce the risks for specific regions or agricultural producers, boosting their confidence in participating in production activities and market competition. Consequently, this contributes to the stable operation of the agricultural product market and the sustainable development of the regional economy [77].

3. Hypothesis

3.1. Hypothesis 1

Geographical distance creates impediments to regional trade, resulting in market segmentation [78]. According to the resource allocation theory, this market segmentation significantly impedes the economic development efficiency of both the local region and neighboring areas [15]. Studies on the market segmentation of agricultural products and its impact on China's economic growth reveal a substantial negative correlation between this segmentation and the actual per capita agricultural GDP and per capita GDP. This leads to the conclusion that market segmentation is detrimental to the coordinated development of the economy [79]. Furthermore, when examining 12 Chinese cities and the prices of six major commodity categories, it has been demonstrated that interprovincial market segmentation hampers the economic growth rates of respective provinces [80]. Synthesizing existing research findings, market segmentation has been identified as having a severe negative impact on economic growth [81]. This is primarily due to its reduction in resource allocation efficiency, resulting in increased production costs while simultaneously diminishing market transaction efficiency [82]. Ultimately, market segmentation affects the well-being of the population and imposes constraints on regional economic development [83].

In the late 1980s, China initiated its reform efforts by introducing the land contracting system in rural areas. Subsequently, the agricultural industry underwent a gradual process of marketization [84]. Specifically, the introduction of market mechanisms to agricultural product transactions enabled farmers and businesses to participate in free trade [85]. As transportation infrastructure was constructed and improved, the rapid development of China's integrated agricultural product market contributed significantly to the increase in agricultural output [86]. The domestic integration of markets, encompassing agricultural products, in China has been verified to have a noteworthy impact on the rate of economic growth [87]. More importantly, it consistently contributes to enhancing income levels for residents in the regions, particularly fostering accelerated income growth for farmers in less developed areas [88]. This not only alleviates poverty but also mitigates economic imbalances between regions [86]. Based on these observations, we propose the following hypothesis:

Hypothesis 1. *The agricultural market integration has a positive impact on regional economic development.*

3.2. Hypothesis 2

In the context of free trade, the inclination and policy decisions of local governments to establish trade barriers are influenced by the current level of international trade in the region [89]. Common indicators of international trade levels encompass total imports, total exports, and total trade volume. In diverse conditions of international trade in agricultural products, the positive impact of China's integration of the domestic agricultural product market on regional economic development varies [90]. The theory of comparative advantage suggests that the international trade level of agricultural products reflects both the competitive advantage of the agricultural product industry and the ability to leverage international market resources [91]. Conversely, risk theory posits that the international trade level of agricultural products mirrors the degree of regional economic dependence on foreign or non-regional economies [92]. Specifically, this dependence manifests in varying degrees of impact on regional economic development due to international trade in agricultural products.

According to the economies of scale theory, a large scale of international trade in agricultural products implies a high degree of dependence of the agricultural industry on foreign markets [93]. When closely connected to foreign economies, the domestic economy's economic development relies on the economies of scale gained from foreign markets [94]. This explains why, when the level of foreign trade in agricultural products is high, the role of domestic agricultural product market integration in promoting regional economic development is relatively weak. In a sense, international trade "substitutes" the role of promoting regional economic development through domestic agricultural product market integration [93]. However, in regions lacking a competitive advantage to enter foreign agricultural product markets, it is necessary to strengthen market connections with other domestic regions. This is done to enhance the level of agricultural product market integration, aiming to gain spatial spillover effects [95] and scale effects [96] from domestic agricultural product market integration. This, in turn, facilitates the sustained development of regional agriculture and the agricultural product industry. Based on these observations, the following hypothesis is proposed:

Hypothesis 2. *In different regions, the "substitution relationship" wherein domestic agricultural market integration is replaced by agricultural foreign trade to promote regional economic development varies.*

3.3. Hypothesis 3

In addition to the consumer utility theory in economics, the "substitution effect" theory finds widespread application in agricultural markets and international trade, albeit with a unique interpretation. This interpretation encompasses three aspects: firstly, a substitution effect exists between the international agricultural product market and the domestic agricultural product market [97,98]. The development and changes in the international agricultural product market can lead to corresponding changes in the domestic market, and vice versa. Specifically, businesses and consumers must make choices between the domestic agricultural product market and the international market. If a particular agricultural product becomes more competitive or experiences a lower price in the international market, it will impact the demand in the domestic market. Secondly, changes in international agricultural product trade can have an impact on the import and export of domestic agricultural products [99]. If there is an increased demand for domestic products in the international market, domestic farmers or producers may ramp up exports, thereby reducing the supply to the domestic market. Conversely, if the demand for domestic agricultural products rises in the domestic market, producers may decrease exports and increase domestic supply [100]. Thirdly, changes in international trade policies related to agricultural products can also influence the substitution relationship between the international and domestic markets [101]. For example, trade protectionist policies may restrict the import of foreign agricultural products, ensuring a demand for domestic agricultural products.

Due to geographical, climatic, and historical factors, China's administrative regions are categorized into the eastern, central, and western parts. The levels of economic, social, and trade development generally align with this regional division [102]. Specifically, the eastern region is the most developed, the western region is the least developed, and the central region falls in between. Research indicates that the level of agricultural product market integration in these three major regions corresponds to these developmental differences [83]. Using the price index method, the calculated agricultural product market integration indices for the eastern, central, and western regions in 2019 were approximately 2.3, 1.8, and 1.0, respectively [103]. Further analysis reveals that, owing to varying levels of economic and trade development, the impact of agricultural product market integration on regional economic development exhibits regional differences [104]. Domestic agricultural product market integration can effectively promote regional economic growth [105]. However, in the economically advanced eastern regions, the promotional role of the domestic agricultural product market on the local economy is often overlooked due to their close ties to international markets.

The eastern region, being in proximity to overseas markets, incurs lower export trade costs, facilitating easier access to international markets, and exhibits a higher dependence on foreign trade, as promptly reflected in the international market environment [106]. Provinces with higher levels of economic openness, such as Shanghai, Jiangsu, and Shandong, tend to utilize foreign trade as a lever for promoting the development of the local agricultural industry [107]. Comparative studies suggest that international trade plays a more significant role in the economic growth of coastal regions, while domestic market integration has a more pronounced impact on inland areas [108]. The western inland regions, characterized by lower levels of external openness, are unable to substitute domestic market demand with international markets. Consequently, they rely more on domestic market integration to propel economic development [109]. This implies that the role of domestic agricultural product market integration is primarily concentrated in the less externally open western regions. In the eastern regions, international markets substitute a portion of the domestic agricultural product market, reducing the demand for agricultural products from other domestic regions. In this sense, it may adversely affect the economic development of other domestic regions. Based on these observations, the following hypotheses are proposed:

Hypothesis 3a. *Regional heterogeneity exists in the promoting effect of agricultural product market integration on regional economic development.*

Hypothesis 3b. *The substitutive role of agricultural product foreign trade in promoting economic development through regional agricultural product market integration exhibits regional heterogeneity.*

4. Materials and Methods

The empirical research framework of this article encompasses three key aspects. Firstly, we provided a precise estimation of the agricultural market integration index and analyzed its temporal trends. Secondly, we constructed a dynamic spatial Durbin model to investigate the spatial effects of agricultural market integration on regional economic development and regional heterogeneity. Finally, we introduced variables related to international trade of agricultural products to examine their "substitution effects" in promoting regional economic development through agricultural market integration, with a specific focus on differences between the eastern and western regions. Furthermore, to ensure that the model accurately describes the relationships between variables, we addressed estimation biases caused by endogeneity issues within the model. Additionally, we utilized officially published statistical data from 31 provinces, municipalities, and autonomous regions in China from 2010 to 2019 to ensure the scientific validity and accuracy of the results. These three aspects represent the marginal contributions of this article compared to existing research.

4.1. Data Source

We utilized annual data from 2010 to 2019 for 31 provinces in China. The data are primarily sourced from the "China Statistical Yearbook" and the China Stock Market & Accounting Research Database. Data not available in the aforementioned statistical yearbooks are obtained from various provincial statistical yearbooks and the official website of the Ministry of Commerce of the People's Republic of China. The empirical analysis of the paper was conducted using Stata (16.0, StataCorp, College Station, TX, USA).

4.2. Variables

4.2.1. The Dependent Variable

Regional economic development ($\ln pergdp$) encompasses various indicators, such as total GDP, economic growth rate, industrial added value, per capita disposable income, and the level of social welfare. The per capita disposable income indicator not only captures the economic growth trend but also reflects the relative prosperity of each resident and their ability to utilize social resources. Hence, the natural logarithm of per capita disposable income is employed as the proxy variable for regional economic development.

4.2.2. Explanatory Variable

Agricultural market integration ($together$) contrasts with the segmentation of agricultural markets, and its measurement involves an initial assessment of the agricultural market segmentation index. In this study, the agricultural market segmentation index for 31 provinces is constructed using the relative price method [110]. The relative price method refers to using the relative prices of agricultural products as a tool to measure the degree of agricultural market integration. After obtaining the prices of agricultural products, the volatility of relative price differences is calculated to estimate the level of agricultural market integration between regions. When agricultural products can freely move, the prices in various regions will eventually converge. Therefore, using the relative price method to measure the degree of agricultural market integration is an effective approach.

Before calculating the relative price index of agricultural products, it is necessary to construct a three-dimensional ($t \times m \times k$) panel dataset, where t represents the years, m represents the regions and k represents the various types of agricultural products. The original data are sourced from the annual Provincial Statistical Yearbooks of China, covering the years 2010–2019. This dataset spans a decade and includes data from 31 provinces, municipalities, and autonomous regions nationwide (excluding Hong Kong, Macao, and Taiwan due to missing data). It encompasses seven major categories of agricultural products: cereals, oils and fats, meat and poultry and their products, eggs, aquatic products, vegetables, and fresh and dried fruits (classified according to the China Statistical Yearbook). The utilized dataset comprises three dimensions: time, location, and type of agricultural product ($10 \times 31 \times 7$).

a.   Calculate the absolute value of relative prices

We utilize the retail price index of agricultural products for each province to calculate relative prices, as shown in Formulas (1) and (2):

$$\Delta Q_{ijt}^k = \ln\left(\frac{p_{it}^k}{p_{jt}^k}\right) - \ln\left(\frac{p_{it-1}^k}{p_{jt-1}^k}\right) = \ln\left(\frac{p_{it}^k}{p_{it-1}^k}\right) - \ln\left(\frac{p_{jt}^k}{p_{jt-1}^k}\right) \tag{1}$$

$$\left|\Delta Q_{ijt}^k\right| = \left|\ln\left(\frac{p_{it}^k}{p_{it-1}^k}\right) - \ln\left(\frac{p_{jt}^k}{p_{jt-1}^k}\right)\right| \tag{2}$$

In Equation (1), ΔQ_{ijt}^k denotes the relative price of any two provinces for the k_{th} agricultural product in year t. Here, i and j represent any two provinces, t denotes the year, and k signifies the specific agricultural product. Additionally, p_{it}^k stands for the retail price index of the k_{th} agricultural product in province i in year t, p_{it-1}^k represents the retail price index of the same product in province i in the previous year, p_{jt}^k signifies the retail price index of the k_{th} agricultural product in province j in year t, and p_{jt-1}^k denotes the retail price index of the same product in province j in the previous year. Equation (2) encompasses 465 pairs of any two provinces or municipalities for the years 2010–2019, resulting in the computation of 32,550 different differentials in the form of relative prices $\left|\Delta Q_{ijt}^k\right|$. The utilization of absolute values in this context ensures that the variance of relative prices remains unaffected by the order of provinces or municipalities in the combinations.

b.   Eliminate the mean

Given the heterogeneity in agricultural products between two provinces, the fact that $\left|\Delta Q_{ijt}^k\right|$ is not solely influenced by interprovincial market characteristics, we employ a method of demeaning to address this heterogeneity in agricultural products [111]. The procedure is as follows: assuming that $\left|\Delta Q_{ijt}^k\right|$ comprises two components α^k and ε_{ijt}^k, where α^k is related to the characteristics of the k_{th} agricultural product itself and ε_{ijt}^k reflects the market features of provinces i and j. To eliminate the influence of α^k, we calculate the average relative price for the k_{th} agricultural product for 465 pairs of provincial combinations in year t, denoted as $\left|\Delta \overline{Q_t^k}\right|$. Subsequently, we subtract this mean from $\left|\Delta Q_{ijt}^k\right|$ to obtain $q_{ijt}^k = \left|\Delta Q_{ijt}^k\right| - \left|\Delta \overline{Q_t^k}\right| = (\alpha^k - \overline{\alpha^k}) + (\varepsilon_{ijt}^k - \overline{\varepsilon_{ijt}^k})$, thereby obtaining the relative price variation component used for calculating the variance, denoted as q_{ijt}^k. It is solely associated with the market characteristics of the agricultural products in the two provinces and some random factors.

c.   Market segmentation index

Var (q_{ijt}) represents the variance in the relative price fluctuations for the seven categories of agricultural products. By summing the relative price variances for each province with any other province and dividing the result by 30 (as each province has 30 different combinations), we obtain the agricultural market segmentation index for each province: Var $(q_{nt}) = (\sum_{i \neq j} \text{Var}\left(q_{ijt}\right))/30$. Since there is an inverse relationship between agricultural market segmentation and market integration, the reciprocal of the square root of the agricultural market segmentation index can be used as the agricultural market integration index for each province, denoted as $together = \sqrt{1/\text{Var}(q_{nt})}$. It is important to note once again that existing research calculates the national average index of agricultural integration for the years 2010 to 2019 to be approximately 1.3, which slightly differs from the calculation method employed in this study [103]. Specifically, after obtaining the reciprocal of the agricultural segmentation index, there is no multiplication by 1/1000; instead, the square root of this reciprocal is taken directly. Both calculation methods, when subjected to the corresponding inverse operation, can be converted into indices with a consistent measurement scale.

4.2.3. Control Variables

According to previous studies, the following control variables have been selected [86,112–115]; the government intervention level variable (*govern*), calculated using the proportion of general public budget expenditures to GDP; The level of agricultural development (*agri*), measured by the proportion of the added value of the primary industry to GDP; the level of urbanization (*urban*), measured by the proportion of the urban population to the total population; agricultural labor supply (*labor*), measured by the number of people engaged in agriculture, forestry, animal husbandry, and fishery; agricultural fixed asset investment (*invest*), measured by the proportion of total fixed asset investment in agriculture, forestry, animal husbandry, and fishery to GDP; population density (*density*), reflected by dividing the year-end population by the administrative area.

4.2.4. Instrumental Variables

Firstly, the transportation network density (*traffic*) is calculated by dividing the total mileage of highways, railways, and inland waterways by the administrative area of each province, drawing on existing research [87]. Secondly, the interaction term between the annual sunshine hours and the agricultural market integration index for each province (*sun*) is employed as an instrumental variable. Annual sunshine hours are highly correlated with crop yield and diversity, and they are mutually independent of the agricultural market integration index, making them a suitable choice as an instrumental variable.

4.2.5. Moderating Variables

The foreign trade volume of agricultural products is measured by the total import and export volumes, including separate assessments for imports and exports.

The specific meanings and descriptive statistics of each variable are shown below (Table 1).

Table 1. Introduction and descriptive statistics of regional economic development, agricultural market integration, and related control variables in China from 2010 to 2019.

Variables	Variables Interpretation	Mean	Standard Deviation	Minimum	Maximum
ln *pergdp*	Per capita GDP	10.75	0.463	9.482	12.01
together	Relative price index method	32.72	6.176	16.51	45.97
govern	Government budget expenditure/GDP (%)	27.93	20.83	10.58	137.9
agri	Primary industry value added/GDP (%)	9.459	4.957	0.281	25.28
urban	Urban population/total population (%)	56.09	13.39	22.67	89.60
labor	Labor supply in the agricultural category (persons)	791.6	594.1	23.61	2559
invest	Social fixed asset investment in agriculture (CNY)	3.407	2.548	0.00520	15.16
density	Year-end resident population/administrative district land area (persons/square kilometers)	517.7	690.5	3.376	3913
traffic	Total mileage of roads, railroads and inland waterways/land area of administrative region (ten thousand kilometers/square kilometers)	1.298	0.875	0.0695	5.344
sun	*together* ∗ annual sunshine hours (hours)	1996	577.0	257.1	3163

Data source: calculated by the author.

4.3. Trend of Agricultural Market Integration

By averaging the agricultural market integration index across the 31 provinces and municipalities in China, we derive the agricultural market integration index for China (nation) for the decade preceding the COVID-19 pandemic, along with its temporal trends. Concurrently, Figure 1 illustrates the average changes over time in this index for the eastern (east), central (middle), and western (west) regions.

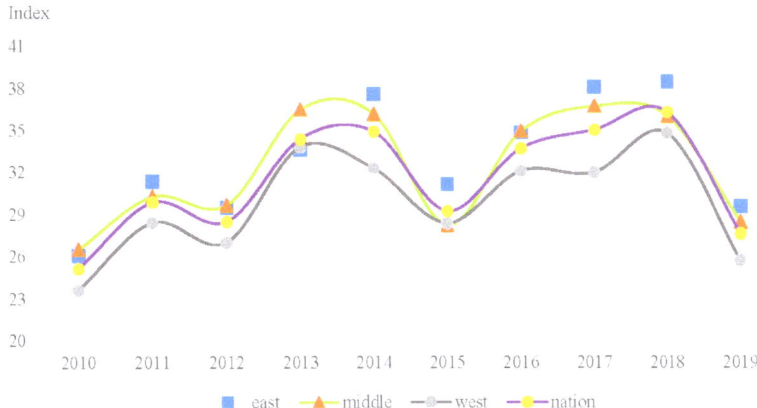

Figure 1. Agricultural market integration trends in China as a whole and in the eastern, central, and western regions from 2010 to 2019. Note: east (eastern regions): Beijing, Tianjin, Hebei, Liaoning, Shandong, Shanghai, Jiangsu, Zhejiang, Fujian, Guangdong, Hainan; Middle (central regions): Shanxi, Jilin, Heilongjiang, Anhui, Jiangxi, Henan, Hubei, Hunan; west (western regions): Inner Mongolia, Guangxi, Chongqing, Sichuan, Guizhou, Yunnan, Tibet, Shaanxi, Gansu, Qinghai, Ningxia, Xinjiang.

It is evident that from 2010 to 2019, the national agricultural market integration index exhibits a cyclical trend with a periodicity of approximately 5 years (Figure 1). Analyzing regional differences, the cyclical fluctuation trends in the eastern and central regions are essentially similar to the national trend. In contrast, the western region has the lowest agricultural market integration index, and its fluctuations show no clear pattern.

4.4. Spatial Econometric Models

In accordance with the resource allocation theory and economies of scale theory discussed in Section 2.3, a higher level of agricultural market integration, denoted by an elevated agricultural market integration index, implies a stronger degree of integration for resources related to agricultural products [67,68]. This heightened integration is associated with a greater impact on regional economic development [73]. The role of agricultural market integration in regional economic development is reflected through the construction of a panel econometric model (Equation (3)).

$$\ln pergdp_{it} = \beta_1 together_{it} + \beta_2 govern_{it} + \beta_3 agri_{it} + \beta_4 urban_{it} + \beta_5 labor_{it} + \beta_6 invest_{it} + \beta_7 density_{it} + \lambda_t + \alpha_i + \varepsilon_{it} \quad (3)$$

$lnpergdp_{it}$ represents regional economic development, $together_{it}$ stands for the agricultural market integration index, β_2–β_7 are control variables, representing the 31 provinces ($i = 1, 2, \ldots, 31$), t is the year ($t = 2010, 2012, \ldots, 2019$), and λ_t and α_i denote time and regional effects, respectively. ε_{it} represents the random error term.

The level of agricultural market integration in one province affects the economic development of other provinces, indicating the presence of spatial spillover effects. The examination of such spatial effects can be conducted using a spatial Durbin model. Additionally, due to the time trend characteristics of regional economic development, it is necessary to establish a dynamic spatial Durbin model to investigate the relationship between agricultural market integration and regional economic development. The formula is as follows:

$$\ln pergdp_{it} = \varphi \ln pergdp_{i,t-1} + \vartheta \sum_{j=1}^{n} W_{ij} \ln pergdp_{j,t-1} + \rho \sum_{j=1}^{n} W_{ij} \ln pergdp_{jt} +$$
$$\beta_1 together_{it} + \beta_2 govern_{it} + \beta_3 agri_{it} + \beta_4 urban_{it} + \beta_5 labor_{it} + \beta_6 invest_{it} + \beta_7 density_{it} +$$
$$\gamma_1 \sum_{j=1}^{n} W_{ij} together_{jt} + \gamma_2 \sum_{j=1}^{n} W_{ij} govern_{jt} + \gamma_3 \sum_{j=1}^{n} W_{ij} agri_{jt} + \gamma_4 \sum_{j=1}^{n} W_{ij} urban_{jt} \tag{4}$$
$$+ \gamma_5 \sum_{j=1}^{n} W_{ij} labor_{jt} + \gamma_6 \sum_{j=1}^{n} W_{ij} invest_{jt} + \gamma_7 \sum_{j=1}^{n} W_{ij} density_{jt} + \varepsilon_{it}$$

φ and ϑ represent the coefficients of time and spatiotemporal lag terms for regional economic development, W_{ij} is the spatial weight matrix, ρ is the coefficient for spatial spillover effects on regional economic development, and $\gamma_1 - \gamma_7$ represent coefficients for the spatial lag terms. The meanings of the other variables remain consistent with the previous context.

4.5. Spatial Weight Matrix

The interdependence and correlation among spatial units can be represented using a spatial weight matrix. Various types of spatial weight matrices include matrices based on proximity distance, geographical distance, economic distance, and economic–geographic embedding. The proximity-based spatial weight matrix enumerates other provinces that are adjacent to it and assigns corresponding weight values to reflect the strength of their relationship. The geographical distance spatial weight matrix calculates the actual geographical distances between different provinces and assigns weight values. Typically, provinces that are closer have higher weights, while those that are farther away have lower weights. The economic distance spatial weight matrix is based on economic factors to measure the correlation between different provinces and then assign weight values. The economic–geographical embedding spatial weight matrix integrates the influences of economic distance and geographical distance to represent the complex relationships among different provinces. Essentially, it reflects the spatial connections between regions in terms of both economic and geographical aspects. Selecting the appropriate spatial weight matrix is crucial for studying the relationship between the agricultural market integration index and regional economic development.

The spatial weight matrix based on proximity distance refers to designating the k ($k = 1, 2, \ldots, 31$) nearest provinces as neighboring regions. In this matrix, elements corresponding to pairs of adjacent provinces are assigned a value of 1, while others are set to 0. Considering that although Guangdong and Hainan provinces are not contiguous on land, their maritime areas are connected, and they share close agricultural trade activities and other economic and social connections. Therefore, the model assigns a value of 1 to the elements of the spatial weight matrix between these two provinces. The formula is as follows:

$$W_1 = \begin{cases} 1 \text{ region } i \text{ and } j \text{ are adjacent} \\ 0 \text{ region } i \text{ and } j \text{ are not adjacent} \end{cases} \quad i = 1, 2, \cdots, 31; \quad j = 1, 2, \cdots, 31 \tag{5}$$

In the spatial weight matrix based on geographical distance, the distance relationship between province i and province j adheres to the following conditions: when $i \neq j$, $W_2 = 1/d_{ij}$; when $i = j$, $W_2 = 0$. d_{ij} represents the spherical distance between the two provinces, calculated based on latitude and longitude data. The formula is as follows:

$$W_2 = \begin{cases} 1/d_{ij} & i \neq j \\ 0 & i = j \end{cases} \quad i = 1, 2, \cdots 31; \quad j = 1, 2, \cdots, 31 \tag{6}$$

The spatial weight matrix based on economic distance uses the reciprocal of the difference in economic levels between two provinces, where $\overline{X_i}$ and $\overline{X_j}$ represent the mean per capita GDP for provinces i and j in the sample period, respectively. The formula is as follows:

$$W_3 = \frac{1}{\left| \overline{X_i} - \overline{X_j} \right|} (i \neq j) \quad i = 1, 2, \cdots, 31; j = 1, 2, \cdots, 31 \tag{7}$$

The economic–geographic spatial weight matrix not only takes into account the spatial influence of geographical distance between provinces but also reflects the factual connections and spatial spillover effects in their economic development. Therefore, adopting the economic–geographic spatial weight matrix is essential for accurately capturing the spatial relationships among provinces. The diagonal elements of the economic–geographic spatial weight matrix are the product of the spatial weight based on geographical distance and the proportion of GDP for each region, where $\overline{X}$ represents the mean per capita GDP for all provinces during the sample period. The formula is as follows:

$$W_4 = \frac{1}{d_{ij}} * diag(\frac{\overline{X_1}}{\overline{X}}, \frac{\overline{X_2}}{\overline{X}}, \cdots \frac{\overline{X_n}}{\overline{X}})\ (i \neq j)\ i = 1, 2, \cdots, 31; \quad j = 1, 2, \cdots, 31 \quad (8)$$

4.6. Global Moran's Index

We employ the Global Moran's Index to examine the spatial correlation of regional economic development. In this context, n represents the number of administrative divisions nationwide (n = 1, 2,..., 31), where x_i and x_j denote the per capita GDP of provinces i and j, respectively. $\overline{y}$ represents the national per capita GDP, S^2 is the sample variance, and W_{ij} is the spatial weight matrix. The calculation formula is as follows:

$$GlobalMoran'sIndex = \frac{\sum_{i=1}^{n} \sum_{j=1}^{n} D_{ij}(x_i - \overline{y})(y_j - \overline{y})}{S^2 \sum_{i=1}^{n} \sum_{j=1}^{n} W_{ij}} \quad (9)$$

5. Results

5.1. Spatial Correlation

Equation (9) illustrates the Global Moran's Index for regional economic development under four distinct spatial weights (Table 2). It is noteworthy that the Global Moran's Index for regional economic development, spanning from 2010 to 2019, exhibits positivity across all four spatial weights, and all values successfully pass the 1% significance test. This observation implies that, irrespective of the spatial weight matrices employed, a discernible spatial correlation exists in the economic development of diverse provinces and regions.

Table 2. Global Moran's Index.

	Adjacent Distance Spatial Weight Matrix		Geographical Distance Spatial Weight Matrix		Economic Distance Spatial Weight Matrix		Nested Economic Geospatial Weight Matrix	
Year	Moran's I	Z Value	Moran's I	Z Value	Moran's I	Z Value	Moran's I	Z Value
2010	0.4440 ***	4.0080	0.1600 ***	5.5640	0.4970 ***	5.7580	0.1810 ***	5.5440
2011	0.4360 ***	3.9350	0.1580 ***	5.4890	0.4910 ***	5.6920	0.1770 ***	5.4560
2012	0.4170 ***	3.7760	0.1520 ***	5.3300	0.4920 ***	5.6950	0.1730 ***	5.3510
2013	0.3990 ***	3.6170	0.1460 ***	5.1510	0.5000 ***	5.7760	0.1690 ***	5.2320
2014	0.3750 ***	3.4160	0.1370 ***	4.8740	0.5110 ***	5.8880	0.1600 ***	4.9960
2015	0.3630 ***	3.3150	0.1290 ***	4.6660	0.5180 ***	5.9680	0.1530 ***	4.8270
2016	0.3730 ***	3.4090	0.1240 ***	4.5171	0.5150 ***	5.9530	0.1480 ***	4.7070
2017	0.3970 ***	3.6190	0.1230 ***	4.5010	0.5020 ***	5.8290	0.1480 ***	4.6960
2018	0.3570 ***	3.3040	0.1080 ***	4.0940	0.4670 ***	5.4870	0.1200 ***	3.9990
2019	0.3571 ***	3.3110	0.1080 ***	4.1090	0.4610 ***	5.4140	0.1190 ***	3.9760

Note: The significance level of Moran's I index is tested according to the Monte Carlo simulation method (999 times). ***, **, and * represent the significance levels of 1%, 5%, and 10%, respectively. Data source: calculated by the author.

5.2. Model Testing

The dynamic spatial panel Durbin model provides an accurate estimation of the spatial spillover effects on regional economic development resulting from agricultural market integration. However, its application necessitates the use of empirical data and rigorous statistical scrutiny. The Lagrange Multiplier Test (LM) and Likelihood Ratio Test (LR) are two major testing methods in econometrics, both of which can be used to examine the model selection. The test results are shown below (Table 3). Upon conducting statistical tests on the data, LM tests and their robust LM forms for spatial lag models and spatial error models did not achieve statistical significance. Consequently, spatial error models and spatial lag models are deemed unsuitable for this study. Employing the LR test to scrutinize the model allows verification of whether the impact of agricultural market integration on regional economic development is best analyzed using the spatial Durbin model. The estimation results reveal that the LR test decisively rejects the null hypothesis, signifying the dismissal of the assumption that the spatial Durbin model can be simplified into spatial lag models and spatial error models. The Wald test further corroborates this conclusion, affirming the appropriateness of choosing the spatial panel Durbin model.

Table 3. LM and LR tests: examining the suitability of the model for assessing the impact of agricultural market integration on regional economic development in China from 2010 to 2019.

	Test	***p*-Value**
LM spatial lag	0.1650	0.6850
Robust LM spatial lag	0.1810	0.6710
LM spatial error	0.5660	0.4520
Robust LM spatial error	0.5500	0.4590
LR spatial lag	46.240 ***	0.0000
LR spatial error	83.9700 ***	0.0000
Hausman chi2	77.3000 ***	0.0000
Wald test	1499.3563 ***	0.0000

Note: ***, **, and * represent the significance levels of 1%, 5%, and 10%, respectively. Data source: calculated by the author.

Considering the temporal continuity and spatial transmission of regional economies, neglecting these characteristics can lead to biases in model estimation. By setting up a dynamic spatial panel Durbin model, we can control for the factors influencing regional economic development in both the temporal and spatial dimensions, yielding robust estimation results [116]. The estimated time lag term for regional economic development indicates the impact of the previous period's economic development on the current period [117]. The estimated spatial spillover term for regional economic development signifies the impact of economic development in other provinces on the economic development of the focal province [118]. The estimated spatiotemporal lag term for regional economic development reflects the influence of economic development in other provinces in the previous period on the economic development of the focal province in the current period [119]. The estimated spatial spillover terms for key explanatory variables and control variables indicate the impact of these variables in other provinces on the economic development of the focal province [120]. This research further undertakes a comparative analysis of the influence of agricultural market integration on regional economic development, employing both non-spatial panel models and static spatial panel Durbin models. Given the diverse natural resource endowments and varying levels of socio-economic development across provinces, the use of a fixed-effects model is deemed appropriate. The results of the Hausman test also affirm that both the non-spatial panel model and the dynamic spatial panel Durbin model in this study are suitable for estimation using fixed effects.

5.3. Regression Analysis

We specified the models in the following forms, including a pooled regression model, a fixed-effects regression model, a static spatial Durbin model, a dynamic spatial panel Durbin model based on time lag, and a dynamic spatial panel Durbin model based on both time and spatial double lags. The results of each model are sequentially presented in Table 4, labeled as (1)–(5).

Table 4. Regression results of the impact of agricultural market integration on regional economic development in China from 2010 to 2019.

Variables	(1) Pooled Regression Model	(2) Fixed-Effects Regression Model	(3) Static Spatial Durbin Model	(4) Dynamic Spatial Durbin Model (Time Lag)	(5) Dynamic Spatial Durbin Model (Time-Space Lag)
together	0.00396 ** (0.00155)	0.00176 ** (0.00088)	0.00182 ** (0.00079)	0.00060 (0.00052)	0.00156 *** (0.00053)
govern	0.00223 *** (0.00063)	−0.01828 *** (0.00199)	−0.01664 *** (0.00185)	−0.00148 *** (0.00025)	−0.00101 *** (0.00025)
agri	−0.01528 *** (0.00267)	−0.02860 *** (0.00437)	−0.02987 *** (0.00390)	0.01085 *** (0.00107)	0.00427 *** (0.00108)
urban	0.02878 *** (0.00179)	0.01655 *** (0.00379)	0.01742 *** (0.00407)	−0.02643 *** (0.00102)	−0.01218 *** (0.00103)
labor	0.00009 *** (0.00002)	−0.00003 (0.00007)	−0.00010 (0.00006)	−0.00008 *** (0.00001)	−0.00004 *** (0.00001)
invest	−0.01551 *** (0.00482)	0.00737 ** (0.00309)	0.00329 (0.00292)	0.00409 *** (0.00158)	0.00219 (0.00159)
density	−0.00005 *** (0.00002)	0.00080 *** (0.00023)	0.00108 *** (0.00022)	0.00012 *** (0.00001)	0.00007 *** (0.00001)
$W \times together$			0.00208 (0.00209)	0.01876 *** (0.00146)	0.01158 *** (0.00147)
$L.\ln pergdp$				2.39660 *** (0.02526)	1.41647 *** (0.02621)
$L.W \ln pergdp$					0.61438 *** (0.13705)
Spatial rho (lambda)			0.67575 *** (0.05644)	0.26735 *** (0.06252)	0.30633 ** (0.12378)
N	310.00000	310.00000	310.00000	279.00000	279.00000
R^2	0.88894	0.92808	0.61377	0.76800	0.80374

Note: The values in () are the heteroskedasticity robust standard errors. ***, **, and * represent the significance levels of 1%, 5%, and 10%, respectively. Data source: calculated by the author.

In models (1) and (2), where spatial effects are not considered, it is evident that agricultural market integration exerts a positively facilitating influence on regional economic development, with all regression coefficients passing the 5% significance test. In models (3) to (5), incorporating the economic distance spatial weight matrix, a comparable positive promoting effect of agricultural market integration on regional economic development is identified. Notably, in model 5, the coefficients of the time lag and spatial lag terms for regional economic development are both significantly positive at the 1% level. The ensuing discussion will center on the estimation results of model 5.

5.3.1. Testing Hypothesis 1

Drawing upon the regression results from model 5 (Table 4), the coefficient associated with agricultural market integration is found to be 0.00156, successfully passing the 1% significance level test. This observation signifies that agricultural market integration plays a positive role in promoting regional economic development, thereby substantiating Hypothesis 1. The specific interpretation of this coefficient suggests that a one-unit increase in the agricultural market integration index corresponds to a 0.156% rise in the level of economic development. The rationale behind this finding, rooted in theory and empirical evidence, lies in the ability of agricultural market integration to enhance the efficiency of the agricultural supply chain, broaden the market for agricultural product sales, optimize agricultural product prices, encourage specialization in agricultural production, mitigate the risk of agricultural trade defaults, reduce the overall cost of agricultural products from production to consumption, and ultimately stimulate economic development.

5.3.2. Spatial Spillover Effects of Agricultural Market Integration

The coefficient of the spatial lag term for agricultural market integration is 0.01158, passing the 1% significance test. This suggests that a one-unit increase in the level of agricultural market integration in neighboring provinces would lead to a 1.158% increase in the level of regional economic development in the province under consideration. It is evident that the spatial spillover effects of agricultural market integration are significant. This can be primarily attributed to the frequent agricultural trade between provinces and the close interconnection of the upstream and downstream sectors of the agricultural production chain, which optimizes the rational allocation of agricultural and production factors. This allows provinces to engage in production and sales based on cost and technological comparative advantages, while simultaneously gaining access to new markets and consumers, thereby substantially promoting economic development.

5.3.3. Spatiotemporal Effects of Regional Economic Development

Model 5 reveals that regional economic development manifests both temporal lag effects and spatial spillover effects. Firstly, the coefficient of the time lag term for regional economic development is significantly positive at the 1% level, signifying that if the regional economic development level of the province in the previous period was high, the economic level in the next period will continue to increase. This growth trend intensifies over time, demonstrating a pronounced "time effect". Secondly, the coefficient of the spatial lag term for regional economic development is positive and passes the 1% significance test, indicating that the economic development level of neighboring provinces exerts a promotional effect on the economic development of the province under consideration. This implies the presence of spatial connections between provinces in economic development. On average, a 1% increase in the economic development level of neighboring provinces promotes the economic development level of the province under consideration by approximately 0.614%, as elucidated by the analysis.

5.4. Robustness Test

To ensure the robustness of the estimated parameters, three methods were employed in this study for robustness testing.

5.4.1. Weight Transformation

In the regression analysis, the economic distance spatial weight matrix was primarily utilized. For the robustness testing in this section, three additional matrices were constructed: proximity distance, geographical distance, and economic–geographical spatial nesting. The results indicate that, whether in the short term or long term, the promoting effect of agricultural market integration on regional economic development remains robust when employing these three spatial weight matrices (Table 5). In the short term, with the adoption of these three weight matrices, the regression coefficients are 0.01108, 0.05603, and 0.05164, respectively, with the geographical distance spatial weight matrix having the largest coefficient. This suggests that, in the short-term impact of agricultural market integration on regional economic development, the actual geographical distance between two provinces exerts the greatest influence. In the long term, the regression coefficients are 0.00915, 0.01310, and 0.01617, respectively, with the economic–geographical spatial weight matrix having the largest coefficient. This indicates that, in the long-term impact, the economic connection between two provinces and the actual geographical distance both play crucial roles. In conclusion, while there are differences in the magnitude of coefficients between the short term and long term, the signs and significance levels remain fundamentally unchanged, demonstrating the robustness of the research results.

Table 5. Regression results of the impact of agricultural market integration on regional economic development in China from 2010 to 2019 based on weight matrix transformation.

Variables	Short-Term			Long-Term		
	Adjacent Distance Spatial Weight Matrix	Geographical Distance Spatial Weight Matrix	Nested Economic Geospatial Weight Matrix	Adjacent Distance Spatial Weight Matrix	Geographical Distance Spatial Weight Matrix	Nested Economic Geospatial Weight Matrix
together	0.01108 *** (0.00363)	0.05603 *** (0.01981)	0.05164 *** (0.01489)	0.00915 *** (0.00293)	0.01310 *** (0.00280)	0.01617 *** (0.00306)
govern	0.00495 *** (0.00192)	−0.02562 ** (0.01101)	−0.01872 ** (0.00877)	0.00409 *** (0.00154)	−0.00598 *** (0.00184)	−0.00584 ** (0.00232)
agri	−0.04546 *** (0.00929)	−0.18295 *** (0.05977)	−0.16620 *** (0.04569)	−0.03756 *** (0.00723)	−0.04265 *** (0.00569)	−0.05193 *** (0.00792)
urban	0.03206 *** (0.00718)	0.05915 ** (0.02422)	0.03414 ** (0.01461)	0.02647 *** (0.00557)	0.01375 *** (0.00372)	0.01064 *** (0.00368)
labor	0.00005 (0.00006)	−0.00103 ** (0.00041)	−0.00115 *** (0.00040)	0.00004 (0.00005)	−0.00024 *** (0.00006)	−0.00036 *** (0.00009)
invest	−0.00241 (0.01020)	0.12934 *** (0.04873)	0.07688 ** (0.03606)	−0.00199 (0.00838)	0.03028 *** (0.00772)	0.02405 ** (0.00961)
density	0.00011 * (0.00007)	0.00059 ** (0.00025)	0.00029 ** (0.00012)	0.00009 * (0.00006)	0.00014 *** (0.00004)	0.00009 *** (0.00003)
N	310.00000	310.00000	310.00000	310.00000	310.00000	310.00000

Note: The values in () are the heteroskedasticity robust standard errors. ***, **, and * represent significance levels of 1%, 5%, and 10%, respectively. Data source: calculated by the author.

5.4.2. Model Transformation

To further validate the promoting effect of agricultural market integration on regional economic development, spatial lag models (SAR) and spatial error models (SEM) were chosen for data estimation and analysis. In the spatial lag model (SAR), the addition of the spatial lag term for regional economic development as an explanatory variable signifies the spatial impact of economic development in neighboring provinces on the economic development of the province under consideration. This approach, distinct from the dynamic spatial panel Durbin model, can once again verify the role of agricultural market integration in promoting regional economic development. Column (1) reveals that the coefficient for agricultural market integration is 0.00318 and passes the 5% significance test (Table 6). In the spatial error model (SEM), unobserved factors influencing the regional economic development of neighboring provinces are incorporated as spatial error terms. Introducing this spatial error term as an explanatory variable into the model serves the same purpose as the spatial lag model (SAR). Column (2) shows that the coefficient for agricultural market integration is 0.00447 and passes the 1% significance test (Table 6). In conclusion, by analyzing the results of the spatial lag model (SAR) and spatial error model (SEM), Hypothesis 1 is once again confirmed.

Table 6. Robustness test of the impact of agricultural market integration on regional economic development in China from 2010 to 2019.

Variables	Model Transformation Test		Variable Transformation Test	Endogeneity Test
	(1) SAR Model	(2) SEM Model	(3) Dynamic Spatial Durbin Model	(4) 2SLS Estimation
together	0.00318 ** (0.00134)	0.00447 *** (0.00142)		0.00309 *** (0.00105)

Table 6. *Cont.*

Variables	Model Transformation Test		Variable Transformation Test	Endogeneity Test
	(1) SAR Model	**(2) SEM Model**	**(3) Dynamic Spatial Durbin Model**	**(4) 2SLS Estimation**
seg			−0.00256 *** (0.00064)	
$W \times seg$			−0.00949 *** (0.00189)	
L. ln *pergdp*			1.39564 *** (0.02512)	
Spatial rho	0.51365 *** (0.07051)		0.12300 ** (0.06270)	
Spatial lambda		0.48326 *** (0.09334)		
Control variables	Yes	Yes	Yes	Yes
N	310.00000	310.00000	279.00000	310.00000
R^2	0.87055	0.80516	0.94763	0.97972
D–W–H Endogenous test				4.00021 [0.0455]
Kleibergen–Paap rk LM statistic				76.81 [0.0000]
Kleibergen–Paap rk Wald F statistic				58.85 {19.93}
Anderson–Rubin Wald statistic				15.39 [0.0005
Sargen–Hansen statistic				2.659 [0.1030]

Note: The values in () are the heteroskedasticity robust standard errors, the values in [] are the *p*-values of the corresponding test statistics, and the values in {} are the critical values at the 10% level of the Stock–Yogo test. ***, **, and * represent the significance levels of 1%, 5%, and 10%, respectively. Data source: calculated by the author.

5.4.3. Variable Transformation

By appropriately transforming the explanatory variables through mathematical formulas, the influence of outliers in the data can be effectively mitigated, thereby verifying the robustness of the results (Table 4). In this research, the reciprocal of the square root of the agricultural market integration index represents the fragmentation index of the agricultural product market. From an economic perspective, when the level of agricultural market integration is low, local regional markets are relatively closed, indicating a high degree of fragmentation in the agricultural product market. According to the theories of resource allocation and economies of scale, it can be inferred that a high degree of fragmentation in the agricultural product market will adversely affect regional economic development. Conducting a robustness test with the fragmentation index of the agricultural product market as the core explanatory variable provides a reverse verification of the impact of agricultural market integration on regional economic development, as demonstrated in column (3) (Table 6). The analysis of these results indicates that when the fragmentation index of the agricultural product market in a province increases by one unit, the level of regional economic development will decrease by 0.256%, and this passes the 1% significance test. This once again confirms the robustness of the results presented in Table 4.

5.5. Endogeneity Test

From a practical standpoint, the elevation in agricultural market integration fosters regional economic development. Conversely, regional economic development may also contribute to an increased level of agricultural market integration, suggesting a potential bidirectional causal relationship between them. This bidirectional causal relationship among variables can lead to endogeneity in the estimation model, resulting in biased outcomes. To address this endogeneity concern, instrumental variable methods can be employed to control for the relationships between variables. We select traffic network density (*traffic*) and the interaction term between annual sunshine hours and agricultural market integration (*sun*) as two instrumental variables for agricultural market integration.

This selection is based on two considerations: first, traffic network density (*traffic*) reflects the impact of geographical factors on agricultural market integration, satisfying the exogeneity condition. The more developed the transportation infrastructure in each province, the lower the transportation costs for agricultural products. Consequently, consumers can obtain the required products at a lower cost in the agricultural market. Therefore, this variable fulfills the conditions of exogeneity and correlation with agricultural market integration. Second, annual sunshine hours, as a natural condition closely related to agricultural production, also satisfies the exogeneity and correlation conditions with agricultural market integration. Considering that the annual sunshine hours in each province generally remain stable, using a time-invariant instrumental variable would not yield meaningful estimation results. Therefore, we construct an interaction term as an instrumental variable, namely the product of the annual agricultural market integration index and annual sunshine hours. This effectively addresses the endogeneity issue in the regression model, ensuring the correctness and robustness of the estimation results.

Column (4) reports the Two-Stage Least Squares (2SLS) estimation results using the traffic network density and annual sunshine hours as instrumental variables (Table 6). To examine the effectiveness of the instrumental variables, several statistical tests were conducted. First, the *Durbin–Wu–Hausman (D–W–H)* endogeneity test results indicate rejection of the exogeneity assumption for agricultural market integration at a 5% significance level. Second, the *Kleibergen–Paap rk LM statistic* for the underidentification test rejects the assumption of insufficient instrument identification at a 1% significance level. Furthermore, the *Kleibergen-Paap rk Wald F statistic* is 58.85, exceeding the critical value of 19.93 for the *Stock–Yogo* test at a 10% significance level, thereby rejecting the weak instrument assumption. Simultaneously, the associated probability of the *Sargan–Hansen* overidentification test is 0.103, indicating no rejection of the null hypothesis that the instrumental variables are not overidentified at a 10% significance level. This suggests that there is no issue of overidentification with the instrumental variables. Lastly, the *Anderson–Rubin Wald statistic* at a 1% significance level substantiates the reasonableness of the instrumental variable assumption. In summary, the aforementioned statistical tests affirm the rationality of selecting traffic network density and annual sunshine hours as instrumental variables. The 2SLS regression results reveal a coefficient of 0.00309 for agricultural market integration, signifying its promoting effect on regional economic development.

6. Extensibility Analysis: Substitution Relationships and Regional Heterogeneity

6.1. Substitution Relationships: Testing Hypothesis 2

The foundational regression results have conclusively affirmed the positive impact of domestic agricultural product market integration on regional economic development under closed conditions (Table 4). In this context, we introduce the agricultural product foreign trade volume (import volume, export volume, and total trade volume of agricultural products) as a moderating variable. By incorporating the levels of agricultural product market integration, regional economic development, and agricultural product foreign trade volume into the moderation effects model framework for analysis, we aim to explore the variations in the role of domestic agricultural product market integration in regional economics under conditions of an open economy. To delve into this, we employ a dynamic panel spatial

Durbin model with time lag terms, as depicted in column (4) (Table 4). This model allows us to scrutinize the interaction terms between agricultural product import trade volume and agricultural product market integration, as well as the interaction terms between agricultural product export trade volume and agricultural product market integration. We assess the magnitude and significance of these coefficients, and the outcomes are elucidated in columns (1) and (2) (Table 7). Subsequently, we utilize the dynamic panel spatial Durbin model with time and space lag terms from column (5) (Table 4), introducing the interaction term between the total volume of agricultural product import and export trade and agricultural product market integration. Once again, we examine the size and significance of the interaction term coefficients, and the findings are presented in column (3) (Table 7).

Table 7. Effect of agricultural market integration on regional economic development in China from 2010 to 2019 under the influence of agricultural product foreign trade.

Variables	(1)	(2)	(3)
together $\times$ *import*	−0.00097 *** (0.00026)		
together $\times$ *export*		−0.00120 *** (0.00041)	
together $\times$ *overall*			−0.00380 *** (0.00036)
$W \times$ *together* $\times$ *import*	−0.00377 *** (0.00110)		
$W \times$ *together* $\times$ *export*		−0.01266 *** (0.00180)	
$W \times$ *together* $\times$ *overall*			−0.02201 *** (0.00157)
$L.\ln pergdp$	1.55856 *** (0.02697)	1.65591 *** (0.02588)	1.47874 *** (0.02689)
$L.W \ln pergdp$			2.00517*** (0.13973)
Spatial rho (lambda)	0.28222 *** (0.06602)	0.27936 *** (0.06870)	0.57710 *** (0.12446)
Control variables	Yes	Yes	Yes
N	279.00000	279.00000	279.00000
R^2	0.45805	0.96205	0.95019

Note: The values in () are the heteroskedasticity robust standard errors. ***, **, and * represent the significance levels of 1%, 5%, and 10%, respectively. Data source: calculated by the author.

In columns (1) and (2) (Table 7), the interaction terms between the volume of agricultural product imports and agricultural product market integration, as well as between the volume of agricultural product exports and agricultural product market integration, are both negative, measuring −0.00097 and −0.0012, respectively. Importantly, both of these values pass significance tests in econometrics. Analyzing the results in column (3) (Table 7), we find that the coefficient for the interaction term between the total volume of agricultural product imports and exports and domestic agricultural product market integration is −0.0038, passing the significance test at the 1% level. This indicates that as the level of foreign trade in agricultural products increases, the promoting effect of agricultural product market integration on regional economic development will decrease. Conversely, when the level of foreign trade in agricultural products decreases, the role of domestic agricultural product market integration in regional economic development will increase. Clearly, if the speed of regional economic development remains constant (at equilibrium), a "substitute relationship" exists between international trade of agricultural products and domestic agricultural product market integration in promoting regional economic development. In summary, all three coefficients for the interaction terms are negative, suggesting that foreign trade in agricultural products can indeed substitute for some of the effects of domestic agricultural product market integration on economic development, supporting the validity of Hypothesis 2. However, it is crucial to note that the regulation of the types of agricultural products in the domestic market differs between agricultural product imports and exports, influencing their respective impacts on domestic agricultural product market integration.

The distinct structures of agricultural product imports and exports can yield varying degrees of substitution effects in international trade [121]. The top three categories in the 2019 agricultural product import trade, by proportion, were edible oils and fats (35%), meat and poultry and their products (33%), and aquatic products (17%) (Figures 2 and 3). This signals that the demand for these three types of agricultural products surpasses domestic supply, necessitating imports to meet consumption needs within China. In the 2019 agricultural product export trade, the leading categories were aquatic products (38%), vegetables (30%), and dried and fresh melons and fruits (14%). This implies that the demand for these three types of agricultural products in the domestic market is lower than the supply, indicating a competitive advantage and effective demand in the international market. It is crucial to note that China predominantly imports aquatic products such as fish fry, fresh or chilled freshwater products, and dried fish. On the other hand, exported aquatic products encompass fish fillets, mollusks, frozen fish, and crustaceans. In conclusion, a precise analysis of the types of agricultural products in imports and exports is essential for a thorough evaluation of the opportunities and risks posed by agricultural foreign trade to diverse agricultural sectors and provincial economic development. Such an analysis also facilitates the development of appropriate response policies.

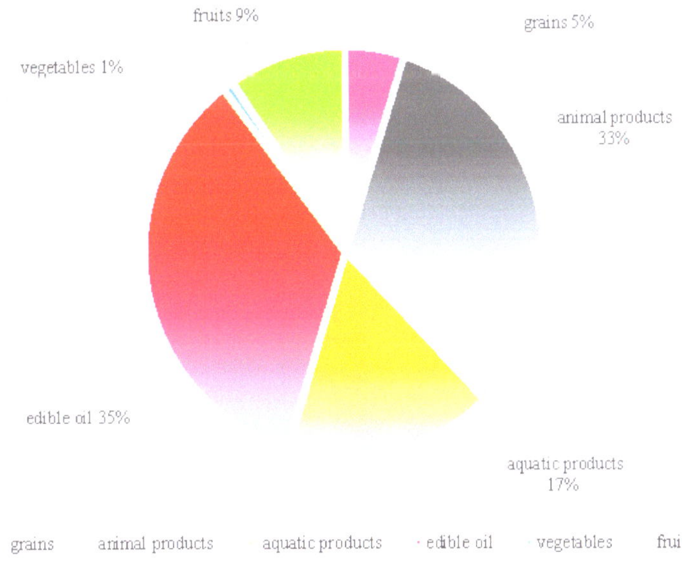

Figure 2. The import proportion of major agricultural products in China for the year 2019. Data source: General Administration of Customs of the People's Republic of China. http://stats.customs.gov.cn/ (accessed on 30 June 2020).

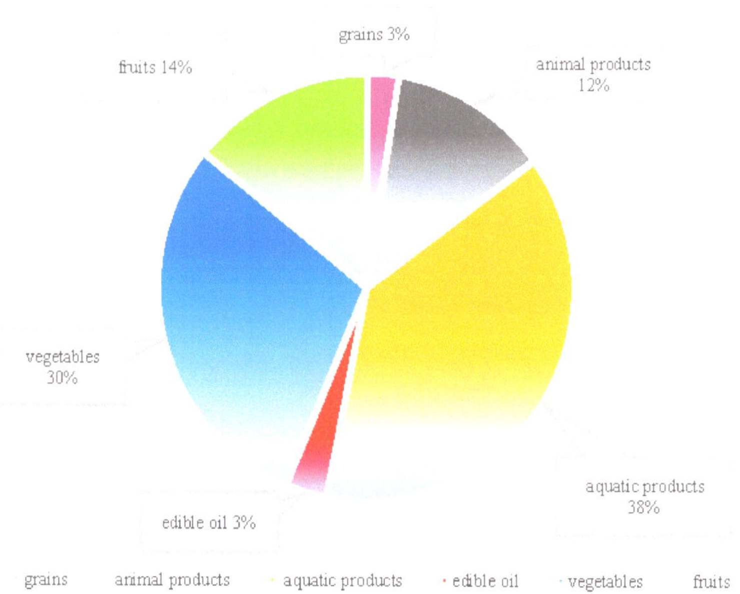

Figure 3. The export proportion of major agricultural products in China for the year 2019. Data source: General Administration of Customs of the People's Republic of China. http://stats.customs.gov.cn/ (accessed on 30 June 2020).

6.2. Heterogeneity Analysis: Testing Hypothesis 3

The economic development disparity between the eastern and western regions of China is substantial, marked by variations in the sophistication of agricultural market mechanisms and distinct levels of agricultural foreign trade development. In contrast to the western region, the eastern region exhibits advanced economic development, characterized by a considerable total volume of agricultural imports and exports. According to data released by the National Bureau of Statistics of China, the average total volume of agricultural imports and exports in the eastern region from 2010 to 2019 amounted to USD 13.826 billion, while in the western region, it stood at USD 1.411 billion, representing a nearly tenfold difference. In this context, drawing on the theories of resource allocation efficiency and specialization, the role of agricultural product market integration in economic development is likely to diverge between the two regions. Consequently, it becomes imperative to scrutinize and ascertain the impact of different levels of agricultural international trade on the relationship between these regions. The central region's diverse development indices fall within the spectrum delineated by the eastern and western regions. Thus, it is adequate to conduct a comparative analysis of the disparities between the eastern and western regions to meet the testing requirements for Hypothesis 3.

The regression coefficient for agricultural product market integration in the eastern region is 0.01804, while in the western region, it is 0.02354 (Table 8). This suggests that a one-unit increase in the level of agricultural product market integration corresponds to a 1.804% and 2.354% increase in the economic development levels of the two regions, respectively. The regional impact of agricultural product market integration on economic development exhibits heterogeneity, with a more pronounced effect in the relatively closed western region, confirming Hypothesis 3a. The coefficient of the variable *"together × overall"* signifies the magnitude of the positive effect of agricultural international trade substituting for regional agricultural product market integration on economic development. The results show that the interaction term coefficients for the eastern and western regions are -0.00099 and -0.00217, respectively, indicating differing substitution effects of agricultural international trade in these regions and confirming Hypothesis 3b. This discrepancy is attributed to the high overall openness of the eastern region, limiting the positive impact of agricultural international trade as a substitute for agricultural product market integration in promoting economic development. Conversely, the relatively closed western region has significant potential for increased openness, and agricultural international trade can play a crucial role in deepening economic openness, thereby substantially promoting economic development in the western region.

Table 8. Heterogeneity of agricultural market integration on regional economic development in China from 2010 to 2019 under the influence of agricultural product foreign trade.

Variables	(1) Eastern Region	(2) Western Region
together	0.01804 (0.01528)	0.02354 *** (0.00535)
overall	0.05344 (0.04161)	0.05761 *** (0.01573)
together × *overall*	−0.00099 (0.00112)	−0.00217 *** (0.00047)
$L. \ln pergdp$	1.44172 *** (0.07188)	1.22098 *** (0.03602)
W × *together*	0.30812 *** (0.11839)	0.13842 *** (0.02886)
W × *overall*	0.85030 *** (0.26373)	0.44810 *** (0.09133)
W × *together* × *overall*	0.02134 ** (0.00846)	0.01322 *** (0.00257)
Spatial rho (lambda)	0.40622 ** (0.17840)	1.85951 *** (0.21567)
Control variables	Yes	Yes
N	99.00000	108.00000
R^2	0.78104	0.07547

Note: The values in () are the heteroskedasticity robust standard errors. ***, **, and * represent the significance levels of 1%, 5%, and 10%, respectively. Data source: calculated by the author.

7. Discussions

According to data released by the Food and Agriculture Organization of the United Nations, China's Gross Domestic Product accounted for approximately 16% of the global total in 2019. Against the backdrop of economic globalization, the level of agricultural product market integration among China's provinces has intensified, fostering close supply chain connections for agricultural production and consumption across different regions [122]. Unlike industries such as domestic tourism, manufacturing, and energy, the agricultural product market is more severely impacted by disruptions in the supply chain and a reduction in effective demand. This is partly due to the intricate nature of the agricultural product supply chain, involving producers, consumers, inputs for agriculture and fisheries across regions, as well as various stages such as processing, storage, transportation, and marketing. Furthermore, the complexity of the agricultural product value chain and its dependence on trade and transportation render agricultural production and supply highly susceptible to external shocks. Such disruptions not only affect the daily lives and food supply security of ordinary citizens but also pose a threat to social stability [123].

In late 2019, the sudden outbreak of the COVID-19 pandemic profoundly impacted agricultural markets across various provinces in China. Stringent nationwide traffic lockdowns played a crucial role in curbing the spread of the disease. However, a substantial volume of agricultural, livestock, and horticultural products could not be promptly sold within local and cross-regional markets due to these measures, leading to severe stockpiling and missing the optimal market entry timing [124]. This not only increased the storage costs of agricultural products but also diminished their freshness and edible value, resulting in significant economic losses for farmers and other producers and hindering economic development in different regions [125]. Various restrictive measures implemented by local governments to control the COVID-19 pandemic adversely affected the development of agricultural product market integration. During this period, various indicators of agricultural market development failed to reflect its normal state and trends, rendering them unsuitable as reference standards for the recovery and improvement of agricultural production and sales supply chains in the post-pandemic era. Therefore, assessing the pre-pandemic level of agricultural product market integration and conducting comparative studies among different regions and provinces can provide valuable theoretical support for formulating effective policies to restore the damage caused by the pandemic to regional economic development.

7.1. Analysis of Positive Effects Based on Different Theories

The research findings of this study suggest that, in a closed state without considering agricultural import and export trade, agricultural product market integration positively contributes to regional economic development. Through calculations, it was determined that a one-unit increase in the agricultural product market integration index corresponds to a 0.156% rise in economic development. This conclusion aligns with previous research outcomes [126]. Analyzing from the perspective of economies of scale theory, agricultural product market integration directly introduces new cross-regional markets to farmers and other producers [127]. The substantial consumer demand arising from these new markets inherently motivates them to expand their operations. The adoption of new technologies supports larger-scale production operations, leading to reduced unit costs, ultimately boosting profits for producers and enhancing overall industry productivity [128].

According to the Resource Allocation Efficiency Theory, agricultural product market integration helps rectify the inefficient allocation of agricultural resources caused by the COVID-19 pandemic, distributing resources reasonably across different regions to meet diverse production and consumption needs [30]. With optimal utilization of agricultural resources such as land, water sources, and labor, producers can more effectively manufacture agricultural products and enhance the efficiency of the agricultural supply chain. Specifically, agricultural product market integration can guide the optimal utilization of agricultural resources throughout the entire industry chain, spanning production, transportation, and consumption, thus preventing the wastage of limited agricultural resources. Consequently, agricultural products from different regions can be concentrated in areas most suitable for their production, reducing unnecessary redundancy and resource wastage [129]. In summary, agricultural product market integration creates favorable conditions for the swift recovery of regional economies after the pandemic and improves the operational efficiency of the economy.

7.2. Analysis of Substitution Effects Based on Different Theories

When contemplating international trade in agricultural products or, in other words, opening the domestic agricultural product market, the international arena may replace a portion of the domestic market. The regression coefficients for the three interaction terms between agricultural product import and export trade (import volume, export volume, and total trade volume) and the agricultural product market integration index are -0.00097, -0.0012, and -0.0038, respectively. This suggests that as the level of agricultural product foreign trade increases, the promoting effect of agricultural product market integration on regional economic development will diminish. The opposite holds true as well. In summary, a "substitute relationship" exists between international agricultural product trade and domestic agricultural product market integration in fostering regional economic development. In simpler terms, agricultural product foreign trade substitutes for some of the effects of domestic agricultural product market integration on economic development. Drawing on the theories of specialization and comparative advantage, different regions within the country and other nations will focus on producing specific types of agricultural products to improve overall production efficiency in the agricultural industry. If certain agricultural products on the international market possess technological and cost advantages, they will partially substitute for domestic agricultural product markets based on the fundamental principles of a market economy.

In order to balance production efficiency and ensure a stable market supply, it is necessary to encourage different regions to specialize in agricultural production, while also meeting the diverse demands for various agricultural products across regions [130]. This necessitates an increase in the level of domestic agricultural product market integration. Moreover, when a particular region engages in the import and export trade of agricultural products, it will no longer solely rely on the production and sales of agricultural products from other domestic regions. This shift implies that the economic development of the local region is no longer solely dependent on interregional trade of agricultural products within the country; in fact, to some extent, the international market is "diluting" the promoting effect of domestic agricultural product market integration on regional economic development. Some studies suggest that during the COVID-19 pandemic, difficulties in cross-regional domestic agricultural product supply were encountered, and the international agricultural product market played a positive role in ensuring a stable food supply in certain domestic regions [131]. Therefore, it is evident that in the post-pandemic era, along with the vigorous restoration of domestic agricultural product market integration and economic development in various regions, there is a need to enhance the level of openness to the outside world, promoting the sustained development of international agricultural product trade.

7.3. Analysis of Heterogeneity Based on Different Theories

In an open economy context, this analysis examines the variations in the role of agricultural product market integration in promoting regional economic development across different regions in China. The regression coefficients for agricultural product market integration in the eastern and western regions are 0.01804 and 0.02354, respectively, indicating a stronger impact in the western region, consistent with previous research findings [104]. The theory of economies of scale suggests that agricultural product market integration incentivizes farmers and other producers to achieve economies of scale in production and engage in cross-regional or international trade. As the first region in China to embrace international trade, the eastern region, situated near economically vibrant areas like Japan, South Korea, and ASEAN countries, enjoys convenient shipping and inherent advantages for fostering international agricultural trade [132]. Over time, the agricultural product markets in these East and Southeast Asian countries have become intricately linked with the sale of agricultural products in the eastern region of China. For instance, in 2019, agricultural product exports from Shandong province to Japan and ASEAN countries constituted 42.3% of the total agricultural product exports. Conversely, the western region, being relatively closed, primarily sells its agricultural products to other provinces and cities within the country. The economic development in the western region is more significantly influenced by domestic agricultural product market integration than in the eastern region.

In-depth analysis reveals that the international agricultural product market can substitute for a portion of the domestic market, but the strength of this substitution effect varies regionally. The research findings in this paper indicate that the substitution effect coefficients in the eastern and western regions are −0.00099 and −0.00217, respectively. This further confirms that, in the less open western region, developing foreign agricultural product trade holds significant potential for promoting regional economic development.

However, the impact of the COVID-19 pandemic, leading to disruptions in global supply chains and transportation difficulties, has resulted in increased production and sales costs for agricultural products [133], reversing the effects of economies of scale. Faced with uncertainty and risks, risk theory suggests that farmers and businesses tend to adopt conservative strategies. Governments and enterprises worldwide are more inclined to produce and supply agricultural products locally to reduce dependence on international trade. This, in turn, results in a reduction in trade volume, which is detrimental to global economic development [134]. Addressing concerns about global agricultural supply chain stability and preventing further deglobalization in agricultural trade is crucial in the post-pandemic era.

8. Conclusions

8.1. Key Findings

This paper employs a dynamic spatial Durbin model to analyze the impact of agricultural market integration on regional economic development across 31 provinces in China during the ten years preceding the onset of the COVID-19 pandemic. The research findings indicate the following: firstly, under closed economic conditions, agricultural market integration effectively promotes the development of the domestic economy in various regions. Secondly, under open economic conditions, domestic agricultural market integration continues to foster regional economic development, although this promoting effect is somewhat diminished. Thirdly, the role of agricultural market integration in promoting economic development is greater in the western region compared to the eastern region. Lastly, there are significant regional variations in the strength and weakness of the substitutive effect of agricultural import and export trade on domestic agricultural market integration.

8.2. Policy Recommendations

In the post-pandemic era, countries worldwide are keenly focused on issues such as fluctuations in agricultural import and export prices and the vulnerability of single-supplier supply chains. There is a heightened emphasis on the localization of agricultural production and supply. This has resulted in two prominent features in the international agricultural market: a weakening of complementary capabilities and an increase in trade restrictive measures. Against this backdrop, China, in order to better address global challenges like pandemics, should underscore the integrated development of regional agricultural markets while leveraging the strengths of various domestic regions. To achieve this, the following three policy measures are proposed.

Firstly, strengthening connections in the domestic agricultural market. Attention should be given to eliminating policy and regulatory barriers to market access between regions. Further efforts should be made to plan a rational layout of transportation infrastructure connecting the entire nation, and the construction of an efficient agricultural cold chain transportation network and storage facilities. Secondly, encouraging international trade of agricultural products. In alignment with WTO rules and regulations of other economic cooperation organizations, tariffs should be reduced, and efforts should be intensified to strengthen trade connections with countries worldwide in the agricultural sector. Thirdly, emphasizing the development of the agricultural market in the western region. Promoting the development of internet-based smart agriculture in the western region will help overcome the geographical constraints of this area.

8.3. Limitations of this Research

This paper has contributed novel findings to the field of agricultural market research, yet certain limitations persist. Firstly, while the level of agricultural market integration influences regional economic development, the specific pathways through which it affects regional economic development require meticulous examination through the acquisition of substantial real-world data and empirical validation. Secondly, constrained by the difficulty in systematically obtaining micro-level agricultural market data, the authors were unable to investigate the specific impacts of agricultural market integration on the income and costs of farmers, agricultural enterprises, and other producers at the level of individual cities or more granular entities.

8.4. Future Research Prospects

The elevation of the level of agricultural market integration may also be accompanied by adverse effects. Subsequent research should focus on two aspects: firstly, in conjunction with policy incentive mechanisms, efforts should be made to prevent agricultural enterprises and individuals from pursuing profit as the sole objective, prioritizing the supply of high-quality agricultural products to the most economically developed regions. This practice may impact the food and nutritional health of the population in less developed regions. Secondly, considering the productivity differences between the less developed western regions and the developed eastern regions, differentiated subsidy measures should be introduced. This aims to encourage agricultural enterprises in less developed regions to adopt clean energy and low-carbon production methods, preventing environmental damage and safeguarding public health. In conclusion, in the post-pandemic era, research on the relationship between agricultural market integration, international trade of agricultural products, and economic development should place greater emphasis on a micro-level analysis, considering production efficiency, environmental factors, social aspects, and economic development concurrently.

Author Contributions: Conceptualization, X.M. and S.W.; methodology, J.H.; writing—original draft preparation, X.M.; writing—review and editing, Z.R., T.M. and H.X. All authors have read and agreed to the published version of the manuscript.

Funding: This study was funded by the Postgraduate Research & Practice Innovation Program of Jiangsu Province (Funder: Jiangsu Education Department; Funding number: KYCX21_0554).

Institutional Review Board Statement: Not applicable.

Informed Consent Statement: Not applicable.

Data Availability Statement: Data are contained within the article.

Conflicts of Interest: The authors declare no conflicts of interest.

References

1. Cariappa, A.A.; Acharya, K.K.; Adhav, C.A.; Sendhil, R.; Ramasundaram, P. Impact of COVID-19 on the Indian agricultural system: A 10-point strategy for post-pandemic recovery. *Outlook Agric.* **2021**, *50*, 26–33. [CrossRef]
2. Barman, A.; Das, R.; De, P.K. Impact of COVID-19 in food supply chain: Disruptions and recovery strategy. *Current Res. Behav. Sci.* **2021**, *2*, 100017. [CrossRef]
3. Thilmany, D.; Canales, E.; Low, S.A.; Boys, K. Local food supply chain dynamics and resilience during COVID-19. *Appl. Econ. Perspect. Policy* **2021**, *43*, 86–104. [CrossRef]
4. Yang, B.; Asche, F.; Li, T. Consumer behavior and food prices during the COVID-19 pandemic: Evidence from Chinese cities. *Econ. Inq.* **2022**, *60*, 1437–1460. [CrossRef]
5. Ali, J.; Khan, W. Impact of COVID-19 pandemic on agricultural wholesale prices in India: A comparative analysis across the phases of the lockdown. *J. Public. Aff.* **2020**, *20*, e2402. [CrossRef]
6. Gruère, G.; Brooks, J. Characterising early agricultural and food policy responses to the outbreak of COVID-19. *Food Policy* **2021**, *100*, 102017. [CrossRef] [PubMed]
7. Zant, W. How is the liberalization of food markets progressing? Market integration and transaction costs in subsistence economies. *World Bank. Econ. Rev.* **2013**, *27*, 28–54. [CrossRef]
8. Baffes, J.; Kabundi, A. Commodity price shocks: Order within chaos? *Resour. Policy* **2023**, *83*, 103640. [CrossRef]
9. Akter, S.; Basher, S.A. The impacts of food price and income shocks on household food security and economic well-being: Evidence from rural Bangladesh. *Global Environ. Chang.* **2014**, *25*, 150–162. [CrossRef]
10. Dawe, D.; Timmer, C.P. Why stable food prices are a good thing: Lessons from stabilizing rice prices in Asia. *Glob. Food Secur.* **2012**, *1*, 127–133. [CrossRef]
11. Martin, W. Economic growth, convergence, and agricultural economics. *Agric. Econ.* **2019**, *50*, 7–27. [CrossRef]
12. Beghin, J.C.; Schweizer, H. Agricultural trade costs. *Appl. Econ. Perspect. Policy* **2021**, *43*, 500–530. [CrossRef]
13. Banerjee, A.; Duflo, E.; Qian, N. On the road: Access to transportation infrastructure and economic growth in China. *J. Dev. Econ.* **2020**, *145*, 102442. [CrossRef]

14. Moser, C.; Barrett, C.; Minten, B. Spatial integration at multiple scales: Rice markets in Madagascar. *Agric. Econ.* **2009**, *40*, 281–294. [CrossRef]
15. Fan, X.; Song, D.; Zhao, X. Does Infrastructure Construction Break up Domestic Market Segmentation? *Econ. Res. J.* **2017**, *52*, 20–34.
16. Robinson, A.L. Internal Borders: Ethnic-based market segmentation in Malawi. *World Dev.* **2016**, *87*, 371–384. [CrossRef]
17. Luckmann, J.; Ihle, R.; Kleinwechter, U.; Grethe, H. World market integration of Vietnamese rice markets during the 2008 food price crisis. *Food Sec.* **2015**, *7*, 143–157. [CrossRef]
18. Liu, Z.; Xu, Y. Local Government Behavioral Patterns and Industrial Structure Evolution in the Process of Marketization. *Econ. Manag. J.* **2014**, *36*, 12–23. [CrossRef]
19. Hang, L.; Ullah, I.; Kim, D.H. A secure fish farm platform based on blockchain for agriculture data integrity. *Comput. Electron. Agric.* **2020**, *170*, 105251. [CrossRef]
20. Oladele, O.I. Effect of information communication technology on agricultural information access among researchers, extension agents, and farmers in southwestern Nigeria. *J. Agric. Food Inf.* **2011**, *12*, 167–176. [CrossRef]
21. Engotoit, B.; Kituyi, G.M.; Moya, M.B. Influence of performance expectancy on commercial farmers' intention to use mobile-based communication technologies for agricultural market information dissemination in Uganda. *J. Syst. Inform. Tech.* **2016**, *18*, 346–363. [CrossRef]
22. Escobal, J.A.; Cavero, D. Transaction costs, institutional arrangements and inequality outcomes: Potato marketing by small producers in rural Peru. *World Dev.* **2012**, *40*, 329–341. [CrossRef]
23. Cirera, X.; Arndt, C. Measuring the impact of road rehabilitation on spatial market efficiency in maize markets in Mozambique. *Agric. Econ.* **2008**, *39*, 17–28. [CrossRef]
24. Zakari, S.; Ying, L.; Song, B. Market integration and spatial price transmission in Niger grain markets. *Afr. Dev. Rev.* **2014**, *26*, 264–273. [CrossRef]
25. Zhao, X.; Calvin, K.V.; Wise, M.A.; Iyer, G. The role of global agricultural market integration in multiregional economic modeling: Using hindcast experiments to validate an Armington model. *Econ. Anal. Policy* **2021**, *72*, 1–17. [CrossRef]
26. Gluschenko, K. Market integration in Russia during the transformation years. *Econ. Transit.* **2003**, *11*, 411–434. [CrossRef]
27. Yang, H.; Qi, J.; Xu, X.; Yang, D.; Lv, H. The regional variation in climate elasticity and climate contribution to runoff across China. *J. Hydrol.* **2014**, *517*, 607–616. [CrossRef]
28. Hobbs, J.E. Food supply chains during the COVID-19 pandemic. *Can. J. Agric. Econ.* **2020**, *68*, 171–176. [CrossRef]
29. Elavarasan, R.M.; Pugazhendhi, R.; Shafiullah, G.M.; Kumar, N.M.; Arif, M.T.; Jamal, T.; Chopra, S.S.; Dyduch, J. Impacts of COVID-19 on Sustainable Development Goals and effective approaches to maneuver them in the post-pandemic environment. *Environ. Sci. Pollut. Res.* **2022**, *29*, 33957–33987. [CrossRef]
30. Khedhiri, S. The impact of COVID-19 on agricultural market integration in Eastern Canada. *Reg. Sci. Policy Pract.* **2023**, *15*, 371–386. [CrossRef]
31. Zhao, Y.; Ni, J. The border effects of domestic trade in transitional China: Local Governments' preference and protectionism. *Chin. Econ.* **2018**, *51*, 413–431. [CrossRef]
32. Halder, P.; Pati, S. A need for paradigm shift to improve supply chain management of fruits & vegetables in India. *Asian J. Agric. Rural Dev.* **2011**, *1*, 1–20. [CrossRef]
33. Olper, A.; Raimondi, V. Agricultural market integration in the OECD: A gravity-border effect approach. *Food Policy* **2008**, *33*, 165–175. [CrossRef]
34. Barrett, C.B. Measuring integration and efficiency in international agricultural markets. *Appl. Econ. Perspect. Policy* **2021**, *23*, 19–32. [CrossRef]
35. Takeshima, H.; Nagarajan, L. Minor millets in Tamil Nadu, India: Local market participation, on-farm diversity and farmer welfare. *Environ. Dev. Econ.* **2012**, *17*, 603–632. [CrossRef]
36. Gluschenko, K. Analysing changes in market integration through a cross-sectional test for the law of one price. *Int. J. Financ. Econ.* **2004**, *9*, 135–149. [CrossRef]
37. Cecchetti, S.G.; Mark, N.C.; Sonora, R.J. Price index convergence among United States cities. *Int. Econ. Rev.* **2002**, *43*, 1081–1099. [CrossRef]
38. Fan, S.; Wei, X. The law of one price: Evidence from the transitional economy of China. *Rev. Econ. Stat.* **2006**, *88*, 682–697. [CrossRef]
39. Lan, Y.; Sylwester, K. Does the law of one price hold in China? Testing price convergence using disaggregated data. *China Econ. Rev.* **2010**, *21*, 224–236. [CrossRef]
40. Young, A. The Razor's edge: Distortions and incremental reform in the People's Re-public of China. *Q. J. Econ.* **2000**, *115*, 1091–1135. [CrossRef]
41. Bai, C.; Du, Y.; Tao, Z.; Tong, Y. Local protectionism and industrial concentration in China: Overall trend and important factors. *Econ. Res. J.* **2004**, *4*, 29–40. (In Chinese)
42. Calderón, C.; Servén, L. Infrastructure and economic development in Sub-Saharan Africa. *J. Afr. Econ.* **2010**, *19*, i13–i87. [CrossRef]
43. Mohsin, M.; Yongtong, M.; Hussain, K.; Mahmood, A.; Zhaoqun, S.; Nazir, K.; Wei, W. Contribution of fish production and trade to the economy of Pakistan. *Int. J. Mar. Sci.* **2015**, *5*, 1–7. [CrossRef]
44. Islam, S.M.; Clarke, M. The relationship between economic development and social welfare: A new adjusted GDP measure of welfare. *Soc. Indic. Res.* **2002**, *57*, 201–229. [CrossRef]
45. Lin, S. Resource allocation and economic growth in China. *Econ. Inquiry* **2000**, *38*, 515–526. [CrossRef]

46. Luo, W.; Timothy, D.J. An assessment of farmers' satisfaction with land consolidation performance in China. *Land Use Policy* **2017**, *61*, 501–510. [CrossRef]
47. Saleem, H.; Khan, M.B.; Mahdavian, S.M. The role of green growth, green financing, and eco-friendly technology in achieving environmental quality: Evidence from selected Asian economies. *Environ. Sci. Pollut. Res.* **2022**, *29*, 57720–57739. [CrossRef] [PubMed]
48. Liu, Y.; Ma, X.; Shu, L.; Hancke, G.P.; Abu-Mahfouz, A.M. From Industry 4.0 to Agriculture 4.0: Current status, enabling technologies, and research challenges. *IEEE Trans. Ind. Inform.* **2020**, *17*, 4322–4334. [CrossRef]
49. Surya, B.; Salim, A.; Hernita, H.; Suriani, S.; Menne, F.; Rasyidi, E.S. Land use change, urban agglomeration, and urban sprawl: A sustainable development perspective of Makassar City, Indonesia. *Land* **2021**, *10*, 556. [CrossRef]
50. Magesa, M.M.; Michael, K.; Ko, J. Access and use of agricultural market information by smallholder farmers: Measuring informational capabilities. *Electron. J. Inf. Syst. Dev. Ctries.* **2020**, *86*, e12134. [CrossRef]
51. Kofi, O.M.; Biekpe, N. Agricultural commodity supply response in Ghana. *J. Econ. Stud.* **2008**, *35*, 224–235. [CrossRef]
52. Ahlers, C.; Broll, U.; Eckwert, B. Information and output in agricultural markets: The role of market transparency. *Agric. Econ.* **2013**, *1*, 15. [CrossRef]
53. Schipmann, C.; Qaim, M. Supply chain differentiation, contract agriculture, and farmers' marketing preferences: The case of sweet pepper in Thailand. *Food Policy* **2011**, *36*, 667–677. [CrossRef]
54. Gardas, B.B.; Raut, R.D.; Narkhede, B. Modeling causal factors of post-harvesting losses in vegetable and fruit supply chain: An Indian perspective. *Renew. Sust. Energ. Rev.* **2017**, *80*, 1355–1371. [CrossRef]
55. Staal, S.; Delgado, C.; Nicholson, C. Smallholder dairying under transactions costs in East Africa. *World Dev.* **1997**, *25*, 779–794. [CrossRef]
56. Birthal, P.S.; Joshi, P.K.; Gulati, A. *Vertical Coordination in High-Value Commodities: Implications for Smallholders*; (No. 596-2016-40047); AgEcon Search: St. Paul, MN, USA, 2005. [CrossRef]
57. Ruben, R.; Boselie, D.; Lu, H. Vegetables procurement by Asian supermarkets: A transaction cost approach. *Supply Chain Manag. Int. J.* **2007**, *12*, 60–68. [CrossRef]
58. Ao, G.; Liu, Q.; Qin, L.; Chen, M.; Liu, S.; Wu, W. Organization model, vertical integration, and farmers' income growth: Empirical evidence from large-scale farmers in Lin'an, China. *PLoS ONE* **2021**, *16*, e0252482. [CrossRef]
59. Zheng, X.; Huang, Q.; Zheng, S. The Identification and Applicability of Regional Brand-Driving Modes for Agricultural Products. *Agriculture* **2022**, *12*, 1127. [CrossRef]
60. Urruty, N.; Tailliez-Lefebvre, D.; Huyghe, C. Stability, robustness, vulnerability and resilience of agricultural systems. A review. *Agron. Sustain. Dev.* **2016**, *36*, 15. [CrossRef]
61. Acosta, I.C.G.; Cano, L.A.F.; Peña, O.D.L.; Rivera, C.L.; Bravo, B.J.J. Design of an inventory management system in an agricultural supply chain considering the deterioration of the product: The case of small citrus producers in a developing country. *J. Appl. Eng. Sci.* **2018**, *16*, 523–537. [CrossRef]
62. Alabi, M.O.; Ngwenyama, O. Food security and disruptions of the global food supply chains during COVID-19: Building smarter food supply chains for post COVID-19 era. *Brit. Food J.* **2023**, *125*, 167–185. [CrossRef]
63. Rizou, M.; Galanakis, I.M.; Aldawoud, T.M.; Galanakis, C.M. Safety of foods, food supply chain and environment within the COVID-19 pandemic. *Trends Food Sci. Technol.* **2020**, *102*, 293–299. [CrossRef]
64. Streimikis, J.; Baležentis, T. Agricultural sustainability assessment framework integrating sustainable development goals and interlinked priorities of environmental, climate and agriculture policies. *Sustain. Dev.* **2020**, *28*, 1702–1712. [CrossRef]
65. Molden, D.; Oweis, T.; Steduto, P.; Bindraban, P.; Hanjra, M.A.; Kijne, J. Improving agricultural water productivity: Between optimism and caution. *Agric. Water Manag.* **2010**, *97*, 528–535. [CrossRef]
66. Doran, J.W. Soil health and global sustainability: Translating science into practice. *Agric. Ecosyst. Environ.* **2002**, *88*, 119–127. [CrossRef]
67. Li, M.; Zhao, L.; Zhang, C.; Liu, Y.; Fu, Q. Optimization of agricultural resources in water-energy-food nexus in complex environment: A perspective on multienergy coordination. *Energy Convers. Manag.* **2022**, *258*, 115537. [CrossRef]
68. Komilova, N.K.; Haydarova, S.A.; Xalmirzaev, A.A.; Kurbanov, S.B.; Rajabov, F.T. Territorial structure of agriculture development in Uzbekistan in terms of economical geography. *J. Adv. Res. Law Econ.* **2019**, *10*, 2364.
69. Jambor, A. Comparative advantages and specialisation of the Visegrad countries agri-food trade. *Acta Oecon. Inform.* **2013**, *16*, 22–34. [CrossRef]
70. Ma, M.; Steinbach, S.; Wu, J. A study on regional specialization of China's agricultural production: Recent trends and drivers. *Asian J. Agric. Rural Dev.* **2014**, *4*, 113–127. [CrossRef]
71. Self, S.; Grabowski, R. Economic development and the role of agricultural technology. *Agric. Econ.* **2007**, *36*, 395–404. [CrossRef]
72. Hadley, D.; Irz, X. Productivity and farm profit–A microeconomic analysis of the cereal sector in England and Wales. *Appl. Econ.* **2008**, *40*, 613–624. [CrossRef]
73. Key, N. Farm size and productivity growth in the United States Corn Belt. *Food Policy* **2019**, *84*, 186–195. [CrossRef]
74. Asche, F.; Cojocaru, A.L.; Roth, B. The development of large scale aquaculture production: A comparison of the supply chains for chicken and salmon. *Aquaculture* **2018**, *493*, 446–455. [CrossRef]
75. De Roest, K.; Ferrari, P.; Knickel, K. Specialisation and economies of scale or diversification and economies of scope? Assessing different agricultural development pathways. *J. Rural Stud.* **2018**, *59*, 222–231. [CrossRef]
76. Valencia, V.; Wittman, H.; Blesh, J. Structuring Markets for Resilient Farming Systems. *Agron. Sustain. Dev.* **2019**, *39*, 25. [CrossRef]

77. Gaupp, F.; Pflug, G.; Hochrainer-Stigler, S.; Hall, J.; Dadson, S. Dependency of crop production between global breadbaskets: A copula approach for the assessment of global and regional risk pools. *Risk Anal.* **2017**, *37*, 2212–2228. [CrossRef] [PubMed]
78. Roberts, M.; Goh, C.C. Density, distance and division: The case of Chongqing municipality, China. *Camb. J. Reg. Econ. Soc.* **2011**, *4*, 189–204. [CrossRef]
79. Poncet, S. Measuring Chinese domestic and international integration. *China Econ. Rev.* **2003**, *14*, 1–21. [CrossRef]
80. Zhao, Y.; Liu, D. Market discrimination, regional boundary effect and economic growth. *China Ind. Econ.* **2008**, *12*, 27–37. (In Chinese) [CrossRef]
81. Pan, W. Regional linkage and the spatial spillover effects on regional economic growth in China. *Econ. Res. J.* **2012**, *47*, 54–65. (In Chinese)
82. Yang, J.; Luo, L. Industrial policy, regional competition and distortion of resources spatial allocation. *China Ind. Econ.* **2018**, *12*, 5–22. (In Chinese) [CrossRef]
83. Liu, Z.; Kong, L. From segmentation to integration: The resistance and countermeasures to promote the construction of a unified domestic market. *China Ind. Econ.* **2021**, *8*, 20–36. [CrossRef]
84. Huang, S. The evolution of China's rural development strategy and its theoretical basis over the past 40 years of reform and opening-up. *Econ. Res. J.* **2018**, *53*, 4–19. (In Chinese)
85. Lin, Y. Main issues and prospects of rural reform in China in the 1990s. *Manag. World* **1994**, *3*, 139–144. (In Chinese) [CrossRef]
86. Dong, Y.; Gu, Y.; Yang, K. Agricultural products brands, market integration and agricultural income growth. *J. Cap. Univ. Econ. Bus.* **2021**, *1*, 70–80. (In Chinese) [CrossRef]
87. Liu, S.; Hu, A. Transport infrastructure and economic growth: Perspective from China's regional disparities. *China Ind. Econ.* **2010**, *4*, 14–23. (In Chinese) [CrossRef]
88. Du, Y. The impact of agricultural vertical integrating and farmer's organizational innovation to the farmer's income. *China Rural Survey* **2005**, *3*, 9–18+80. (In Chinese)
89. Lu, M.; Chen, Z. Fragmented growth: Why economic opening may worsen domestic market segmentation? *Econ. Res. J.* **2009**, *44*, 42–52. (In Chinese)
90. Zheng, F.; Cheng, Y. From agricultural industrialization to agricultural industrial clusters: Feasibility analysis of the development of competitive agricultural industrialization. *Manag. World* **2005**, *7*, 64–73+93. (In Chinese) [CrossRef]
91. Liu, L.; Zhou, L. Comparative advantage, FDI and international competitiveness of China's agricultural products industry. *J. Int. Trade* **2011**, *12*, 39–54. (In Chinese) [CrossRef]
92. Li, X.; Li, H. The probable impacts of appreciation of RMB on Chinese agriculture: A case of soybean. *Probl. Agric. Econ.* **2005**, *1*, 31–36+79. (In Chinese)
93. Qiang, W.; Niu, S.; Wang, X.; Zhang, C.; Liu, A.; Cheng, S. Evolution of the Global Agricultural Trade Network and Policy Implications for China. *Sustainability* **2020**, *12*, 192. [CrossRef]
94. Moon, W. Are there dynamic productivity gains from agricultural trade? *China Agric. Econ. Rev.* **2022**, *14*, 32–46. [CrossRef]
95. Cai, Z. The formation path of regional economic layout with high-quality development: Based on the perspective of regional complementary advantages. *Reform* **2020**, *8*, 132–146. (In Chinese)
96. Li, X.; Zhang, Y.; Sun, B. Does regional integration promote economic growth efficiency? An empirical analysis based on the Yangtze River Economic Belt. *China Popul. Resour. Environ.* **2017**, *27*, 10–19. (In Chinese)
97. Banik, N.; Das, K.C. The location substitution effect: Does it apply for China? *Global Bus. Rev.* **2014**, *15*, 59–75. [CrossRef]
98. Erokhin, V.; Ivolga, A.; Heijman, W.J.M. Trade liberalization and state support of agriculture: Effects for developing countries. *Agric. Econ.* **2014**, *60*, 524–537. [CrossRef]
99. Kastner, T.; Erb, K.H.; Nonhebel, S. International wood trade and forest change: A global analysis. *Global Environ. Chang.* **2011**, *21*, 947–956. [CrossRef]
100. Kumar, A. Exports of livestock products from India: Performance, competitiveness and determinants. *Agric. Econ. Res. Rev.* **2010**, *23*, 57–68. [CrossRef]
101. MacLaren, D. Agricultural trade policy analysis and international trade theory: A review of recent developments. *J. Agric. Econ.* **1991**, *42*, 250–297. [CrossRef]
102. Song, Y. An analysis of the relationship between administrative divisions and the types of administrative region and regional administration in modern China. *Peking. Univ. J. Philos. Soc. Sci.* **1999**, *4*, 55–62. (In Chinese)
103. Miao, X.; Han, J.; Wang, S.; Li, X. Spatial effects of market integration on the industrial agglomeration of agricultural products: Evidence from China. *Environ. Sci. Pollut. Res.* **2023**, *30*, 84949–84971. [CrossRef]
104. Lou, W. Measurement and comparison of regional economic integration in the Beijing-Tianjin-Hebei, Yangtze River Delta, and Pearl River Delta regions. *Stat. Decis.* **2014**, *2*, 90–92. (In Chinese) [CrossRef]
105. Ma, Y. Agricultural product trade, agricultural technological progress and interregional income gap among farmers in China. *J. Int. Trade* **2018**, *6*, 41–53. (In Chinese) [CrossRef]
106. Démurger, S. Infrastructure development and economic growth: An explanation for regional disparities in China? *J. Comp. Econ.* **2001**, *29*, 95–117. [CrossRef]
107. Wang, D.; Abula, B.; Lu, Q.; Liu, Y.; Zhou, Y. Regional business environment, agricultural opening-up and high-quality development. Dynamic empirical analysis from China's agriculture. *Agronomy* **2022**, *12*, 974. [CrossRef]

108. Sheng, B.; Mao, Q. Trade openness, domestic market integration, and inter-provincial economic growth in China: 1985–2008. *J. World Econ.* **2011**, *11*, 44–66. (In Chinese) [CrossRef]
109. Yang, S.; Gou, L.; Mao, Y. Market segmentation, openness to the outside world, and economic growth of urban agglomerations: A case study of the Guangdong-Hong Kong-Macao Greater Bay area. *Econ. Issues Explor.* **2019**, *11*, 125–133. (In Chinese)
110. Gui, Q.; Chen, M.; Lu, M.; Chen, Z. Is the domestic goods market in China tending toward segmentation or integration: An analysis based on the relative price method. *J. World Econ.* **2006**, *2*, 20–30. (In Chinese)
111. Wu, J.; Ge, Z.; Han, S.; Xing, L.; Zhu, M.; Zhang, J.; Liu, J. Impacts of agricultural industrial agglomeration on China's agricultural energy efficiency: A spatial econometrics analysis. *J. Clean. Prod.* **2020**, *260*, 121011. [CrossRef]
112. Zhao, L.; Huang, G.; Wang, X. Labor market segmentation, factor allocation efficiency and agricultural product circulation industry growth- An investigation of the moderated mediation effects. *J. Agrotech. Econ.* **2021**, *3*, 4–19. (In Chinese) [CrossRef]
113. Zou, B.; Cao, Y.; Xiao, Y. Effect of market integration on farmers' increasing income from the perspective of common prosperity. *Res. Financ. Econ. Issues* **2023**, *11*, 115–129. (In Chinese) [CrossRef]
114. Song, D.; Gao, X.; Fan, X. Is supply-side structural reform in agriculture conducive to inclusive growth? *J. East. China Norm. Univ.* **2020**, *52*, 146–161+199–200. (In Chinese) [CrossRef]
115. Zhu, H.; Zheng, X. Digital economy, urban-rural integration and agricultural economic resilience. *Stat. Decis.* **2023**, *39*, 22–27. (In Chinese) [CrossRef]
116. Shao, S.; Zhang, K.; Dou, J. Effects of economic agglomeration on energy saving and emission reduction: Theory and empirical evidence from China. *Manag. World* **2019**, *35*, 36–60+226. (In Chinese) [CrossRef]
117. Bian, Y.; Wu, L.; Bai, J. Market segmentation and high-quality economic development: Evidence from the perspective of green growth. *J. Environ. Eco.* **2019**, *4*, 96–114. (In Chinese) [CrossRef]
118. Huang, J. Domestic trade, spatial spillovers and provincial business cycle synchronization: 1987-2011. *Finance Trade Res.* **2014**, *25*, 18–27. (In Chinese) [CrossRef]
119. Cui, J.; Shi, L.; Xie, B. Can the convergence of technology and industry drive the development of regional economy—A dynamic spatial Durbin model based study. *J. Harbin Univ. Commer.* **2022**, *2*, 15–25. (In Chinese)
120. Nie, Y.; Yao, Q.; Zhou, Z. Industrial collaborative agglomeration and high-quality economic development in the Yangtze River Delta. *East. China Econ. Manag.* **2022**, *10*, 16–30. (In Chinese) [CrossRef]
121. Podoba, Z.S.; Moldovan, A.A.; Faizova, A.A. The import substitution of agricultural products in Russia. *Probl. Econ. Transit.* **2020**, *62*, 707–720. [CrossRef]
122. Fu, S.; Zhan, Y.; Ouyang, J.; Ding, Y.; Tan, K.; Fu, L. Power, supply chain integration and quality performance of agricultural products: Evidence from contract farming in China. *Prod. Plan. Control* **2021**, *32*, 1119–1135. [CrossRef]
123. Zhou, J.; Fei, H.; Kai, L.; Yu, W. Vegetable production under COVID-19 pandemic in China: An analysis based on the data of 526 households. *J. Integr. Agric.* **2020**, *19*, 2854–2865. [CrossRef]
124. Zhao, H.; Guo, X.; Peng, N. What catalyzes the proactive recovery of peasants from the COVID-19 pandemic? A livelihood perspective in Ningqiang County, China. *Int. J. Disaster Risk Reduct.* **2022**, *73*, 102920. [CrossRef] [PubMed]
125. Zhou, W. Meta-analysis of the effect of the COVID-19 epidemic on China's food supply chain. *Open J. Bus. Manag.* **2022**, *11*, 303–312. [CrossRef]
126. Zeng, G.; Ding, Y. Does agglomeration externalities promote the growth of county-level industries? Based on the empirical evidence of the agricultural and sideline food processing industry in Hubei Province. *World Agric.* **2021**, *3*, 120–130. [CrossRef]
127. Liu, J.; Bailey, D. Examining economies of scale for farmer cooperatives in China's Shanxi Province. *J. Rural. Coop.* **2013**, *41*, 147–175. [CrossRef]
128. Pan, D.; Tang, J.; Zhang, L.; He, M.; Kung, C. The impact of farm scale and technology characteristics on the adoption of sustainable manure management technologies: Evidence from hog production in China. *J. Clean. Prod.* **2021**, *280*, 124340. [CrossRef]
129. Tan, Y.; Hai, F.; Popp, J.; Oláh, J. Minimizing waste in the food supply chain: Role of information system, supply chain strategy, and network design. *Sustainability* **2022**, *14*, 11515. [CrossRef]
130. Lu, S.; Liu, Y.; Long, H.; Xing, H. Agricultural production structure optimization: A case study of major grain producing areas, China. *J. Integr. Agric.* **2013**, *12*, 184–197. [CrossRef]
131. Cao, L.; Li, T.; Wang, R.; Zhu, J. Impact of COVID-19 on China's agricultural trade. *China Agric. Econ. Rev.* **2021**, *13*, 1–21. [CrossRef]
132. Francois, J.F.; Wignaraja, G. Economic implications of Asian integration. *Glob. Econ. J.* **2008**, *8*, 850139. [CrossRef]
133. Coluccia, B.; Agnusdei, G.P.; Miglietta, P.P.; De Leo, F. Effects of COVID-19 on the Italian agri-food supply and value chains. *Food Control* **2021**, *123*, 107839. [CrossRef] [PubMed]
134. Gu, H.; Wang, C. Impacts of the COVID-19 pandemic on vegetable production and countermeasures from an agricultural insurance perspective. *J. Int. Agric.* **2020**, *19*, 2866–2876. [CrossRef]

 *sustainability*

Article

Relationship of Arable Land Scale and High-Quality Development of Farmers' Cooperatives: Evidence from Grain Production Cooperatives in China

Yang Xu [1], Yujia Huo [2] and Xiangyu Guo [1,*]

[1] College of Economics and Management, Northeast Agricultural University, Harbin 150030, China; b210801009@neau.edu.cn

[2] School of Economics and Management, Anqing Normal University, Anqing 246133, China; huoyujia152311@aqnu.edu.cn

* Correspondence: guoxy@neau.edu.cn; Tel.: +86-0451-8713-213

Abstract: Sustainable agricultural development relies significantly on the high-quality progression of farmers' cooperatives. While growing in number, farmers' cooperatives are still facing the dilemma of improving the quality of their development. Land endowment is the foundation of agricultural production and the farmers' cooperatives. Clarifying the correlation between arable land scale and the high-quality development of farmers' cooperatives is conducive to the optimization of land use and the adoption of scientific land management measures to improve the quality of the development of farmers' cooperatives. Based on the micro-survey data of 448 farmers' cooperatives in three major grain-producing provinces, namely Heilongjiang, Henan, and Shandong in China, this paper constructs an evaluation index system for the high-quality development of farmers' cooperatives and theoretically and empirically explores the impact mechanism of arable land scale on the high-quality development of farmers' cooperatives. The results suggest the following: (1) there exists a significant "inverted U-shaped" association between the arable land scale and the development quality of cooperatives, and this result remains robust after testing through substitution variable and instrumental variable methods; (2) further research on the "inverted U-shaped" association reveals that the impact of arable land scale on the high-quality development of cooperatives undergoes four stages: "weak impact—rapid improvement—diminished growth effect—decline in development quality"; and (3) mechanism tests suggest that the "inverted U-shaped" association between the arable land scale and the development quality of cooperatives is mainly constrained by industrial development input, and arable land scale and industrial development show a strong complementary relationship. Therefore, in the course of enhancing the quality of farmers' cooperatives, it is crucial to select appropriate land management strategies based on to their stage, paying special attention to the compatibility between arable land scale and industrial development.

Keywords: arable land scale; high-quality development; farmers' cooperatives; agricultural industry

Citation: Xu, Y.; Huo, Y.; Guo, X. Relationship of Arable Land Scale and High-Quality Development of Farmers' Cooperatives: Evidence from Grain Production Cooperatives in China. *Sustainability* **2024**, *16*, 2389. https://doi.org/10.3390/su16062389

Academic Editors: Fotios Chatzitheodoridis, Efstratios Loizou and Achilleas Kontogeorgos

Received: 29 January 2024
Revised: 10 March 2024
Accepted: 11 March 2024
Published: 13 March 2024

1. Introduction

According to data from COPAC (The Committee for the Promotion and Advancement of Cooperatives), approximately 800 million people worldwide are employed in cooperatives, accounting for about 10% of the global workforce. This statistic underscores the significance of cooperatives as a crucial organizational form. The Rochdale Cooperative, established in Manchester, UK, in 1844, is widely recognized as the world's first successful cooperative. Its values of openness, integrity, and respect for members continue to be fundamental principles endorsed by the International Cooperative Alliance [1]. Examining the historical development of cooperatives, Europe stands as their place of origin, with the United States and Japan subsequently adopting this organizational form. Developing countries, such as China and Brazil, gradually witnessed the formation of large-scale cooperatives several decades later [2].

Over the past century, agricultural cooperatives in various countries have developed distinctive characteristics. In several European countries, cooperatives hold a market share exceeding half, primarily concentrated in dairy, fruit, and vegetable cooperatives [3]. Cooperatives in the United States often involve corporate investments and rely on large family farms, forming integrated agricultural organizations encompassing production, processing, and sales [4]. Japan, on the other hand, has established agricultural associations top–down through the central government, characterized by a high degree of centralization and management [5]. In contrast, cooperatives in developing countries like China and Brazil are primarily composed of farmer members, tend to be smaller in scale, exhibit lower levels of industrial development, and focus on the production of staple crops [6,7]. However, regardless of the country, cooperatives share the eternal goals of enhancing cooperative efficiency, increasing product quality, and achieving sustainable, high-quality development.

In 1987, the World Commission on Environment and Development introduced the concept of sustainable development, defining it as "the ability to meet the needs of the present without compromising the ability of future generations to meet their own needs". Scholars gradually recognized that sustainable economic development must consider certain factors, such as culture, society, and the environment [8,9]. The study of development issues in economics began to shift from simply pursuing the quantity of economic growth to focusing on the quality of economic growth [10]. In 2017, the Chinese government put forward the concept of high-quality development based on sustainable development, the core objective of which is to satisfy the people's growing needs for a better life [11]. Promoting high-quality development cannot overlook the agricultural sector, and the enhancement of the development of the quality of farmers' cooperatives emerges as a crucial topic given their significant role as key players in agricultural operations.

The high-quality development of farmers' cooperatives is built on sustainable economic growth [12], utilizing innovative and environmentally friendly production methods [13,14], adhering to cooperative principles [15], and fostering development that aligns with the shared interests of producers and consumers [16]. The connotations of high-quality development for farmers' cooperatives are extensive, and merely measuring their development based on economic performance is inadequate [17]. As agricultural organizations with broad farmer participation, the development status of farmers' cooperatives requires consideration not only of farmers' economic income and living needs but also of consumers' expectations regarding the quality of agricultural products [11]. To comprehensively assess the development quality of farmers' cooperatives and strategically promote their enhancement, it is essential to construct a comprehensive evaluation indicator system for high-quality development. This will facilitate further research into the factors influencing the development of farmer cooperatives.

Land has consistently played a crucial role in the agricultural development process, significantly impacting economic growth, environmental protection, and the formation of rural community relations [18]. Numerous studies indicate that differences in arable land scale are a major factor restricting the market competitiveness of agricultural entities [19,20]. There are generally two perspectives on the impact of arable land scale on cooperative development. First is the resource constraint view, which posits that the scarcity of arable land resources is one of the limiting factors for cooperative development. In comparison with farmers' cooperatives in developed countries, a prominent characteristic of farmers' cooperatives in developing countries is relatively smaller scale [21,22]. The small scale hinders the production and service capabilities of cooperatives while also limiting their ability to assume more social responsibilities [23]. With the development of rural revitalization and agricultural modernization, large-scale production of arable land in rural areas has become the mainstream trend in current agricultural production, and farmers' cooperatives are essential entities in the large-scale management of arable land [24–26]. Therefore, many developing countries have implemented policies encouraging cooperatives to expand their scale. During the rapid development stage of cooperatives in developing countries, expanding operational scale is often the primary development approach.

However, another perspective is the resource curse, which suggests that an excess of arable land resources can also impact cooperative development. Continuously expanding scale is an inefficient cooperative model that leads to a collective decision-making dilemma for cooperatives. Larger operational scales can also result in uneven resource allocation and increased environmental pollution [27,28]. Studies indicate that large-scale cultivation by cooperatives has led to serious pesticide abuse and a decline in the quality of agricultural products [29], making it challenging to meet consumer market demands. For groups composed of farmers, blindly expanding scale also exposes issues of insufficient management capability [30]. Existing studies suggest that the economic growth of cooperatives primarily comes from the added value and precision management of agricultural products [31]. This is because consumer demands for the quality of agricultural products are increasing, leading to higher expectations for the quality of arable land cultivation [32]. Furthermore, cases of farms' technical efficiency from Hungary and Ecuador illustrate that the spatial potential for increasing productivity by expanding land scale is limited unless there is a change in technology [33,34]. In sub-Saharan African countries, medium-sized farms are more prevalent than large-scale farms [35].

Both of the above perspectives emphasize the decisive impact of arable land on organizational development. However, with technological advancements and changes in market demands, agricultural production has shifted from singular production of agricultural products to integrated development encompassing production, processing, and sales [36]. This shift has resulted in a transformation of the role of arable land in agricultural production. With the maturity of agricultural processing technologies and the development of rural e-commerce, the income growth of cooperatives is no longer reliant on expanding scale [37], leading to a continuous reduction in the dependence on arable land. So, in the context of the high-quality development of farmers' cooperatives, does the determinant role of arable land resources still exist? Is arable land still a foundational element influencing the high-quality development of farmers' cooperatives? Moreover, what are the characteristics of arable land's influence on the high-quality development of cooperatives? What factors constrain this influence? This paper aims to provide answers to these questions.

In contrast to prior research, this paper's marginal contributions are as follows. Firstly, based on micro-level data from major grain-producing provinces in China, this paper assesses the impact of arable land on the high-quality development of cooperatives from a cross-sectional perspective, enriching the analytical framework of arable land's impact on the high-quality development of cooperatives. Secondly, it constructs a more comprehensive indicator system to measure the high-quality development of farmers' cooperatives, broadening the analytical framework for evaluating the development level of cooperatives. Finally, it analyzes the characteristics of arable land's impact on the high-quality development of farmers' cooperatives at different stages and delves into the underlying reasons for the influence of land endowment on the high-quality development of cooperatives. Building upon the above analysis, this paper lays out theoretical and practical foundations to facilitate the government's efforts in fostering the advancement of high-quality development within farmers' cooperatives.

The remaining sections of this paper are organized as follows. Section 2 elucidates the theoretical basis of how arable land impacts the high-quality development of cooperatives. Section 3 presents the methodology, including empirical procedures, research scope, data collection, considered variables, and analysis procedures. Section 4 provides a detailed overview of the empirical results of the study. Section 5 concludes the paper and offers policy insights.

2. Theoretical Analysis

Based on the theory of economies of scale, there exists a close relationship between arable land scale and the high-quality development of cooperatives. The theory of economies of scale emphasizes that with the expansion of production scale, the average cost per unit product gradually decreases. In the agricultural sector, large-scale agricultural production can better utilize modern agricultural technology and improve land utilization efficiency, thus reducing production costs [38,39]. For farmers' cooperatives, the expansion of arable land scale implies that cooperatives can implement more detailed division of labor, which is conductive to reducing average costs and increasing profit levels, leading to an enhancement in the quality of development.

On the one hand, cooperatives with larger arable land scale can better integrate resources and enhance organizational and managerial capabilities [40,41]. By expanding arable land scale, cooperatives can achieve better economies of scale in agricultural production, procurement, and sales. Larger cooperatives are better equipped to organize members' production activities, coordinate the procurement and sales of agricultural products, and respond more flexibly to market fluctuations [42]. Simultaneously, cooperatives with a larger arable land scale often possess more comprehensive management teams, enabling them to plan and organize agricultural production more professionally [43,44]. The professionalism and efficiency in management can also lead to more refined production and operations, assisting cooperatives in better adapting to market demands and changes and promoting the green transformation of agriculture [45]. This efficient organizational and management level contributes to enhancing the overall efficiency of the cooperative, laying the foundation for its high-quality development.

On the other hand, cooperatives with a greater arable land size can reduce transaction costs, thereby enhancing synergy among members. In agricultural production, various cooperative and coordinated activities, including resource integration, production coordination, and product sales, are involved [46]. Cooperatives with a larger arable land scale have stronger negotiation power and market bargaining capabilities, allowing them to collaborate more effectively with various stakeholders in the agricultural value chain. This reduces information asymmetry and lowers transaction costs for cooperation [47]. This enables members to fully share the overall benefits of the cooperative, fostering a closer sense of community and synergistic effects [48]. Cooperation and coordination among members will be smoother, propelling the cooperative towards high-quality development.

In summary, the larger the arable land scale, the more likely it is for the cooperative to achieve high-quality development, thereby enhancing overall competitiveness and sustainable growth. Based on this, Hypothesis 1 is proposed.

Hypothesis 1. *The arable land scale has a markedly positive impact on the development quality of farmers' cooperatives. The larger the arable land scale, the higher the development quality of farmers' cooperatives.*

With the trend of rural arable land scale becoming the mainstream in current agricultural production, cooperatives also exhibit distinct characteristics in the variation of arable land scale. When exploring the impact of arable land scale on the high-quality development of cooperatives, it is essential not only to focus on the direct impact of arable land scale on the high-quality development of cooperatives but also to consider the possibility of diminishing marginal returns resulting from the expansion of arable land scale [49–51].

Firstly, the expansion of scale leads to management and decision-making challenges. On one hand, as the size of the cooperative grows, the difficulty of management increases exponentially. Large-scale cooperatives require a more complex and efficient management system, and the improvement in management level requires corresponding time and resources. If the management level does not keep pace with the increase in scale, the cooperative faces issues, such as internal organizational disarray and delayed decision making, thereby reducing the overall level of high-quality development [11]. On the other hand, blindly expanding scale can lead to a collective decision-making dilemma for the cooperative. In a large organization, the transmission of decisions becomes slow and cumbersome, posing challenges for the cooperative in adapting to market changes and formulating flexible decisions [52]. Excessive scale can weaken the flexibility and adaptability of the cooperative, thereby reducing its agility in market competition.

Secondly, uneven distribution of resources is also a potential issue. Large-scale cooperatives concentrate more resources on core operations, leading to an uneven distribution of resources in terms of the interests and needs of some grassroots members [53]. This can result in dissatisfaction among members and the estrangement of cooperative relationships, impacting the formation of synergistic effects.

Additionally, environmental issues are also aspects that need consideration. Large-scale agricultural production triggers more environmental pollution issues, including soil contamination, excessive water resource utilization, and the misuse of fertilizers and pesticides [54]. These issues not only impose a burden on the environment but also have negative impacts on the quality and safety of agricultural products, subsequently affecting the reputation and sustainable development of the cooperative in the market. Drawing from the above arguments, this paper posits Hypothesis 2.

Hypothesis 2. *There is an "inverted U-shaped" non-linear association between arable land scale and the development quality of cooperatives. As the arable land scale increases, the development quality of farmers' cooperatives initially improves and then declines.*

Existing studies have indicated that the expansion of arable land scale, especially the fragmentation of plots, is a significant factor leading to the enlargement of marginal costs and a decline in development quality [55]. Additionally, with the maturity of agricultural processing technologies and the development of rural e-commerce, the improvement in the development quality of cooperatives comes from the value addition and refined management of agricultural products rather than blindly pursuing the expansion of arable land scale. When cooperatives have a certain arable land scale, they typically allocate funds for investment in industrial development [56]. At this point, the income growth of cooperatives no longer solely depends on the expansion of arable land scale but relies on agricultural industrialization development [57]. According to David Ricardo's perspective, when additional units of variable inputs, such as labor or land, are employed with fixed inputs, like capital or technology, the marginal productivity of the variable input will eventually diminish. In the context of agricultural cooperatives, this implies that while enlarging the scale of arable land may initially boost productivity and reduce average costs, there will eventually be a point where further expansion leads to diminishing returns, thus limiting the extent to which economies of scale can be realized.

To illustrate the changes in the development quality of cooperatives, this paper has designed a simple Cooperative Profit Model. Assuming other input factors remain constant and if the configuration of input factors does not affect input prices (i.e., land rent and equipment costs), only arable land area and industrial equipment investment are considered. Among these, arable land area is a continuously variable factor, while industrial equipment investment is a long-term input factor for cooperatives, which remains constant in the short term and is considered a non-continuous factor.

$$\pi = PA^{\alpha}K^{\beta} - wA - rK \tag{1}$$

π represents the total profit of the cooperative, P is the price of agricultural products, A and K represent the quantities of inputs of arable land and industrial equipment, respectively, α and β are the output elasticities of arable land and industrial equipment, and $\alpha + \beta = 1$, w, and r are the land rent and equipment costs. According to Equation (1), the first-order conditions for the marginal returns of arable land and industrial equipment, respectively, can be derived:

$$\frac{\partial \pi}{\partial A} = \alpha PA^{\alpha-1}K^{\beta} - w = 0 \tag{2}$$

$$\frac{\partial \pi}{\partial K} = \beta PA^{\alpha}K^{\beta-1} - r = 0 \tag{3}$$

Based on Equations (2) and (3), the marginal output of arable land and industrial equipment can be further derived:

$$MPRA = \alpha PA^{\alpha-1}K^{\beta} = w \tag{4}$$

$$MPRK = \beta PA^{\alpha}K^{\beta-1} = r \tag{5}$$

From Equations (4) and (5), it can be observed that at profit maximization, the marginal output of arable land and industrial equipment equals their respective costs (i.e., the marginal output of arable land equals the unit cost of arable land, and the marginal output of industrial equipment equals the unit cost of industrial equipment). In this scenario, to determine the changes in cooperative profits, it is necessary to identify the cost inputs of arable land and industrial equipment, further considering the budget constraint l faced by the cooperative.

$$wA + rK \leq l \tag{6}$$

To provide a more intuitive analysis of how cooperatives maximize output under budget constraints, we can adopt the approach of equilibrium diagrams.

In the equilibrium configuration diagram of arable land area and industrial equipment investment (Figure 1), the vertical axis A represents the arable land area, and the horizontal axis K represents industrial equipment investment. Contour lines π_1, π_2, π_3, and π_4 represent different profit levels for the cooperative, where profit level $\pi_2 > \pi_3 > \pi_1 > \pi_4$. Moreover, on the same contour line of profit level, different combinations of arable land area and industrial equipment can enable the cooperative to achieve the same profit. Additionally, due to the complementary relationship between arable land area and industrial equipment investment, the initial budget constraint line l_1 is perpendicular to the 45-degree line. When industrial equipment investment is K_1, the optimal configuration is achieved with arable land area input A_1, and the equilibrium point is M_1, located on the 45° line.

According to Figure 1a, it can be observed that in consideration of long-term input expansion, the cooperative's budget constraint line shifts to l_2, resulting in a new profit-maximizing intersection at M_2. At this point, the arable land input is A_2, and the industrial equipment input is K_2. Due to the complementary relationship between arable land and industrial equipment, the cooperative can process all output from the arable land, leading to all profits being value-added profits.

However, when considering short-term input variations for the cooperative, according to Figure 1b, the industrial equipment input remains unchanged in the short term, as the capital required for industrial equipment input exceeds the increase in arable land scale. Consequently, as the relative price of arable land decreases for the cooperative, the budget constraint line shifts to l_3. At this juncture, the profit-maximizing intersection becomes M_3, surpassing the processing capacity of industrial equipment A_1. As a result, the produced agricultural products cannot all be processed and sold. The shaded portion of agricultural products must be sold at a lower market price, thereby reducing the cooperative's profit.

Additionally, considering the scenario where the cooperative reduces its inputs, according to Figure 1c, the budget constraint line shifts leftward to l_4. At this point, the cooperative's arable land area input to A_4, the processing capacity of the cooperative's industrial equipment, remains unchanged. The intersection point M_4 satisfies profit maximization. However, due to the small output of agricultural products, it cannot meet the processing needs of the industrial equipment. This leads to the underutilization of a portion of processing capacity, resulting in wastage of fixed costs for agricultural product industrial equipment. Additionally, it increases the unit output cost, thereby resulting in the cooperative's inability to achieve optimal profit.

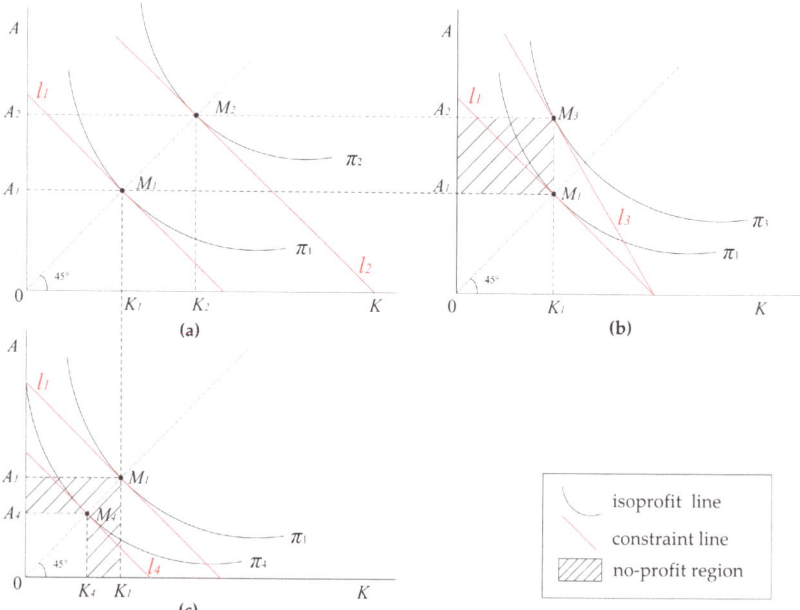

Figure 1. Equilibrium diagram of cooperatives' arable land area and industrial equipment input allocation. (**a**) Increased inputs; (**b**) Relative price changes; (**c**) Reduced inputs.

In summary, when the arable land area is too large or too small, influenced by the short-term constancy of industrial equipment investment, it will prevent the cooperative from obtaining optimal profits. Moreover, the profit level is just one evaluation criterion for the high-quality development of the cooperative. After considering certain factors, such as social benefits, green production, and standardization, the "inverted U-shaped" association between arable land scale and cooperative high-quality development becomes more pronounced. Drawing on the above arguments, this paper posits Hypothesis 3.

Hypothesis 3. *Industrial development is the key factor in the "inverted U-shaped" association between arable land and the development quality of the cooperative.*

3. Materials and Methods

3.1. Research Methods

3.1.1. Evaluation Model

The commonly used weighting methods in evaluative research include expert weighting, the analytic hierarchy process (AHP), the entropy method, and factor analysis. The first two methods rely to some extent on subjective judgment, while the latter two, though more objective, may lead to unreasonable weight assignments when dealing with unevenly distributed data. Therefore, adopting a single weighting method can easily lead to a contradiction between the weight of indicators and their actual importance. Therefore, many scholars in existing evaluative research have employed a combination weighting method, achieving favorable results [58,59]. On one hand, combining weighting methods reduces the subjective bias generated during expert judgment. On the other hand, it helps avoid the objective bias caused by differences in data quality. Therefore, this paper ultimately adopts a combination of the analytic hierarchy process and entropy method. It utilizes the comprehensive weights calculated by these two methods to measure the development quality of farmers' cooperatives.

1. Analytic hierarchy process (AHP). Firstly, based on the evaluation indicator system for the high-quality development of farmers' cooperatives, construct a hierarchical structure model with four levels. Secondly, experts score the matrix using a 1~9 scale and calculate the weights. Finally, obtain consistency test results by calculating the characteristic roots, eigenvector values, and the corresponding average random consistency index (RI). After passing the consistency test, AHP indicator weights can be obtained.

2. Entropy method. Considering the different attributes, units, and ranges of the evaluation indicators, the data are first standardized. Secondly, compute the information entropy and information utility values of each indicator using the standardized data. Finally, after normalization, obtain the entropy method weights for each evaluation indicator.

$$e_j = -\frac{1}{\ln n} \sum_{i=1}^{n} \frac{X'_{ij}}{\sum_{i=1}^{n} X'_{ij}} \ln \frac{X'_{ij}}{\sum_{i=1}^{n} X'_{ij}} \tag{7}$$

$$d_j = 1 - e_j \tag{8}$$

$$\beta_j = d_j / \sum_{j=1}^{m} d_j \tag{9}$$

e_j represents entropy, n is the number of sample cooperatives, and X'_{ij} represents the standardized values; d_j represents the coefficient of variation for the j indicator, also known as information utility or information entropy redundancy; β_j is the weight value, and m is the number of indicators.

3. Comprehensive Weight Calculation. Based on the principle of minimum relative entropy, we use the Lagrange multiplier method to compute composite weights, ensuring the accuracy and scientific validity of the weight calculation results.

The calculation formula is as follows:

$$W_f = \sqrt{\alpha_j \cdot \beta_j} / \sum_{j=1}^{m} \alpha_j \cdot \beta_j \tag{10}$$

α_j represents the analytic hierarchy process (AHP) weight result, β_j represents the entropy method weight result, and W_f is the comprehensive weight.

3.1.2. Regression Model

This paper mainly explores the linear relationship and "inverted U-shaped" association between land scale and cooperative development quality. Based on this, linear and nonlinear function models are constructed.

The calculation formula is as follows:

$$CDQ_i = \alpha + \beta ALS_i + \theta control_i + \varepsilon_i \tag{11}$$

$$CDQ_i = \alpha + \beta_1 ALS_i + \beta_2 ALS^2_i + \theta control_i + \varepsilon_i \tag{12}$$

The dependent variable CDQ_i represents the development quality index of cooperative i and the core explanatory variables, ALS_i and ALS^2_i, respectively, represent the land scale and its squared term for the cooperative. In addition, $control_i$ represents a series of control variables affecting the development quality of the cooperative, and ε_i represents the residual term.

3.2. Variable Definitions and Descriptions

3.2.1. Dependent Variable

The high-quality development of farmers' cooperatives is a crucial component of China's high-quality economic development. It extends traditional performance evaluations of cooperatives under the premise of sustainable development and represents a key link in the transformation of cooperatives into high-level agricultural organizations. Therefore, the high-quality development of farmers' cooperatives should encompass four dimensions. Firstly, it should be premised on the continuous creation of economic value. Secondly, it should prioritize agricultural innovation and the production of green products as means. Thirdly, it should adhere to operational and financial standards to ensure effective governance and management. Lastly, it should aim to drive the development of rural communities and farmers as its ultimate goal.

Specifically, the achievement of any single goal can contribute to the improvement of high-quality development in cooperatives. In the assessment system, a deficiency in any indicator may impact the overall development evaluation. For instance, if a cooperative solely pursues economic profit growth at the expense of environmental sustainability, the quality of its development may be questioned. Conversely, if a cooperative focuses only on sustainable product production but sacrifices operational profits, it can also affect its development quality. We aim to highlight cooperatives that prioritize sustainable development, even if it means sacrificing some profit, and actively support their communities. This is because solely using economic growth as a criterion may overlook the significant contributions of these cooperatives. However, we do not encourage neglecting cooperatives that prioritize their economic development and the interests of their members. Therefore, the best practice is to conduct a comprehensive assessment of high-quality development to thoroughly measure the performance of cooperatives.

In selecting specific indicators, we drew upon the three dimensions of the economic, social, and ecological performance evaluation system of cooperatives [7]. Additionally, we expanded these indicators to encompass aspects of innovation and standardization, aligning with the principles of high-quality development (refer to Table 1). Through the establishment of an assessment system for high-quality development, our aim is to furnish cooperatives in other countries with enhanced understanding and practical insights to attain sustainable development goals, thereby fostering global advancement and expansion of cooperative movements.

Due to the importance of indicators being dependent on their assigned weights, we employed a comprehensive evaluation method that combines both subjective and objective considerations in assigning weights. According to the results in Table 1, among the five primary indicators, the comprehensive weight for the economic foundation is the highest, indicating that the economy remains the cornerstone of high-quality development for cooperatives. Following in descending order of weights are social value, green development, standardization, and innovation capacity, reflecting the varying degrees of significance across different aspects in the process of high-quality development for cooperatives.

Table 1. Evaluation indicator system for high-quality development of farmers' cooperatives.

Primary Indicator	Comp. Wt.	Secondary Indicator	Comp. Wt.	Tertiary Indicator	AHP Wt.	Entropy Wt.	Comp. Wt.
Economic Basis (EB)	0.2622	Profitability	0.1799	Average Income per Member (CNY 10,000)	0.1489	0.0165	0.0579
				Operating Profit (CNY 10,000)	0.1489	0.0733	0.1220
		Industry Integration	0.0823	Processing Proportion (%)	0.0695	0.0128	0.0349
				New Sales Methods (%)	0.1290	0.0128	0.0474
Innovation Capability (IC)	0.1026	Technological Innovation	0.0321	Standards/Patents (pcs)	0.0147	0.0233	0.0216
				Application of Fine Seeds (%)	0.0098	0.0081	0.0104
		Cooperative Branding	0.0705	Registered Trademarks (pcs)	0.0086	0.0774	0.0301
				Brand Coverage (1~6)	0.0159	0.0751	0.0404

Table 1. *Cont.*

Primary Indicator	Comp. Wt.	Secondary Indicator	Comp. Wt.	Tertiary Indicator	AHP Wt.	Entropy Wt.	Comp. Wt.
Green Development (GD)	0.2161	Ecological Protection	0.0686	Proportion of Reduced Chemical Area (%)	0.0421	0.0638	0.0606
				Recycling Rate of Agricultural Waste (%)	0.0227	0.0021	0.0080
		Product Safety	0.1475	Quality Certification Standards (1~5)	0.0583	0.0816	0.0806
				Traceability Proportion (%)	0.0388	0.0845	0.0669
Standardization Level (SL)	0.1366	Operational Standards	0.0771	Frequency of Member Meetings/Director Meetings (times/year)	0.0158	0.0407	0.0296
				Proportion of Distributable Surplus Returned (%)	0.0369	0.0447	0.0474
		Financial Standards	0.0596	Frequency of Financial Report Disclosure (times/year)	0.0151	0.0301	0.0249
				Frequency of Accounting (times/year)	0.0280	0.0315	0.0347
Social Value (SV)	0.2825	Social Participation	0.1409	Investment in Village Collective Construction (CNY 10,000)	0.0355	0.1328	0.0802
				Number of Cooperative Enterprises/Other Cooperatives (pcs)	0.0532	0.0507	0.0607
		Training and Employment	0.1416	Number of People Trained in Farmer Training Projects (ppl)	0.0379	0.0672	0.0590
				Number of Jobs Created by the Cooperative (pcs)	0.0705	0.0710	0.0827

Note: The specific explanation of indicators, data, and the original questionnaire can be found in Sections A–D.

3.2.2. Independent Variables

This paper takes "cooperative arable land scale" as the core explanatory variable, which includes the total area of land contributed by cooperative members and the land area acquired by the cooperative through leasing. Cooperative farm management involves three types of land. The first is land contributed by cooperative members, where members invest their land as capital in cooperative production and they receive profit dividends. The second is leased land, where the cooperative acquires land from farmers or other organizations by paying rent and then manages it uniformly. The third is land management, where the cooperative provides services for arable land owned by farmers or other organizations and only charges service fees. The land scale discussed in this paper is based on the cooperative's input and output; land management does not reflect this characteristic.

3.2.3. Control Variables

1. Characteristics of the cooperative chairman. Referring to previous research, the chairman of a farmer's cooperative, as a crucial leader of the organization, plays a significant role in the decision making for the cooperative's development. This paper controls for the chairman's gender, age, educational level, and position.
2. Basic characteristics of the cooperative. The basic characteristics of a farmers' cooperative include fixed assets, the number of members, and the external environment

of the cooperative, which have fundamental effects on the development quality of the cooperative.

3. Operational and management characteristics of the cooperative. This mainly includes the operational system and advantageous resources of the cooperative.

By comprehensively considering the above variables, a more accurate analysis of the relationship between the cooperative's land scale and high-quality development can be achieved, eliminating the potential impact of other factors on the research results and enhancing the credibility and scientific rigor of the study. The definition and description of variables are shown in Table 2.

Table 2. Variable definition and description.

Variable Type	Variable Name	Variable Definition and Description
Dependent Variable	Cooperative Development Quality (CDQ)	Evaluated Through the Evaluation Index System for High-Quality Development of farmers' cooperatives (Table 1)
Independent Variable	Arable Land Scale (ALS)	Total Area of Land Invested by Cooperative Members and Land Leased by The Cooperative (kha)
Control Variable	Characteristics of Cooperative Chairman (CCC)	Gender (1 = Male, 0 = Female) Age (years, logarithm) Educational Level (5 = Bachelor's and above, 4 = College, 3 = High School or Technical School, 2 = Junior High School, 1 = Elementary School and below) Village Cadre (1 = Yes, 0 = No)
	Basic Characteristics of Cooperative (BCC)	Cooperative Total Assets (CNY 10,000, logarithm) Number of Cooperative Members (ppl, logarithm) Large Agricultural Machinery Quantity (units) Demonstration Level (5 = National, 4 = Provincial, 3 = Municipal, 2 = County, 1 = None) Distance to County Town (kilometers) Distance to The Nearest Formal Financial Institution (kilometers)
	Operational and Management Characteristics of Cooperative (OMCC)	Second Rebate System (1 = Yes, 0 = No) One-Person-One-Vote System (1 = Yes, 0 = No) Number of Full-time Employees (ppl) Proportion of Social Relationship Expenses to Cooperative Surplus (%)

3.3. Data Sources

The data for this study were obtained via a questionnaire survey conducted from December 2022 to September 2023 in the top three grain-producing provinces in China (Heilongjiang, Henan, and Shandong), covering 14 cities and 70 counties. In the sample questionnaire survey, approximately 7 cooperatives were randomly selected in each county, with a total of 500 questionnaires distributed and 487 successfully collected, of which 448 met the research requirements, achieving an effective rate of 89.6%.

Due to significant differences in resource endowments among different types of farmers' cooperatives, it is not possible to conduct a high-quality development assessment under a single standard. Therefore, based on the core issue of this study, the final choice was to investigate production and operation-oriented cooperatives with grain production as the main business. Livestock and poultry farming cooperatives and service cooperatives are not within the scope of this study. This choice aims to ensure an in-depth investigation into the mechanism of how land scale influences the high-quality development of farmers' cooperatives while avoiding interference from differences between different types of cooperatives.

3.4. Descriptive Statistics

Table 3 presents the descriptive statistical analysis results of the data obtained in this study. Firstly, among the 448 sampled farmers' cooperatives, the average development quality of farmers' cooperatives is 0.1408, with a maximum value of 0.7670 and a minimum value of 0.0060. It can be observed that there is significant variation in the development quality among the sampled cooperatives, and, overall, the development quality of cooperatives is relatively low, which is consistent with the conclusions drawn from the literature review and on-site investigations. Secondly, the mean value of the explanatory variable, land scale, is 0.4544, with a maximum value of 5.6000 and a minimum value of 0.0191. This indicates substantial individual heterogeneity in land scale among the sampled cooperatives, implying significant differences in land scale among different cooperatives potentially facing various management and operational challenges. Descriptive statistical analysis results for other control variables are presented below.

Table 3. Descriptive statistical analysis results of sample cooperatives.

Variable	Observations	Mean	SD	Min	Max
Cooperative Development Quality (CDQ)	448	0.1408	0.1589	0.0060	0.7670
Arable Land Scale (ALS)	448	0.4439	0.6323	0.1910	4.2
Chairman's Gender	448	0.7656	0.4241	0	1
Chairman's Age	448	3.8022	0.1643	3.3322	4.2047
Chairman's Educational Level	448	3.3527	0.7177	2	5
Chairman's Village Cadre	448	0.3013	0.4594	0	1
Cooperative Total Assets	448	4.8042	1.1102	3.3499	8.5348
Number of Cooperative Members	448	4.1358	0.9327	2.9444	6.9217
Large Agricultural Machinery Quantity	448	1.3359	0.9707	0	4.8903
Demonstration Cooperative Level	448	1.8147	1.4092	1	5
Distance to County Town	448	19.9634	8.9498	5.5	53
Distance to Financial Institutions	448	3.4196	2.7343	0.5	20
Second Rebate System	448	0.0759	0.2651	0	1
One-Person-One-Vote System	448	0.1272	0.3336	0	1
Number of Full-time Employees	448	1.8585	1.3420	0	5.1985
Proportion Social Relationship Expenses	448	0.6775	1.4832	0	10

Note: The data in the table are organized based on the content of on-site investigations.

4. Results

4.1. Empirical Results

4.1.1. Inverted U-Shaped Relationship

Table 4 analyzes the impact of arable land scale on cooperative development quality (CDQ). From Model 1-1 to Model 1-4, the core explanatory variable land scale (ALS), land scale squared term (ALS^2), and other control variables are gradually introduced. The results show that land scale has a significant impact on the development quality of cooperatives in all models, with positive coefficients. This indicates that, on average, as the land scale operated by cooperatives increases, the development quality of cooperatives also increases. This conclusion verifies H1 stated earlier and is consistent with conclusions drawn from existing research.

It is worth noting this when considering the nonlinear relationship between land scale and the development quality of cooperatives. In Model 1-2 and Model 1-4, the regression coefficients of the land scale squared term are -0.067 and -0.031, with p-values less than 0.01, indicating a significant "inverted U-shaped" association. This implies that as the land scale expands, the rate of improvement in the development quality of cooperatives gradually slows down. There is also a threshold effect, where beyond a certain critical point, the expansion of land scale may lead to a decrease in the development quality of cooperatives. This result validates H2 mentioned earlier.

Table 4. Arable land scale and cooperative development quality.

Variable	Model 1-1	Model 1-2	Model 1-3	Model 1-4
Arable Land Scale (ALS)	0.197 *** (0.006)	0.409 *** (0.009)	0.054 *** (0.006)	0.201 *** (0.013)
Square of Arable Land Scale (ALS^2)		−0.067 *** (0.002)		−0.031 *** (0.003)
Chairman's Gender			−0.006 (0.006)	−0.005 (0.005)
Chairman's Age			−0.028 * (0.016)	−0.014 (0.014)
Chairman's Educational Level			0.006 (0.004)	0.005 (0.003)
Chairman's Village Cadre			0.013 ** (0.005)	0.011 ** (0.004)
Cooperative Total Assets			−0.024 *** (0.005)	−0.021 *** (0.004)
Number of Cooperative Members			0.023 *** (0.006)	0.021 *** (0.005)
Large Agricultural Machinery Quantity			0.014 *** (0.004)	0.003 (0.003)
Demonstration Cooperative Level			0.015 *** (0.003)	0.010 *** (0.003)
Distance to County Town			0.001 *** (0.000)	0.001 *** (0.000)
Distance to Financial Institutions			−0.003 ** (0.001)	−0.001 (0.001)
Second Rebate System			0.032 *** (0.012)	0.023 ** (0.010)
One-Person-One-Vote System			−0.017 (0.011)	−0.010 (0.010)
Number of Full-time Employees			0.026 *** (0.004)	0.016 *** (0.003)
Proportion Social Relationship Expenses			0.049 *** (0.003)	0.036 *** (0.003)
Constant	0.051 *** (0.005)	−0.067 (0.004)	0.084 (0.066)	0.032 (0.057)
Sample Size	448	448	448	448
Adjusted R^2	0.703	0.889	0.927	0.945

Note: *** indicates $p < 0.01$, ** indicates $p < 0.05$, * indicates $p < 0.1$; values in parentheses are standard errors, etc.

4.1.2. Categorized Cooperative Development Quality

Cooperative development quality is a comprehensive indicator. To further explore the relationship between ALS and the categorized development quality of cooperatives, this study conducted tests using five models, with EB, IC, GD, SL, and SV as the dependent variables. The results indicate that there is a significant "inverted U-shaped" association between ALS and the development quality of cooperatives in all categorized models, consistent with the overall development quality results. This implies that with the expansion of ALS, the quality in each category follows the pattern of first increasing and then decreasing.

Specifically, the impact of changes in ALS on the categorized development quality of cooperatives is as follows: GD > SV > EB > IC > SL. Among them, the impact of ALS on GD is the most significant. This indicates that with the expansion of cooperative ALS, it is more likely to cause environmental issues, requiring the formulation of more reasonable strategies for green development to ensure environmental sustainability. Changes in ES and SV are also relatively pronounced, indicating that in the process of expanding ALS, cooperatives may face issues, such as declining profits and neglecting social responsibilities. Therefore, cooperatives need to focus on maintaining economic stability, enhancing social responsibility, and caring for community development. The impact of ALS on IC and SL is relatively small, indicating that these two aspects are relatively stable and less affected by changes in scale.

Overall, with the expansion of ALS, the categorized development quality of cooperatives exhibits a significant "inverted U-shaped" association. However, the impact of ALS on the development quality of different cooperative classifications varies. Therefore, when cooperatives pursue high-quality development, it is necessary to consider the development quality of different classifications. The management of ALS needs to comprehensively consider various influences to balance economic benefits, social responsibility, and environmental sustainability (Table 5).

Table 5. Arable land scale for grain crops and categorized development quality of cooperatives.

Variable	EB	IC	GD	SL	SV
	Model 2-1	Model 2-2	Model 2-3	Model 2-4	Model 2-5
ALS	0.035 ***	0.032 ***	0.065 ***	0.025 ***	0.044 ***
	(0.005)	(0.002)	(0.006)	(0.003)	(0.004)
ALS^2	−0.008 ***	−0.005 ***	−0.010 ***	−0.004 ***	−0.005 ***
	(0.001)	(0.000)	(0.001)	(0.001)	(0.001)
CCC	Control	Control	Control	Control	Control
BCC	Control	Control	Control	Control	Control
OMCC	Control	Control	Control	Control	Control
Sample Size	448	448	448	448	448
Adjusted R^2	0.862	0.912	0.856	0.892	0.904

Note: *** indicates $p < 0.01$. "Control" indicates variables that have been controlled, etc.

4.2. Robustness Tests

4.2.1. Estimation Results with Replacement of Independent Variables

In the process of exploring the impact of ALS on CDQ, this paper uses the total area of cooperative ALS as the core explanatory variable. To further validate the robustness of the regression results mentioned earlier, this paper divides the total area of ALS into two categories, invested arable land and transferred arable land, and conducts a regression analysis again. Similarly, the models control for CCC, BCC, and OMCC. The regression results in Table 6 are consistent with the previous findings, indicating the robustness of the research conclusions.

According to Models 3-1 and 3-2, it can be observed that whether it is invested arable land or transferred arable land, with the expansion of ALS, the development quality of cooperatives shows an increasing trend. However, the impact of the area of invested arable land on development quality is significantly greater than that of transferred land. One important reason is that land transfer requires priority payment of rent and is not conducive to the long-term arable land infrastructure construction of cooperatives, thereby restricting the high-quality development of cooperatives to a certain extent.

Table 6. The impact of varied sources of arable land scale on cooperative development quality.

Variable	Model 3-1	Model 3-2	Model 3-3	Model 3-4
ALS (invested)	0.328 *** (0.020)		0.708 *** (0.033)	
ALS (transferred)		0.041 *** (0.007)		0.106 *** (0.015)
ALS^2 (invested)			−0.297 *** (0.022)	
ALS^2 (transferred)				−0.019 *** (0.004)
CCC	Control	Control	Control	Control
BCC	Control	Control	Control	Control
OMCC	Control	Control	Control	Control
Sample Size	448	448	448	448
Adjusted R^2	0.946	0.920	0.962	0.924

Note: *** indicates $p < 0.01$.

Furthermore, the results from Models 3-3 and 3-4, with the introduction of squared terms, indicate a significant "inverted U-shaped" association for both invested arable land and transferred arable land. This suggests that both land investment and transfer need to be controlled within a certain scale, as exceeding a critical point will lead to a decline in development quality. Therefore, cooperatives need to balance the expansion of scale with the improvement of development quality in arable land management, thus preventing the negative impact of excessive scale expansion.

4.2.2. Estimation Results after Addressing Endogeneity Issues

Although this paper thoroughly discusses the regression results of different explanatory variables and the explained variable in the preceding sections, arriving at consistent conclusions, there may still be endogeneity issues in the process of econometric regression, primarily in the following aspects. First, there might be a problem of reverse causality. The more superior the development quality of farmers' cooperatives, the more extensive the demand for arable land, and the more robust the capacity to expand ALS. Second, there is an issue of measurement errors. Although the questionnaire was corrected through pre-survey before the investigation and the interviewers were trained, to adhere to the principle of random sampling, the sample cooperatives were relatively dispersed, and some questionnaires were conducted via telephone interviews, leading to potential measurement errors. Third, there is a problem of omitted variables. To avoid potential endogeneity issues arising from the omission of important variables, this paper, referring to the existing literature and considering practical data acquisition, selected control variables from three aspects, characteristics of cooperative chairpersons, basic features of cooperatives, and operational and managerial features of cooperatives, possibly neglecting factors from other aspects. Therefore, to address the potential endogeneity issues mentioned above, this paper adopts the Two-Stage Least Squares (2SLS) method for instrumental variable regression.

The instrumental variable selected in this paper is "Average Household Arable Land (IV1)", which represents the ratio of the arable land area owned by cooperative members to the total number of members. The reasons for choosing this instrumental variable are twofold. Firstly, the Average Household Arable Land in cooperative societies is not directly related to the quality of cooperative development. However, generally, the larger the per capita arable land area, the more extensive the ALS that cooperatives can acquire through members' equity participation or land transfer. Therefore, Average Household Arable Land can indirectly influence cooperative development quality by directly affecting the cooperative ALS. Secondly, the Average Household Arable Land is determined by land tenure rights and remains constant over the long term, and the sample survey point in this paper is the year 2022. Therefore, the instrumental variable is not significantly correlated with other control variables for that year. Combining these two reasons, the paper considers the selection of "Average Household Arable Land" as an appropriate instrumental variable. Next, regression and testing of the instrumental variable will be conducted. Considering that this paper's core explanatory variables consist of two variables, ALS and ALS^2, in the empirical test, the "Square of Average Household Arable Land (IV2)" is additionally included as the second instrumental variable.

Table 7 displays the regression outcomes of the instrumental variable. Models 4-1 and 4-2 present the test results of the correlation between the instrumental variable and the ALS along with ALS^2, and both model results are significant. Model 4-3 displays the test results of the ALS and ALS^2 on CDQ after applying the instrumental variable. The signs of model coefficients and statistical significance are consistent with the regression results mentioned earlier, further supporting the theoretical assumptions of the study. The instrumental variable test results indicate that the absolute values of the coefficients of key explanatory variables in the regression equation are larger than those in the baseline regression model (Model 1-4). This aligns with the convention of instrumental variable regression results. The R-squared of this model is 0.936, indicating a relatively high level of model fit.

Furthermore, using a 10% deviation as the maximum range criterion for the instrumental variable [60], the study's results show that the minimum eigenvalue statistic is significantly greater than the critical value within the 10% deviation range of the instrumental variable. Therefore, there is no issue of weak instrumental variables.

Finally, based on the results of the Durbin–Wu–Hausman test (DWH test), it indicates the presence of certain endogeneity issues in the model. Therefore, the estimates based on the instrumental variable (Model 4-3) should be considered, and it still concludes a significant "inverted U-shaped" impact of land scale on the development quality of cooperatives.

Table 7. Instrumental variable test regression results.

Variable	ALS	ALS^2	CDQ
	Model 4-1	Model 4-2	Model 4-3
IV1	0.255 *** (0.061)		
IV2		1.449 *** (0.204)	
ALS			0.320 *** (0.030)
ALS^2			−0.055 *** (0.010)
CCC	Control	Control	Control
BCC	Control	Control	Control
OMCC	Control	Control	Control
Minimum Eigenvalue Statistic			100.718
DWH Test			13.191 ***
Sample Size	448	448	448
R^2	0.974	0.908	0.936

Note: *** indicates $p < 0.01$.

4.3. Mechanism Analysis

4.3.1. Stage-Specific Characteristics

According to the analysis in the previous sections, the impact of ALS on the CDQ exhibits an "inverted U-shaped" pattern. Based on existing research findings, fitting a scatter plot to assess the "inverted U-shaped" relationship is considered an effective method [61]. According to the distribution of ALS and the CDQ shown in Figure 2, the impact of ALS on the CDQ can be divided into four main stages. (1) Initial Impact Stage (ALS 0–0.4 kha): At this point, cooperatives have a relatively small land scale. Although expanding the land scale would enhance development quality, the impact is relatively weak. (2) Rapid Improvement Stage (ALS 0.4–0.8 kha): Once cooperatives achieve a certain land scale, further expansion rapidly improves the development quality of the cooperative. (3) Diminishing Growth Effect Stage (ALS 0.8–1.2 kha): When the land scale exceeds a certain threshold, the rate of improving development quality through expanding the land scale significantly slows down. (4) Declining Development Quality Stage (ALS above 1.2 kha): When the cooperative's land scale reaches a threshold where development quality starts to decline, continuous expansion of the land scale leads to a decrease in development quality.

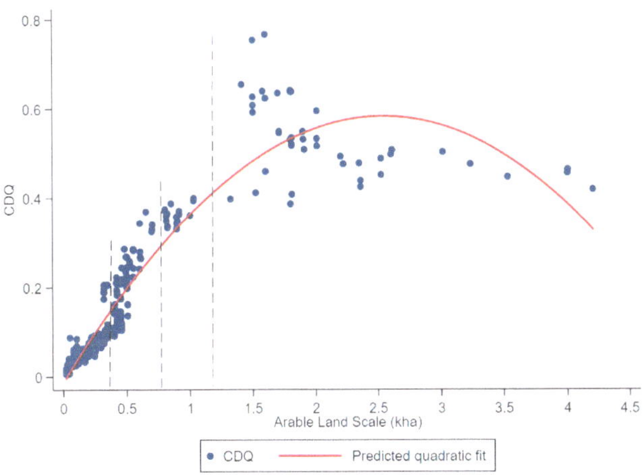

Figure 2. Scatter plot of ALS and CDQ distribution.

According to the regression results in Table 8, it is evident that in Models 5-1 to 5-3, the coefficient of ALS on the CDQ undergoes an initial increase followed by a decrease. However, during this stage, the coefficient value remains positive. In Model 5-4, when the ALS exceeds 1.2 kha, the impact of ALS on the CDQ becomes significantly negative. This indicates that it has entered the fourth stage of development, where further increases in land scale led to a decline in the CDQ. It is important to note that the delineated turning points for land scale represent statistical patterns observed in the sampled cooperative societies and may vary in reality regarding the turning points between ALS and CDQ.

Table 8. Revised regression results by segments.

Variable	0–0.4 kha	0.4–0.8 kha	0.8–1.2 kha	1.2+ kha
	Model 5-1	Model 5-2	Model 5-3	Model 5-4
ALS	0.257 ***	0.561 ***	0.284 ***	−0.056 ***
	(0.012)	(0.065)	(0.059)	(0.015)
CCC	Control	Control	Control	Control
BCC	Control	Control	Control	Control
OMCC	Control	Control	Control	Control
Sample Size	296	90	18	44
Adjusted R^2	0.801	0.798	0.743	0.431

Note: *** indicates $p < 0.01$.

4.3.2. Arable Land Fragmentation and Cooperative Industrialization

After clarifying the four-stage characteristics of how ALS influences the high-quality development of cooperatives, this paper focuses on the reasons behind the threshold of high-quality development in cooperatives. Based on the theoretical analysis in the previous sections, industrialization development is a key factor influencing the inverted U-shaped relationship between arable land scale and the quality of farmers' cooperatives development. Therefore, this paper adopts three methods to analyze the influence mechanism of industrialization development.

(1) Grouping cooperatives based on whether they undergo industrialization development and then separately examining the relationship between arable land scale and the quality of cooperative development within each group. If the inverted U-shaped relationship between arable land scale and cooperative development quality is more pronounced in the groups undergoing industrialization, it indirectly supports the complementary mechanism of industrialization development.

(2) Constructing a complementary function between arable land scale and industrialization development to explore the impact of this complementary function on the quality of cooperative development.

(3) Introducing a moderation effects model of industrialization development. Interacting the linear and quadratic terms of arable land scale with the indicator of industrialization development to investigate how industrialization influences the nonlinear impact of arable land scale on the quality of cooperative development.

The specific formula settings are as follows:

$$COMP_i = \frac{IDI_i - IDI_{min}/IDI_{max} - IDI_{min}}{ALS_i - ALS_{min}/ALS_{max} - ALS_{min}} \tag{13}$$

$$CDQ_i = \alpha + \beta COMP_i + \theta control_i + \varepsilon_i \tag{14}$$

$$CDQ_i = \alpha + \beta_1 ALS_i + \beta_2 ALS^2_i + \beta_3 IDI_i + \beta_4 ALS_i \times IDI_i + \beta_5 ALS^2_i \times IDI_i + \theta control_i + \varepsilon_i \tag{15}$$

$COMP_i$ represents the complementarity of the cooperative and IDI_i and ALS_i, respectively, denote the amount of investment in industrial development and the arable land scale of the cooperative. CDQ_i represents the development quality of the cooperative, *control$_i$* represents a series of control variables influencing the development quality of the cooperative, and ε_i represents the residual term.

According to the empirical results in Table 9, Model 6-1 represents cooperatives undergoing industrialization development, and its regression results are consistent with the overall findings. Model 6-2 represents cooperatives that have not undergone industrialization development. From the regression results, it can be observed that the inverted U-shaped relationship between ALS and CDQ does not hold in these cooperatives. In cooperatives without industrialization development, there is a linear correlation between ALS and CDQ. This clearly indicates that industrialization development is a crucial factor influencing the establishment of the inverted U-shaped relationship between ALS and CDQ.

Model 6-3 demonstrates the impact of COMP on CDQ, revealing a significant positive influence of COMP on the CDQ. This validates the crucial role of industrialization development discussed earlier. Only when IDI is aligned with ALS can it promote the improvement of cooperative development quality.

Table 9. The mechanism analysis of the impact of arable land scale on cooperative development quality.

Variable	Model 6-1	Model 6-2	Model 6-3	Model 6-4	Model 6-5
ALS	0.205 *** (0.032)	0.167 *** (0.043)		0.311 *** (0.018)	0.379 *** (0.018)
ALS^2	−0.024 *** (0.004)	0.322 *** (0.104)		−0.041 *** (0.003)	−0.070 *** (0.004)
COMP			0.118 *** (0.018)		
IDI					0.023 *** (0.006)
ALS × IDI					−0.053 *** (0.018)
ALS^2 × IDI					0.014 *** (0.003)
CCC	Control	Control	Control	Control	Control
BCC	Control	Control	Control	Control	Control
OMCC	Control	Control	Control	Control	Control
Sample Size	133	315	448	448	448
Adjusted R^2	0.936	0.877	0.829	0.955	0.964

Note: *** indicates $p < 0.01$.

Finally, Models 6-4 and 6-5 analyze the moderating effects of industrial development. In Model 6-5, the coefficient of ALS × IDI is significantly negative, indicating that in the initial stages of smaller arable land scale, as the level of industrial development increases, the positive impact of ALS on CDQ diminishes. Possible reasons include the inability of smaller arable land scale to fully leverage the advantages brought by industrialization (such as technology, market access, etc.) or the increased costs associated with industrialization that are harder to spread over smaller arable land scale. Additionally, the coefficient of ALS2 × IDI is significantly positive, suggesting that with further expansion of ALS, the higher level of IDI intensifies the marginal positive impact of ALS on CDQ. This may indicate that at a certain level of industrialization, larger arable land scale can benefit more from industrialization, such as through economies of scale, more efficient resource utilization, and better market access.

Combining the research findings, simply pursuing an expansion of arable land scale is not always the optimal strategy in an industrialized context. Cooperatives need to identify the point where arable land scale matches the level of industrialization to achieve the best development outcomes.

5. Discussion

Cooperative development assessment is a globally discussed topic, with many scholars selecting cooperatives as their research focus, albeit with varying emphases [62,63]. Some studies primarily analyze the economic benefits of cooperatives, providing clear insight into their economic development while often neglecting other aspects [64,65]. In recent years, an increasing number of scholars have begun to emphasize comprehensive performance, advocating for sustainable performance systems that encompass economic, social, and ecological dimensions [7,66]. Building upon this approach and incorporating China's experiences [8], this study constructs a comprehensive assessment system for the high-quality development of farmers' cooperatives. The aim is to comprehensively evaluate the development situation of cooperatives, considering economic, social, and ecological factors.

Due to variations in evaluation systems, there are differing conclusions regarding the impact of arable land on the development of farmers' cooperatives. Some studies, particularly those emphasizing the economic benefits of cooperatives, often assert that arable land has a positive influence on the economic growth of cooperatives [67,68]. However, as assessment systems become more comprehensive in evaluating cooperatives, many studies suggest that the impact of arable land on the overall development of cooperatives gradually diminishes [69]. This aligns with the findings of our research. Furthermore, for cooperatives engaged in non-grain production, the role of arable land in their development also tends to weaken over time [70].

It is well-established that arable land has a positive impact on the quality development of cooperatives, meaning that as arable land scale expands, farmers' cooperatives will possess a richer resource endowment, thereby enhancing the development quality of the cooperative [42]. However, as the role of arable land continues to weaken, it is more important to explore the factors that influence the role of arable land [71,72]. As cooperatives enter a new stage of high-quality development, industrial development is considered an effective means to improve production efficiency, reduce labor costs, reduce agricultural production risks, and increase agricultural added value [73]. Xu and Guo (2022) provide a comprehensive analysis of cooperative development, emphasizing the interaction and underlying mechanisms between industrialization and cooperative performance and revealing the multifaceted dynamics provided by industrialization for cooperative development [74]. In addition, existing studies suggest that industrial development serves as a substitute for arable land, allowing cooperatives to reduce agricultural production and even transform into agricultural companies after undergoing industrial transformation and upgrading [75].

However, field research reveals that this theory does not align with the actual situation. The "inverted U-shaped" association between arable land and the high-quality development of cooperatives indicates that the cooperative's arable land scale and industrial development are complementary rather than substitutive. This is because cooperatives are fundamentally agricultural organizations deeply rooted in rural communities and among farmer populations, making it challenging to achieve fully enterprise-oriented development [76]. Therefore, in the early stages of cooperative development, expanding arable land scale can fully leverage economies of scale. More importantly, as cooperatives increase investment in industrial development, expanding into processing, sales, and other aspects, the key to development lies in balancing the cooperative's arable land output and industrial development capabilities. This ensures that processing and sales capabilities match production capabilities, avoiding resource wastage.

Therefore, in the context of high-quality development, cooperatives should not blindly expand their production and operation scale. It is essential to ensure that industrialization investment maximally leverages the advantages of arable land scale, preventing the suppression of high-quality development due to scale expansion.

6. Conclusions and Implications

6.1. Conclusions

In the dual context of increasingly scarce arable land resources and the high-quality development of agriculture, exploring the mechanism of the impact of arable land on the high-quality development of farmers' cooperatives is important for promoting the quality of farmers' cooperatives and sustainable agricultural development. This paper conducts on-site investigations on 448 farmers' cooperatives. Through theoretical analysis and statistical analysis of sample data, it is found that both excessively large and excessively small arable land scale will lead to a decline in the development quality of farmers' cooperatives. Even among cooperatives with similar arable land scale, there is considerable variability in development quality. To further explore the complex relationship between them, this paper utilizes survey data for empirical analysis and draws the following conclusions:

(1) There is a significant positive relationship between arable land scale and the development quality of cooperatives, as the larger the arable land scale, the higher the development quality of cooperatives.

(2) After adding the squared term, there is an "inverted U-shaped" association between arable land scale and the development quality of cooperatives. As arable land scale continuously expands, the development quality of cooperatives first increases and then decreases.

(3) Upon classifying cooperative development quality into five categories (economic foundation, innovation capability, green development, standardization, and social value), regression results for each category still show a significant "inverted U-shaped" association. However, there is strong variability in the impact on development quality in different categories.

(4) Analyzing the phased characteristics of the "inverted U-shaped" association between arable land scale and high-quality development of cooperatives reveals four stages, "weak impact—rapid improvement—diminishing growth effect—decline in development quality", as cooperatives' arable land scale increases.

(5) Industrial development is a conditional factor affecting the establishment of the "inverted U-shaped" association between arable land scale and high-quality development in cooperatives. The intrinsic mechanism is that industrial development and arable land scale in cooperatives exhibit a strong complementary relationship rather than a substitution relationship.

6.2. Policy Implications

Arable land is an indispensable basic resource for the high-quality development of farmers' cooperatives. However, considering the heterogeneity among different cooperatives, it is crucial to pay attention to the moderate expansion of arable land scale. The research findings of this paper have enlightening implications for the government to guide the allocation of arable land in cooperatives and thereby promote the high-quality development of farmers' cooperatives. Firstly, the government should guide cooperatives to consider their own situations. For cooperatives with smaller arable land scale, utilizing land transfer to achieve economies of scale can rapidly improve development quality. Secondly, for cooperatives with larger arable land scale engaged in industrial development, the government should guide these cooperatives to align their industrial development capabilities with arable land production capacity. Additionally, reinforcing standardized and refined management can leverage the complementary effects of resources. Finally, for cooperatives with relatively abundant arable land resources, especially during stages where saturation in industrial development may occur in the short term, it is essential to fully leverage human resources and other innovative inputs. This will better promote the high-quality development of farmers' cooperatives, ensuring their competitiveness in the ever-changing market environment.

6.3. Research Limitations and Prospects

This study has certain limitations, which can be further expanded upon in future research. While the paper extensively analyzes the impact of arable land scale on the high-quality development of farmers' cooperatives and identifies industrial development as a key factor influencing the inverted U-shaped relationship, future research could explore the significant role of certain factors, such as human capital and social capital, in the high-quality development of farmers' cooperatives. To better understand the role of arable land scale, the paper only utilizes data from grain-production-type cooperatives in three major grain-producing regions in China. Future researchers may consider larger sample sizes and conduct comparative analyses using data from various types of cooperatives.

Author Contributions: Conceptualization, Y.X. and X.G.; methodology, Y.X. and Y.H.; software, Y.X.; validation, Y.X. and Y.H.; formal analysis, X.G.; investigation, X.G. and Y.X.; resources, X.G.; data curation, Y.X.; writing—original draft preparation, Y.X.; writing—review and editing, Y.H. and X.G.; supervision, Y.H.; funding acquisition, X.G. All authors have read and agreed to the published version of the manuscript.

Funding: This research was funded by the National Social Science Fund of China, grant number 23AJY016.

Institutional Review Board Statement: Not applicable.

Informed Consent Statement: Not applicable.

Data Availability Statement: The data that support our research findings are available from the corresponding author upon request.

Acknowledgments: The authors are grateful for the patient review and helpful suggestions from the editor of this journal, as well as the anonymous referees. Additionally, the authors express our gratitude to the farmers' cooperatives surveyed in the three major grain-producing regions of Shandong, Henan, and Heilongjiang in China for their support and assistance in obtaining research data.

Conflicts of Interest: The authors declare no conflicts of interest.

Abbreviations

Acronyms and Symbols	Description
EB	Economic Basis
IC	Innovation Capability
GD	Green Development
SL	Standardization Level
SV	Social Value
CDQ	Cooperative Development Quality
CCC	Characteristics of Cooperative Chairman
BCC	Basic Characteristics of Cooperative
OMCC	Operational and Management Characteristics of Cooperative
ALS	Arable Land Scale
ALS^2	Square of Arable Land Scale
IV1	Average Household Arable Land
IV2	Square of Average Household Arable Land
ALD	Arable Land Distribution
IDI	Industrial Development Investment

Appendix A. Survey Questionnaire on Farmers' Cooperative Development

We are the Agricultural Economics Research Team from Northeastern Agricultural University. We sincerely invite you to participate in this research. The name of the research project is the relationship between land endowment and the high-quality development of farmers' cooperatives. Before you decide whether to do the questionnaire or not, please read the following carefully.

Research Purpose:

The objective of this study is to delve into the relationship between land endowment and the high-quality development of farmers' cooperatives. We will investigate the basic characteristics of the chairperson of farmers' cooperative, the scale of arable land, and the overall development of farmers' cooperative. The aim is to gain crucial insights into the nexus between land endowment and the development of farmers' cooperatives.

Privacy and Confidentiality:

Your personal information will be strictly confidential, unless otherwise required by applicable laws and regulations. The research findings will solely be utilized for academic research purposes and will not be employed for any other purposes.

Rights and Voluntary Participation:

You have the right to refuse participation or withdraw from the study at any point without facing any adverse consequences. You may raise concerns or questions at any time, and we will provide satisfactory explanations.

Consent to Participate:

I have read and understood the information provided above. I voluntarily agree to participate in this study and acknowledge my right to withdraw at any time.

Researcher's Contact Information:

Name:
E-mail:
Survey Code: Survey Location: Survey Date: Surveyor:
Respondent's Name: Phone Number: E-mail:

1. Basic Information of the Farmers' Cooperative

(1) Cooperative Name: [Fill in the blank]
(2) Location of the Cooperative: [Fill in the blank]
(3) Registration Date of the Cooperative: [Fill in the blank]
(4) Number of Cooperative Members (people): [Fill in the blank]
(5) Total Assets of the Cooperative (thousands of RMB): [Fill in the blank]
(6) Number of Large Agricultural Machinery Owned by the Cooperative, where large agricultural machinery refers to those with a power of 50 horsepower or above (units): [Fill in the blank]
(7) Main Agricultural Products Operated by the Cooperative: [Multiple-choice]
1) Rice 2) Wheat 3) Corn 4) Other
(8) Demonstration Level of the Cooperative: [Single-choice]
1) Non-demonstration cooperative 2) County-level 3) City-level 4) Provincial-level 5) National-level
(9) Does the Cooperative Have a One-Person-One-Vote System for Major Decision-Making? [Single-choice]
1) Yes 2) No
(10) Does the Cooperative Have a Rebate System? [Single-choice]
1) Yes 2) No
(11) Distance from the Cooperative to the Nearest County Town (km): [Fill in the blank]
(12) Distance from the Cooperative to the Nearest Financial Institution (km): [Fill in the blank]
(13) Average Attendance Rate of Members in the Cooperative's 2022 Annual Meeting (%): [Fill in the blank]
(14) Average Participation Rate of Members in the Cooperative's 2022 Member Skills Training Activities (%): [Fill in the blank]

2. Basic Information of the Chairperson of the Cooperative

(1) Gender of the Chairperson of the Cooperative: [Single-choice]
1) Male 2) Female
(2) Age of the Chairperson of the Cooperative: [Fill in the blank]
(3) Educational Level of the Chairperson of the Cooperative: [Single-choice]
1) Elementary school and below 2) Junior high school 3) High school/Technical secondary school 4) College 5) Bachelor's degree and above
(4) Is the Chairperson of the Cooperative a Village Official? [Single-choice]
1) Yes 2) No

3. Relevant Information on the High-Quality Development of the Cooperative

(1) Cooperative's Operating Income in 2022 (CNY 10,000): [Fill in the blank]
(2) Cooperative's Operating Profit in 2022 (CNY 10,000): [Fill in the blank]
(3) Cropland Area Sown by the Cooperative in 2022 (hectares): [Fill in the blank]
(4) Total Agricultural Production of the Cooperative in 2022 (tons): [Fill in the blank]
(5) Quantity of Processed Agricultural Products by the Cooperative in 2022 (tons): [Fill in the blank]
(6) Quantity of Agricultural Products Sold through Grain Merchants by the Cooperative in 2022 (tons): [Fill in the blank]
(7) Quantity of Agricultural Products Sold by the Cooperative through Enterprise Specialty Stores, Online Sales, Order Production, Supermarket Connections, and Rural Tourism in 2022 (tons): [Fill in the blank]
(8) Number of Agricultural Production Standards Formulated and Adopted by the Cooperative in 2022 (units): [Fill in the blank]
(9) Number of Patents Applied for by the Cooperative in 2022 (units): [Fill in the blank]
(10) Cropland Area Sown with Superior Crop Varieties Adopted by the Cooperative in 2022 (hectares): [Fill in the blank]
(11) Number of Registered Trademarks Owned by the Cooperative as of 2022 (units): [Fill in the blank]
(12) Brand Influence Range of the Cooperative: [Single-choice]
1) No independent brand 2) County-level 3) City-level 4) Provincial-level 5) National-level 6) Export sales
(13) Production Area in 2022 with Reduced Chemical Fertilizers and Pesticides by the Cooperative (hectares): [Fill in the blank]
(14) Percentage of Comprehensive Treatment and Utilization of Crop Straw, Agricultural Plastic Film, Packaging Bags for Fertilizers and Pesticides, and Related Waste by the Cooperative in 2022 (%): [Fill in the blank]
(15) Number of Agricultural Product Certifications (including "Pollution-Free Agricultural Products," "Green Agricultural Products," "Organic Agricultural Products," and "Geographical Indication Agricultural Products"): [Single-choice]
1) No certification 2) One certification 3) Two certifications 4) Three certifications 5) Four certifications
(16) Quantity of Agricultural Products Clearly Displayed on Cooperative's Product Packaging or Labels in Terms of Production, Processing, or Sales Processes (tons): [Fill in the blank]
(17) Number of Times the Cooperative's General Meeting for Members was Held in 2022 (times/year): [Fill in the blank]
(18) Number of Times the Cooperative's Board of Directors was Convened in 2022 (times/year): [Fill in the blank]
(19) Proportion of Provident Fund Withdrawal by Cooperative Members in 2022 (%): [Fill in the blank]
(20) Distributable Surplus Amount in 2022 by the Cooperative (CNY 10,000): [Fill in the blank]

(21) Amount of Surplus Refunded by the Cooperative in 2022 (CNY 10,000): [Fill in the blank]

(22) Number of Times in 2022 the Cooperative Publicly Disseminated Financial Information through Postings, Mass Distribution, etc., for All Members to Understand Financial Information (times/year): [Fill in the blank]

(23) Number of Times in 2022 the Cooperative Consolidated Income and Expenditure Accounts by Accounting (e.g., monthly frequency is 12, quarterly frequency is 4) (times/year): [Fill in the blank]

(24) Total Amount of Direct Economic Support Provided by the Cooperative for Village Construction and Development, such as Cooperative Contributions to Village Road Repair, etc. (CNY 10,000): [Fill in the blank]

(25) Number of Cooperative Partnerships with Other Cooperatives and Enterprises, with Cooperation Duration Generally Exceeding 1 Year (units): [Fill in the blank]

(26) Number of Participants in Training Opportunities Provided by the Cooperative for Members and Other Villagers in 2022 (people): [Fill in the blank]

(27) Number of Long-Term Employed Workers by the Cooperative, Considering the Seasonal Nature of Agriculture (defined as workers employed for more than 4 months in the entire year) (people): [Fill in the blank]

4. Current Situation of Cooperative Resource Input

(1) Cooperative's Arable Land Scale (hectares): [Fill in the blank]
Total Scale: Scale Contributed by Members: Leased Land Scale:
(2) Number of Villages Where the Cooperative Operates Arable Land: [Fill in the blank]
(3) Does the Cooperative Engage in Industrial Development? [Single-choice]
1) Yes 2) No
(4) If yes, what is the investment in industrial development (CNY 10,000)? [Fill in the blank]
(5) Number of Full-time Staff in the Cooperative: [Fill in the blank]
(6) Does the Cooperative Hire Professional Managers? [Single-choice]
1) Yes 2) No
(7) Does the Cooperative Have Agricultural Technology Introduction? [Single-choice]
1) Yes 2) No
(8) If yes, how many types of agricultural technology have been introduced (types)? [Fill in the blank]
(9) Does the Cooperative Have Unified Purchase of Agricultural Insurance? [Single-choice]
1) Yes 2) No
(10) Amount Spent by the Cooperative on Hospitality Expenses (CNY 10,000): [Fill in the blank]
(11) Percentage of Cooperative Members' Sales through the Cooperative (%): [Fill in the blank].

Appendix B. Interview Outline for Cooperative Chairperson

1. Background and Introduction

(1) Please introduce your cooperative, including the establishment time, main business, and development history.

(2) How do you define the development quality of the cooperative? Which factors do you consider most important for development quality?

(3) How competitive do you think the current agricultural product market is?

(4) How do you perceive the demand for the agricultural products you operate in the current market among relevant consumers?

(5) Have there been significant changes in the development plan of the cooperative in recent years? If so, what are the reasons?

2. Cooperative Resource Input Situation

(1) How is the demand for arable land scale in your cooperative? Does the cooperative have a significant demand for arable land? How much land does the cooperative need to obtain through leasing? Will expanding the arable land face obstacles?

(2) Does your cooperative process agricultural products? Does the cooperative have its own sales channels?

(3) Will the arable land scale of your cooperative consider the level of industrialization investment?

(4) Does your cooperative lack professional management, accounting, technical, and sales personnel? What channels are typically used when hiring professional staff?

(5) How intense is the focus of your cooperative on technology introduction? Has the technology introduced by your cooperative been fully utilized? What impact does technology introduction have on the cooperative's development?

(6) Does your cooperative pay attention to maintaining social relationships? How is the expenditure on social relationships? How do these relationships contribute to the development of your cooperative?

(7) How do you assess the value and contribution of the above resources to the cooperative? Considering the development history of your cooperative, do you find the demand for resources to be the same at different stages? In which stages do the mentioned resources play a crucial role?

(8) How do you determine the proportion of investment in various resources for the cooperative? Does these proportions vary based on different business and market demands?

(9) How do you coordinate and balance different types of resources to meet the overall needs of the cooperative?

(10) How do you assess the success of the cooperative's resource allocation strategy? What are your plans and expectations for future resource allocation?

3. Development Challenges

(1) How do you assess the economic benefits of the cooperative? Which indicators best reflect the financial condition and profitability of the cooperative?

(2) How do you assess the cooperative's contribution to society? What impact do you think the cooperative has on the community and local residents?

(3) How do you assess the ecological environmental impact of the cooperative? Has your cooperative taken measures to protect the environment?

(4) Do the products produced by your cooperative meet market demands? How do you assess and improve product quality?

4. Challenges and Opportunities

(1) What do you consider the biggest challenge currently facing your cooperative? How do you plan to address it?

(2) What do you see as the biggest opportunities for the development of the cooperative in the coming years? How do you plan to leverage these opportunities?

(3) How do you view the government's support for farmer cooperatives and the policy environment? Do you believe that the policy environment has a significant impact on the development of cooperatives?

(4) Are you familiar with the digital economy? What impact do you think the digital economy has on the development of cooperatives?

(5) In the context of high-quality development, what aspects do you think the cooperative will pay more attention to in the future?

Appendix C. The Descriptive Statistical Analysis of Indicators for the High-Quality Development of Cooperatives

Due to space constraints and the specific focus of our research, some details have not been fully elaborated upon in the main text of the manuscript. To ensure comprehensive coverage and clarity, these details have been included in the appendix. Specifically, Table A1 comprises a descriptive statistical analysis of indicators pertaining to the high-quality development of cooperatives. This supplemental information offers additional data and insights to complement the findings discussed in the paper.

Table A1. The descriptive statistical analysis of indicators for the high-quality development of cooperatives.

Primary Indicator	Secondary Indicator	Tertiary Indicator	Mean	SD	Min	Max
Economic Basis (EB)	Profitability	Average Income per Member (CNY 10,000)	1047.370	2039.968	30	15,000
		Operating Profit (CNY 10,000)	138.487	419.284	3.2	4000
	Industry Integration	Processing Proportion (%)	41.529	26.605	0	100
		New Sales Methods (%)	41.632	26.536	0	100
Innovation Capability (IC)	Technological Innovation	Standards/Patents (pcs)	1.908	1.906	0	10
		Application of Fine Seeds (%)	81.384	15.065	50	100
	Cooperative Branding	Registered Trademarks (pcs)	0.725	1.681	0	12
		Brand Coverage (1~6)	1.717	1.417	1	6
Green Development (GD)	Ecological Protection	Proportion of Reduced Chemical Area (%)	6.549	11.647	0	50
		Recycling Rate of Agricultural Waste (%)	74.580	16.843	10	100
	Product Safety	Quality Certification Standards (1~5)	1.460	1.053	1	5
		Traceability Proportion (%)	5.795	13.436	0	80
Standardization Level (SL)	Operational Standards	Frequency of Member Meetings/Director Meetings (times/year)	2.288	1.863	1	12
		Proportion of Distributable Surplus Returned (%)	18.150	24.890	0	101
	Financial Standards	Frequency of Financial Report Disclosure (times/year)	2.844	3.292	0	12
		Frequency of Accounting (times/year)	4.007	3.509	1	24
Social Value (SV)	Social Participation	Investment in Village Collective Construction (CNY 10,000)	22.165	120.646	0	2000
		Number of Cooperative Enterprises/Other Cooperatives (pcs)	2.156	3.457	0	18
	Training and Employment	Number of People Trained in Farmer Training Projects (ppl)	55.647	111.930	0	583
		Number of Jobs Created by the Cooperative (pcs)	6.924	13.795	0	70

Appendix D. Specific Explanation of Tertiary Indicators for the High-Quality Development of Cooperatives

In Table A2, we provide detailed explanations of the tertiary indicators for the high-quality development of cooperatives. While these explanations were not extensively discussed in the main body of the manuscript due to space constraints, we have included them here for clarity and completeness.

Table A2. Specific explanation of tertiary indicators for the high-quality development of cooperatives.

Tertiary Indicator	Meaning
Average Income per Member (CNY 10,000)	Operating Income per Member Household
Operating Profit (CNY 10,000)	Cooperative's Operating Profit
Processing Proportion (%)	Quantity of Processed Agricultural Products to Total Agricultural Products Production

Table A2. *Cont.*

Tertiary Indicator	Meaning
New Sales Methods (%)	Quantity of Agricultural Products Sold through Specialty Stores, Online Sales, Order Production, Supermarket Connections, and Rural Tourism by the Cooperative to Total Agricultural Products Production
Standards/Patents (pcs)	Sum of the Number of Standards Formulated by the Cooperative and the Number of Patents Applied
Application of Fine Seeds (%)	Proportion of Cultivated Area for Superior Crop Varieties to Total Crop Sowing Area
Registered Trademarks (pcs)	Number of Registered Trademarks Owned by the Cooperative
Brand Coverage (1~6)	1 = No independent brand; 2 = County-level; 3 = City-level; 4 = Provincial-level; 5 = National-level; 6 = Export sales
Proportion of Reduced Chemical Area (%)	Production Area with Reduced Use of Fertilizers and Pesticides to Total Crop Sowing Area
Recycling Rate of Agricultural Waste (%)	Percentage of Comprehensive Treatment and Utilization of Crop Straw, Agricultural Plastic Film, Packaging Bags for Fertilizers and Pesticides, Agricultural Machinery, and Related Waste to Total Waste
Quality Certification Standards (1~5)	Number of Chinese Government-Certified Pollution-Free Agricultural Products, Green Agricultural Products, Organic Agricultural Products, and Geographical Indication Agricultural Products
Traceability Proportion (%)	Proportion of Agricultural Products Displaying the Cooperative's Name Clearly on Packaging or Labels to the Total Agricultural Product Quantity
Frequency of Member Meetings/Director Meetings (times/year)	Sum of the Number of General Meetings for Members and the Number of Board of Directors Meetings in the Cooperative
Proportion of Distributable Surplus Returned (%)	Amount of Surplus Refunded to Distributable Surplus Amount
Frequency of Financial Report Disclosure (times/year)	Number of Times Financial Statements Were Publicized to All Members through Postings, Mass Distribution, etc.
Frequency of Accounting (times/year)	Number of Times the Unified Summary of Income and Expenditure Accounts by the Accountant
Investment in Village Collective Construction (CNY 10,000)	Total Direct Economic Support Provided by the Cooperative for Village Construction and Development
Number of Cooperative Enterprises/Other Cooperatives (pcs)	Number of Partnerships
Number of People Trained in Farmer Training Projects (ppl)	Number of Participants in Training Opportunities Provided for Members and Other Villagers
Number of Jobs Created by the Cooperative (pcs)	Number of Job Positions Offering Employment for More Than 4 Months

References

1. Hilson, M. A Consumers' international? The international cooperative alliance and cooperative internationalism, 1918–1939: A Nordic Perspective. *Int. Rev. Soc. Hist.* **2011**, *56*, 203–233. [CrossRef]
2. Ribas, W.P.; Pedroso, B.; Vargas, L.M.; Picinin, C.T.; de Freitas Júnior, M.A. Cooperative organization and its characteristics in economic and social development (1995 to 2020). *Sustainability* **2022**, *14*, 8470. [CrossRef]
3. Bijman, J.; Iliopoulos, C. Farmers' cooperatives in the EU: Policies, strategies, and organization. *Ann. Public Coop. Econ.* **2014**, *85*, 497–508. [CrossRef]
4. Pokharel, K.P.; Regmi, M.; Featherstone, A.M.; Archer, D.W. Examining the financial performance of agricultural cooperatives in the USA. *Agric. Finance Rev.* **2019**, *79*, 271–282. [CrossRef]
5. Maclachlan, P.L.; Shimizu, K. Japanese farmers in flux: The domestic sources of agricultural reform. *Asian Surv.* **2016**, *56*, 442–465. [CrossRef]
6. Hairong, Y.; Yiyuan, C. Debating the rural cooperative movement in China, the past and the present. *J. Peasant. Stud.* **2013**, *40*, 955–981. [CrossRef]
7. Marcis, J.; de Lima, E.P.; Gouvea da Costa, S.E. Model for assessing sustainability performance of agricultural cooperatives'. *J. Clean. Prod.* **2019**, *234*, 933–948. [CrossRef]
8. Zhang, R.; Aljumah, A.I.; Ghardallou, W.; Li, Z.; Li, J.; Cifuentes-Faura, J. How economic development promotes the sustainability targets? Role of natural resources utilization. *Resour. Policy* **2023**, *85*, 103998. [CrossRef]

9. Xu, F.-L.; Zhao, S.-S.; Dawson, R.W.; Hao, J.-Y.; Zhang, Y.; Tao, S. A triangle model for evaluating the sustainability status and trends of economic development. *Ecol. Model.* **2006**, *195*, 327–337. [CrossRef]
10. Chang, Y.; Wang, S. China's pilot free trade zone and green high-quality development: An empirical study from the perspective of green finance. *Environ. Sci. Pollut. Res.* **2023**, *30*, 88918–88935. [CrossRef]
11. Liu, X.; Zhang, X. The impact of the digital economy on high-quality development of specialized farmers' cooperatives: Evidence from China. *Sustainability* **2023**, *15*, 7958. [CrossRef]
12. Kalogiannidis, S. Economic cooperative models: Agricultural cooperatives in Greece and the need to modernize their operation for the sustainable development of local societies. *Int. J. Acad. Res. Bus. Soc. Sci.* **2020**, *10*, 452–468. [CrossRef]
13. Futemma, C.; De Castro, F.; Brondizio, E.S. Farmers and social innovations in rural development: Collaborative arrangements in Eastern Brazilian Amazon. *Land Use Policy* **2020**, *99*, 104999. [CrossRef]
14. Candemir, A.; Duvaleix, S.; Latruffe, L. Agricultural cooperatives and farm sustainability—A literature review. *J. Econ. Surv.* **2021**, *35*, 1118–1144. [CrossRef]
15. Ajates, R. An integrated conceptual framework for the study of agricultural cooperatives: From repolitisation to cooperative sustainability. *J. Rural Stud.* **2020**, *78*, 467–479. [CrossRef]
16. Gonzalez, R.A. Going back to go forwards? From multi-stakeholder cooperatives to Open Cooperatives in food and farming. *J. Rural Stud.* **2017**, *53*, 278–290. [CrossRef]
17. Zhong, Z.; Zhang, C.; Jia, F.; Bijman, J. Vertical coordination and cooperative member benefits: Case studies of four dairy farmers' cooperatives in China. *J. Clean. Prod.* **2018**, *172*, 2266–2277. [CrossRef]
18. Yan, J.; Chen, C.; Hu, B. Farm size and production efficiency in Chinese agriculture: Output and profit. *China Agric. Econ. Rev.* **2018**, *11*, 20–38. [CrossRef]
19. Brümmer, B.; Glauben, T.; Lu, W. Policy reform and productivity change in Chinese agriculture: A distance function approach. *J. Dev. Econ.* **2006**, *81*, 61–79. [CrossRef]
20. Yang, H.; Huang, K.; Deng, X.; Xu, D. Livelihood capital and land transfer of different types of farmers: Evidence from panel data in Sichuan Province, China. *Land* **2021**, *10*, 532. [CrossRef]
21. Geng, L.; Yan, S.; Lu, Q.; Liang, X.; Li, Y.; Xue, Y. A rural land share cooperative system for alleviating the small, scattered, and weak dilemma in agricultural development: The cases of Tangyue, Zhouchong, and Chongzhou. *Agriculture* **2023**, *13*, 1675. [CrossRef]
22. Lai, Z.; Chen, M.; Liu, T. Changes in and prospects for cultivated land use since the reform and opening up in China. *Land Use Policy* **2020**, *97*, 104781. [CrossRef]
23. Wiggins, S.; Kirsten, J.; Llambí, L. The future of small farms. *World Dev.* **2010**, *38*, 1341–1348. [CrossRef]
24. Cortner, O.; Garrett, R.D.; Valentim, J.F.; Ferreira, J.; Niles, M.T.; Reis, J.; Gil, J. Perceptions of integrated crop-livestock systems for sustainable intensification in the Brazilian Amazon. *Land Use Policy* **2019**, *82*, 841–853. [CrossRef]
25. Deininger, K.; Jin, S. The potential of land rental markets in the process of economic development: Evidence from China. *J. Dev. Econ.* **2005**, *78*, 241–270. [CrossRef]
26. Shen, M.; Shen, J. Evaluating the cooperative and family farm programs in China: A rural governance perspective. *Land Use Policy* **2018**, *79*, 240–250. [CrossRef]
27. Li, F.; Zhao, W.; Yeh, E.T. The locally managed agrarian transition in China: Land shareholding cooperatives and the agricultural co-management system in Chongzhou, Sichuan. *Eurasian Geogr. Econ.* **2022**, *64*, 732–757. [CrossRef]
28. Ye, J. Land transfer and the pursuit of agricultural modernization in China. *J. Agrar. Change* **2015**, *15*, 314–337. [CrossRef]
29. Li, M.; Wang, J.; Zhao, P.; Chen, K.; Wu, L. Factors affecting the willingness of agricultural green production from the perspective of farmers' perceptions. *Sci. Total Environ.* **2020**, *738*, 140289. [CrossRef]
30. Zhang, Z.; Paudel, K.P. Small-scale forest cooperative management of the grain for Green Program in Xinjiang, China: A SWOT-ANP analysis. *Small-Scale For.* **2021**, *20*, 221–233. [CrossRef]
31. Brandão, J.B.; Breitenbach, R. What are the main problems in the management of rural cooperatives in Southern Brazil? *Land Use Policy* **2019**, *85*, 121–129. [CrossRef]
32. Lu, H.; Zhang, P.; Hu, H.; Xie, H.; Yu, Z.; Chen, S. Effect of the grain-growing purpose and farm size on the ability of stable land property rights to encourage farmers to apply organic fertilizers. *J. Environ. Manag.* **2019**, *251*, 109621. [CrossRef]
33. Baráth, L.; Fertö, I. Heterogeneous technology, scale of land use and technical efficiency: The case of Hungarian crop farms. *Land Use Policy* **2015**, *42*, 141–150. [CrossRef]
34. Yarzábal, L.A.; Chica, E.J. Microbial-based technologies for improving smallholder agriculture in the Ecuadorian Andes: Current situation, challenges, and prospects. *Front. Sustain. Food Syst.* **2021**, *5*, 617444. [CrossRef]
35. Wineman, A.; Jayne, T.S.; Modamba, E.I.; Kary, H. Characteristics and spillover effects of medium-scale farms in Tanzania. *Eur. J. Dev. Res.* **2021**, *33*, 1877–1898. [CrossRef]
36. Bijman, J. Exploring the sustainability of the cooperative model in dairy: The case of the Netherlands. *Sustainability* **2018**, *10*, 2498. [CrossRef]
37. Sala-Ríos, M. What are the determinants affecting cooperatives' profitability? Evidence from Spain. *Ann. Public Coop. Econ.* **2023**, *95*, 85–111. [CrossRef]
38. Adamopoulos, T.; Brandt, L.; Leight, J.; Restuccia, D. Misallocation, selection, and productivity: A quantitative analysis with panel data from China. *Econometrica* **2022**, *90*, 1261–1282. [CrossRef]

39. Key, N. Farm size and productivity growth in the United States Corn Belt. *Food Policy* **2019**, *84*, 186–195. [CrossRef]
40. Munnangi, A.K.; Lohani, B.; Misra, S.C. A review of land consolidation in the state of Uttar Pradesh, India: Qualitative approach. *Land Use Policy* **2020**, *90*, 104309. [CrossRef]
41. Yin, Q.; Sui, X.; Ye, B.; Zhou, Y.; Li, C.; Zou, M.; Zhou, S. What role does land consolidation play in the multi-dimensional rural revitalization in China? A research synthesis. *Land Use Policy* **2022**, *120*, 106261. [CrossRef]
42. Liao, Y.; Zhang, B.; Kong, X.; Wen, L.; Yao, D.; Dang, Y.; Chen, W. A cooperative-dominated model of conservation tillage to mitigate soil degradation on cultivated land and its effectiveness evaluation. *Land* **2022**, *11*, 1223. [CrossRef]
43. Cheng, Y.; Hu, Y.; Zeng, W.; Liu, Z. Farmer heterogeneity and land transfer decisions based on the dual perspectives of economic endowment and land endowment. *Land* **2022**, *11*, 353. [CrossRef]
44. Franco, D.; Singh, D.R.; Praveen, K. Evaluation of adoption of precision farming and its profitability in banana crop. *Indian J. Econ. Dev.* **2018**, *14*, 225–234. [CrossRef]
45. Ran, G.; Wang, G.; Du, H.; Lv, M. Relationship of cooperative management and green and low-carbon transition of agriculture and its impacts: A case study of the Western Tarim River Basin. *Sustainability* **2023**, *15*, 8900. [CrossRef]
46. Trang, N.T.N.; Nguyen, T.-T.; Pham, H.V.; Cao, T.T.A.; Thi, T.H.T.; Shahreki, J. Impacts of collaborative partnership on the performance of cold supply chains of agriculture and foods: Literature review. *Sustainability* **2022**, *14*, 6462. [CrossRef]
47. Huo, Y.; Wang, J.; Guo, X.; Xu, Y. The collaboration mechanism of agricultural product supply chain dominated by farmer cooperatives. *Sustainability* **2022**, *14*, 5824. [CrossRef]
48. Zhang, H.; Wu, D. The impact of agricultural factor inputs, cooperative-driven on grain production costs. *Agriculture* **2023**, *13*, 1952. [CrossRef]
49. Helfand, S.M.; Taylor, M.P. The inverse relationship between farm size and productivity: Refocusing the debate. *Food Policy* **2021**, *99*, 101977. [CrossRef]
50. Khan, N.; Ray, R.L.; Kassem, H.S.; Ihtisham, M.; Siddiqui, B.N.; Zhang, S. Can cooperative supports and adoption of improved technologies help increase agricultural income? Evidence from a recent study. *Land* **2022**, *11*, 361. [CrossRef]
51. Sheng, Y.; Chancellor, W. Exploring the relationship between farm size and productivity: Evidence from the Australian grains industry. *Food Policy* **2019**, *84*, 196–204. [CrossRef]
52. Pilny, A.; Poole, M.S.; Reichelmann, A.; Klein, B. A structurational group decision-making perspective on the commons dilemma: Results from an online public goods game. *J. Appl. Commun. Res.* **2017**, *45*, 413–428. [CrossRef]
53. Aragón, F.M.; Restuccia, D.; Rud, J.P. Are small farms really more productive than large farms? *Food Policy* **2022**, *106*, 102168. [CrossRef]
54. Ronner, E.; van de Ven, G.; Nowakunda, K.; Tugumisirize, J.; Kayiita, J.; Taulya, G.; Uckert, G.; Descheemaeker, K. What future for banana-based farming systems in Uganda? A participatory scenario analysis. *Agric. Syst.* **2023**, *209*, 103669. [CrossRef]
55. Zhang, J.; Chen, M.; Huang, C.; Lai, Z. Labor endowment, cultivated land fragmentation, and ecological farming adoption strategies among farmers in Jiangxi Province, China. *Land* **2022**, *11*, 679. [CrossRef]
56. Guo, F.; Lai, I.K.W.; Zhang, T.; Zhong, Y. Profit coordination and optimization of agricultural product brand promotion lead by farmer cooperative organizations. *Complexity* **2023**, *2023*, 1536341. [CrossRef]
57. Liu, H. The tripartite evolutionary game of green agro-product supply in an agricultural industrialization consortium. *Sustainability* **2022**, *14*, 11582. [CrossRef]
58. Fu, L.; Li, J. Comprehensive evaluation and research on China's public culture service system based on AHP method and entropy weight method. *J. Chem. Pharm. Res.* **2014**, *6*, 230–238.
59. Luo, W.; Li, Y.; Chen, D.; Luo, H. The evaluation model of a country's health care system based on AHP and entropy weight method. *Int. J. Appl. Math. Stat.* **2014**, *52*, 70–83.
60. Stock, J.H.; Yogo, M. Testing for weak instruments in linear IV regression. *Natl. Bur. Econ. Res.* **2002**. [CrossRef]
61. Maertens, R. Adverse Rainfall Shocks and Civil War: Myth or Reality? *J. Confl. Resolut.* **2020**, *65*, 701–728. [CrossRef]
62. Marcis, J.; Bortoluzzi, S.C.; de Lima, E.P.; da Costa, S.E.G. Sustainability performance evaluation of agricultural cooperatives' operations: A systemic review of the literature. *Environ. Dev. Sustain.* **2019**, *21*, 1111–1126. [CrossRef]
63. Peng, X.; Liang, Q.; Deng, W.; Hendrikse, G. CEOs versus members' evaluation of cooperative performance: Evidence from China. *Soc. Sci. J.* **2020**, *57*, 219–229. [CrossRef]
64. Lauermann, G.J.; Moreira, V.R.; Souza, A.; Piccoli, P.G.R. Do cooperatives with better economic–financial indicators also have better socioeconomic performance? *Voluntas* **2020**, *31*, 1282–1293. [CrossRef]
65. Donkor, E.; Hejkrlik, J. Does commitment to cooperatives affect the economic benefits of smallholder farmers? Evidence from rice cooperatives in the Western province of Zambia. *Agrekon* **2021**, *60*, 408–423. [CrossRef]
66. Xu, Y.; Liang, Q.; Huang, Z. Benefits and pitfalls of social capital for farmer cooperatives: Evidence from China. *Int. Food Agribus. Manag. Rev.* **2018**, *21*, 1137–1152. [CrossRef]
67. Deininger, K.; Byerlee, D. The rise of large farms in land abundant countries: Do they have a future? *World Dev.* **2012**, *40*, 701–714. [CrossRef]
68. Manjunatha, A.V.; Anik, A.R.; Speelman, S.; Nuppenau, E.A. Impact of land fragmentation, farm size, land ownership and crop diversity on profit and efficiency of irrigated farms in India. *Land Use Policy* **2013**, *31*, 397–405. [CrossRef]
69. Hong, W. Initial natural endowment and farmers' land abandonment behavior: Based on the investigation of the scale of contracted land. *J. Nanjing Agric. Univ. (Soc. Sci. Ed.)* **2022**, *22*, 124–135. (In Chinese) [CrossRef]

70. Kontogeorgos, A.; Sergaki, P.; Kosma, A.; Semou, V. Organizational models for agricultural cooperatives: Empirical evidence for their performance. *J. Knowl. Econ.* **2018**, *9*, 1123–1137. [CrossRef]
71. Nyanga, A.; Kessler, A.; Tenge, A. Key socio-economic factors influencing sustainable land management investments in the West Usambara Highlands, Tanzania. *Land Use Policy* **2016**, *51*, 260–266. [CrossRef]
72. Veronica, P.; Victor, M.-G.; Elena, M.-M.; Jose-Maria, G.-A. Drivers of joint cropland management strategies in agri-food cooperatives. *J. Rural. Stud.* **2021**, *84*, 162–173. [CrossRef]
73. Liu, S.; Wang, B. The decline in agricultural share and agricultural industrialization—Some stylized facts and theoretical explanations. *China Agric. Econ. Rev.* **2022**, *14*, 469–493. [CrossRef]
74. Xu, Y.; Guo, X. Mechanism of three-industry integration on the performance of farmers' cooperatives: Experience from 254 farmers' cooperatives in Heilongjiang Province. *J. China Agric. Univ.* **2022**, *27*, 265–278. (In Chinese)
75. Li, J.; Chen, Y. Agricultural corporatization is the only way to agricultural modernization in China. *China Rural Econ.* **2022**, *38*, 52–69. (In Chinese)
76. Chuanmin, S.; Falla, J.S. Agro-Industrialization: A comparative study of China and developed countries. *Outlook Agric.* **2006**, *35*, 177–182. [CrossRef]

Article

Innovations in Agricultural Bio-Inputs: Commercial Products Developed in Argentina and Brazil

Gabriel da Silva Medina [1,*], Rosana Rotondo [2] and Gustavo Rubén Rodríguez [2,3,*]

[1] Faculty of Agronomy and Veterinary Medicine, Campus Universitário Darcy Ribeiro, University of Brasília, ICC Sul, Brasília 70800-100, Brazil
[2] Faculty of Agricultural Sciences, Campo Experimental Villarino, National University of Rosario, Zavalla S2125ZAA, Argentina; rotondorosana@gmail.com
[3] Institute for Agricultural Research of Rosario (IICAR-CONICET-UNR), Campo Experimental Villarino, National University of Rosario, Zavalla S2125ZAA, Argentina
* Correspondence: gabriel.medina@unb.br (G.d.S.M.); grodrig@unr.edu.ar (G.R.R.)

Abstract: Innovations in agricultural bio-inputs can lead to sustainable alternatives to replace synthetic fertilizers and pesticides. However, there is no clear understanding of what technologies can become available to farmers as commercial products, particularly in developing countries. This study summarizes the innovations used in commercial products in Argentina and Brazil based on the countries' official data and on in-depth surveys conducted with 14 bio-input private companies. The results reveal ongoing development efforts to improve traditional products, such as inoculants that help plants fix nitrogen. There is also progress in mastering the formulation of new bio-inputs, such as bio-fertilizers that promote plant growth and bio-pesticides for pest control. Lastly, the next generation of bio-inputs composed of phytovaccines promises to help prepare plants' immune systems against the attack of pathogenic fungi and bacteria, while bio-herbicides can potentially reduce the use of synthetic herbicides to prepare fields for harvest. Domestic companies based in Argentina and Brazil play an important role in these innovations that can underpin bio-economy growth in developing countries.

Keywords: bio-fertilizer; bio-pesticide; phytovaccine; bio-herbicide; bio-economy; horizon scanning

Citation: da Silva Medina, G.; Rotondo, R.; Rodríguez, G.R. Innovations in Agricultural Bio-Inputs: Commercial Products Developed in Argentina and Brazil. *Sustainability* **2024**, *16*, 2763. https://doi.org/10.3390/su16072763

Academic Editors: Fotios Chatzitheodoridis, Efstratios Loizou and Achilleas Kontogeorgos

Received: 11 February 2024
Revised: 1 March 2024
Accepted: 19 March 2024
Published: 27 March 2024

1. Introduction

Decreasing the use of synthetic agricultural inputs is a target shared by countries worldwide due to the negative impacts agrochemicals have on human health and the environment [1,2]. One of the main potential alternatives to agrochemicals is the use of bio-inputs whenever innovations provide solutions to supersede synthetic fertilizers, pesticides, and herbicides [3]. Agricultural bio-inputs are biological products developed from enzymes, extracts (from plants or microorganisms), microorganisms, macro-organisms (invertebrates), and secondary metabolites, which are intended for biological control, nutrition, and abiotic and biotic stress relief [4].

There is a growing market demand for agricultural bio-inputs, leading to important technological developments in this field. The global market for bio-pesticides was estimated at US $6.51 billion in 2022, with an estimated growth of 15.7% by 2029 [5]. The market for bio-fertilizers in 2022 represented US $2.02 billion, with an estimated growth of 12.1% by 2029, while the market for bio-stimulants in 2022 represented US $3.14 billion, with an estimated growth of 11.4% by 2029 [5].

Innovations range from improving the efficacy of existing bio-inputs such as inoculants, bio-fertilizers, and bio-pesticides [6–9] to developing new products based on modern technologies such as nanotechnology and molecular biology [10,11]. There are also efforts to combine different microorganisms into single products [12–14] that solubilize nutrients available in the soil and help plants resist pests and drought [15,16]. Industrial processes such as bioreaction and lyophilization are also applied to prepare long-lasting and easy-to-apply bio-products [17,18].

After long processes of product development in public research centers and private companies, bio-inputs are becoming largely available to farmers in the marketplace. Inoculants were the first generation of bio-inputs to have a large-scale market, particularly for soybeans in the 1990s [6]. After 2015, a new generation of commercial products such as biological fertilizers and biological fungicides became available on the market, which resulted in the current growth in the bio-inputs sector [1]. Scientists expect that the next generation of bio-inputs will result from the growing market demand for bio-herbicides [19] and from developments in the field of biotechnology [11]. This recent market growth for large-scale crops took place alongside the steadily growing demand for peri-urban and organic areas, where synthetic inputs are not allowed [3].

Innovation in agricultural bio-inputs represents an opportunity for sustainable development in countries like Argentina and Brazil, which have a large demand for agricultural inputs [20,21]. Both countries have agricultural-related industries, support from public innovation centers, and dedicated policies promoting the sector of bio-inputs [22–24]. Strategic investments in this sector may also benefit from the growing market for bio-economy-related products [25].

However, despite the potential for innovation in agricultural bio-inputs and the intensive research by the public and private sectors reported by the scientific community, there is no clear understanding of what innovations are now available in the market or can potentially make it to the market in the coming years and what solutions they are expected to deliver [26]. We lack information from a horizon scanning approach to what is happening at the end of the product development pipeline, where products are either being developed and registered for commercial use or are becoming available in the marketplace [17,27].

This study aims to characterize the state-of-the-art in commercial agricultural bio-input development that can lead developing countries to play an important global role in sustainable agriculture. Specifically, we aim to:

- Summarize the innovations registered by governmental agencies for commercial use in Argentina and Brazil;
- Characterize the commercial products made available in the market or under development by private companies;
- Analyze the main market-related innovative pathways based on the products' and companies' characteristics.

2. Theoretical Framework

While the potential uses of microorganisms for agricultural purposes are well covered by the scientific literature [6,13], there are fewer studies on market dynamics, particularly on the innovations that are being made available in the market [21]. The existing literature on commercial agricultural bio-inputs highlights the importance of biotechnology to foster sustainable agriculture [28,29], its potential for bio-economy [25], as well as its limitations [30]. However, the academic literature still lacks a clear understanding of the existing innovations as a means to assess how the bio-inputs sector can eventually underpin bio-economy growth in developing countries.

The rising global demand for agricultural products presents opportunities for investments in innovative agro-industrial sectors in developing countries [31]. Innovation is a new or improved product or process (or a combination thereof) that differs significantly from the previous products or processes and that has been made available to potential users [32]. In other words, innovation can be defined as an invention that becomes a product available to consumers in the marketplace.

Developing countries often lack investments by industrial sectors for product development [33]. However, this is not the case for the sector of bio-input, which receives investments from both private companies (including startups) and public research centers [34].

Cooperation for innovation between private companies, universities, and research centers occurs moderately with domestic high-tech companies in developing countries [35,36].

Public research centers include mainly the Instituto Nacional de Tecnología Agropecuaria (INTA) in Argentina and the Empresa Brasileira de Pesquisa Agropecuária (Embrapa) in Brazil, but there are also universities and other important research centers in both countries [37]. These centers often develop new technologies that are then licensed to private companies that transform them into commercial products [23,24].

The development of a commercial bio-product begins with basic research to isolate and select the strains of beneficial microorganisms and laboratory tests to assess its efficacy. The next phases include trials in greenhouses to validate the product prototype in a controlled environment and field trials on farms that go along with the registration of the product for commercial use [17]. Once the product is registered, it can be made available on the market for farmers. The whole process of registering a product requires years of field testing with diverse crops and in distinct regions. The main types of agricultural bio-inputs available on the market are inoculants, bio-stimulators (bio-fertilizers), and products for plant therapeutics (bio-pesticides) [24]. Industrial agricultural bio-inputs are a new and growing sector with potential solutions for sustainable agriculture, which is also the case for other sectors such as organic fertilizers that use agricultural residues [38,39].

3. Methods

This study is based on official data on commercial bio-inputs and their active ingredients that are registered in Argentina and Brazil and on interviews with private companies that invest in research to develop new bio-inputs in these countries. The official data provided a comprehensive overview of the main technologies (extracts from live organisms or micro- and macro-organisms) used in both countries based on the products registered by private and public companies by January 2023. The interviews added information on the private companies' new developments and perspectives for the future based on the products these companies were expecting to have on the market in the coming years.

For Argentina, we used official data published online by the Argentinean Servicio Nacional de Sanidad y Calidad Agroalimentaria (National Service for Agrifood Health and Quality—SENASA) that registers the agricultural products used in Argentina. The study is based on two databases, which are the registered fertilizers (Registro de Productos fertilizantes, enmiendas y otros) [40] and the registered pesticides (Registro nacional de terapéutica vegetal) [41]. From the list of registered fertilizers, all bio-fertilizers under the titles fertilizante biológico and enmienda biológica were selected. From the list of registered pesticides, all bio-pesticides based on fungi, bacteria, or viruses were selected. While the second database provides the list of active ingredients for each product, the same information is not available in the first list.

For Brazil, we used data published by the Ministry of Agriculture Livestock and Procurement (Ministério da Agricultura Pecuária e Abastecimento—MAPA). In the option "Produtos formulados" of the Agrofit website [42], we selected all the classes that were listed as biological, i.e., acaricide, biological agent, bactericide, fungicide, insecticide, and nematicide. These lists provided the names of the registered products, the active principles, and the companies that registered these products.

We contacted all the companies with products registered that had a website in Argentina and Brazil by e-mail and/or WhatsApp provided on their websites. From the 22 companies that replied to our contacts, we interviewed 14 domestic companies that have capital either from Argentinean or Brazilian groups and that have a strong product development component in their business model. Between October 2022 and March 2023, interviews were carried out with representatives of seven companies from Argentina and seven companies from Brazil (Table 1). While most of the companies are small and family-owned, the sample also included medium-to-large corporations such as Nova (owned by two families from Argentina and one US investor), Protergium (part of the Argentinean group Terragene), Rizobacter (part of the Argentinean group Bioceres, which is listed in the NASDAQ), Biagro (part of the Brazilian holding GoExper that also owns Simbiose and Bioma), Biotrop (part of the Brazilian private equity firm Aqua Capital), and Vittia (a Brazilian group listed in the Brazilian stock market, B3).

Table 1. Interviewed companies and their webpages.

Headquarters	Company (and Business Group)	Webpage
Argentina	Ceres Demeter	https://ceresdemeter.com.ar/
	Fragaria	http://fragaria.com.ar/
	Microvidas	https://www.microvidas.com.ar/
	Nova	https://laboratorios-nova.com/web/
	Protergium (Terragene)	https://protergium.com/es/
	Rizobacter (Bioceres Crop Solutions Corp)	https://www.rizobacter.com.ar/
	Summabio	https://summabio.com.ar/
Brazil	Biagro (Go Exper holding—Biagro/Simbiose/Bioma)	https://www.biagro.com.br
	Biotrop/Total (Aqua Capital private equity) *	https://biotrop.com.br/
	Forbio (Forus Group—Forquímica)	https://for.bio/
	Krilltech	https://krilltech.com.br/
	Moara	https://moara.agr.br/
	Vital Force (products registered as Vital Brasil)	https://vitalforce.com.br/
	Vittia	https://vittia.com.br/

* By September 2023, after this study was finished, Biotrop was sold to the Belgian company Biobest. Websites accessed on 18 January 2023.

Interviews were conducted in person in nine of the cases, and there were five interviews conducted by online video calls. Five of the in-person interviews included a visit to the company's plant and lasted for around two hours, while the other interviews lasted for around one hour. The interviews were not recorded, and no personal information became available. The interviews covered the characteristics of the commercial products, their development process, and the products that were being registered or in the stage of field validation and are expected to be made available in the market in the coming years.

A horizon scanning approach was deployed to make a first evaluation of the new technologies. Horizon scanning is the first stage of the subject-wide synthesis of evidence since it involves listing all the known options for addressing a particular problem [43]. While a complete review of the evidence base would be preferable, the scale and duration of such reviews are often impractical [44].

A cluster analysis using Ward's method with Jaccard distances was conducted to determine the companies performing the main technological developments that can lead to innovative pathways in the field of bio-inputs. For this, we used 28 variables in the survey that represent the technologies adopted by the interviewed companies. All variables were binary, i.e., either the company had or did not have the specific technology in its portfolio. The technologies were coded by segments composed of inoculants, bio-fertilizers, bio-pesticides, phytovaccines, and other developments. The software INFOSTAT version 2011 (Universidad Nacional de Córdoba, Córdoba, Argentina) was used to perform the cluster analysis [45].

4. Results

4.1. Products Registered by Governmental Agencies

The official data revealed that by January 2023, Argentina had registered 824 inoculants or bio-fertilizers and 39 bio-pesticides registered for plant therapeutics (Figure 1).

Companies in Argentina reported that some products used as both bio-fertilizers and bio-pesticides are often registered as bio-fertilizers since registering bio-fertilizers is less time- and resource-consuming than registering bio-pesticides for plant therapeutics. However, in the case of bio-pesticides, the registration must be made as plant therapeutics. It can be the reason why Argentina has many more products registered as bio-fertilizers when compared with plant therapeutics (or bio-pesticides).

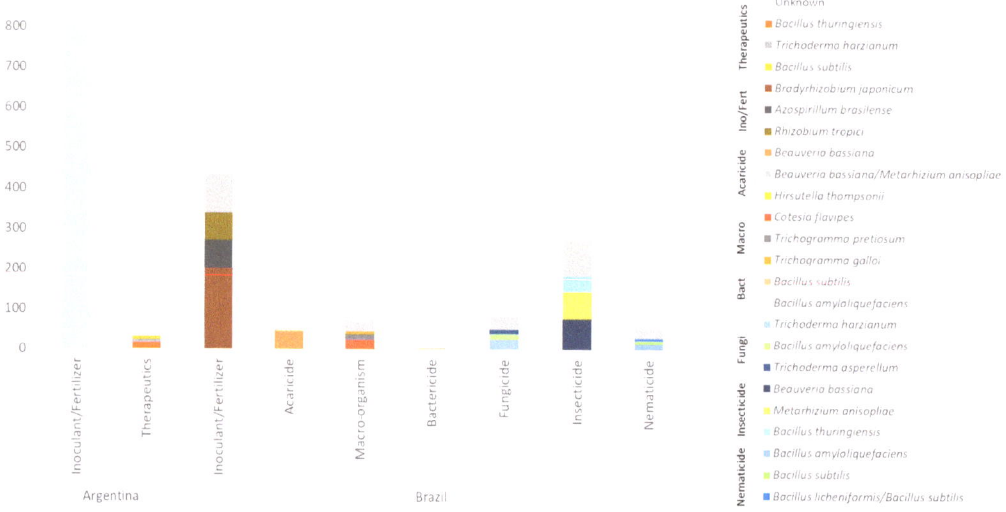

Figure 1. Total number of products and the most common technologies (microorganisms) registered in Argentina and Brazil by January 2023. Source: Adapted from SENASA for Argentina and MAPA for Brazil [41,42]. Note: While Argentina registers products either as bio-fertilizers or plant therapeutics, Brazil registers bio-pesticides (or plant therapeutics) as acaricide, biological agent, bactericide, fungicide, insecticide, and nematicide.

Most of the products for plant therapeutics in Argentina were based on the bacteria *Bacillus thuringiensis*, but there were also products based on the fungus *Trichoderma* and one virus. Whereas the active ingredients for inoculants and bio-fertilizers were not made available by SENASA, the interviewed companies mentioned that their products are mostly based on bacteria such as *Bradyrhizobium japonicum* for inoculants and *Bacillus subtilis* and *Pseudomonas fluorescens* for bio-fertilizers.

In Argentina, 123 companies had registered the 824 inoculants or bio-fertilizers, and 22 companies had registered the 39 products for plant therapeutics by January 2023, adding up to 129 companies in total since both types of products were registered by the same companies in some cases. The leading company in inoculants and bio-fertilizers is Rizobacter (part of the Argentinean Bioceres Crop Solution holding), while six other companies have three products each in plant therapeutics. As for the public companies, INTA registered one product for plant therapeutics, and YPF Sociedad Anonima registered two bio-fertilizers [41].

By January 2023, Brazil had 436 inoculants or bio-fertilizers and 546 bio-pesticides registered by the MAPA, half of which were registered as insecticides, but the list also included acaricides and other products (Figure 1). The vast majority of the acaricides were made of the fungus *Beauveria bassiana*, but there were also products based on other fungi and mites. Most of the agents for biological control were macro-organisms such as wasps of the genus *Cotesia* and *Trichogramma*, but there were also bugs and mites. The few bactericides were made of bacteria of the genus *Bacillus*, while the fungicides were made of fungi of the genus *Trichoderma* or bacteria of the genus *Bacillus*, either as isolated microorganisms or as pools of different organisms. Finally, bio-insecticides were mostly made of the fungi *Beauveria bassiana* and *Metarhizium anisopliae*, and the nematicides were made of bacteria of the genus *Bacillus*.

Bio-pesticides (such as fungicides, insecticides, etc.) are registered in accordance with the Pesticides Law (lei 7802/89) in Brazil, but biological products can also be registered as inoculants or as phytosanitary products approved for use in organic agriculture. Since phytosanitary products approved for use in organic agriculture have low toxicity, their registration is faster than registration by the Pesticides Law.

In Brazil, 45 companies had registered the 436 inoculants and bio-fertilizers and 114 companies had registered the 546 bio-pesticides by January 2023, adding up to 137 companies with bio-inputs registered in Brazil since both types of products were registered by the same companies in some cases. The leading company in inoculants and bio-fertilizers was Biotrop/Total (Biotrop is the trade name of a company that is part of the Brazilian Aqua Capital private equity fund that registered products as Total). The leading private companies in bio-pesticides were Koppert (a Dutch company) for acaricides, Promip (a Brazilian company) for macro-organisms, Biotrop/Total for fungicides and nematicides, and Simbiose (part of the Brazilian GoExper holding) for insecticides. As for the public research centers, Embrapa registered 18 inoculants, and Embrapa and Ceplac registered one insecticide each [42].

4.2. Developments Reported by Companies

The interviews with companies revealed many commercial development fronts underway, including products that are available on the market and products that are still being developed or registered. These developments bring recent improvements in traditional products such as inoculants, efforts to master the formulation of pools of microorganisms in bio-fertilizers and bio-pesticides, and the ongoing development of new products such as phytovaccines and bio-herbicides (Figure 2).

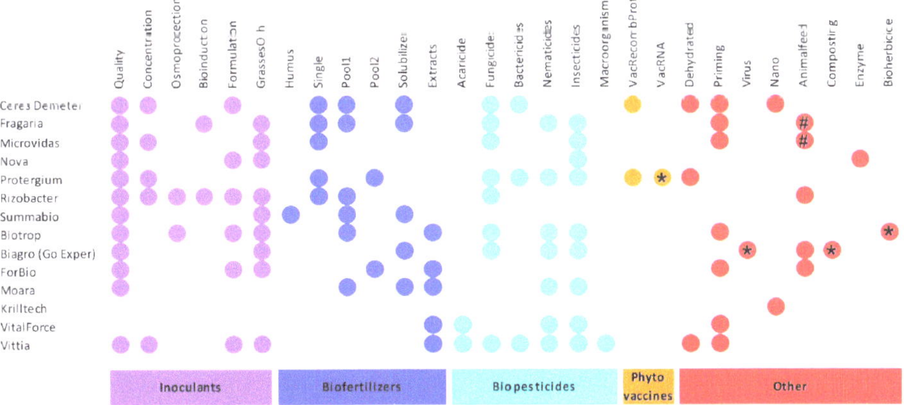

Figure 2. Number of cases of innovations in bio-inputs reported by the interviewed companies. Source: Interviews with companies' representatives. Note: Innovations listed in the figure are available in the market except for the cases where the products are still being developed (noted as *) or registered (noted as #).

4.2.1. More Efficient Inoculants

Large-scale commercial use of inoculants in Argentina and Brazil began with the bacteria *Bradyrhizobium japonicum* used in soybean seeds to improve the plants' capacity to fix nitrogen (Table 2). Over time, different technological developments were made available in the market, such as the selection of more efficient strains of bacteria, formulations in high concentrations, osmoprotection, use of bio-inductors to anticipate the exchange of signals between the root and the microorganisms, and improved formulations with better efficacy (e.g., long-living bacteria, resistance to abiotic stress, and organisms with climate change mitigation effects). Additionally, inoculants were developed for cereals such as corn and wheat based on the bacteria *Azospirillun brasilense* and for other crops such as fruits and vegetables based on the bacteria *Rhizobium tropici* and others.

Table 2. Generations of technological developments in bio-products in Argentina and Brazil.

Generation	Uses (Most Common)	Characteristics	Active Ingredients (Most Common)	Main Technological Development
First	Inoculants to promote nitrogen-fixing by plants	Isolated conventional bacteria	*Bradyrhizobium japonicum*	Isolating, growing, and preventing contamination
Second	Bio-fertilizers for plant growth and nutrient solubilization	Non-conventional PGPR bacteria (isolated or in the pool)	*Bacillus subtilis* and *Pseudomonas fluorescens*	Keeping the stability and levels of different organisms over time
Third	Bio-pesticides for pest control	Pool of isolated, non-conventional bacteria, fungi, and viruses	Bacteria + fungus such as *Trichoderma harzianun*	Isolating spore-producing organisms and stabilizing the fungi
Fourth	Phytovaccines that activate plants' immune systems	Molecular biology to produce recombinant proteins, peptides, and ribonucleic acid (RNA)	Bio-elicitors made of pathogenic fungi or bacteria	Know-how in molecular biology and industrial processes

Source: Interviews with companies' representatives.

All but two interviewed companies have inoculants in their portfolio, and some are focused on developing more efficient inoculants given the size of their market. For example, the Argentinean Rizobacter is a global leader in inoculants for soybeans, and the company is also developing new products on the Rizoliq platform. Biagro (Go Exper) in Brazil also has important developments in inoculants, including Biagro N2 with *Azospirillum brasilense* for grasses and Biagro beans with *Rhizobium tropici* for beans. Almost all companies in Argentina use the *Bradyrhizobium japonicum* strain E109, isolated by the public agency INTA, and companies in Brazil also use strains isolated by different public agencies.

4.2.2. Bio-Fertilizers Made of a Pool of Organisms

Commercial bio-fertilizers began with liquid humus containing non-isolated microorganisms obtained from roundworms. The next generations had single-isolated bacteria, and the recent formulations focus on pools of different bacteria (Pool 1) or pools of bacteria and fungi (Pool 2) (Figure 2). The limited number of companies associated with Pool 2 is explained by the challenges of stabilizing mixes of different organisms, a technology that is still being mastered by companies. Developments already available on the market include the use of plant growth-promoting rhizobacteria (PGPR), such as *Bacillus* and other genera, to promote plant growth. Ongoing developments are related to trying out different organisms, assessing their efficacy, and reaching stable formulations of organism pools.

An important recent commercial development in this segment was the use of soil-nutrient-solubilizing bacteria such as *Pseudomonas fluorescens*. Further developments resulted in pools of other organisms, such as *Bacillus megaterium* and *Bacillus subtilis*, that resulted in the leading phosphorous (P) solubilizer product in the Brazilian market. Innovation centers and companies continue working with different pools of organisms to solubilize different nutrients, such as potassium (K) and others.

Bio-extracts are another area for the development of bio-fertilizers, which include extracts of algae, amino acids, and biopolymers produced by the cells of living organisms and used to protect seeds and bacteria in inoculants. The Argentinean Ceres Demeter is registering Biopol P1 as a mix of natural polymers and sugars for coating peanut seeds, capable of protecting the shell and increasing the adherence and survival of growth-promoting microorganisms on it, improving fluidity compared to untreated seeds. In Brazil, Biotrop has Bionatus (based on extracts of the algae *Ascophyllum nodosum*) positioned in the market as a renewable biodegradable organo-mineral fertilizer; Forbio has Adiplus SojaFort NOD as a biopolymer to protect bacteria inoculated into soybean seeds; and Vital Force has Stimulus Root, which contains amino acids from algae.

All but two of the companies interviewed have bio-fertilizers in their portfolios, but some have a greater focus on bio-fertilizer development. For example, the Argentinean company Fragaria is now combining the bacteria *Azospirillum brasiliense* and *Pseudomonas rhodesiae* to prepare the fertilizer Acqua Duo, which is a PGPR bio-stimulator that solubilizes phosphorus. The Brazilian company Moara developed the product Bioprosolo using three species of *Bacillus* to solubilize P, stimulate root growth, and induce drought resistance. Go Exper is a leading company in P solubilization with a pool of *Bacillus megaterium* and *Bacillus subtilis*, developed in collaboration with Embrapa and registered as Biagro Energia by Biagro and BiomaPhos by Bioma.

4.2.3. Bio-Pesticide Formulations for Bio-Control

Commercial bio-control began with the use of some species of non-pathogenic fungi that control other species of pathogenic fungi. It then evolved to products combining different fungi and bacteria, resulting in products labeled on the market as bio-acaricides, bio-fungicides, bio-bactericides, bio-nematicides, and bio-insecticides. Ongoing developments are related to improving the long-term stability of the formulations, particularly those mixing bacteria and fungi. Such effort often implies having dedicated zones (or industrial plants) to work with microorganisms that produce spores, such as *Bacillus* and *Trichoderma*, and other zones to work with microorganisms that do not produce spores, as a means of preventing contamination by spores.

Developments in this sector also include isolating strains of new microorganisms to be used as bio-controls for different pathogenic organisms. For example, while the fungus *Beauveria bassiana* is well known by different companies, some companies invested in species of the genera *Metarhizium* (fungus) and *Streptomyces* (bacteria) to develop new fungicides. Formulations are also important to guarantee the quality of the product; while some companies use solid subtracts (such as rice), others multiply the microorganisms in liquid subtracts in bioreactors as a means of inducing the production of metabolites.

All but three interviewed companies have bio-pesticides in their portfolio, but with a distinct focus on fungicides, bactericides, nematicides, and insecticides (Figure 2). For example, in Argentina, Microvidas has the fungicide Trichovidas (*Trichoderma harzianum*) as its main commercial product, while Ceres Demeter is developing ISR Wheat and ISR Peanuts, both based on different strains of *Bacillus*, to control fungi, mainly *Fusarium* and *Sclerotinia*. In Brazil, Vital Force has BioBVB based on *Beauveria bassiana* registered as an insecticide but also with acaricide effects, and the insecticide BioScap Liq based on *Beauveria bassiana* and *Metarhizium anisopliae* as its leading commercial product. Biotrop has the fungicides Bombardeiro and Biomagno, the nematicide Furatrop, and the insecticides Biokato and Olimpo on the market.

Macro-organisms are also deployed for bio-control in commercial products used as parasitoids or predators of insects and other pests. For example, Vittia has the product GALLOI-VIT (*Trichogramma galloi*), which is a biological control agent used to destroy the eggs of *Diatraea saccharalis* in sugar cane plantations. Companies are also beginning to work with viruses, and Biagro mentioned that the Go Exper holding is developing the virus-based product VirControl.

4.2.4. Phytovaccines as Molecules for Specific Targets

Applied biotechnology has allowed the development of active principles through the use of recombinant proteins, peptides, or RNA interference that can be used as phytovaccines to prepare the plants' immune systems to resist the attack of pathogenic organisms such as fungi or bacteria. Recombinant proteins can promote efficient absorption of minerals and tolerance for stressful situations. The technique uses an isolated deoxyribonucleic acid (DNA) sequence from a pathogenic fungus or bacteria that is introduced in a bacteria or yeast that works as a factory to reproduce a target protein. Once the product is homogenized, the host bacteria are destroyed, and the resulting recombinant protein is used as an active ingredient for preparing bio-inputs. Similar biotechnology routes are used for peptides and ribonucleic acid (RNA), but the use of RNA interference requires encapsulating the molecule as a means of providing it with environmental stability. The resulting product is used as a bio-elicitor that activates the immune systems of the plants, playing the role of a phytovaccine.

The large-scale use of protein in pest control dates back to the application of *Bacillus thuringiensis* (BT) crystal protein to control lepidopterous larvae. Dating back to the 1920s, this innovation continued to be used as a commercial product even after the development of genetically modified BT corn. The phytovaccine differential is that it acts on the plant to prepare its immune system against potential pathogenic pests and does not act on the pests themselves.

Only two companies mentioned they were developing phytovaccines as bio-elicitors. Protergiun is registering two products as biological conditioners (one for soybeans and the other for cereals) based on molecular biology, whose active ingredient is a recombinant protein, and is also developing a product based on RNA interference. Ceres Demeter also mentioned investments in a metabolite of *Streptomyces* recombinant that can be used as priming, whose timely application increases the plants' capacity to resist stress. The only competitor in this segment in Argentina is Tropfen, which is marketing Tropbio Pro, a product that is made of the Harpin $\alpha\beta$ protein extracted from the bacteria *Escherichia coli* using technology developed by a US company. This technology is not listed in Table 3 since it is dedicated to innovations conducted both in Argentina and Brazil.

Table 3. Examples of state-of-the-art technologies of the main types of bio-inputs that are either available in the market or are under development in Argentina and Brazil (name of the commercial product and description).

Use	Company	Developments in the Market	Ongoing Developments
Inoculants	Rizobacter	Rizoliq—*Bradyrhizobium* sp. in high concentration in the container and on the seed	Rizoliq Dakar—*Bradyrhizobium japonicum* and *Bradyrhizobium diazoefficiens* for better biological fixation in environments of water deficiency and high temperatures
	Biagro (Go Exper)	Biagro N2 with *Azospirillum brasilense* for grasses and Biagro Beans with *Rhizobium tropici* for beans	Liquid and peaty (turfoso) inoculants for 21 days of pre-inoculation
Bio-fertilizers	Fragaria	Acqua Duo—*Azospirillum brasiliense* and *Pseudomonas rhodesiae* as a PGPR and P solubilizer	Bio-fertilizers based on *Bacillus* since it grows faster than conventional bacteria and has both PGPR and therapeutic effects
	Moara	-	Bioprosolo (Registering)—Three species of *Bacillus* to solubilize P, stimulate root growth, and induce drought resistance
	Summabio	Summabalance and Summaroot (liquid humus + Isolated *Pseudomona* and *Bacillus* + micronutrients)	Include *Trichoderma* in the platform
Bio-pesticides	Microvidas	Trichovidas—*Trichoderma harzianum* as a bio-fungicide	Studying the potential of lyophilization for having solid instead of liquid products (focusing on *Trichoderma*)

Table 3. *Cont.*

Use	Company	Developments in the Market	Ongoing Developments
Bio-pesticides	Biotrop	Fungicides Bombardeiro and Biomagno, nematicide Furatrop, and insecticides Biokato and Olimpo	Solid bio-pesticides (Olimpo, the first commercial product)
	Vital Force	Insecticide BioScap Liq based on *Beauveria bassiana* and *Metarhizium anisopliae*	Bio-nematicides and plant resistance induction
Phytovaccines	Protergium	Phytovaccine based on recombinant proteins—Products of the molecular biology platform registered as two bio-conditioners, one for soybeans and the other for cereals	Phytovaccine based on RNA interference (developing)
	Ceres Demeter	-	Metabolite of *Streptomyces* recombinant
Other—Priming	Vittia	Priming—Bio-Immune has endospores that induce plant resistance	Gene silencing in specific microorganisms and investments to increase the portfolio of macro-organisms
Livestock feeding	Forbio	-	Inoculant Forsilo made of *Lactobacillus plantarum* to accelerate the anaerobic fermentation of silage to improve the quality of the feed offered to cows
Nano	Krilltech	Arbolina, which is based on organic carbon nanoparticles	Exploring possibilities to combine microorganisms and organic nanoparticles
Enzymes	Nova	Enzymes Xilanasa and Fitasa used to improve livestock digestion	Peptides from bacterial fermentation to be used as biofungicides

Source: Interviews with companies' representatives. Note: The companies and products listed in this table are examples of technological developments. There are other companies in each sector, and companies also have other products in addition to the ones mentioned in this table.

4.2.5. Other Ongoing Developments

Interviewed companies are investing in a variety of new technologies, ranging from new products to new industrial processes, as summarized below.

- Priming—Priming is the timely use of products to increase the plants' capacity to resist biotic stress, such as fungi attack, and abiotic stress, such as drought. Different types of bio-inputs used at the right moment can have priming effects. The Argentinean companies Protergium and Ceres Demeter mentioned developing products to be used for priming. In Brazil, Vittia mentioned that its commercial fungicide and bactericide, Bio-Immune, based on *Bacillus subtilis*, has endospores that induce plant resistance. They also mentioned working with gene silencing in specific microorganisms.
- Dehydrated products—Industrial processes such as lyophilization using low temperatures and sprays using heat are deployed to have solid/powder products (noted as wettable powder—WP) that reduce the costs of logistics. In Argentina, Fragaria uses lyophilization to prepare Biosilo, marketed under its own brand, and a probiotic produced for a third party. In Brazil, Biotrop uses lyophilization to prepare the insecticide Bioolimpo, sold as a powder-based product.
- Virus—Insect viruses are disease-causing organisms that reproduce within a host insect and can control a variety of insects that attack crops. Biagro mentioned that the Go Exper holding is developing the product VirControl based on viruses.
- Nanotechnology—Nanoparticles can be used as nanopesticides or nanofertilizers that enhance the capacity of plants to absorb nutrients. The Argentinean Ceres Demeter is developing a bio-input that includes microorganisms and organic nanoparticles, while

the Brazilian Krilltech already has on the market the product Arbolina, which is based on organic carbon nanoparticles that function as physiological promoters.

- Inoculants for livestock feeding—Bacteria can be used to improve the quality of livestock feeding. In Argentina, Fragaria has Biosilo on the market as an inoculant for silage made of lactic acid bacteria (*Lactobacillus plantarum*, *Pediococcus acidilactici*, and *Lactobacillus buchneri*) and cellulolytic enzymes, and Microvidas is developing products to be inoculated in cattle feed using bacteria of the genus *Pseudomonas*. The Brazilian Forbio is developing the product Forsilo made of *Lactobacillus plantarum* to accelerate the anaerobic fermentation of silage and improve the quality of the feed offered to cows.
- Composting—Bio-inputs can be used to accelerate the process of bio-composting. Biagro mentioned isolating an inoculant for composting.
- Enzymes—Enzymes are proteins that act as biological catalysts by accelerating chemical reactions. The Argentinean company Nova has an industrial plant dedicated to the production of enzymes, where they use different biological platforms to manufacture Xilanasa and Fitasa, which are enzymes used to improve digestion by monogastric livestock such as chickens and pigs (Table 3).
- Bio-herbicides—Bio-herbicides can reduce the use of chemical herbicides to prepare fields for mechanical harvest. Biotrop mentioned developing a bio-herbicide for broad leaf weed in Brazil.

4.3. Innovation Pathways

The description of the innovations in Section 4.2 revealed different efforts for technological development and levels of exclusivity in the types of commercial products, ranging from simpler technologies such as inoculants that are mastered by a large number of companies to new developments such as phytovaccines based on recombinant proteins that are being developed by a few companies. By crossing these two variables (technological developments and levels of exclusivity in the market), there are different pathways for technological development:

- Improving traditional products such as inoculants to compete in a large and consolidated market that has many competitors, small profit margins, but a large scale of sales. Some companies stand out in this sector, such as the Argentinean Rizobacter, which became a global leader in inoculants;
- Mastering new products such as bio-pesticides and bio-fertilizers to meet growing demand from more exclusive markets that have large profit margins but are still niche markets in some cases. For example, while nematicides used in soybean fields are mostly biological (since biological products are often more effective than chemical products to control nematodes), bio-insecticides are often used alongside chemical products to reduce the number of applications of chemical products. Targeting this market, the Brazilian company Vittia specialized itself in bio-pesticides based on micro- and macro-organisms.
- Developing innovations to conquer new and more exclusive technology-based markets by meeting existing demands for specific products. The Argentinean Protergium stands out in the segment of phytovaccines with two products based on recombinant proteins that are being registered.

Figure 3 also reveals that, considering the types of innovations developed by the interviewed companies, Argentinean companies tend to be grouped in the first cluster of companies (blue lines), while Brazilian companies tend to be grouped in a second cluster (green lines). Since the countries have different natural conditions (Brazil is a tropical country while Argentina ranges from subtropical in the north to polar in the south), this may reveal that the domestic companies focus their R&D on local conditions.

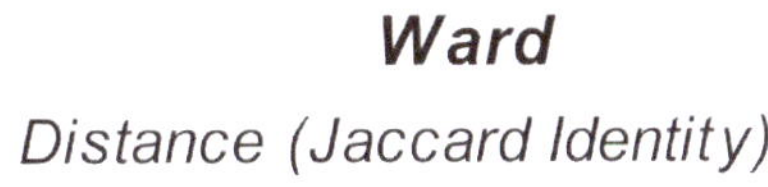

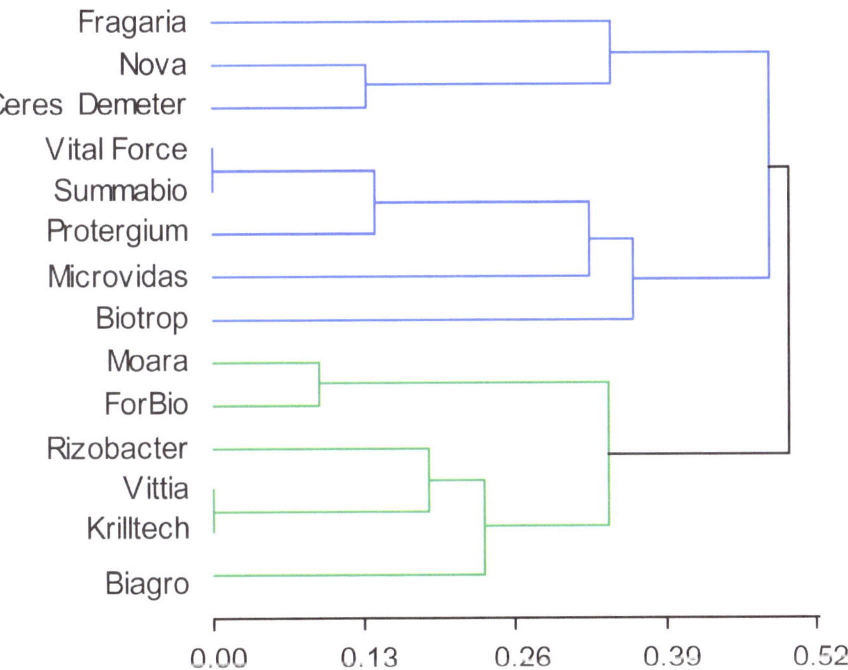

Figure 3. Cluster analysis of innovations developed by interviewed companies in Argentina and Brazil. The two main formed groups are in different colors.

5. Discussion

The publicly available official data revealed a large number of inoculants, bio-fertilizers, and bio-pesticides registered in Argentina and Brazil. However, since the process of having a product registered takes around three years, official data provide an important but not updated understanding of the existing innovations in the field of bio-inputs. Up-to-date information on commercial products' developments was provided by the interviews with companies investing in bio-input innovation in Argentina and Brazil. In addition to efforts to develop more efficient inoculants and to master the formulation of bio-fertilizers and bio-pesticides, the interviews revealed ongoing developments of new commercial bio-inputs such as phytovaccines and bio-herbicides, as well as specific metabolites isolated from microorganisms (instead of the alive microorganisms), products for livestock, and the use of industrial processes such as lyophilization in product formulation (Figure 4).

The literature reveals different generations of technological development in bio-inputs, ranging from inoculants based on a single isolated bacteria [6] to more recent products using a pool of microorganisms or technologies such as molecular biology [11]. The scientific literature also publicizes experimental developments on products such as bio-herbicides that can become the next generation of bio-inputs [19].

The results for Argentina and Brazil reveal that while some companies and research centers focus on improving the quality of existing products, such as inoculants, to compete in a large-scale existing market, others invest in developing new products targeting specific markets where exclusivity may guarantee greater profit margins. Therefore, by looking at the end of the product development pipeline, we found different parallel and contemporaneous technological races for product development that can lead to important progress in the bio-economy [27].

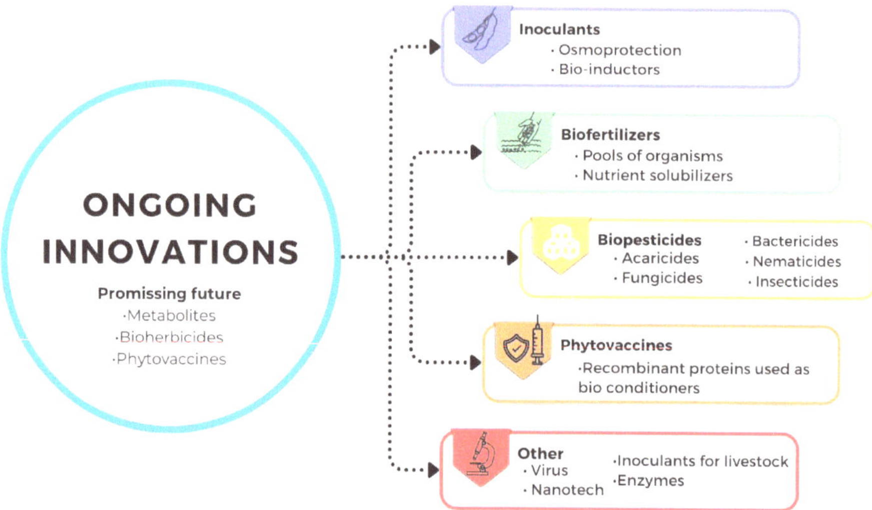

Figure 4. Summary of ongoing innovation fronts in agricultural bio-inputs taking place in Argentina and Brazil.

Consequently, we do not have new generations of bio-inputs replacing the old generations as if in a technological continuum of new technologies taking the place of the old ones, but we rather have parallel races for the improvement of existing products and the development of new products for different markets happening at the same time. The value for money and the efficacy in different field situations of these innovations in bio-inputs [6,12,15] will define the products that may remain on the market over time. Since any technological development and its adoption can be associated with potential risks, especially when living microorganisms are released into the environment, corporate accountability and continued academic research are fundamental for this sector in the long term.

These parallel races may potentially result in innovations that address market demands and can lead domestic companies in Argentina and Brazil to establish themselves in the market for agricultural bio-inputs. The rising global demand for agricultural products creates opportunities for investments in innovative agro-industrial sectors [31]. Since cooperation for innovation between private companies and research institutes is particularly beneficial for domestic high-tech companies [35], the bio-inputs sector can largely benefit from investments made by both private companies and public research centers [34].

The limitations of this study include the relatively small number of companies interviewed. Despite the importance of the bio-inputs sector in Argentina and Brazil, there are still no country-wide updated sources of information on the main sector trends that can guide assertive investments in product development and market expansion. Key market-related missing information includes:

- Main public innovation centers—List of the main research teams alongside the most important research centers (organizations/laboratories/researchers) and their recent developments on bio-inputs. This information is available for Argentina [24], but not for Brazil;

- Companies with commercial products—Description of companies with products registered and technologies currently available and under development (with their technology readiness levels—TRL);
- Bio-inputs market—Survey of the volume and values sold by type or group of products (such as bionematicides) and regions;
- Current adoption by farmers—Interview with a sample of farmers nationwide on adoption levels;
- Potential adoption by farmers—In-depth survey with farmers in a region where there are high adoption levels to identify the ideal/possible level of adoption per crop given the existing solutions in bio-inputs;
- On-farm adoption by farmers—Assessment of the on-farm production (on-farm bio-factories) sector, including its size (number of companies), distribution (by state), and adopting farmers' profiles.

6. Conclusions

The results of this study reveal ongoing development efforts to improve traditional agricultural bio-inputs such as inoculants, to master the formulation of new products such as bio-fertilizers and bio-pesticides, and to develop the next generation of bio-inputs such as phytovaccines and bio-herbicides. Private companies, in collaboration with public research centers, are leading innovations on these different fronts by targeting different market opportunities with potential for important contributions in terms of bio-economy growth. Taken together, these developments already provide more efficient commercial inoculants to improve the biological fixation of nitrogen in soybeans, as well as new inoculants for cereals such as wheat and corn and in fruits and vegetables. Bio-fertilizers made of pools of microorganisms resulted in commercial products to induce plant growth and solubilize soil nutrients. Bio-pesticides are also available on the market as bio-acaricides, bio-fungicides, bio-bactericides, bio-nematicides, and bio-insecticides.

Ongoing developments promise the next generation of agricultural bio-inputs, such as phytovaccines to activate the plant's immune system and bio-herbicides to potentially replace synthetic chemical herbicides in preparing fields to be harvested. The use of industrial processes such as lyophilization also allows the formulation of powder instead of liquid commercial products, which reduces the logistical costs.

Domestic stakeholders have an important role in these technological developments. The official records show that both foreign multinationals and domestic private companies lead in the number of products registered in Argentina and Brazil. The survey conducted with the domestic companies reveals their leading role in the different innovation fronts described in this study that can lead companies in Argentina and Brazil to establish themselves in the global market for agricultural bio-inputs.

Potentially disruptive innovations such as bio-herbicides, phytovaccines, and isolated metabolites extracted from microorganisms can underpin bio-economy growth globally. The sector of agricultural bio-inputs can be seen as an opportunity for developing countries to go beyond the primary production of agricultural commodities and build new agro-industrial capabilities for sustained agro-industrial growth.

Author Contributions: G.d.S.M.—Conceptualization; Data curation; Formal analysis; Funding acquisition; Investigation; Methodology; Validation; Visualization; Writing—original draft; Writing—review and editing. R.R.—Conceptualization; Formal analysis; Investigation; Methodology; Supervision. G.R.R.—Conceptualization; Formal analysis; Investigation; Methodology; Software; Supervision; Writing—review and editing. All authors have read and agreed to the published version of the manuscript.

Funding: This research received external funding from the Fundação de Apoio à Pesquisa do Distrito Federal (FAPDF), call DPG/UnB Nº 0004/2022, Programa de Bolsas de Pós-Doutorado no Exterior/FAPDF.

Institutional Review Board Statement: Not applicable since interviewees were informed that the interviews would not be recorded nor their names would be made available. All interviewees provided informed consent when taking part in the survey, following Argentina's and Brazil's ethics protocols for research with human beings.

Informed Consent Statement: Informed consent was obtained from all subjects involved in the study.

Data Availability Statement: The data presented in this study are available on request.

Conflicts of Interest: The authors declare no conflicts of interest.

References

1. Jacquet, F.; Jeuffroy, M.-H.; Jouan, J.; Le Cadre, E.; Litrico, I.; Malausa, T.; Reboud, X.; Huyghe, C. Pesticide-free agriculture as a new paradigm for research. *Agron. Sustain. Dev.* **2022**, *42*, 8. [CrossRef]
2. Bierbaum, R.; Leonard, S.A.; Rejeski, D.; Whaley, C.; Barra, R.O.; Libre, C. Novel entities and technologies: Environmental benefits and risks. *Environ. Sci. Policy* **2020**, *105*, 134–143. [CrossRef]
3. Palmisano, T. Narratives and practices of pesticide removal in the Andean valleys of Chile and Argentina. *Environ. Sci. Policy* **2023**, *139*, 149–156. [CrossRef]
4. Kaushik, B.; Kumar, D.; Shamim, M. *Biofertilizers and Biopesticides in Sustainable Agriculture*; Apple Academic Press: New York, NY, USA, 2019. [CrossRef]
5. Forbes Agro. O Mercado de Bioinsumos. Available online: https://forbes.com.br/forbesagro/2023/01/o-mercado-de-bioinsumos-vai-para-onde-no-brasil/ (accessed on 18 January 2023).
6. Elnahal, A.S.M.; El-Saadony, M.T.; Saad, A.M.; Desoky, E.-S.M.; El-Tahan, A.M.; Rady, M.M.; AbuQamar, S.F.; El-Tarabily, K.A. The use of microbial inoculants for biological control, plant growth promotion, and sustainable agriculture: A review. *Eur. J. Plant Pathol.* **2022**, *162*, 759–792. [CrossRef]
7. Amaresan, N.; Sankaranarayanan, A.; Kumar, M.; Druzhinina, I. *Advances in Trichoderma Biology for Agricultural Applications*; Springer International Publishing: Berlin/Heidelberg, Germany, 2022. [CrossRef]
8. Rojas, R.; Ávila, G.C.G.M.; Contreras, J.A.V.; Aguilar, C.N. *Biocontrol Systems and Plant Physiology in Modern Agriculture*; Apple Academic Press: New York, NY, USA, 2022. [CrossRef]
9. Mącik, M.; Gryta, A.; Frąc, M. Biofertilizers in agriculture: An overview on concepts, strategies and effects on soil microorganisms. *Adv. Agron.* **2020**, *162*, 31–87. [CrossRef]
10. Kumar, Y. Nanofertilizers and their role in sustainable agriculture. *Ann. Plant Soil Res.* **2021**, *23*, 238–255. [CrossRef]
11. Ke, J.; Wang, B.; Yoshikuni, Y. Microbiome Engineering: Synthetic Biology of Plant-Associated Microbiomes in Sustainable Agriculture. *Trends Biotechnol.* **2021**, *39*, 244–261. [CrossRef]
12. Garcia, M.V.C.; Nogueira, M.A.; Hungria, M. Combining microorganisms in inoculants is agronomically important but industrially challenging: Case study of a composite inoculant containing Bradyrhizobium and Azospirillum for the soybean crop. *AMB Express* **2021**, *11*, 71. [CrossRef]
13. Bercovich, B.A.; Villafañe, D.L.; Bianchi, J.S.; Taddia, C.; Gramajo, H.; Amalia Chiesa, M.; Rodríguez, E. Streptomyces eurocidicus promotes soybean growth and protects it from fungal infections. *Biol. Control* **2022**, *165*, 104821. [CrossRef]
14. Martins, S.J.; Pasche, J.M.; Silva, H.A.; Selten, G.G.; Savastano, N.; Abreu, L.; Bais, H.; Garrett, K.A.; Kraisitudomsook, N.; Pieterse, C.M.J.; et al. The Use of Synthetic Microbial Communities (SynComs) to Improve Plant Health. *Phytopathology* **2023**, *113*, 1369–1379. [CrossRef]
15. Patil, S.V.; Borase, H.P.; Salunkhe, J.D.; Suryawanshi, R.K. Isolation and Screening of Zinc Solubilizing Microbes: As Essential Micronutrient Bio-Inputs for Crops. In *Practical Handbook on Agricultural Microbiology*; Springer Protocols Handbooks; Springer: Berlin/Heidelberg, Germany, 2022; pp. 181–186. [CrossRef]
16. Hakim, S.; Naqqash, T.; Nawaz, M.S.; Laraib, I.; Siddique, M.J.; Zia, R.; Mirza, M.S.; Imran, A. Rhizosphere Engineering With Plant Growth-Promoting Microorganisms for Agriculture and Ecological Sustainability. *Front. Sustain. Food Syst.* **2021**, *5*, 617157. [CrossRef]
17. Polack, L.A.; Lecuona, R.E.; López, S.N. *Control Biológico de Plagas en Horticultura: Experiencias Argentinas de Las Últimas Tres Décadas*; Ediciones INTA: Buenos Aires, Argentina, 2020.
18. Sarkar, D.; Singh, S.; Parihar, M.; Rakshit, A. Seed bio-priming with microbial inoculants: A tailored approach towards improved crop performance, nutritional security, and agricultural sustainability for smallholder farmers. *Curr. Res. Environ. Sustain.* **2021**, *3*, 100093. [CrossRef]
19. Pouresmaeil, M.; Sabzi-Nojadeh, M.; Movafeghi, A.; Aghbash, B.N.; Kosari-Nasab, M.; Zengin, G.; Maggi, F. Phytotoxic activity of Moldavian dragonhead (*Dracocephalum moldavica* L.) essential oil and its possible use as bio-herbicide. *Process Biochem.* **2022**, *114*, 86–92. [CrossRef]
20. Hodson de Jaramillo, E.; Trigo, E.J.; Campos, R. The Role of Science, Technology and Innovation for Transforming Food Systems in Latin America and the Caribbean. In *Science and Innovations for Food Systems Transformation*; Springer International Publishing: Cham, Switzerland, 2023; pp. 737–749 [CrossRef]

21. Medina, G.d.S.; Rotondo, R.; Rodríguez, G.R. Agricultural Bio-Inputs as an Innovative Area of Opportunity for Agro-Industrial Growth in Developing Countries: Lessons from Argentina. *World* **2023**, *4*, 709–725. [CrossRef]
22. Goulet, F.; Hubert, M. Making a Place for Alternative Technologies: The Case of Agricultural Bio-Inputs in Argentina. *Rev. Policy Res.* **2020**, *37*, 535–555. [CrossRef]
23. Meyer, M.; Bueno, A.; Mazaro, S.; Silva, J. *Bioinsumos na Cultura da Soja*; Embrapa Soja: Londrina, Brazil, 2022.
24. Starobinsky, G.; Monzón, J.; Broggi, E.; Braude, H. *Bioinsumos para la Agricultura Que Demandan Esfuerzos de Investigación y Desarrollo*, 1st ed.; Ministerio de Desarrollo Productivo Argentina: Buenos Aires, Argentina, 2021.
25. Laibach, N.; Börner, J.; Bröring, S. Exploring the future of the bioeconomy: An expert-based scoping study examining key enabling technology fields with potential to foster the transition toward a bio-based economy. *Technol. Soc.* **2019**, *58*, 101118. [CrossRef]
26. Keswani, C.; Dilnashin, H.; Birla, H.; Singh, S.P. Obstacles in the Adaptation of Biopesticides in India. In *Bio#Futures*; Springer International Publishing: Cham, Switzerland, 2021; pp. 301–318. [CrossRef]
27. Oguntuase, O.J. Advancing a Framework for Entrepreneurship Development in a Bioeconomy. In *Handbook of Research on Nascent Entrepreneurship and Creating New Ventures*; IGI Global: Hershey, PA, USA, 2021; pp. 295–315. [CrossRef]
28. Das, S.; Ray, M.K.; Panday, D.; Mishra, P.K. Role of biotechnology in creating sustainable agriculture. *PLoS Sustain. Transform.* **2023**, *2*, e0000069. [CrossRef]
29. Ayilara, M.S.; Adeleke, B.S.; Babalola, O.O. Bioprospecting and Challenges of Plant Microbiome Research for Sustainable Agriculture, a Review on Soybean Endophytic Bacteria. *Microb. Ecol.* **2023**, *85*, 1113–1135. [CrossRef]
30. Goulet, F. Characterizing alignments in socio-technical transitions. Lessons from agricultural bio-inputs in Brazil. *Technol. Soc.* **2021**, *65*, 101580. [CrossRef]
31. Cruz, J.E.; Medina, G.d.S.; Júnior, J.R.d.O. Brazil's Agribusiness Economic Miracle: Exploring Food Supply Chain Transformations for Promoting Win–Win Investments. *Logistics* **2022**, *6*, 23. [CrossRef]
32. OECD. The Measurement of Scientific, Technological and Innovation Activities. In *Oslo Manual 2018*; OECD: Paris, France, 2018. [CrossRef]
33. Leonardi, V.; García Casal, I.; Cristiano, G. Desempeño innovador de un grupo de Mipymes agroindustriales argentinas. *Econ. Soc.* **2009**, *14*, 45–64.
34. Sili, M.; Dürr, J. Bioeconomic Entrepreneurship and Key Factors of Development: Lessons from Argentina. *Sustainability* **2022**, *14*, 2447. [CrossRef]
35. Tessarin, M.; Suzigan, W.; Guilhoto, J.J.M. Cooperação para inovar no Brasil: Diferenças segundo a intensidade tecnológica e a origem do capital das empresas. *Estud. Econ.* **2020**, *50*, 671–704. [CrossRef]
36. Morceiro, P.C.; Tessarin, M.S.; Guilhoto, J.J.M. Produção e uso setorial de tecnologia no Brasil. *Econ. Apl.* **2022**, *26*.
37. OECD. Agricultural Policy Monitoring and Evaluation. In *Agricultural Policy Monitoring and Evaluation 2022*; OECD: Paris, France, 2022. [CrossRef]
38. Rana, S.; Kapoor, S.; Rana, A.; Dhaliwal, Y.S.; Bhushan, S. Industrial Apple Pomace as a Bioresource for Food and Agro Industries. In *Sustainable Agriculture Reviews 56*; Springer: Cham, Switzerland, 2021; pp. 39–65. [CrossRef]
39. Ekielski, A.; Żelaziński, T.; Mishra, P.K.; Skudlarski, J. Properties of Biocomposites Produced with Thermoplastic Starch and Digestate: Physicochemical and Mechanical Characteristics. *Materials* **2021**, *14*, 6092. [CrossRef]
40. Senasa. Registro de Productos Fertilizantes, Enmiendas y Otros. Servicio Nacional de Sanidad y Calidad Agroalimentaria—Dirección de Tecnología de la Información. Available online: https://aps2.senasa.gov.ar/vademecumFertilizantes/app/publico (accessed on 18 January 2023).
41. Senasa. Registro Nacional de Terapéutica Vegetal. Servicio Nacional de Sanidad y Calidad Agroalimentaria—Dirección de Tecnología de la Información. Available online: https://aps2.senasa.gov.ar/vademecum/app/publico/formulados (accessed on 18 January 2023).
42. MAPA. Agrofit—Sistema de Agrotóxicos Fitossanitários. MAPA (Ministério da Agricultura Pecuária e Abastecimento). Available online: https://agrofit.agricultura.gov.br/agrofit_cons/principal_agrofit_cons (accessed on 17 January 2023).
43. Petrovan, S.O.; Aldridge, D.C.; Bartlett, H.; Bladon, A.J.; Booth, H.; Broad, S.; Broom, D.M.; Burgess, N.D.; Cleaveland, S.; Cunningham, A.A.; et al. Post COVID-19: A solution scan of options for preventing future zoonotic epidemics. *Biol. Rev.* **2021**, *96*, 2694–2715. [CrossRef]
44. Orr, A.; Ahmad, B.; Alam, U.; Appadurai, A.; Bharucha, Z.P.; Biemans, H.; Bolch, T.; Chaulagain, N.P.; Dhaubanjar, S.; Dimri, A.P.; et al. Knowledge Priorities on Climate Change and Water in the Upper Indus Basin: A Horizon Scanning Exercise to Identify the Top 100 Research Questions in Social and Natural Sciences. *Earth's Futur.* **2022**, *10*, e2021EF002619. [CrossRef]
45. Di Rienzo, J.A.; Casanoves, F.; Balzarini, M.G.; Gonzalez, L.; Tablada, M.; Robledo, C. *INFOSTAT Version 2011*; Universidad Nacional de Córdoba: Córdoba, Spain, 2016.

 *sustainability*

Article

The Spatiotemporal Impact of Digital Economy on High-Quality Agricultural Development: Evidence from China

Qi Li and Zhijiao Liu *

School of Economics and Business Administration, Heilongjiang University, Harbin 150080, China; 2220082@s.hlju.edu.cn
* Correspondence: 2018035@hlju.edu.cn; Tel.: +86-132-1280-7882

Abstract: China's high-quality economic development is strongly supported by the high-quality development of agriculture, and the digital economy has emerged as a key driver for promoting shared prosperity and high-quality economic development. Against this backdrop, investigating the connection between high-quality agricultural development and the digital economy holds significant importance. This study utilized the entropy-weighted TOPSIS model to evaluate comprehensive evaluation indicators of the two according to panel data from 30 provinces in China between 2011 and 2021. Subsequently, GIS spatial analysis and exploratory spatial data analysis (ESDA) were employed to investigate the spatiotemporal evolution features and spatial correlations. Finally, the spatiotemporal geographically weighted regression (GTWR) model was constructed to examine the spatiotemporal impact of the digital economy on the advancement of high-quality agricultural growth. The results indicate that: (1) from 2011 to 2021, China's high-quality agricultural development and digital economy both demonstrated a general increasing trend. In terms of spatial distribution, there were significant spatial variations, with a general trend of "Southeast is higher, whereas the Northwest is lower". The regions with significant value were primarily clustered in the coastal areas in the east and several provincial capitals. (2) Both of the two exhibited significant global spatial self-correlation, and there were also significant spatiotemporal clustering effects in high-quality agricultural growth, gradually forming a high-value cluster centered around Shanghai and a low-value cluster centered around western provinces. (3) The digital economy positively influences the enhancement of high-quality agricultural development, demonstrating notable spatial and temporal heterogeneity. In contrast to the southeastern areas, the influence is more pronounced in the northern and central-western areas.

Keywords: digital economy; high-quality agricultural development; spatiotemporal effects; GTWR model

1. Introduction

China stands as a prominent global agricultural force, with agriculture holding considerable importance in the nation's economic landscape. "The White Paper on China's Food Security" pointed out that China has undergone a significant transition from hunger to sufficiency and then to moderate prosperity. In 2022, China achieved an overall grain output of 687 million tons [1], and some key agricultural products have consistently held the top position globally in terms of production. Despite China's remarkable achievements in agriculture, it finds itself in a dual predicament of low efficiency, low innovation drive, and environmental pollution. The high agricultural inputs are accompanied by low grain yields which are far below those of European and American developed nations [2]. The application of synthetic fertilizers not only seriously pollutes the environment but also accelerate the depletion of resources [3]. Traditional production factors, agricultural production techniques, and agricultural development models can no longer provide sustainable momentum for achieving high-quality agricultural development. Insufficient innovation

Citation: Li, Q.; Liu, Z. The Spatiotemporal Impact of Digital Economy on High-Quality Agricultural Development: Evidence from China. *Sustainability* **2024**, *16*, 2814. https://doi.org/10.3390/su16072814

Academic Editors: Fotios Chatzitheodoridis, Efstratios Loizou and Achilleas Kontogeorgos

Received: 15 February 2024
Revised: 24 March 2024
Accepted: 25 March 2024
Published: 28 March 2024

drive in agriculture and inadequate international competitiveness of agricultural products have gradually become important factors restricting the modernization of agriculture and the overall revitalization of the countryside. It is imperative to investigate new paths to foster high-quality agricultural growth.

The digital economy has emerged as a key driver for promoting shared prosperity and high-quality economic development. The Chinese government has repeatedly emphasized the necessity of implementing rural digital development initiatives, expediting the creation of digital application scenarios, and accelerating the growth of "smart agriculture". In the last few years, China has been committed to implementing the digital agricultural growth strategy, and modern agriculture is moving towards digitization. Agricultural digitization has emerged as a significant factor contributing to high-quality agricultural growth and providing a new impetus for rural economic development [4,5]. Against this backdrop, studies that explore the connection between the digital economy and high-quality agricultural development are both theoretically and practically significant.

2. Literature Review

There is a relatively abundant body of studies related to this topic, which mainly focuses on three aspects. One aspect is the connotation, spatial patterns, and economic effects of the digital economy. "Digital economy" is a recently emerged economy that has been shaped by advancements in information and communication technologies (ICTs). Since the formal proposal of the idea of the "digital economy" by Tapscott [6], scholars both domestically and internationally have engaged in plenty of discussions on different aspects, leading to a substantial body of studies. In relation to the connotation, Tapscott suggests that its prominent features include digitization, the fact that it is knowledge-based, virtualization, and interconnectedness. In terms of spatial patterns, there are notable regional disparities, primarily because of variations in available resources and digital infrastructure [7–10]. Research has found that China's digital economy has evolved from an initial stage of agglomeration to a more balanced and diffused development pattern. Regarding economic effects, Zhang proposed that it promotes the modernizing of the industrial framework and enhances industrial economy's technological content [11]. Zhang and colleagues suggested that it is beneficial for top-notch economic growth and shared prosperity. It can enhance resource allocation efficiency to promote top-notch economic growth [12,13]. Cao proposed that it can encourage top-notch industrial growth through new models such as networked collaboration and personalized customization [14]. Su and colleagues pointed out that it offers strong support for environmentally conscious development of the tourism industry and for promoting energy-saving, low-carbon, and top-notch development of the tourism sector [15].

The second aspect involves the investigation of high-quality agricultural development. The existing literature has explored it from two perspectives. Firstly, researchers have primarily focused on constructing evaluation systems and measuring indicators. A key metric for assessing the rise of a nation's wealth is total factor productivity (TFP), which is especially important in developing nations [16–18]. It has become a crucial metric for measuring superior agricultural development. However, TFP in agriculture primarily considers capital and labor factors while neglecting environmental factors related to the growth of sustainable agriculture. Therefore, it is unable to adequately capture the environmental advantages of high-quality agricultural development [19]. Secondly, consider the natural motivations and pathways leading to high-quality agricultural development. Currently, there is abundant research in the theoretical field. This research includes studies on the incremental inputs of key resources like land and labor [20], technological innovation, and digital finance [21]. These factors are all important in influencing high-quality agricultural development. Furthermore, innovation in agricultural technology is essential for achieving high-quality agricultural advancement [22,23]. Informatization is a key driver of technological innovation, and agricultural informatization helps to improve the TFP in agriculture [24].

The third aspect involves studying the connection between the digital economy and advanced agricultural development. With the swift advancement of new-generation information technologies, the digital economy is constantly empowering agriculture and becoming a fresh source of power and momentum for high-quality agricultural development [25,26]. Scholars such as Zhou and Li have suggested that the digital economy has the potential to overcome the constraints of time and space in exchanging information, thus facilitating the integration of agricultural resources and thereby improving agricultural productivity and development [27,28]. Zhang and colleagues suggested that advancements in agricultural technology within the digital economy could enhance overall green productivity in agriculture [29]. Wang and colleagues discovered that digital finance could enhance the process of high-quality agricultural development, but there are dual thresholds for the impact of the former on the latter [30]. Tang Y. highlighted the importance of agricultural digitalization in achieving the goals of high-quality agricultural advancement [31].

The significance of leveraging the digital economy to enhance the quality of agricultural development has been acknowledged and underscored. A comprehensive examination of the relationship between the two from an empirical perspective has been fairly thorough. However, the current body of literature is deficient in novel methodologies and innovative viewpoints for investigating the relationship between the two entities. Currently, there has been no utilization of ArcGIS spatial visualization techniques and GTWR models to investigate the spatiotemporal impact effects of the two. Hence, examining the influence of the digital economy on the advancement of high-quality agriculture from both temporal and spatial viewpoints, bears significant practical and theoretical implications. The study will utilize data collected from 30 provinces in China between 2011 and 2021. The study aims to investigate the spatiotemporal distribution patterns and effects of the two in China by utilizing the entropy-weighted TOPSIS model, geographic information system (GIS) spatial analysis, and the GTWR model. The objective is to further clarify the relationship between the two and provide valuable insights for achieving high-quality agricultural development in the context of agricultural and rural modernization.

3. Model Construction and Indicator Description

3.1. Entropy-Weighted TOPSIS

The entropy-weighted TOPSIS method is used to evaluate the levels of different subsystems and the overall levels of digital economy and high-quality agricultural development. This method integrates the entropy weight method with the TOPSIS model to offer a comprehensive evaluation approach. The integration of both methods can mitigate the challenges of exclusively relying on the entropy method for ranking evaluation outcomes and the subjectivity concerns linked to the exclusive use of the TOPSIS model. This integration enhances the objectivity and comparability of research findings. The principle entails the application of the entropy weight method to standardize indicators of each subsystem, determining the weights of each evaluation indicator in the overall indicator system. Subsequently, the TOPSIS method is employed to calculate the comprehensive scores of the evaluation subjects. It is a frequently employed multi-criteria decision-making analysis technique. It is well-suited for comparative studies involving multiple scenarios and objects, aiming to determine the optimal choice or the most competitive object. The evaluation process involves ranking objects by assessing the distance between each object and the ideal and worst solutions, thereby providing a comprehensive score. An object is deemed optimal when it is in proximity to the ideal solution and distant from the worst solution; otherwise, it is considered non-optimal. The subsequent stages outline the exact steps involved in the calculation.

(1) The entropy weighting technique is employed to determine the weights of the indicators. The variable m denotes the sample size, while n represents the number of evaluation indicators. Additionally, $k = 1/ln(m)$. Initially, the entropy values of each indicator are calculated using Equation (1), where "y_i" represents the value of the j-th indicator in the i-th sample, and e_j denotes the entropy value of each indicator, with

$0 \leq e_j \leq 1$. Subsequently, the weights of each indicator are calculated based on the entropy values e_j of each indicator, as shown in Equation (2), where W_j represents the weight of the j-th indicator.

$$e_j = -k \sum_{i=1}^{m} \left[\left(y_{ij} / \sum_{i=1}^{m} y_{ij} \right) \times \ln \left(y_{ij} / \sum_{i=1}^{m} y_{ij} \right) \right] \tag{1}$$

$$W_j = (1 - e_j) / \sum_{j=1}^{n} (1 - e_j) \tag{2}$$

(2) The Euclidean distance is calculated, representing the distance of evaluation indicators from the ideal solution D_i^+ and the worst solution D_i^-, as shown in Equations (3) and (4), where v_{ij} denotes the weighted normalized matrix of each indicator ($v_{ij} = W_j y_{ij}$), v_j^+ denotes the ideal solution ($v_{ij} = W_j y_{ij}$), and v_j^- denotes the worst solution ($v_{ij} = W_j y_{ij}$).

$$D_i^+ = \sqrt{\sum_{j=1}^{m} \left(v_{ij} - v_j^+ \right)^2} \tag{3}$$

$$D_i^- = \sqrt{\sum_{j=1}^{m} \left(v_{ij} - v_j^- \right)^2} \tag{4}$$

(3) The comprehensive evaluation index is calculated, specifically measuring the relative closeness of each evaluation object to the optimal solution.

$$N_i = D_i^- / \left(D_i^+ + D_i^- \right) \tag{5}$$

In the equation, N_i denotes the comprehensive score, which refers to the comprehensive evaluation index. The value of N_i ranges from 0 to 1; if N_i approaches 1, it signifies that the indicator is closer to the optimal value. Conversely, if N_i approaches 0, it suggests that the indicator is closer to the worst level.

3.2. Exploratory Spatial Data Analysis (ESDA)

ESDA is used to observe the correlation and variation of phenomena in space and involves both global and local aspects. The Global Moran's I is used to measure the dispersion or clustering degree among all spatial units in the entire study area. The formula is as follows:

$$I_G = \frac{\sum_{i=1}^{n} \sum_{j=1}^{n} w_{ij}(x_i - \bar{x})(x_j - \bar{x})}{S^2 \sum_{i=1}^{n} \sum_{j=1}^{n} w_{ij}} \tag{6}$$

In the equation, I_G represents the Global Moran's index, n represents the number of cities in the study, x_i and x_j, respectively, represent the attribute values of cities i and j, $\bar{x}$ represents the average value of the attribute, S^2 represents the sample variance of the attribute values of the research units, and w_{ij} represents the inverse distance spatial weight matrix.

In the local autocorrelation test, the $LISA$ statistic is introduced to measure the spatial clustering features of different samples in space, as follows:

$$LISA = \frac{X_i - \bar{X}}{S^2} \sum_{j=1}^{n} \left[w_{ij}(X_j - \bar{X}) \right] \tag{7}$$

3.3. GTWR Model

Huang [32] first proposed the use of the spatiotemporal geographic weighted regression model (GTWR) for modeling spatiotemporal changes in house prices. Compared with the traditional geographic weighted regression model [33], the GTWR model incorporates a time dimension, which can better handle the non-stationarity of "time–space" and produce more effective estimation results [34]. Therefore, this paper adopts the GTWR model to analyze the impact in different spatiotemporal dimensions. The model is as follows:

$$Y_i = \beta_0(u_i, v_i, t_i) + \sum_{k=1}^{p} \beta_k(u_i, v_i, t_i) X_{ik} + \varepsilon_i \tag{8}$$

In the equation, Y_i represents the observed value. (u_i, v_i) represents the latitude and longitude of the i-th observation point. t_i represents the time parameter. (u_i, v_i, t_i) represents the spatiotemporal coordinates of the i-th observation point. β_0 represents the regression constant. $\beta_k(u_i, v_i, t_i)$ symbolizes the k-th independent variable's regression coefficient at the i-th observation point. ε_i represents the residual.

The core elements of the GTWR model are the selection of the spatiotemporal weight matrix and bandwidth. The spatiotemporal weight matrix provides an estimation for each spatiotemporal location of the observation point "i" and the independent variable "k", as illustrated in the following formula:

$$\hat{\beta}(u_i, v_i, t_i) = \left[X^T W(u_i, v_i, t_i) X^{-1} \right] X^T W(u_i, v_i, t_i) Y \tag{9}$$

In the equation, $\hat{\beta}(u_i, v_i, t_i)$ symbolizes the estimated value of $\beta_k(u_i, v_i, t_i)$; $W(u_i, v_i, t_i)$ symbolizes the spatiotemporal weight matrices; X^T presents the transposition of the matrices; and Y represents the matrices composed of observed values.

This study employs the bi-square spatial weighting function, a finite Gaussian function, to mitigate the "long tail effect" resulting from data dispersion. The model is presented as follows:

$$W_{ij}^{ST} = \begin{cases} \left[1 - \left(\frac{d_{ij}^{ST}}{b_i} \right)^2 \right]^{-2}, d_{ij}^{ST} \leq b_i \\ 0, d_{ij}^{ST} > b_i \end{cases} \tag{10}$$

In the equation, W_{ij}^{ST} represents the spatiotemporal weight matrices produced from the b$_i$-square spatial weight function, and d_{ij}^{ST} represents the spatiotemporal distance between observation point "i" and observation point "j". The formula is as follows: $\sqrt{\delta\left[(u_i - u_j)^2 + (v_i - v_j)^2 \right] + \mu(t_i - t_j)^2}$. Clearly, the selection of the bandwidth "b" in the formula will have a big impact on how the spatiotemporal weight matrix is established, and this paper utilizes the AICc criterion method to determine an adaptive bandwidth.

3.4. Indicator Description

3.4.1. Dependent Variable

The comprehensive assessment index system for promoting high-quality agricultural development is built upon five key development principles: creativity, collaboration, sustainability, openness, and innovation [35], as well as the viewpoints of improving quality and efficiency and integrating industries, drawing from relevant research. The system consists of five dimensions: basic development, quality improvement, integrated development, new quality development, and shared development (Table 1). Subsequently, the level of high-quality agricultural growth was assessed using the TOPSIS approach for 30 Chinese provinces between 2011 and 2021.

Table 1. System of evaluation indicators for high-quality development of agriculture.

Primary Indicators	Secondary Indicators	Indicator Description	Characteristics	Weight of Indicators
Basic development	Grain yield per unit area	Total area of grain cultivation/seeding area	+	0.01509
	Level of agricultural mechanization	Total agricultural machinery power/arable land area	+	0.04114
Quality improvement	Labor productivity	The combined value of animal husbandry, forestry, and agriculture/number of primary industry employees	+	0.03242
	Land productivity	Total value of agricultural production/crop sown area	+	0.04593
	Rural per capita income	Average income per person in rural family	+	0.0441
	Index of agricultural industrial structure adjustment	Value of agricultural products/the combined value of animal husbandry, forestry, and agriculture	+	0.00605
	Engel coefficient of rural residents	Engel coefficient of rural populations	−	0.01028
	Urban/rural income ratio	Urban dwellers' per capita disposable income/rural dwellers' per capital disposable income	−	0.01192
	Urban/rural consumption ratio	Urban dwellers' per capita consumption expenses/rural dwellers' per capita consumption expenses	−	0.01106
Integrated development	The proportion of added value of the tertiary industry	Value added of the tertiary industry/value added of the primary industry	+	0.31082
	Proportion of agricultural product processing industry	The agricultural product manufacturing sector's operating income/value of agricultural products	+	0.11777
	Proportion of agriculture, forestry, animal husbandry	The combined value of animal husbandry, forestry, and agriculture/the combined value of primary sector	+	0.07435
New quality development	Degree of plastic film usage in agriculture	Amount of plastic film used in agriculture/crop sown area	−	0.0058
	Level of fertilizer usage in agriculture	Amount of fertilizer used in agriculture/crop sown area	−	0.00937
	Level of pesticide usage	Quantity of pesticide applied/crop sown area	−	0.01644
	Forest coverage rate	Forest area/cultivated land area	+	0.04131
	Rate of efficient irrigation	Area of efficient irrigation/cultivated land area	+	0.03811
Shared development	Local financial education level	Local fiscal education expenditure	+	0.04487
	Financial support for agriculture	Expenditure on agricultural, forestry, and water affairs	+	0.03527
	Level of agricultural openness	Total agricultural product import and export trade/GDP	+	0.08789

Note: "+" indicates a positive indicator, "−" indicates a negative indicator.

In terms of the basic development dimension, a solid agricultural foundation is an essential and important support. The stability, efficiency, and sustainable development of the agricultural sector cannot be achieved without the development and application of revolutionary cutting-edge technologies. With the trend of modern agricultural production being dominated by large-scale land use, modern mechanical equipment has begun to be widely utilized [36]. Therefore, this study integrates the degree of mechanization in addition to the per unit area grain yield to jointly measure the level of agricultural basic development.

In terms of the quality improvement dimension, development must consider both quality and efficiency. This dimension includes improving the living standards of rural residents and promoting agricultural innovation. Therefore, this study utilizes rural residents' net income and the rural Engel coefficient to depict the production and living conditions of farmers. Additionally, the urban/rural income ratio and urban/rural consumption ratio are chosen to assess the urban/rural development gap [37]. The study also incorporates the agricultural industry structure adjustment index, labor productivity, and land productivity to assess agricultural innovation.

In terms of the integrated development dimension, the integration of agricultural industries is indispensable [38]. The deep integration of agriculture with industries such as ecology, culture, tourism, and elderly care represents an inevitable choice. Deep integration refers to efficient integration across various agricultural stages by widening the industrial chain, extending vertically, making the industry multifunctional, and aggregating factors based on agricultural production [39,40]. It enhances the effectiveness, sustainability, and competitiveness of agricultural production by combining the output of the primary sector, the processing and sales of the secondary industry, and the services of the tertiary sector. This study employs the ratio of the output value of the agriculture, forestry, animal husbandry, and fishery service industry, the output value proportion of the agricultural product processing industry, and the added value of the tertiary industry as indicators to quantify agricultural industry integration.

In terms of the new quality development dimension, this study integrates agricultural green and sustainable development and introduces the concept of "new quality development". The precondition for the green development of agriculture is to avoid excessive consumption of resources and environmental pollution. Hence, the study employs the per unit area of plastic film usage as a metric to assess resource consumption levels. (Agricultural plastic films require significant amounts of raw materials such as petroleum, natural gas, and water in their production processes, potentially resulting in considerable resource depletion. During the utilization phase, inadequate management practices may result in wastage and the frequent replacement of plastic films, leading to resource depletion). Additionally, the measurement of environmental pollution levels includes the assessment of fertilizer and pesticide usage per unit area. The sustainable development of agriculture emphasizes the effective management and sustainable utilization of natural resources. This approach aims to prevent the overuse and wastage of resources, thereby guaranteeing the long-term stability of resource availability. Hence, this study opts to utilize forest coverage and the effective irrigation rate as indicators to assess the degree of sustainable agricultural resource utilization.

The shared development dimension includes both domestic shared development and international shared development. For domestic agricultural shared development, local financial resources directed towards agricultural education expenditure and providing financial support to the agricultural sector are considered. The intensity of financial support for agriculture reflects the strength of national financial investment [41], and it is an important policy for supporting agricultural development, as well as a significant approach to addressing the shortage of funds for agricultural production among farmers. For international shared development, the level of agricultural foreign trade is used to reflect the influence of a country's agriculture and its level of openness. This paper selects the proportion of total agricultural import and export trade to measure this.

3.4.2. Explanatory Variables

According to previous studies, considering data availability and continuity, this study has chosen three dimensions—digital industrial development, digital infrastructure, and digital inclusive finance—to establish a comprehensive evaluation system for the digital economy (Table 2). This study employed the TOPSIS approach to obtain the thorough index of the digital economy for every province and city. The Digital Inclusive Finance Index incorporates the Inclusive Finance Index, a collaborative effort between Peking University Digital Finance Research Center and Ant Financial Group [42]. This dataset provides three secondary indicators: coverage breadth, depth of usage, and level of digitization. It is widely acknowledged as a reputable source of data in China.

Table 2. System of evaluation indicators for the digital economy.

Primary Indicators	Secondary Indicators	Characteristics	Weight of Indicators
Digital infrastructure	Ports available for broadband internet connectivity (in tens of thousands)	+	0.02861
	Long-distance optical cable length (in ten thousand kilometers)	+	0.01764
	Rate of mobile phone ownership per one hundred individuals	+	0.01178
	Domain name count (in ten thousands)	+	0.06488
Digital industrialization	Total postal business volume (in CNY 100 millions)	+	0.09503
	Total telecommunications business volume (in CNY ten billions)	+	0.05837
	Scale of software product revenue	+	0.09107
	Number of workers in the information service sector (in ten thousands)	+	0.05252
	Expenditure on research and development (in CNY ten billions)	+	0.48609
	Number of patent applications (in thousands)	+	0.06043
Inclusive digital finance	Extent of inclusive digital finance coverage	+	0.01268
	Depth of inclusive digital finance usage	+	0.01113
	Degree of digitization of inclusive digital finance	+	0.00978

Note: "+" indicates a positive indicator.

3.4.3. Selection of Control Variables

The following control variables are included: (1) urbanization level (ur), determined by the ratio of urban population to total population; urbanization provides support for the advancement of agriculture in terms of market, technology, and funding, creating objective conditions for the process of modernizing and improving agriculture. (2) Rural education level (edu), determined by the average number of years of education among rural dwellers. (3) The industrial structure level (jgh) is indicated by the ratio of value added from the tertiary industry to the value added from the secondary industry. Upgrading the industrial structure promotes the optimization and adjustment of the agricultural production structure and increases the demand for higher-quality agricultural products. (4) Level of openness (open) is represented by the proportion of the actual total import and export volume of each province to the regional gross domestic product. It can promote regional economic prosperity and accelerate rural agricultural modernization. (5) The old-age dependency (old) refers to the severe shortage of human capital caused by the aging of the rural workforce, which will hinder the further development of rural industries.

3.4.4. Source of Data

Given the importance of data accessibility and timeliness, this study opted to utilize panel data from 30 Chinese provinces between 2011 and 2021 as the sample for analysis. The data primarily originate from the annual publications such as the *China Rural Statistical Yearbook, China Agricultural Products Processing Industry Yearbook, China Statistical Yearbook, China Agricultural Yearbook,* as well as from various provincial statistical yearbooks. Missing data were approximated through the utilization of the mean annual growth rate and interpolation.

4. Results and Analysis

4.1. The Spatiotemporal Evolution of the Digital Economy and the High-Quality Development of Agricultural Economy

4.1.1. Time Series Feature Analysis

The time series chart (Figure 1) depicted in this study illustrates the comprehensive evaluation index of the digital economy and high-quality agriculture for 30 provinces. The index was calculated using the entropy-weighted TOPSIS method. The line graph illustrates the aggregated composite scores calculated by averaging the composite scores of the 30 provinces, while the bar graph depicts the composite scores of different sub-regions obtained by averaging the composite scores of the corresponding provincial regions. The data presented in Figure 1 illustrate a continuous upward trend in China's digital economy from 2011 to 2021. The comprehensive score of the digital economy increased from 0.053 in 2011 to 0.187 in 2021, indicating a relatively rapid growth rate. Furthermore, a consistent upward trend has been noted in various regions. Particularly noteworthy is the higher level in the eastern regions compared to other areas annually, indicating a geographical distribution pattern of "East > Central > West > Northeast". The phenomenon can be attributed to the larger economic scale and faster economic growth rate in the eastern region coupled with policy incentives aimed at fostering the digital revolution and expanding the digital industry.

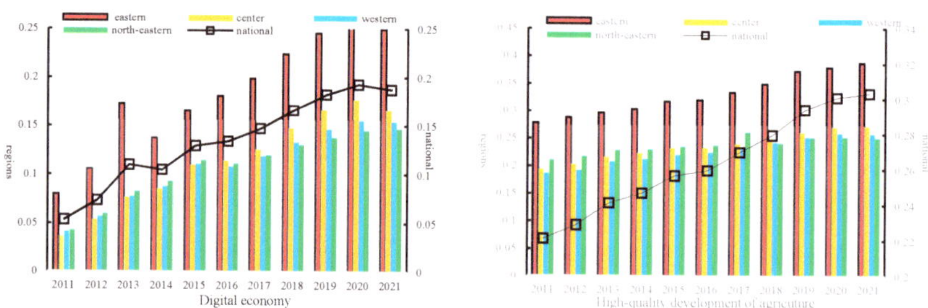

Figure 1. Value of the digital economy and high-quality development of agriculture, 2011–2021.

From Figure 1, it is evident that China's level of high-quality agricultural development has been steadily rising between 2011 and 2021. The level in various regions has also shown an overall stable upward trend; however, there are significant regional variations: the level in the eastern region surpasses that in other regions, suggesting a high quality of agricultural development. Conversely, the western and northeastern regions exhibit levels below the national average, indicating a lower quality of agricultural development. Possible reasons for this include the following: (1) the accelerated pace of agricultural advancement in the eastern regions has resulted in a more pronounced reduction in arable land area compared to the western regions. This situation necessitates an improvement in efficiency and yield to compensate for losses, consequently fostering a higher standard of agricultural development; (2) provinces with high-quality agricultural development have received better financial and technological support, enabling more efficient use of resources, accelerating agricultural transformation, promoting industrial structure adjustment, and continuously improving industrial efficiency and productivity; (3) provinces with low-quality agricultural development face different limiting factors, with some provinces lacking in environmental resources, while others, despite abundant resources, have relatively weak production efficiency and technological innovation.

Through comparative analysis, it was found that there is a high degree of consistency in the temporal characteristics of digital economy and high-quality development of agriculture between the country and various regions. Therefore, a preliminary prediction can be made: the digital economy possesses the possibility to drive agriculture towards high-quality development.

4.1.2. Spatial Distribution Characteristics Analysis

Utilizing ArcGIS and the natural breaks method to visualize the levels of digital economy and high-quality agricultural development in 2011, 2016, and 2021 allows for an intuitive observation of the spatial agglomeration characteristics and evolutionary trends in different provinces, as depicted in Figures 2 and 3.

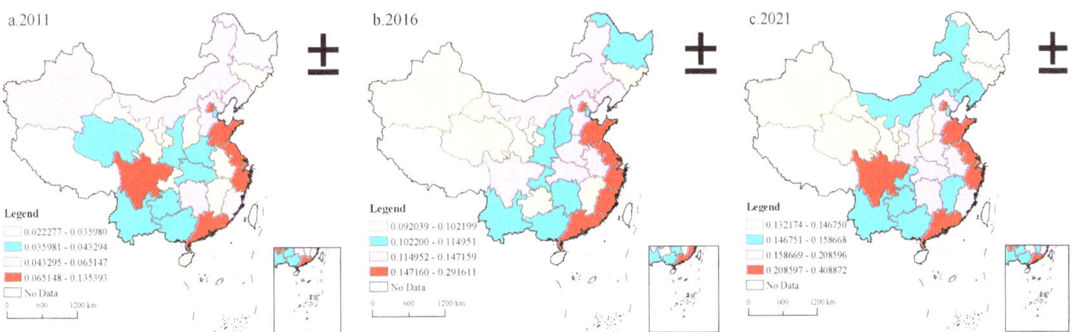

Figure 2. Spatial distribution of the digital economy.

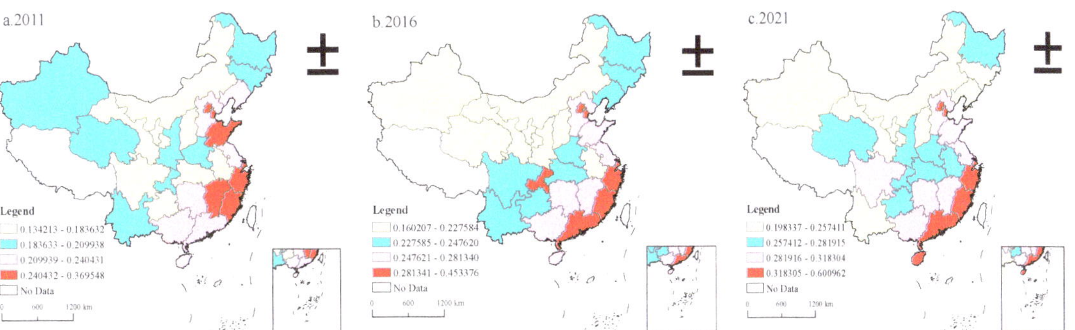

Figure 3. Spatial distribution of the high-quality development of agriculture.

In terms of the spatial distribution of the digital economy, significant regional disparities are evident among the eastern, central, and western regions. In 2011, it showed significantly greater progress in the eastern areas compared to other areas, with the northeastern and northwestern areas surpassing the central and western areas. Areas with low levels of it were primarily concentrated in the central and western regions. From 2016 to 2021, there was a consistent pattern of gradual decline observed from the coastal areas in the east towards the northwestern and northeastern regions. Specifically, high-value areas were primarily concentrated in the eastern coastal areas and sporadically dispersed in the central and western areas. Shanghai, Beijing, Zhejiang Province, and Fujian Province consistently maintained their positions in the top five rankings. Low-value regions were mostly centered on the northwestern and northeastern areas, like Xinjiang Province, Gansu Province, Qinghai Province, Jilin Province, and Heilongjiang Province, which consistently maintained relatively low levels.

High-quality agricultural development showed a spatial distribution pattern with high levels in the southeastern coastal areas and low levels in the central, western, and northeastern regions, and the distribution of high- and low-value areas demonstrated relatively stable changes over time. In terms of geographical distribution, Shanghai, Beijing, Fujian Province, Zhejiang Province, and Tianjin consistently maintained their status as high-value areas. The surrounding radiation areas, centered around Hunan Province and Shandong Province, were considered secondary high-value areas. The majority of the central, western, and northeastern regions had levels of high-quality agricultural development below the national average, classifying them as low-value areas. Specifically, Shanghai consistently maintained the highest level, while Gansu Province and Shanxi Province consistently remained at the lowest level.

4.2. Spatial Correlation Analysis

4.2.1. Global Spatial Autocorrelation

From a spatial perspective (Table 3), the Global Moran's index is positive and has passed the significance test at the 1% level. This indicates the presence of significant global spatial autocorrelation.

Table 3. Global Moran index.

Year	High-Quality Agricultural Development			Digital Economy		
	Moran's I	Z Value	*p* Value	Moran's I	Z Value	*p* Value
2011	0.350	5.132	0.000	0.204	3.240	0.001
2012	0.340	5.080	0.000	0.260	3.981	0.000
2013	0.385	5.628	0.000	0.212	5.108	0.000
2014	0.374	5.480	0.000	0.198	3.210	0.001
2015	0.374	5.599	0.000	0.192	3.151	0.002
2016	0.360	5.529	0.000	0.262	4.078	0.000
2017	0.335	5.330	0.000	0.341	5.093	0.000
2018	0.322	5.177	0.000	0.269	4.237	0.000
2019	0.317	5.158	0.000	0.215	3.492	0.000
2020	0.321	5.220	0.000	0.135	2.389	0.017
2021	0.335	5.417	0.000	0.186	3.061	0.002

4.2.2. Local Spatial Autocorrelation

However, local spatial autocorrelation allows for the assessment of spatial correlation among individual spatial entities within a specific region and their neighboring entities, a capability not provided by the Global Moran's index. Hence, this study opted to analyze the local spatial autocorrelation in 2011 and 2021. Building on the Global Moran's index, LISA cluster maps were generated to visually represent the spatial clustering of heterogeneous characteristics, as depicted in Figures 4 and 5. As shown in Figure 4, the region characterized by high–high agglomeration in terms of digital economy development in 2011 is situated in Shanghai. In contrast, by 2021, the high–high aggregation area has extended from Shanghai to Zhejiang Province, while the low–low aggregation region was primarily concentrated in Xinjiang Province. As depicted in Figure 5, high and low levels of agglomeration are observed in the high-quality agricultural development in China, and the high–high agglomeration areas are mainly concentrated in the eastern coastal areas. A comparison reveals a significant overlap of high-value cluster areas between the two during the study period. Therefore, it is inferred that by enhancing the digital economy development in these areas and leveraging the driving and overflow impacts of the digital economy, high-quality development can be promoted.

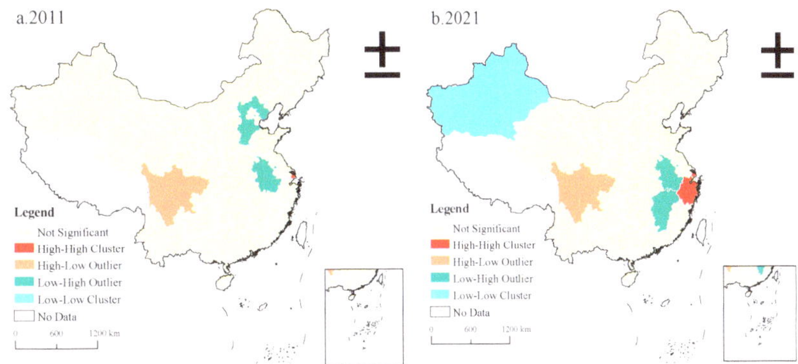

Figure 4. LISA diagram of the digital economy, 2009 and 2021.

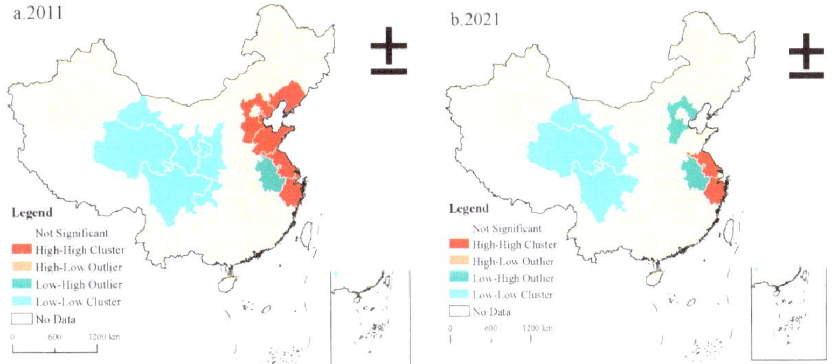

Figure 5. LISA diagram of the high-quality agricultural development, 2009 and 2021.

4.3. Spatiotemporal Non-Stationarity Analysis

4.3.1. Prior Estimation of the Model

To avoid multicollinearity issues between variables and the problem of spurious regression, tests for multicollinearity and panel data stationarity were conducted, and Table 4 presents the findings. The value of VIF for each variable is less than 3, suggesting the absence of multicollinearity issues among the driving factors. The variables of the LLC test show that the panel data are steady at the 1% significance level and there are no issues of spurious regression, thus allowing for direct parameter estimation. The industrial structure and degree of openness did not meet the IPS test criteria. After first-order differentiating all variables, they all passed the IPS test, suggesting that all variables exhibit first-order integration. Given that the economic implications of variables may alter following first-order differentiation, panel cointegration tests were performed to optimize the utilization of the original sequence data. The test results indicate that both the Pedroni test and the Westerlund test reject the null hypothesis at a significance level of 1%. This confirms the presence of cointegration among the original sequence data, thereby enabling direct regression analysis.

Table 4. Multicollinearity test and variable unit root test.

Variables	VIF	1/VIF	LLC Test	IPS Test (Level)	IPS Test (1st Difference)
ng	-	-	−12.161 ***	−3.7503 ***	−7.3356 ***
dig	1.83	0.545	−13.939 ***	−5.7094 ***	−6.7229 ***
ur	2.60	0.384	−5.772 ***	−2.4742 ***	−3.3516 ***
edu	1.77	0.565	−13.367 ***	−4.8112 ***	−7.9699 ***
jgh	1.61	0.622	−6.073 ***	−0.4750	−4.8186 ***
open	1.90	0.525	−4.196 ***	0.9432	−5.9407 ***
old	1.24	0.806	−13.104 ***	−4.3520 ***	−7.5757 ***
Mean VIF	1.83	-	-		

Note: *** represents significance at the 1% level.

4.3.2. Model Testing and Selection

Leveraging ArcGIS to compare and contrast the goodness of fit (R-squared), AICc, and RSS for the OLS, TWR, GWR, and GTWR models enhances the scientific rigor of model selection. The model automatically sets the optimal bandwidth based on the AICc criterion. A higher R-squared value indicates a better model fit, while smaller RSS and AICc values indicate better model fit. Table 5 displays the results. The GTWR model demonstrates the lowest RSS and AICc values as well as the highest R-squared value, indicating that the GTWR model's fit is better. Consequently, the GTWR model was chosen.

Table 5. Regression model selection test.

Parameters	OLS	TWR	GWR	GTWR
R-squared	0.763	0.845	0.924	0.961
Adjusted R-squared	-	0.842	0.922	0.960
AICc	−746.374	−846.323	−1025.1	−1153.03
RSS	1.929	1.270	0.624	0.319
Bandwidth	-	0.196	0.115	0.115

Note: bandwidth is the optimal bandwidth of the model, controlling its smoothing level.

4.3.3. Robustness Test

In this study, the explanatory variable "digital economy" was replaced by three variables: digital infrastructure (dig1), industrial digitalization (dig2), and digital inclusive finance (dig3). Subsequently, regression analysis was performed using the OLS and GTWR model. As demonstrated in Table 6, it is clear that the R-squared values for these three variables are higher in the GTWR model, with smaller RSS and AICc, suggesting that the GTWR model's fit is better than the OLS model, thus demonstrating the robustness of the model selection.

Table 6. Robustness test results.

Parameters	dig1		dig2		dig3	
	OLS	GTWR	OLS	GTWR	OLS	GTWR
R-squared	0.769	0.976	0.753	0.945	0.772	0.974
Adjusted R-squared		0.975		0.944		0.973
AICc	−1257.9	−1770.6	−1235.9	−1616.12	−1260.7	1735.5
RSS	0.409	0.042	0.437	0.098	0.406	0.047
Bandwidth		0.011		0.017		0.012

4.3.4. Temporal Non-Stationarity Evolution of the Impact of the Digital Economy

Using ArcGIS 10.8 software, the regression coefficients for the impact of the digital economy under the GTWR model were calculated, and the trend over time (Figure 6) was further visualized. From Figure 6, it is evident that the average value of it is consistently positive, indicating a significant promotional effect on the high-quality development of agriculture. However, because of factors such as the level of urbanization and industrial structure in various provinces, the promotional effect fluctuates over time, showing significant temporal non-stationarity with significant differences in different years. Additionally, the dispersion of the regression coefficients across provinces gradually decreases from 2011 to 2015 and increases from 2015 to 2021, indicating a significant spatial non-stationarity in the promotional effect.

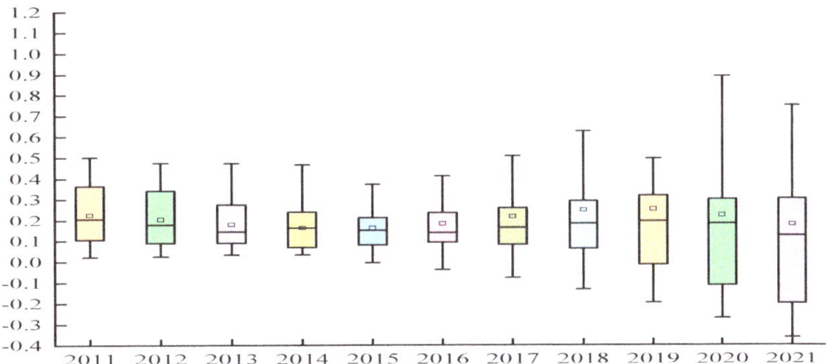

Figure 6. Temporal changes of regression coefficients for the digital economy, 2011–2021.

4.3.5. Spatial Non-Stationarity Evolution of the Impact of the Digital Economy

For the purpose of clearly demonstrating the heterogeneity of the impact of the digital economy on the high-quality development of agriculture in different regions, the regression results of the GTWR model for 2011, 2016, and 2021 were selected for analysis (Figure 7).

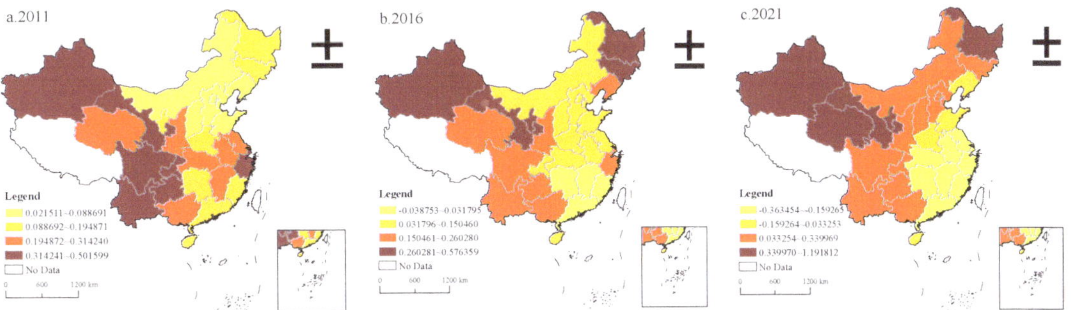

Figure 7. Spatial distribution of regression coefficients for the digital economy.

It can be observed that the promotional effect shows significant spatial differentiation. In 2011, high-value areas were primarily located in the northwest and southwest regions. In 2016 and 2021, there was a gradual decrease from the northwest to the northeast-southeast, with the low-value areas forming in the southeast, suggesting that the promotional effect is weaker in the southeast coastal areas and stronger in the northwest and northeast areas. Specifically, in 2011, high-value areas were mostly centered in Xinjiang, Gansu, Yunnan, Guizhou, Sichuan, and Zhejiang province, which may be attributed to the following reason: the economic level and agricultural skills in the grain production and sales balance areas were relatively weak in 2011, and the digital economy significantly motivated the efficiency of agriculture by changing traditional agricultural production methods.

In 2016 and 2021, the promotional effect was not significant in the economically developed areas along the southeast coast. High-value areas were mostly concentrated in economically underdeveloped areas such as Xinjiang, Gansu, and the three northeastern provinces. The following are some potential explanations for this: firstly, while rural areas in the northwest and northeast regions may be trailing behind the southeast in terms of infrastructure, factor allocation, and locational conditions, the nascent stage of digital economy development in these areas also suggests a higher potential for development. The spatial spillover capacity of the digital economy is playing a significant role in mitigating the inherent developmental resource disparities, thereby alleviating the weaknesses in agricultural development within these regions. Secondly, the developed southeastern coastal areas have responded actively to the demands of rapid economic development and have successfully introduced a large number of other economic policy measures to encourage regional economic development. Therefore, the digital economy has not shown notable effectiveness. In contrast, the economic development of provinces in the northwest and northeast areas has many shortcomings, and the digital economy can bring significant economic dividends to them. Thirdly, in the southeastern coastal areas such as Shanghai, Guangdong, Jiangsu, and Zhejiang, the development of the digital economy has been far ahead of other provinces in recent years. These regions have abundant factor supply and market demand. However, challenges such as heightened competition in the digital technology industry, urgent need for technological innovation, talent shortage, and intense competition of homogenization have gradually emerged, diminishing these areas' capacity to significantly enhance the high-quality development of agriculture.

In conclusion, it is clear that the promotional effect exhibits significant spatial heterogeneity.

5. Conclusions and Policy Implications

5.1. Conclusions

This study primarily examined the spatiotemporal heterogeneity of the impact of China's digital economy on the high-quality development levels of agriculture across 30 provinces and municipalities from 2011 to 2021. (1) In terms of temporal characteristics, from 2011 to 2021, there has been a consistent upward trajectory in the advancement of the digital economy in China alongside a steady annual growth in high-quality agricultural development. In terms of spatial distribution characteristics, both sectors demonstrate a spatial distribution trend characterized by higher levels of development in the southeast region and lower levels in the northwest, highlighting notable spatial disparities. (2) Regarding spatial correlation, the Global Moran's I index indicates that both exhibit significant global spatial autocorrelation, and the LISA cluster map shows a significant spatiotemporal clustering effect of high-quality agricultural development, demonstrating the existence of pronounced "high–high" and "low–low" clustering phenomena. (3) The findings of the GTWR model suggest that the advancement of the digital economy plays a crucial role in fostering the high-quality development of agriculture. Robustness tests reveal that this promotional effect remains significant when the digital economy is tested across different dimensions. (4) This promotional effect exhibits temporal and spatial non-stationarity, with a weaker promotional effect in the southeast coastal areas and a stronger promoting effect in the northwest and northeast regions.

The study makes the following contributions: 1. Innovation in the indicator system for high-quality agricultural development, by constructing a new comprehensive evaluation index system with "basic development, quality improvement, integrated development, new quality development, and shared development" as its core. 2. Innovation in research content, as there are limited publications in the literature exploring the spatiotemporal impact of the digital economy on high-quality agricultural development. 3. Innovation in research methods, by using ArcGIS for spatial visualization to intuitively demonstrate the heterogeneity at the regional and geographical distance levels. Additionally, the study utilizes the GTWR model to analyze spatiotemporal impact. This provides practical evidence for formulating differentiated digital economic policies to serve high-quality agricultural development and offers a new approach for research on high-quality agricultural development, thereby supplementing existing research to a certain extent.

5.2. Policy Implications

Firstly, the rural digital economy should be utilized to stimulate the vitality of rural digital innovation and entrepreneurship. This can be achieved in the following manner: (1) by developing Internet and related communication infrastructure to enhance the digital infrastructure in rural areas. (2) By highlighting innovation in digital technology, cultivating a high-level digital talent team, and fostering agricultural products highly integrated with digital technology. This will accelerate the industrialization and digital transformation of rural digital industries. (3) By promoting the deep integration development of digital inclusive finance with agriculture and rural areas. Emphasis should be placed on innovating and advancing diversified digital financial service models such as credit, insurance, and credit services, enabling better adaptation to the rural areas.

Secondly, enhancement of the market capitalization of agricultural listed companies can be achieved by incorporating digital economy elements and utilizing financial markets. One manner in which to accomplish this is through the implementation of a comprehensive strategy that spans across the stages of production, processing, marketing, distribution, and sales. During production and processing, optimizing digital technology use, fostering innovation in new product development, and embracing value-added processing technologies are essential. In the marketing process, utilizing E-commerce platforms and digital marketing strategies is recommended to expand market reach and target new consumer segments. In the distribution and supply chain stages, it is crucial to streamline distribution and achieve digitalization in the supply chain. The second strategy involves blended finance, which leverages the collaboration of investment funds, securities firms, and angel investors to promote the growth of agricultural finance markets and sustainable agriculture. Blended finance entails leveraging preferential capital from the public sector to attract private capital investment, thereby yielding supplementary development advantages [43]. It can manifest in various forms: 1. equity investment: ownership of shares in agricultural-related companies can be purchased through private equity or public markets (also known as "listed stocks"). 2. Debt investment: various forms of credit can be directly provided to agricultural enterprises or through intermediary entities such as banks or non-bank financial institutions [44]. This blended finance approach leverages public funds to steer private investments towards more high-risk ventures within the agricultural sector, thereby fostering the development of sustainable agricultural practices. This may involve investing in alignment with new financing arrangements or investment counterparts, particularly encouraging the involvement of "mainstream" institutional investors like pension funds [45].

Third, we should aim to optimize agricultural and rural resource allocation and promoting rural transformation and upgrading. On one hand, local governments can improve the efficiency of utilizing rural financial resources by establishing rural financial service centers through the use of digital technologies. The efficiency of land resource allocation can be enhanced by utilizing digital technology to establish land factor circulation platforms. On the other hand, there is a need to facilitate the transformation of traditional financial institutions in rural areas. Cooperation can be encouraged between rural financial institutions and major internet companies in urban areas. The following actions should be carried out: utilizing digital technology to establish cross-regional urban–rural financial service networks, introducing excellent financial products from urban areas to rural areas, and implementing innovative upgrades.

Fourth, we should strengthen inter-provincial communication and cooperation and support for the coordinated development of digital economy and agriculture in various regions. This is recommended to formulate corresponding policy documents to encourage developed provinces to support less developed provinces. The digital economy's driving and overflow effects can be utilized to support comprehensive and coordinated rural development among regions. Additionally, efforts need to be made by the government to strengthen policy enforcement, especially in regions with relatively slower development and lower policy enforcement. As an additional measure, ensuring the diligent implementation of industry assistance policies is essential for the overall success of comprehensive reform.

Fifth, development strategies that are customized, dynamic, and adapted to local conditions should be clarified. The focus should be on leveraging the digital economy to facilitate the superior agricultural development of the northeastern and western regions. Each region must leverage its unique resource endowment and local characteristics to implement tailored development strategies according to the level of agriculture. The eastern region should leverage its advantages to promote agricultural innovation. Midwest and northeastern regions should accelerate to make up for the shortcomings of basic elements in the construction of digital villages.

Author Contributions: Conceptualization, Q.L.; methodology, Q.L.; software, Q.L.; validation, Q.L.; formal analysis, Q.L.; investigation, Q.L.; resources, Q.L.; data curation, Q.L.; writing—original draft preparation, Q.L.; writing—review and editing, Z.L.; visualization, Q.L.; supervision, Z.L.; project administration, Z.L.; funding acquisition, Z.L. All authors have read and agreed to the published version of the manuscript.

Funding: The Fundamental Research Funds for provincial universities in Heilongjiang Province (grant number: 2021-KYYWF-0092, "Research on Financial Innovation Promoting High-Quality Economic Development in Heilongjiang Province").

Institutional Review Board Statement: Not applicable.

Informed Consent Statement: Not applicable.

Data Availability Statement: The data presented in this study are available on request from the corresponding author.

Conflicts of Interest: The authors declare no conflicts of interest.

References

1. China's 2022 Grain Production New Record High, Exceeding 650b Kg Mark for Eight Consecutive Years. Available online: https://www.globaltimes.cn/page/202212/1281648.shtml (accessed on 20 January 2024).
2. Yao, W.; Sun, Z. The Impact of the Digital Economy on High-Quality Development of Agriculture: A China Case Study. *Sustainability* **2023**, *15*, 5745. [CrossRef]
3. West, P.C.; Gerber, J.S.; Engstrom, P.M.; Mueller, N.D.; Brauman, K.A.; Carlson, K.M.; Cassidy, E.S.; Johnston, M.; MacDonald, G.K.; Ray, D.K.; et al. Leverage Points for Improving Global Food Security and the Environment. *Science* **2014**, *345*, 325–328. [CrossRef] [PubMed]
4. Tian, T.; Li, L.; Wang, J. The Effect and Mechanism of Agricultural Informatization on Economic Development: Based on a Spatial Heterogeneity Perspective. *Sustainability* **2022**, *14*, 3165. [CrossRef]
5. Ding, J.; Liu, W.; Wu, J. The Influence Factors and Mechanism of Rural Informationizational Spatial Effect. *Areal Res. Dev.* **2012**, *31*, 151–155.
6. Tapscott, D. The Digital Economy: Promise and Peril in the Age of Networked Intelligence. *Choice Rev. Online* **1996**, *33*, 5199. [CrossRef]
7. Xu, W.; Zhou, J.; Liu, C. The Impact of Digital Economy on Urban Carbon Emissions: Based on the Analysis of Spatial Effects. *Geogr. Res.* **2022**, *41*, 111–129. [CrossRef]
8. Han, Z.; Zhao, J.; Wu, H. A Study on the Scale Measurement, Disequilibrium and Regional Differences of China's Inter-Provincial Digital Economy. *J. Quant. Tech. Econ.* **2021**, *38*, 164–181.
9. Wang, B.; Tian, J.; Cheng, L.; Hao, F.; Han, H.; Wang, S. Spatial Differentiation of Digital Economy and Its Influencing Factors in China. *Sci. Geogr. Sin.* **2018**, *38*, 859–868.
10. Chen, Z.; Wei, Y.; Shi, K.; Zhao, Z.; Wang, C.; Wu, B.; Qiu, B.; Yu, B. The Potential of Nighttime Light Remote Sensing Data to Evaluate the Development of Digital Economy: A Case Study of China at the City Level. *Comput. Environ. Urban. Syst.* **2022**, *92*, 101749. [CrossRef]
11. Zhang, Y. The Development Strategy and Main Tasks of the Digital Economy Driving the Industrial Structure to the Middle and High End. *Econ. Rev.* **2018**, *9*, 85–91. [CrossRef]

12. Zhang, Y.; Wang, M.; Liu, T. Spatial Effect of Digital Economy on High-Quality Economic Development in China and Its Influence Path. *Geogr. Res.* **2022**, *41*, 1826–1844. [CrossRef]

13. Jing, W.; Sun, B. Digital Economy Promotes High-Quality Economic Development: A Theoretical Analysis Framework. *Economist* **2019**, *2*, 66–73. [CrossRef]

14. Cao, Z. Research on the New Manufacturing Model to Promote High-Quality Development of China's Industry under the Background of Digital Economy. *Theor. Investig.* **2018**, *2*, 99–104. [CrossRef]

15. Su, Z.; Zheng, Y.; Guo, L. Impact of the Digital Economy on the Carbon Efficiency of the Tourism Industry and Its Threshold Effect. *China Popul. Resour. Environ.* **2023**, *33*, 69–79. [CrossRef]

16. Ju, X.; Xue, Y.; Xi, B.; Jin, T.; Xu, Z.; Gao, S. Establishing an Agro-Ecological Compensation Mechanism to Promote Agricultural Green Development in China. *J. Resour. Ecol.* **2018**, *9*, 426–433. [CrossRef]

17. Krugman, P. The Myth of Asia's Miracle. *Foreign Aff.* **1994**, *73*, 62–78. [CrossRef]

18. Johnson, D.G. Agriculture and the Wealth of Nations. *Am. Econ. Rev.* **1997**, *87*, 1–12.

19. Feng, Y.; Zhong, S.; Li, Q.; Zhao, X.; Dong, X. Ecological Well-Being Performance Growth in China (1994–2014): From Perspectives of Industrial Structure Green Adjustment and Green Total Factor Productivity. *J. Clean. Prod.* **2019**, *236*, 117556. [CrossRef]

20. Kalirajan, K.P.; Obwona, M.B.; Zhao, S. A Decomposition of Total Factor Productivity Growth: The Case of Chinese Agricultural Growth before and after Reforms. *Am. J. Agric. Econ.* **1996**, *78*, 331–338. [CrossRef]

21. Deng, F.; Jia, S.; Ye, M.; Li, Z. Coordinated Development of High-Quality Agricultural Transformation and Technological Innovation: A Case Study of Main Grain-Producing Areas, China. *Environ. Sci. Pollut. Res.* **2022**, *29*, 35150–35164. [CrossRef]

22. Li, H.; Zhang, J.; Luo, S.; He, K. Impact and Mechanism of Agricultural Technology Innovation on Agricultural Development Quality—Empirical Evidence from the Spatial Perspective. *RD Manag.* **2021**, *33*, 1–15. [CrossRef]

23. Fan, S.; Pardey, P.G. Research, Productivity, and Output Growth in Chinese Agriculture. *J. Dev. Econ.* **1997**, *53*, 115–137. [CrossRef]

24. Zhu, Q.; Bai, J.; Peng, C.; Zhu, C. Do Information Communication Technologies Improve Agricultural Productivity? *Chin. Rural. Econ.* **2019**, *4*, 22–40.

25. Qin, Q.; Guo, H.; Zeng, Y. The Mechanism and Countermeasures of Digital Empowerment to Promote Rural Revitalization. *J. Jiangsu Univ. Soc. Sci. Ed.* **2021**, *23*, 22–33. [CrossRef]

26. Gao, D.; Lyu, X. Agricultural Total Factor Productivity, Digital Economy and Agricultural High-Quality Development. *PLoS ONE* **2023**, *18*, e0292001. [CrossRef] [PubMed]

27. Zhou, Q.; He, A. High Quality Development of the Yellow River Basin Empowered by Digital Economy. *Econ. Probl.* **2020**, *11*, 8–17. [CrossRef]

28. Zhou, Q.; Li, X. Digital Economy and High-Quality Agricultural Development Internal Mechanism and Empirical Analysis. *Reform. Econ. Syst.* **2022**, *6*, 82–89.

29. Zhang, Y.; Ji, M.; Zheng, X. Digital Economy, Agricultural Technology Innovation, and Agricultural Green Total Factor Productivity. *SAGE Open* **2023**, *13*, 21582440231194388. [CrossRef]

30. Wang, S.; Chen, Y. How Does Digital Inclusive Finance Promote the High-Quality Development of Agriculture? On Intermediary and Threshold Mechanism. *J. Manag.* **2022**, *35*, 72–87. [CrossRef]

31. Tang, Y.; Chen, M. The Impact of Agricultural Digitization on the High-Quality Development of Agriculture: An Empirical Test Based on Provincial Panel Data. *Land* **2022**, *11*, 2152. [CrossRef]

32. Huang, B.; Wu, B.; Barry, M. Geographically and Temporally Weighted Regression for Modeling Spatio-Temporal Variation in House Prices. *Int. J. Geogr. Inf. Sci.* **2010**, *24*, 383–401. [CrossRef]

33. Fotheringham, A.S.; Brunsdon, C.; Charlton, M. *Geographically Weighted Regression: The Analysis of Spatially Varying Relationships*; John Wiley & Sons: Hoboken, NJ, USA, 2003; ISBN 0-470-85525-8.

34. Fotheringham, A.S.; Crespo, R.; Yao, J. Geographical and Temporal Weighted Regression (GTWR): Geographical and Temporal Weighted Regression. *Geogr. Anal.* **2015**, *47*, 431–452. [CrossRef]

35. Liu, R.; Guo, T. Construction and Application of the High-quality Development Index——Also on the High-quality Development of Northeast China's Economy. *J. Northeast. Univ. Soc. Sci.* **2020**, *22*, 31–39. [CrossRef]

36. Wang, J.; Shi, M.; Li, Z. Digital Economy Promotes High-Quality Agricultural Development: Internal Mechanism and Empirical Evidence. *J. Univ. Financ. Econ.* **2023**, *36*, 78–89. [CrossRef]

37. Xin, L.; An, X. Construction and Empirical Analysis of Agricultural High-Quality Development Evaluation System in China. *Econ. Rev.* **2019**, *5*, 109–118. [CrossRef]

38. Ye, F.; Qin, S.; Nisar, N.; Zhang, Q.; Tong, T.; Wang, L. Does Rural Industrial Integration Improve Agricultural Productivity? Implications for Sustainable Food Production. *Front. Sustain. Food Syst.* **2023**, *7*, 1191024. [CrossRef]

39. Xiang, W.; Zhu, K.; Teo, B.S.-X.; Zunirah, M.T. Development Experience and Future Prospects of the Integration of Three Rural Industries in China. *Asian J. Soc. Sci. Stud.* **2022**, *7*, 98. [CrossRef]

40. Zhang, L.; Wen, T.; Liu, Y. Integrated Development of Rural Industries and Farmers' Income Growth: Theoretical Mechanism and Empirical Determination. *J. Southwest Univ.* **2020**, *46*, 42–56.

41. Liu, T.; Li, J.; Huo, J. Spatial-Temporal Pattern and Influencing Factors of High-Quality Agricultural Development in China. *J. Arid. Land Resour. Environ.* **2020**, *34*, 1–8. [CrossRef]

42. Sun, Y.; Tang, X. The Impact of Digital Inclusive Finance on Sustainable Economic Growth in China. *Financ. Res. Lett.* **2022**, *50*, 103234. [CrossRef]

43. Attridge, S.; Engen, L. *Blended Finance in the Poorest Countries: The Need for a Better Approach*; ODI Report: London, UK, 2019.
44. Havemann, T. *Landscape Report: Blended Finance for Agriculture*; SAFIN: Zürich, Switzerland, 2019.
45. Havemann, T.; Negra, C.; Werneck, F. Blended Finance for Agriculture: Exploring the Constraints and Possibilities of Combining Financial Instruments for Sustainable Transitions. *Agric. Hum. Values* **2020**, *37*, 1281–1292. [CrossRef] [PubMed]

 *sustainability*

Article

Evaluation of China's Marine Aquaculture Sector's Green Development Level Using the Super-Efficiency Slacks-Based Measure and Global Malmquist–Luenberger Index Models

Deli Yang * and Qionglei Wang

School of Economics and Management, Shanghai Ocean University, Shanghai 201306, China; m210701227@st.shou.edu.cn
* Correspondence: dlyang@shou.edu.cn

Abstract: Given China's rapidly expanding marine aquaculture industry, the associated ecological issues have garnered widespread attention. Therefore, it is crucial to speed up the green growth of marine aquaculture in order to save the environment and use resources sustainably. In order to statically assess and dynamically analyze the green development efficiency levels of marine aquaculture in nine coastal provinces of China from 2012 to 2021, this study uses the non-expected output super-efficiency Slacks-Based Measure model and the Global Malmquist–Luenberger index method. Additionally, it integrates input–output redundancy rates to analyze the causes of efficiency loss. Static efficiency primarily reflects whether a region's inputs and outputs at a given point in time reach an effective efficiency level, while the level of dynamic efficiency mainly gauges the dynamic changes in the efficiency of green production. The results show that, from 2012 to 2021, China's marine aquaculture industry's average static efficiency of green output was 0.705. The southern marine economic zone exhibited the highest static efficiency value in the green development of marine aquaculture, displaying a stepped distribution pattern of "south–north–east" in decreasing order. The input–output redundancy analysis reveals that the primary causes of static efficiency loss in China's marine aquaculture industry are attributed to varying degrees of redundant inputs and carbon emission outputs. Looking through the lens of the GML index, the annual average growth rate of the green total factor productivity in China's marine aquaculture stands at 11.1%, with an annual average change in technical efficiency of 1.8%, while the annual average change in technological progress amounts to 9.1%, suggesting that technological advancement is the primary driver of the rise in green total factor productivity in China's marine aquaculture sector. According to the study, in order to encourage China's marine aquaculture industry to grow sustainably, efforts should be made not only to accelerate technological advancements but also to enhance technical efficiency. Policies that are specifically designed for the local environment should be developed to support the sustainable development of the marine aquaculture sector and to make resource allocation easier.

Keywords: super-efficiency slacks-based measure model; global Malmquist–Luenberger index; marine aquaculture industry; green total factor productivity

Citation: Yang, D.; Wang, Q. Evaluation of China's Marine Aquaculture Sector's Green Development Level Using the Super-Efficiency Slacks-Based Measure and Global Malmquist–Luenberger Index Models. *Sustainability* **2024**, *16*, 3441. https://doi.org/10.3390/su16083441

Academic Editors: Fotios Chatzitheodoridis, Efstratios Loizou and Achilleas Kontogeorgos

Received: 22 March 2024
Revised: 17 April 2024
Accepted: 17 April 2024
Published: 19 April 2024

1. Introduction

People's standard of living has steadily increased since being reformed and opening up. Seafood, characterized by its high protein, low fat, and low calorie content, has become immensely popular among the populace. Marine aquaculture has emerged as a crucial component of fisheries development. As per the "China Fishery Statistical Yearbook 2021", China's total marine aquaculture production reached 22.11 million tons, marking a 3.55% year-on-year increase, accounting for approximately 41% of the nation's total aquaculture output. The rapid expansion of marine aquaculture has significantly contributed to addressing the challenges of seafood scarcity in inland areas, enhancing public health, and promoting coastal economic development. However, alongside the rapid growth of aquaculture, the discharge of exogenous pollutants during the farming process, such as industrial wastewater and solid waste, has adversely affected marine environments, leading to environmental degradation [1]. The "Several Opinions on Accelerating the Green Development of Aquaculture," published in 2019 by the Ministry of Agriculture and Rural Affairs, is the first set of guidelines devoted exclusively to aquaculture since the People's Republic of China was established. This document, approved by the State Council, holds significant importance in promoting the green transformation of aquaculture [2]. The "Five Major Actions for Green and Healthy Aquaculture" were put into action by the Ministry of Agriculture and Rural Affairs in 2020, which is crucial for advancing green aquaculture technologies and protecting the ecological environment [3]. Against this backdrop, the imperative of harmonizing the rapid development of marine aquaculture with environmental conservation to achieve green development has become a pivotal research topic in contemporary discourse.

Through reviewing the relevant literature on the green development of aquaculture, scholars have identified three primary areas of focus within the realm of green aquaculture development research. Firstly, the meaning and implications of "green aquaculture development" are highlighted. According to scholars, the development of aquaculture with consideration for ecological environmental safety and resource conservation is the essence of green aquaculture development. Lu C.C. [4] proposed that green development in aquaculture is established under the dual constraints of aquatic ecological capacity and resource carrying capacity. This is achieved through advanced management concepts, scientific technologies, and material equipment, thus forming a novel development model characterized by efficient resource utilization, ecosystem stability, favorable local environmental conditions, and product quality assurance. Cao J.H. [5] posited that the term "green" emphasizes the protection of the ecological environment and resource conservation, while "development" underscores economic growth and social progress. Therefore, "green development" emphasizes the growth of economic and social welfare under the dual constraints of resource conservation and ecological environmental protection. Yue D.D. [6] suggested that green development in aquaculture aims to achieve a harmonious coexistence among people, aquaculture activities, and the natural environment. This is accomplished through the formulation of plans and standards for green aquaculture development, innovation in aquaculture technologies and mechanisms, and the realization of a whole-process green development mechanism characterized by environmentally friendly aquaculture, efficient technology, safe products, increased income for fishers, and consumer satisfaction.

Secondly, a particular topic of focus for scholarly research in this sector is aquaculture's assessment of its green development level, concentrating primarily on two aspects. One facet involves assessing the green development level of aquaculture through the comprehensive construction of an evaluation index system. Xing Y. [7] established an evaluation system for the green development of aquaculture, selecting 18 indicators based on four dimensions: the input of aquaculture resources, technological advancement and diffusion, managerial oversight, and stage of development. The entropy method was employed for calculations. Yue D.D. [8] created a thorough assessment index system for the environmentally friendly growth of marine aquaculture from four angles: product eco-friendliness, geographic expansion, environmental friendliness, and resource conservation. Furthermore, there is a need to assess the effectiveness of ecofriendly practices in aquaculture by evaluating resource utilization, economic performance, and environmental impact. This evaluation chiefly encompasses metrics like the efficiency of green technologies and overall green productivity. When it comes to research methodologies, primary approaches encompass stochastic frontier analysis (SFA) and data envelopment analysis (DEA), among other relevant methods. Within the domain of SFA methodology, Zhu A.F. [9] employed stochastic frontier analysis (SFA) to gauge the green productivity of China's marine fisheries sector. Yu L. [10] utilized the SFA methodology, incorporating white noise into the computational framework, to assess the efficiency of novel agricultural operators in the aquaculture industry. Indeed, while stochastic frontier analysis (SFA) offers the advantage of accounting for the influence of random errors, it requires a predefined functional form and distinguishes between error and inefficiency terms, which limits its applicability. In the realm of DEA methodology, the main models predominantly focus on the radial paradigm, as demonstrated by the CCR (Charnes, Cooper, and Rhodes) and BCC (Banker, Charnes, and Cooper) models, and the non-radial paradigm mainly represented by the SBM (Slacks-Based Measure) model. Ji J.Y. [11] conducted a comprehensive analysis of green efficiency in Chinese mariculture using a global DEA model. Yang Z.Y. [12] measured the green index of Chinese mariculture using the super-efficient SBM model. Qin H. [13] assessed the ecological–economic efficiency of Chinese marine aquaculture utilizing the SBM model. In the scholarly evaluation of green development in aquaculture using output indicators, the focus is primarily on measuring the anticipated output against the total output value of aquaculture. In terms of non-anticipated output indicators, scholars predominantly take into account pollutants in aquaculture, such as nitrogen (N), phosphorus (P), and chemical oxygen demand (COD).

Lastly, there is the aspect of factors influencing the green development of aquaculture. Xu Y. [14] conducted a study utilizing Feasible Generalized Least Squares (FGLS) estimation, revealing that intertidal aquaculture exerts a negative impact on the overall green productivity of marine aquaculture. Zhang Y. [15] employed the method of Feasible Generalized Least Squares (FGLS) to conclude that there exists a negative correlation between the regulatory level of marine environment and the overall green productivity. Furthermore, investments in marine science and education, the intensity of fisherman training, and dissemination of technological advancements contribute positively to the enhancement in green productivity. Jiang Q.J. [16] utilized the Analytic Hierarchy Process (AHP) to examine the factors that impact the sustainable development of aquaculture. The study revealed that the primary influencers of sustainable aquaculture development are the main stakeholders in aquaculture.

After reviewing the literature, researchers have made significant advancements in the field of sustainable aquaculture development. However, there is room for further optimization in the selection of indicators for evaluating the green development of aquaculture. It has been noted that, during aquaculture operations, both carbon sequestration and carbon emissions are present. Previous studies have tended to focus solely on aquaculture value or production when selecting desirable output indicators, often overlooking the carbon sequestration component. Similarly, in the context of non-desirable output indicators, the emphasis has been predominantly on pollution outputs, neglecting the crucial aspect of carbon emissions. Building upon these observations, this study aims to estimate carbon sequestration and carbon emissions in marine aquaculture in China. These estimations are categorized into desired and undesired outputs. The study assesses the extent of green development in Chinese marine aquaculture by employing the Super-Efficiency Slacks-Based Measure (SBM) model and the Global Malmquist–Luenberger (GML) index methodology. This analytical approach aims to deliver a comprehensive evaluation of the influence of carbon sequestration and emissions on the green development of aquaculture in China.

2. Research Methods and Index Selection

2.1. Research Methods

2.1.1. Superefficient SBM Model with Undesired Outputs

Data Envelopment Analysis (DEA) is a non-parametric efficiency analysis method that provides various advantages, such as avoiding assumptions about indicator weights and uniform measurement units. This approach eliminates the impact of subjective factors on efficiency evaluation and is suitable for assessing the green development level of marine aquaculture, particularly in scenarios involving multiple inputs and outputs. The traditional DEA model solely assesses the efficiency of various radial input–output combinations, overlooking the impacts of slack variables and non-desirable outputs, which may introduce inherent inaccuracies into the results. In contrast, the SBM model comprehensively addresses both radial enhancements in input–output dimensions and slack improvements, mitigating such limitations. Tone [17] proposed the SBM model to address this limitation, but this approach still has its drawbacks. For example, when multiple evaluated objects achieve an effective status simultaneously, it becomes difficult to perform further ranking and comparative analysis. Hence, Tone [18] proposed the super-efficiency SBM model to address the limitation where all evaluated entities reach an efficiency score of 1, thereby impeding further comparative analysis. In the assessment of green development efficiency levels, this model incorporates environmental pollution and carbon emissions as non-desirable outputs. Therefore, this study opts for the super-efficiency SBM model that integrates non-desirable outputs for evaluation, as illustrated in Equations (1) and (2).

$$\min\rho = \frac{1+\frac{1}{m}\sum\limits_{i=1}^{m} s_i^- / x_{ik}}{1-\frac{1}{q_1+q_2}\left(\sum\limits_{r=1}^{q_1} s_r^+ / y_{rk} + \sum\limits_{t=1}^{q_2} s_t^{b-} / b_{tk}\right)} \tag{1}$$

$$s.t.\begin{cases} \sum\limits_{j=1,j\neq k}^{n} x_{ij}\lambda_j - s_i^- \leq x_{ik} \\ \sum\limits_{j=1,j\neq k}^{n} y_{rj}\lambda_j + s_r^+ \geq y_{rk} \\ \sum\limits_{j=1,j\neq k}^{n} b_{tj}\lambda_j - s_t^{b-} \leq b_{tk} \\ 1 - \frac{1}{q_1+q_2}\left(\sum\limits_{r=1}^{q_1} s_r^+ / y_{rk} + \sum\limits_{t=1}^{q_2} s_t^{b-} / b_{tk}\right) > 0 \\ \lambda, s^-, s^+ \geq 0 \\ i = 1,2\ldots,m; r = 1,2\ldots,q; j = 1,2\ldots,n(j \neq k) \end{cases} \tag{2}$$

In the above equations, n represents the number of decision-making units, where each unit consists of (m) types of inputs, q_1 types of expected outputs, and q_2 types of non-expected outputs. The vectors x, y, and b denote the indicators for inputs, expected outputs, and non-expected outputs, respectively. The symbol S represents the slack variable. The symbol λ represents the weight vector, while ρ signifies the green development efficiency value of marine aquaculture. When ρ is lower than 1, it indicates that the efficiency level of green development is relatively inefficient. Conversely, when ρ is greater than 1, it suggests that the efficiency level of green development is relatively effective. A higher value of ρ signifies a higher level of green development efficiency and overall green development performance. This study calculated the green production efficiency of marine aquaculture in coastal provinces of China from 2012 to 2021 by integrating the evaluation index system of green development level in marine aquaculture, using MATLAB2021a software and a super-efficiency SBM model based on non-expected outputs.

2.1.2. Global Malmquist–Luenberger

The super-efficiency SBM model, which incorporates undesired outputs, is utilized for the evaluation of the static level of green development efficiency, lacking the capability to effectively assess its dynamic changes. As a result, integrating the GML index method is essential for analyzing variations in green total factor productivity. This approach enables the measurement of changes in green total factor productivity from period t to t + 1, facilitating a dynamic analysis of green development efficiency. In contrast, the traditional Malmquist index overlooks the consideration of undesired outputs. Chung [19] combined the Malmquist index with directional distance functions and proposed a Malmquist index that takes into account undesired outputs. The issue of linear programing infeasibility during the intertemporal computation of the Malmquist index has been effectively addressed by Oh [20] through the development of the Global Malmquist–Luenberger (GML) index, which integrates a global approach with directional distance functions. This advancement is documented in Equation (3).

$$
\begin{aligned}
GML^{t,t+1}&\left(x^t, y^t, b^t, x^{t+1}, y^{t+1}, b^{t+1}\right) \\[4pt]
&= \frac{1+D^G\left(x^t,y^t,b^t\right)}{1+D^G\left(x^{t+1},y^{t+1},b^{t+1}\right)} = \frac{1+D^t\left(x^t,y^t,b^t\right)}{1+D^{t+1}\left(x^{t+1},y^{t+1},b^{t+1}\right)} \\[6pt]
&\times \frac{\left(1+D^G\left(x^t,y^t,b^t\right)\right)/\left(1+D^t\left(x^t,y^t,b^t\right)\right)}{\left(1+D^G\left(x^{t+1},y^{t+1},b^{t+1}\right)\right)/\left(1+D^{t+1}\left(x^{t+1},y^{t+1},b^{t+1}\right)\right)} \\[6pt]
&= \frac{TE^{t+1}}{TE^t} \times \frac{BPG^{t,t+1}_{t+1}}{BPG^{t,t+1}_t} = EC^{t,t+1} \times BPC^{t,t+1}
\end{aligned}
\tag{3}
$$

In Equation (3), D^G represents the global directional distance function, and the value of $GML^{t,t+1}$ indicates the green total factor productivity change from period t to t + 1. Specifically, when the $GML^{t,t+1}$ index for period t to t + 1 is greater than 1, it indicates an increase in green total factor productivity compared to the previous period. Conversely, if the GML index is lower than 1, it signifies a decrease in green total factor productivity. Furthermore, it can be decomposed into changes arising from technical efficiency (EC) and changes resulting from the gap with best practice (BPC). Technical efficiency gauges the ability to achieve more outputs without increasing resource inputs, whereas technological progress measures the contribution of changes in production technology to outputs. In this context, EC is used to assess the diffusion of green technologies, where $EC^{t,t+1} > 1$ (<1) indicates an improvement (decline) in efficiency over adjacent periods for the decision-making unit. On the other hand, BPC measures the progression of green technological advancements, with $BPC^{t,t+1} > 1$ (<1) indicating progress (regression) in technology over adjacent periods for the decision-making unit.

2.2. Index Selection

During the development of marine aquaculture, the labor force and fixed assets, among other input factors in the industry, are closely correlated with the growth of the total output value in the marine aquaculture sector. This study, considering the characteristics of marine aquaculture and data availability, chose labor force, aquaculture area, fixed assets, training intensity, and intermediate consumption as resource input indicators. Specifically, labor force refers to individuals engaged in production and management, represented by the professional workforce in marine aquaculture; the aquaculture area represents resources formed naturally and through artificial means, quantified by the area dedicated to marine aquaculture; fixed assets denote the aquaculture fishing vessels utilized in marine aquaculture, indicated by the total power of motorized fishing vessels used for aquaculture purposes; training intensity in the marine aquaculture industry measures the level of technical training engagement by practitioners, calculated by multiplying the number of individuals in fisheries training by the proportion of marine aquaculture practitioners in the total workforce in aquaculture; intermediate consumption in marine aquaculture refers to the daily operational expenses, converted from fisheries' intermediate consumption to marine aquaculture intermediate consumption, and adjusted to comparable prices in 2012 using the agricultural production input price index to mitigate the impact of price fluctuations.

In terms of indicator selection, the expected outputs were categorized into economic and environmental aspects. For the economic output, the indicator chosen was the economic output value of marine aquaculture, with relevant data obtainable from the "China Fishery Statistical Yearbook". On the other hand, for the environmental output, the indicator selected was the carbon sequestration capacity of marine aquaculture. Bivalves and algae aquaculture, which do not require the input of feed, have the ability to absorb a significant amount of carbon through carbon fixation. Moreover, they represent the primary species in Chinese marine aquaculture. Consequently, this study primarily focused on bivalves and algae in the calculation of the carbon sequestration capacity within marine aquaculture. The carbon sequestration levels of major bivalves and algae species with carbon sequestration functions can be obtained from the research findings of Li X. [21] and colleagues. The carbon sequestration assessment method for bivalves and algae in marine aquaculture is referred to in He J.B.'s [22] accounting framework, whose specific formula is as follows:

$$C_t = \sum C_s + \sum C_{al}$$
$$\sum C_s = \sum C_{sh} + \sum C_{st}$$
$$\sum C_{sh} = \sum Q_i \cdot \alpha_i \cdot P_i \cdot \beta_i \qquad (4)$$
$$\sum C_{st} = \sum Q_i \cdot \alpha_i \cdot P_j \cdot \beta_j$$
$$\sum C_{al} = \sum Q_k \cdot \beta_k$$

In the above equation, C_t denotes the total carbon sequestration of bivalves and algae in marine aquaculture; C_s and C_{al} represent the carbon sequestration from bivalves and algae, respectively; and C_{sh} and C_{st} denote the carbon sequestration of bivalve soft tissue and shell, respectively. Qi signifies the yield of bivalves; α_i represents the dry weight coefficient of bivalves; Pi and P_j represent the mass proportions of bivalve soft tissue and shell, respectively; β_i and β_j, respectively, denote the carbon fixation capacity of bivalve soft tissue and shell; Q_k signifies the yield of algae; and β_k represents the carbon fixation capacity of algae.

In the context of unintended output indicators, the pollutants generated by marine aquaculture primarily include nitrogen (N), phosphorus (P), and chemical oxygen demand (COD), with emission rates sourced from the data published in the "Announcement of the Second National Pollution Source Census in China" as follows: nitrogen emissions at 2.5 kg/t, phosphorus emissions at 0.33 kg/t, and COD emissions at 13.6 kg/t. Aside from pollutant emissions, carbon emissions during the process of marine aquaculture also bear environmental implications. Carbon emissions from marine aquaculture can be categorized into two components: firstly, carbon emissions resulting from energy combustion, with this study focusing on the carbon emissions generated during the operation of marine aquaculture vessels utilizing diesel fuel; and secondly, indirect carbon emissions stemming from the use of electricity. While most aquaculture methods primarily rely on natural resources, such as the sea area, thereby exhibiting a low energy dependency, pond aquaculture and intensive aquaculture exhibit a higher degree of energy reliance, primarily attributed to carbon emissions arising from activities such as aeration and electrical operation. The calculation methodology for carbon emissions in marine aquaculture can be obtained from the accounting system proposed by Shao G.L. [23], whose specific formula is as follows:

$$C_{ef} = P \cdot \chi \cdot \theta_1 \cdot \omega + \left(P_{pa} \cdot \kappa + P_{ia} \cdot \eta + S_{pa} \cdot \mu + S_{ia} \cdot \rho \right) \cdot \theta_2 \cdot \omega$$

The formula includes various parameters: C_{ef} represents the total carbon emissions from marine aquaculture; P indicates the total power of marine aquaculture vessels; χ denotes the fuel consumption factor for marine aquaculture vessels, whose value of 0.225 tons per kilowatt was obtained from the "Domestic Motorized Fishing Vessel Fuel Subsidy Fuel Consumption Calculation Reference Standard"; P_{pa} and P_{ia}, respectively, represent the output of marine pond aquaculture and marine intensive aquaculture; κ and η represent the single-unit electricity consumption coefficients for marine pond aquaculture and marine intensive aquaculture, determined as 0.37 kilowatts per kilogram and 8.66 kilowatts per kilogram based on the research findings of Xu H. [24] and Yang Z.Y. [25], respectively; θ_1 and θ_2, respectively, indicate the coefficients for diesel and electricity conversion to standard coal, set at 1.46 kg of standard coal per kilogram and 0.12 kg of standard coal per kilowatt according to the "China Energy Statistical Yearbook"; S_{pa} and S_{ia} represent the area of marine pond aquaculture and the volume of marine intensive aquaculture, respectively; μ and ρ, respectively, denote the unit area oxygenation electricity consumption coefficient and the unit volume oxygenation electricity consumption coefficient for marine pond aquaculture and marine intensive aquaculture, with the coefficients determined as 1440.17 kilowatts per hectare and 37.76 kilowatts per cubic meter based on the research findings of Xu H. [26]; and w represents the carbon emission coefficient, set at 0.68 kg per kilogram of standard coal based on the research findings of Xu D.L. [27] and Yue D.D. [28]. In consideration of the aforementioned indicators, this paper established the following evaluation index system for the sustainable development of marine aquaculture, as shown in Table 1.

Table 1. Evaluation index system of the green development level of mariculture industry.

Index	Index Class	Variable	Unit
Input	Resource input	Breeding area	Hectare
		Labor force	Population
		Intermediate consumption	CNY Ten thousand
		Training intensity	Population
		Aquaculture fishing vessel	Kilowatt
Output	Expected output	Economic value of mariculture	CNY Ten thousand
		Carbon sink of mariculture	Ton
	Undesirable output	N, P, and COD emissions from mariculture	Ton
		Carbon emissions from mariculture	Ton

2.3. Data Source

This paper's input–output indicator data mainly originated from the "China Fishery Statistical Yearbook" (2012–2021), "China Statistical Yearbook" (2012–2021), and the "Handbook of Pollution Source Production and Emission Coefficients for Aquaculture in the First National Pollutant Source Census." For the missing data, interpolation methods were utilized for completion, while certain indicators were derived through calculations. Due to substantial data gaps in Tianjin, Shanghai, Hong Kong, Macau, and Taiwan, coupled with the relatively smaller aquaculture scales in these regions, the analysis in this paper focused solely on examining the level of sustainable development in marine aquaculture in 9 provinces, including Liaoning, Hebei, Shandong, Jiangsu, Zhejiang, Fujian, Guangdong, Guangxi, and Hainan. The northern region encompasses three areas: Liaoning, Hebei, and Shandong. The eastern region comprises Jiangsu and Zhejiang. The southern region includes Fujian, Guangdong, Guangxi, and Hainan. Descriptive statistics for the variables are presented in Table 2.

Table 2. Descriptive statistics of input and output variables.

Indicators	Min	Max	Avg	Sd
Breeding area	15,845	942,050	237,788	247,029
Labor force	23,037	226,137	100,159	63,997
Intermediate consumption	115,658	1,519,737	594,078	374,997
Training intensity	1108	108,321	18,420	17,646
Aquaculture fishing vessel	700	323,379	115,544	94,430
Economic value of mariculture	662,215	10,728,457	3,587,035	2,720,635
Carbon sink of mariculture	1807	688,737	201,638	198,289
N, P, and COD emissions from mariculture	6207	155,247	62,163	45,874
Carbon emissions from mariculture	9402	252,097	69,058	57,748

3. Empirical Analysis

3.1. Static Efficiency Analysis of Green Production in the Chinese Mariculture Industry

Table 3 presents the findings derived from computations using panel data. Table 3 presents the marine aquaculture industry's green production efficiency level for the period from 2012 to 2021 in China. Nationally, from 2012 to 2021, China's marine aquaculture industry's green production efficiency only surpassed 1 in 2021; the other years fell short of 1, yielding an average of 0.705, which suggests a moderate level of green production efficiency in China's marine aquaculture industry. The southern marine economic circle had an average green production efficiency of 0.927, higher than the northern and eastern marine economic circles (0.590 and 0.533), respectively, when comparing the average green production efficiency of the three major marine economic circles from 2012 to 2021.

This suggests a stepwise decline in the green production efficiency of China's marine aquaculture industry from south to north to east. Regionally, Fujian predominantly had green production efficiency values exceeding 1 during the calculation period, with values below 1 only in 2012 and 2015, indicating a relatively high level of green development and rational input–output in the marine aquaculture industry of this region. Other regions showed varying levels of fluctuation in green development efficiency, suggesting insufficient stability and significant room for improvement in green development efficiency. In terms of regional averages, Fujian had the highest green production efficiency average of 1.257, while Hebei had the lowest green production efficiency at only 0.294. Although China's green production efficiency in marine aquaculture has seen rapid development in recent years, its overall efficiency remains relatively low, highlighting the considerable potential for raising the green production efficiency of China's marine aquaculture sector and showing areas for development when compared to the frontier.

Table 3. Static efficiency of the green production of mariculture in Chinese provinces from 2012 to 2021.

Province	2012	2013	2014	2015	2016	2017	2018	2019	2020	2021	Mean
Hebei	0.203	0.227	0.242	0.248	0.271	0.277	0.303	0.321	0.383	0.631	0.294
Liaoning	0.476	0.578	0.580	0.570	0.705	1.151	0.719	1.147	1.159	1.353	0.791
Jiangsu	0.244	0.419	0.428	0.384	0.506	0.567	0.767	0.407	1.131	1.124	0.535
Zhejiang	0.379	0.392	0.419	0.453	0.420	0.554	0.560	0.604	0.783	1.025	0.531
Fujian	0.875	1.265	1.270	0.966	1.120	1.127	1.491	1.517	1.533	1.657	1.257
Shandong	0.476	0.505	0.954	0.797	1.143	1.091	1.043	1.049	1.036	1.099	0.882
Guangdong	0.575	0.597	0.583	0.613	0.662	0.904	1.061	1.114	1.076	1.254	0.807
Guangxi	0.542	0.528	0.694	0.680	1.060	0.867	1.152	1.131	1.052	1.049	0.842
Hainan	0.495	0.587	0.839	1.012	1.024	1.015	1.062	1.056	0.682	1.180	0.864
Northern mean	0.358	0.405	0.512	0.483	0.602	0.703	0.610	0.728	0.772	0.979	0.590
Eastern mean	0.304	0.405	0.423	0.417	0.461	0.561	0.655	0.496	0.941	1.074	0.533
Southern mean	0.606	0.695	0.810	0.799	0.947	0.973	1.180	1.192	1.043	1.267	0.927
National mean	0.436	0.515	0.602	0.585	0.694	0.772	0.831	0.832	0.922	1.120	0.705

Figure 1 shows the evolution of green production efficiency from 2012 to 2021 in the national and three key marine economic zones. From 2012 to 2014, the national efficiency grew progressively, reaching a peak of 0.602 before seeing a slight decline. After making a comeback in 2016, it increased gradually. The State Council's 2019 publication of "Several Opinions on Accelerating the Green Development of Marine Aquaculture," which somewhat raised the cost of marine aquaculture, may have contributed to the less noticeable rise in efficiency between 2018 and 2019. A minor peak was observed in 2020, signifying a noteworthy rise in the green production efficiency of China's marine aquaculture sector. This was possibly a result of the release of the "Five Major Actions for Green Aquaculture in 2020" by the Ministry of Agriculture and Rural Affairs of China, which promoted the demonstration of eco-friendly aquaculture technologies and facilitated the application of green and healthy aquaculture technologies to promote the green development of marine aquaculture nationwide. Looking at the evolving trends in green production efficiency in the three major marine economic zones, the southern and northern economic zones exhibited similar patterns of gradual fluctuating improvement. The eastern economic zone experienced a slow increase initially, followed by fluctuations and a decline in 2019, and a subsequent fluctuating increase in 2020. In the initial stages of the study, the green production efficiency of marine aquaculture in the southern economic zone was the highest, followed by the northern economic zone, and the eastern economic zone ranked the lowest. However, by 2020, the green production efficiency of marine aquaculture in the eastern economic zone surpassed that of the northern economic zone.

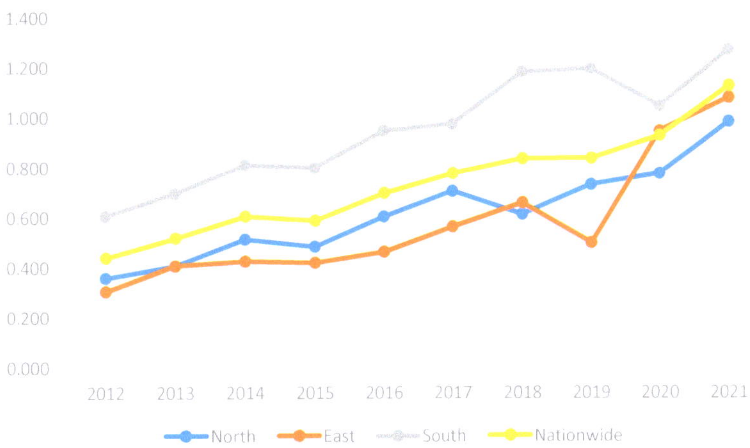

Figure 1. The annual changes in green production efficiency in the national and three major marine economic zones from 2012 to 2021.

Tables 4 and 5, respectively, present the redundancy rates of input and non-desirable output indicators for the national level and each province, as calculated by the super-efficiency SBM model. These rates reflect the distance between each indicator and the green production frontier. A higher value indicates a greater distance from the green production frontier, which exerts a more significant influence on green production efficiency.

Based on the analysis of the redundancy rate of various indicators in Table 4, it is evident from the national average that China's marine aquaculture industry generally involves significant input of resources, leading to considerable waste. Among the input indicators, the average redundancy of training intensity is the largest at 0.400, indicating that excessive investment in training during the sample period is the primary cause of the low green production efficiency in China's marine aquaculture industry. In addition, the aquaculture area and total power of fishing vessels are also important factors contributing to the decrease in green production efficiency, with redundancy degrees of 0.327 and 0.274, respectively. Considering the average redundancy of non-desired outputs, it is observed that there is an issue of excessive output in carbon emissions, nitrogen, phosphorus, and COD, the three major pollutants. Over time, there has been an overall downward trend in the redundancy of both non-desired output indications and input indicators. Only labor redundancies, training intensity, and carbon emissions were zero as of 2021, which explains why China's marine aquaculture sector has been able to continuously increase its green production efficiency. In addition, this draws attention to the crucial areas that need to be addressed in order to improve China's marine aquaculture industry's green production efficiency. These areas include the carbon emissions, aquaculture area, total power of fishing vessels, and intermediate consumption.

According to Table 5, when examining the input–output redundancies of marine aquaculture industry across provinces in China, it is observed that Hebei, Zhejiang, Guangdong, and Guangxi exhibit significant redundancy in labor force. Hebei and Zhejiang have redundancies in all five input factors and two non-desired output factors higher than the national average. Liaoning shows higher redundancies in aquaculture area and training intensity input factors compared to the national average. Jiangsu demonstrates higher redundancies in the total power of fishing vessels, aquaculture area, training intensity as input factors, and carbon emissions compared to the national average. Fujian only has a higher redundancy in training intensity as an input factor, which is lower than the national average. Shandong shows a higher redundancy in the training intensity as an input factor. Guangdong exhibits high redundancies in labor force input and pollutant emission rates. Guangxi has higher redundancies in the total power of fishing vessels and training intensity as input factors. Hainan's redundancy in labor force input is higher than the national average. Therefore, excessive input from a variety of sources and excessive carbon emissions are the primary causes of the decline in green production efficiency in China's marine aquaculture sector. Looking at the longitudinal perspective of inputs and non-desired outputs, the training intensity input redundancy is the highest, followed by aquaculture area. In terms of non-desired output indicators, except for Guangxi and Hainan, the redundancy in carbon emissions is higher than that of pollutant emissions in all other provinces. This suggests that, compared to pollutant emissions, carbon emissions constitute a greater barrier to the improvement in green production efficiency in the marine aquaculture sector. The main causes of China's marine aquaculture industry's decline in production efficiency are high carbon emissions and excessive resource inputs. Thus, the marine aquaculture business may effectively promote the increase in green production efficiency by minimizing non-desired outputs and enhancing input–output efficiency. Therefore, improving aquaculture technologies and reducing carbon emission rates are essential for China's marine aquaculture industry to experience an efficient and sustainable growth in the future.

Table 4. Average input–output redundancy rate of mariculture in China.

Index	Input Redundancy Rate					Redundancy Rate of Undesirable Output	
	Labor Force	Aquaculture Fishing Vessel	Breeding Area	Training Intensity	Intermediate Consumption	Carbon Emissions	Pollutant Discharge
2012	0.383	0.493	0.612	0.779	0.196	0.296	0.206
2013	0.281	0.422	0.537	0.662	0.155	0.279	0.208
2014	0.220	0.365	0.448	0.565	0.144	0.252	0.133
2015	0.184	0.372	0.488	0.593	0.158	0.213	0.155
2016	0.126	0.256	0.404	0.386	0.141	0.175	0.129
2017	0.081	0.277	0.231	0.329	0.116	0.165	0.068
2018	0.119	0.165	0.198	0.265	0.102	0.146	0.054
2019	0.015	0.195	0.223	0.253	0.115	0.155	0.017
2020	0.018	0.108	0.092	0.171	0.070	0.080	0.015
2021	0.000	0.085	0.041	0.000	0.025	0.070	0.000
Mean	0.143	0.274	0.327	0.400	0.122	0.183	0.098

Table 5. Average input–output redundancy rate of Chinese mariculture by province.

Index	Input Redundancy Rate					Redundancy Rate of Undesirable Output	
	Labor Force	Aquaculture Fishing Vessel	Breeding Area	Training Intensity	Intermediate Consumption	Carbon Emissions	Pollutant Discharge
Hebei	0.225	0.817	0.795	0.743	0.473	0.724	0.198
Liaoning	0.143	0.231	0.468	0.212	0.009	0.162	0.138
Jiangsu	0.086	0.550	0.502	0.561	0.284	0.356	0.031
Zhejiang	0.308	0.291	0.485	0.722	0.141	0.250	0.223
Fujian	0.078	0.071	0.059	0.171	0.018	0.013	0.006
Shandong	0.012	0.119	0.193	0.335	0.014	0.113	0.051
Guangdong	0.181	0.077	0.302	0.349	0.099	0.029	0.186
Guangxi	0.107	0.256	0.092	0.256	0.032	0.000	0.053
Hainan	0.144	0.053	0.049	0.254	0.030	0.000	0.000
Mean	0.143	0.274	0.327	0.400	0.122	0.183	0.098

3.2. Dynamic Efficiency Analysis of Green Production in the Chinese Mariculture Industry

The GML index approach was used in this study to examine how China's marine aquaculture industry's overall green production efficiency changed between 2012 and 2021. Based on this, the study divided the analysis into two categories: best practice gap change (BPC) and technical efficiency change (EC). BPC was used to quantify changes in green technological advancement, while EC was used to gauge the spread of green aquaculture technology.

Table 6 displays the GML index and its breakdown for the marine aquaculture sector in China. The findings show that, between 2012 and 2021, China's marine aquaculture industry's green overall production efficiency changed on average by 1.111, growing at an annual rate of 11.1%. This shows that, over the sample period, China's marine aquaculture business had an overall increase trend in green production efficiency. Examining the decomposition of the GML index, it is observed that both the technical efficiency change and technological progress have contributed to the growth in green production efficiency, accounting for 1.8% and 9.1%, respectively. This indicates that the progress of green technology in China's marine aquaculture industry is the driving force behind the growth of the overall green production efficiency, while the contribution of technical efficiency is relatively limited. This is consistent with China's increasing emphasis on green aquaculture technology and the continuous increase in research investment in aquaculture. The GML index mainly experiences positive growth during the sample period, with only a negative growth observed in 2014–2015. Minor variations exist in the growth rates as well. The greatest increase in growth occurred between 2013 and 2014, when it reached an annual growth rate of 18.5%. The State Council published "Several Opinions on Promoting the Sustainable and Healthy Development of Marine Fisheries" at this time, offering recommendations for the advancement of technology and the sustainable growth of the marine aquaculture sector.

Table 6. The changing trends of the GML index, EC index, and BPC index in marine aquaculture in China.

Year/Factor Breakdown	GML	EC	BPC
2012–2013	1.181	1.051	1.123
2013–2014	1.169	0.986	1.185
2014–2015	0.971	0.997	0.975
2015–2016	1.187	0.960	1.236
2016–2017	1.113	1.003	1.110
2017–2018	1.076	1.003	1.072
2018–2019	1.002	1.035	0.968
2019–2020	1.108	1.097	1.010
2020–2021	1.215	1.032	1.178
Mean	1.111	1.018	1.091

The green total factor productivity, technical efficiency, and technological advancement in China's marine aquaculture business showed a varying upward trend between 2012 and 2021. The green total factor production had a notably small fluctuation amplitude, which can be attributable to its breakdown into the product of technological development and technical efficiency. Changes in technical efficiency and advancements in technology work together to produce variations in green total factor productivity. Technology is the main driver of green development in China's marine aquaculture business, as evidenced by the close correlation between the trend of technological advancement and changes in green total factor productivity.

Using the GML index method and MATLAB software, we calculated the level of green total factor productivity. Figure 2 shows the GML index of the marine aquaculture industry in different Chinese regions. When considering changes in green total factor productivity (TFP), the nine provinces that were sampled for the study had an average GML index that was greater than 1, which suggests that, from 2012 to 2021, the green

TFP in the marine aquaculture sector in these provinces continued to improve. Among the provinces, the top five in terms of average GML index were Jiangsu (1.185), Hebei (1.135), Liaoning (1.123), Zhejiang (1.117), and Hainan (1.101), with annual growth rates of 18.5%, 13.5%, 12.3%, 11.7%, and 10.1%, respectively, ranking high in growth rates and exhibiting rapid improvements. Given the constraints of resource and environment, coastal regions have shown an increased attention to green total factor productivity, consistently looking for ways to advance the marine aquaculture sector's green development. The regional heterogeneity along the coastal areas may be attributed to the diverse climate and ecological conditions, as China's maritime territory extends from the Bohai Sea in the north to the South China Sea in the south, spanning 44 degrees of latitude. Consequently, different regions face varying conditions and constraints at different stages of industrial development in the marine aquaculture sector.

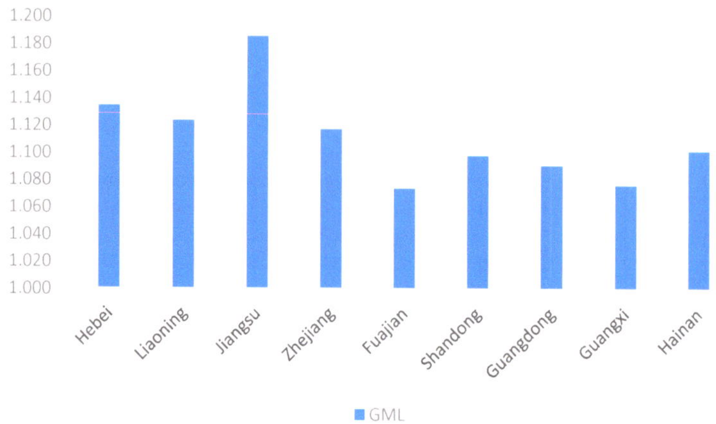

Figure 2. GML index by province.

From 2012 to 2021, the BPC and EC indices of various provinces in China are presented in Figure 3. Looking at the BPC index, the average BPC index of the nine coastal provinces was consistently above 1, indicating progress in the marine aquaculture technology of these regions. Among them, Zhejiang showed the fastest technological progress, with an annual growth rate of 11.3%. Additionally, Liaoning and Hainan also demonstrated notable progress in technology, with annual growth rates of 10.5% and 10%, respectively. However, Fujian exhibited relatively slower growth with an annual average of 3.7%, underscoring the need to enhance the technological level in the aquaculture process and elevate the green aquaculture technology in the marine aquaculture industry. Regarding the EC index, seven provinces had an annual average EC index exceeding 1, indicating a gradual convergence of production efficiency towards the production frontier. The provinces were ranked in descending order of annual average EC index as Jiangsu, Hebei, Liaoning, Fujian, Shandong, Zhejiang, and Hainan; Guangdong and Guangxi had average EC indices below 1, indicating a decline in green production efficiency in the marine aquaculture industry due to an irrational allocation of resource elements. When considering the combined effects of the EC and BPC indices, provinces with both indices exceeding 1 demonstrated continuous progress in the production process while maintaining a rational proportion of input factors. This combined effect has facilitated the improvement in green total factor productivity. For Guangdong and Guangxi, where the BPC index exceeded 1 but the EC index was below 1, it suggests that, while technological progress is occurring in the production process, the ineffective allocation of resources led to a decrease in green total factor productivity in the marine aquaculture industry, despite advancements in production technology.

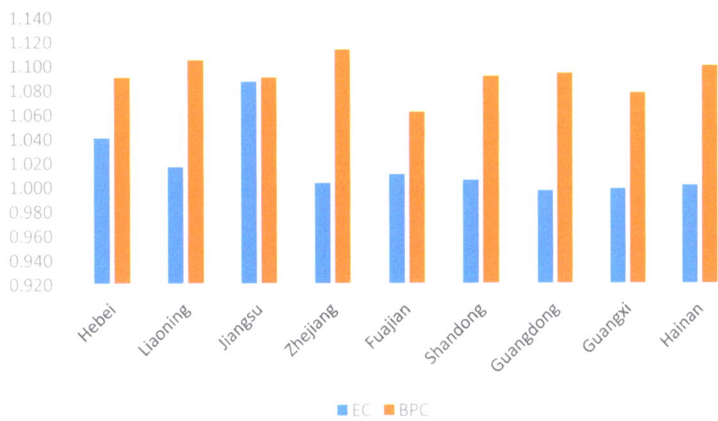

Figure 3. Distribution of EC and BPC indices by province.

Based on the analysis of the GML index, BPC index, and EC index as described above, we can conduct an analysis on the dynamic level of green development in the marine aquaculture industry of various provinces from 2012 to 2021. Solutions are proposed according to the issues faced by each province. In this article, we analyze two provinces that are representative. Firstly, let us focus on Guangdong. With a GTFP index of 1.090, it is positioned at a moderate level nationwide. The BPC index stands at 1.094, also ranking at a moderate level nationally. However, the EC index is below 1, placing Guangdong at the bottom nationally. In the Shandong region, the green total factor productivity in the marine aquaculture industry is mainly influenced by its technical efficiency index, with the progress in catching up to the production frontier significantly slower than other regions. Therefore, to enhance the green total factor productivity of the marine aquaculture industry in Shandong, it is essential to elevate the EC index by actively learning from the management and resource allocation models of other regions.

4. Conclusions and Suggestions

4.1. Conclusions

Using the super-efficiency SBM model, this study measured the green production efficiency in China's marine aquaculture sector from 2012 to 2021. It employed the GML index to examine the trend of the green total factor productivity in China's marine aquaculture sector and integrated the analysis of input–output redundancy rates to pinpoint the reasons behind losses in green production efficiency. The findings suggest that, from 2012 to 2021, China's marine aquaculture industry's average static green production efficiency was 0.705, at a moderate level, exhibiting a stepwise falling pattern of "south–north–east." Province-specific averages for green production efficiency ranged from 0.294 to 1.257, with an apparent rising trend in green production efficiency following 2016. The southern marine economic circle had the highest average green production efficiency during the study period. Excessive unexpected outputs and excessive inputs of factors like total fishing vessel power, aquaculture area, and training were found to be the primary causes of the losses in green production efficiency in China's marine aquaculture industry. These factors also impeded the improvement in green production efficiency. China's marine aquaculture business saw an average annual rise in green total factor production of 11.1%, when seen through the lens of the GML index technique and its decomposition. The technical efficiency grew by 1.8% annually, and the index of changes in the gap to the optimal practice increased by 9.1% annually. This indicates that the growth of green total factor productivity in China's marine aquaculture industry primarily relies on advancements in technology, with the contribution from changes in technical efficiency being relatively limited.

4.2. Suggestions

Based on the research conclusions above, the following strategic recommendations are proposed to enhance the level of green development in China's marine aquaculture industry. Firstly, it is essential to accelerate technological progress while improving technical efficiency in the marine aquaculture industry. The research within the sample period has demonstrated that technological progress is a crucial driver for promoting green total factor productivity, while the contribution from technical efficiency is relatively limited. Therefore, prioritizing the enhancement in technical efficiency is crucial for the green development of marine aquaculture across various provinces. Firstly, it is imperative to improve the communication channels between research institutions and aquaculture markets. If necessary, government support should be leveraged to strengthen the close linkage between research institutions and aquaculture markets, constructing a dynamic linkage mechanism among the government, research institutions, and aquaculture markets to ensure that the technological achievements of research institutions can be applied in aquaculture markets. Secondly, the evaluation of research institutions should be refined, with extended assessment periods and the inclusion of multiple indicators in the evaluation system to promote the technological research and development progress of research institutions. Secondly, efforts should be made to drive the coordinated development of the regional marine aquaculture industry. Coordinated and efficient resource allocation should be planned across provinces to collectively promote the high-quality green development of the marine aquaculture industry. Firstly, the differentiated development of various provinces should be coordinated. This should involve strengthening the close connections of marine aquaculture across different provinces, deepening exchanges and cooperation, and promoting the adoption of advanced aquaculture technologies. Secondly, it is important to expedite the training of large-scale aquaculture enterprises and leading businesses in embracing new knowledge and technologies, as they play a leading role in innovation and serve as exemplary leaders in green development. Lastly, efforts should be focused on enhancing the standardization of the marine aquaculture industry. This involves accelerating the establishment of a standardized system for the green development of the marine aquaculture industry, refining the relevant standards for input factors, pollutants, and carbon emissions, promoting the rational allocation of resources, and advancing the standardized development of green aquaculture.

Author Contributions: Writing—original draft, D.Y.; Writing—review & editing, Q.W. All authors have read and agreed to the published version of the manuscript.

Funding: This research received no external funding.

Institutional Review Board Statement: Not applicable.

Informed Consent Statement: Not applicable.

Data Availability Statement: The data are all from China Fishery Statistical Yearbook, China Statistical Yearbook and Manual of the First National Pollutant Source Census of Aquaculture Pollution Sources, in which carbon sink and carbon emission are calculated by referring to the research of scholars.

Conflicts of Interest: The authors declare no conflicts of interest.

References

1. Yang, Z.Y. On the Non point Source Nature of Endogenous Pollution in Fisheries and the Environmental and Economic Policies for Its Control. *Product. Res.* **2004**, 28–30. [CrossRef]
2. Yu, L.; Jiang, Q.J. The impact of heterogeneous environmental regulation on the green development of fishery in China. *Sci. Geogr. Sin.* **2024**, *44*, 319–328.
3. Chen, Q. Green and healthy aquaculture actions. Selection preferences and sources of heterogeneity of farmers: An analysis based on optimal and worst selection experiment. *J. Agrotech. Econ.* **2023**, 64–79. [CrossRef]

4. Lu, C.C. Thinking on promoting the new pattern of green development of aquaculture in China. *China Fish.* **2018**, *510*, 28–31.
5. Cao, J.H.; Sang, F.E. Thinking on the theory, model and evaluation method of green development of aquaculture industry. *Ecol. Econ.* **2020**, *36*, 101–106+153.
6. Yue, D.D.; Wang, L.M. Study on green development strategy of aquaculture in China. *China Fish.* **2018**, *512*, 34–37.
7. Ying, X.; Ying, P. Analysis of Green Development of Aquaculture in China Based on Entropy Method. *Sustainability* **2023**, *15*, 5585. [CrossRef]
8. Yue, D.D.; Wu, F.X.; Fang, H.; Ruan, W.; Ji, W.W. Evaluation of green development of mariculture in China. *J. Agric. Sci. Technol.* **2021**, *23*, 1–12.
9. Zhu, A.F.; Ping, Y. Spatiotemporal differentiation of green production efficiency of Marine fisheries in China based on SFA model. *Ocean Dev. Manag.* **2023**, *40*, 133–143.
10. Yu, L.; Jiang, Q.J. Research on Total factor productivity of new agricultural management agents in aquaculture industry: Based on micro-agent panel data and SFA method. *Chin. J. Agric. Resour. Reg. Plan.* **2023**, 1–13.
11. Ji, J.Y.; Zeng, Q. Spatiotemporal evolution of green technology efficiency in China's mariculture industry based on global DEA. *Chin. J. Manag. Sci.* **2016**, *24*, 774–778.
12. Yang, Z.Y.; Liu, D.; Peng, L.W. Green development of mariculture in China: Level measurement, regional comparison and development countermeasures. *Ecol. Econ.* **2021**, *37*, 128–135.
13. Qin, H.; Zhang, Y.; Lu, Y.Y. Measurement of eco-economic efficiency of mariculture in China based on SBM model. *J. Agrotech. Econ.* **2018**, *281*, 67–79.
14. Xu, Y.; Ji, J.Y.; Zhou, J.L. Study on the influence of beach culture on the growth of green total factor productivity of mariculture. *Sci. Technol. Manag. Res.* **2022**, *42*, 193–198.
15. Zhang, Y.; Ji, J.Y. Decomposition and influencing factors of green total factor productivity in China's mariculture industry. *Sci. Technol. Manag. Res.* **2022**, *42*, 206–213.
16. Jiang, Q.J.; Li, X.F. Analysis of influencing factors of green development of aquaculture industry based on AHP. *Mar. Econ.* **2021**, *11*, 15–22.
17. Kaoru, T. A slacks-based measure of efficiency in data envelopment analysis. *Eur. J. Oper. Res.* **2001**, *130*, 498–509.
18. Cheng, G. *Data Envelopment Analysis Method and MaxDEA Software*; Intellectual Property Publishing House: Beijing, China, 2014; p. 226.
19. Chung, Y.H.; Färe, R.; Grosskopf, S. Productivity and Undesirable Outputs: A Directional Distance Function Approach. *J. Environ. Manag.* **1997**, *51*, 229–240. [CrossRef]
20. Oh, D.-h. A global Malmquist-Luenberger productivity index. *J. Product. Anal.* **2010**, *34*, 183–197. [CrossRef]
21. Li, X.; Liu, Z.F.; Zhao, M.J.; Xu, L.J.; Sun, H.W. The "double carbon" goal and the realization path of aquaculture and fishing industry in China. *J. Agric. Sci. Technol.* **2022**, *24*, 13–26.
22. He, J.B.; Sun, J.R.; Zhao, Q.; Yang, G.H.; Ke, K.; Liu, P.; Wang, H.; Zhang, Y.H.; Liu, Y.S. Carbon sink contribution and capacity evaluation of shellfish culture in Yantai City. *Trans. Oceanol. Limnol.* **2022**, *44*, 117–122.
23. Shao, G.L.; Kong, H.Z.; Li, C. Net carbon sequestration and its coupling with economy in mariculture in China. *Resour. Sci.* **2019**, *41*, 277–288.
24. Xu, H.; Zhang, Z.L.; Zhang, J.H.; Liu, H.; Zhao, P.; Shi, R.; Wang, J.; He, Y.P. Research and development suggestions on energy conservation and emission reduction of fishery in China. *J. Fish. China* **2011**, *35*, 472–480.
25. Yang, Z.Y.; Peng, L.W.; Yan, Z.F. Dynamic evolution and convergence analysis of green development level of freshwater aquaculture in China: Based on the perspective of green total factor productivity. *J. Agrotech. Econ.* **2022**, 1–21. [CrossRef]
26. Xu, H.; Liu, H.; Zhang, J.H.; Ni, Q.; Shen, J.; Jiang, L. Chinese fishery energy consumption measurement. *China Fish.* **2007**, *384*, 74–76+78.
27. Xu, D.L.; Wang, Y.J. Regional differences and influencing factors of carbon productivity in coastal fisheries in China. *J. China Agric. Univ.* **2015**, *20*, 284–290.
28. Yue, D.D.; Wang, L.M.; Geng, R.; Wang, X.; Dai, Y.Y. Preliminary study on ecological value assessment of algae cultivation in offshore China. *J. Agric. Sci. Technol.* **2014**, *16*, 126–133.

 *sustainability*

Article

Digital Financial Inclusion, Land Circulation and High-Quality Development of Agriculture

Qi Xiong, Xiaoyang Guo * and Jingyi Yang *

Institute of Quantitative Economics, Huaqiao University, Xiamen 361021, China; plove250@hqu.edu.cn
* Correspondence: 21013021005@stu.hqu.edu.cn (X.G.); 22013021014@stu.hqu.edu.cn (J.Y.)

Abstract: With the deep integration of digital technology and inclusive finance, digital inclusive finance has provided a new opportunity for agricultural high-quality development through "overtaking on curves". This article empirically examines the impact of digital inclusive finance on agricultural high-quality development and the dynamic mechanism of land circulation in its transmission process, utilizing panel data from various provinces in China from 2011 to 2021. The research indicates that digital inclusive finance has a significant improvement effect on agricultural high-quality development, and this conclusion remains valid after a series of endogenous treatments and robustness tests. Meanwhile, intelligent manufacturing has a more pronounced role in promoting agricultural high-quality development in China's eastern regions, regions with sound infrastructure, and regions with high environmental regulation intensity. Further research reveals that digital inclusive finance can promote agricultural high-quality development through the mechanism of promoting land circulation. The research conclusions provide important empirical evidence and policy implications for achieving coordinated development of agricultural economic growth and environmental protection, thereby realizing the beautiful vision of comprehensive rural revitalization.

Keywords: digital financial inclusion; high-quality agricultural development; land circulation; environmental regulation

Citation: Xiong, Q.; Guo, X.; Yang, J. Digital Financial Inclusion, Land Circulation and High-Quality Development of Agriculture. *Sustainability* **2024**, *16*, 4775. https://doi.org/10.3390/su16114775

Academic Editors: Stephan Weiler, Fotios Chatzitheodoridis, Efstratios Loizou and Achilleas Kontogeorgos

Received: 12 April 2024
Revised: 14 May 2024
Accepted: 31 May 2024
Published: 4 June 2024

1. Introduction

Since the reform and opening-up, China has made remarkable achievements in economic construction. However, rapid economic growth comes at the cost of an increasing depletion of resources and deterioration of the ecological environment. Facing constraints such as resource shortages, industrial policies, and institutional barriers, the contradictory relationship between agricultural development and ecological environment governance has not been systematically resolved. There is an urgent need to promote quality-oriented agriculture [1]. Furthermore, agriculture, as an inherent ecological carbon sequestration system, manifests unique industrial attributes amidst the impact of climate change. Nevertheless, the present carbon sequestration capacity falls short of balancing the greenhouse gases emitted during production, thereby posing a significant threat to global food security, human health, and the sustainability of economic development [2]. Indeed, agricultural development has reached a crucial period where the development model of "trading the environment for growth" must be changed. Amidst the pursuit of fostering quality development, the report of the 20th National Congress of the Communist Party of China underscores the urgent need to drive agricultural transformation from quantitative expansion towards qualitative enhancement. Specifically, it emphasizes that the entire Party and society must endeavor to comprehensively bolster rural revitalization, expedite agricultural and rural modernization, incessantly enhance agricultural innovation and competitiveness, and resonate with the central theme of "quality agriculture, green agriculture, and brand-building agriculture". Hence, the transformation of agriculture's quantitative growth model, reliant on input factors, can be achieved through the integration of modern agricultural machinery and equipment, expedited technological innovations, and stringent environmental regulations. This approach will enhance the soil quality and carbon sequestration capabilities of farmland, thereby significantly augmenting the level of green production and resource utilization efficiency. Ultimately, it will foster a harmonious alignment of economic, social, and ecological benefits [3]. Additionally, this is imperative for safeguarding China's food security, societal stability, and attaining sustainable economic and environmental development.

Constrained by the marginality and aggregation of rural geographical spaces, weak infrastructure construction, and obstacles such as the "voluntary financial exclusion" of vulnerable groups, traditional inclusive finance is unwilling to penetrate into rural hinterlands, and financial services are difficult to extend to the "long-tail" population [4]. Drawing inspiration from the concept of "internal hematopoiesis" and aligned with the principles of sustainable development, digital inclusive finance maximizes its "multiplier effect" in resource allocation, enabling intelligent analysis and precise delivery of financial services tailored to vulnerable groups, thereby substantially elevating the efficiency of capital matching [5,6]. This enables farmers who have long been excluded by traditional financial services to obtain formal financial support, further contributing to the modernization of traditional agricultural industries, the facilitation of rural public services, the intelligentization of rural governance, the transparency of rural credit systems, and the greening of agricultural industries. Concurrently, while significantly contributing to the traditional financial service system, digital inclusive finance profoundly transforms and revolutionizes agricultural production and economic modalities. In terms of its green attributes and positive environmental externalities, digital inclusive finance consistently assumes the role of green finance, offering an efficacious means to elevate the quality of agricultural development in China [7]. The jointly issued "Key Points for the Development of Digital Villages in 2023" by the Cyberspace Administration of China, the Ministry of Agriculture and Rural Affairs, and four other departments underscores that digital inclusive finance, as an enabler of high-quality agricultural development, serves as the cornerstone for achieving supply-side structural reform and rural revitalization under the "dual carbon" target. Furthermore, it constitutes a fundamental component in China's pursuit of high-quality economic development.

Given this, this article utilizes panel data from various provinces in China from 2011 to 2021 to focus on exploring the impact mechanism of digital inclusive finance on high-quality agricultural development, and analyzes whether land transfer can help tap into the technological and structural dividends of digital inclusive finance in terms of agricultural green and low-carbon transformation, aiming to provide theoretical and empirical evidence for cultivating new competitive advantages and developing new momentum in agriculture.

The research contributions of this article are mainly reflected in the following aspects: Firstly, it focuses on the ecological function of agriculture, establishing a theoretical analysis framework for digital inclusive finance and high-quality agricultural development, and quantitatively assessing their impact effects based on provincial-level data in China. Secondly, unlike heterogeneity analysis based on geographical location, this article divides the samples based on factors such as the degree of digital infrastructure improvement, environmental regulation intensity, and urban location characteristics, identifying the heterogeneous impact of digital inclusive finance on high-quality agricultural development. The conclusions obtained are more conducive to administrative departments and financial institution managers to take precise measures to address deficiencies. At the same time, this article does not simply use the mean of sample data for sample differentiation, but rather relies on the actual economic development status of the region and relevant policy documents, indirectly evaluating the effectiveness of green financial reform pilot zones and national-level big data comprehensive experimental zones in terms of high-quality agricultural development. Thirdly, this article innovatively explores the role of land transfer in the process of digital inclusive finance influencing high-quality agricultural development, thereby expanding the research perspective and content of digital inclusive finance supporting green agricultural development.

The subsequent sections of this scholarly article are structured as follows: The second chapter offers a comprehensive literature review encompassing the essence, evaluation metrics, and contributing factors of high-quality agricultural progression. Additionally, it delves into the mechanisms and avenues through which digital inclusive finance influences this development. The third chapter examines the impact trajectory of digital inclusive finance on high-quality agricultural development, highlighting the pivotal role played by land transfer in this intricate process. This analysis culminates in the formulation of this article's research hypotheses. The fourth chapter introduces the methodological framework and data sources utilized in this article. The fifth chapter employs a range of mathematical and statistical models to validate the stimulatory effect of digital inclusive finance on agricultural advancement, as well as the mediating role of land transfer. The sixth chapter summarizes the key findings, recommends pertinent policies, and discusses the limitations of the present study while outlining potential directions for future research. A visual representation of the research trajectory is provided in Figure 1.

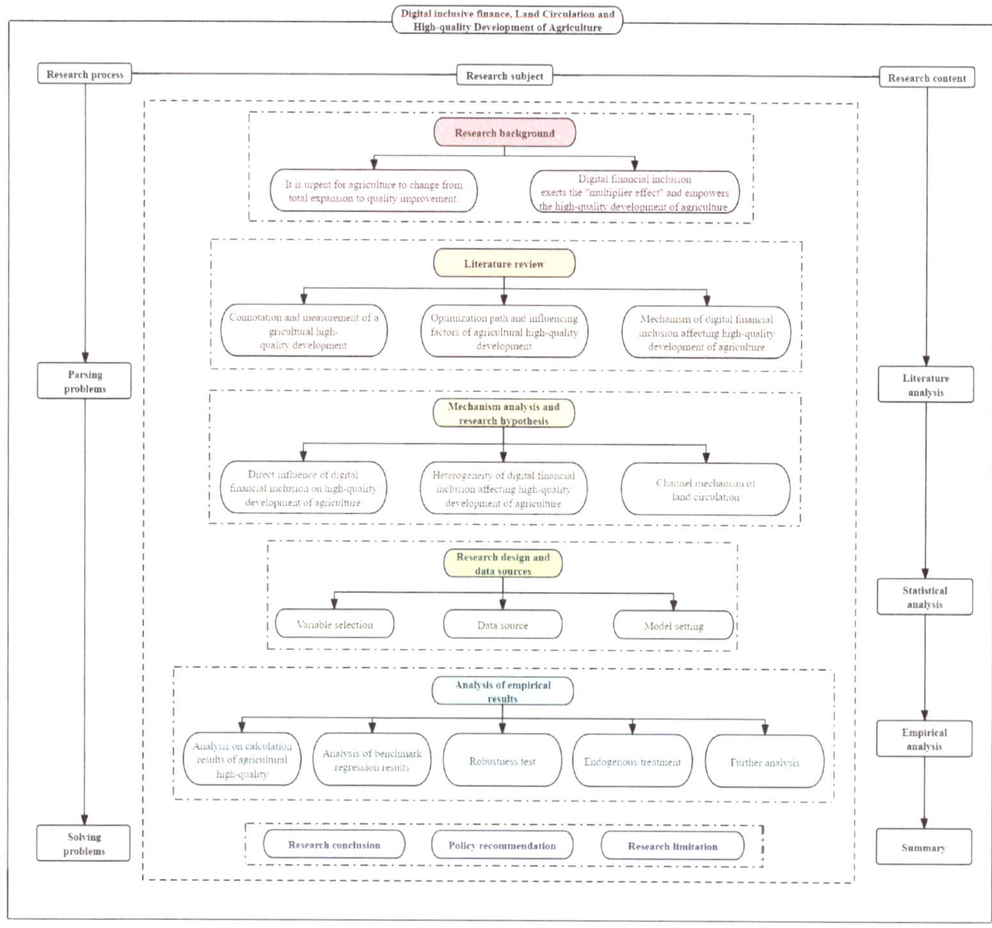

Figure 1. General framework of research.

2. Literature Review

As the product of the integration and innovation of traditional inclusive finance and modern information technology, digital inclusive finance has become a hot topic of concern for governments and scholars in recent years as they explore how to fully unleash its boosting power for high-quality agricultural development, enhance the efficient allocation of production resources, and ensure widespread access to financial digital dividends. A review of the existing literature reveals that research pertaining to this topic can be broadly categorized into the following dimensions:

Firstly, there is the analysis and measurement evaluation of the connotation of high-quality agricultural development. Regarding the conceptual elaboration of high-quality agricultural development, the existing studies generally believe that a sound agricultural production and management system, a diversified supply system of agricultural products, high-level agricultural production efficiency, and high-quality international competitiveness in agriculture are the inherent requirements of high-quality agricultural development [8,9]. Wang et al. (2022) proposed that the essence of high-quality agricultural development is the process of improving agricultural quality and efficiency, upgrading, and modernizing agricultural and rural areas, starting from the power transformation led by advanced technology, focusing on the transformation of agricultural economic efficiency, and aiming to achieve the transformation of agricultural development quality [10]. In terms of scientifically measuring the level of high-quality agricultural development, the academic community has not yet reached a unified standard. Some scholars use multiple indicator-weighting methods such as the multi-objective linear weighted function method, entropy weight method, and fuzzy comprehensive evaluation method to measure it. For example, Huang et al. (2023) constructed a comprehensive evaluation system for high-quality agricultural development from four dimensions: agricultural economic efficiency and stability, optimization of agricultural economic structure, leadership in agricultural green production, and healthy and sustainable agricultural development [11]. They also used the generalized Bonferroni curve to depict its spatial and temporal evolution and convergence. Yang et al. (2024) used a three-stage DEA dynamic analysis model based on spatial heterogeneity and found that high-quality agricultural development level showed a geographic "blockchain" convergence trend with obvious radiation and driving effects [12]. Additionally, a few scholars have used various methods such as Dagum's Gini coefficient decomposition method, kernel density estimation, Markov chain, and spatial Markov chain to analyze the high-quality development status of agriculture in various regions of China. They found that the overall trend is upward, and the regional gap is continuously narrowing [13,14].

The second aspect concerns research on the optimization paths and influencing factors of high-quality agricultural development. In discussing how to promote high-quality agricultural development, many scholars have proposed targeted paths from the perspectives of the industrial system, production system, and management system. For instance, leveraging digital technology to promote deep integration of industries and optimize agricultural industrial structure, strengthening agricultural technological innovation and promoting low-carbon agricultural development, improving the efficiency of land resource allocation, and cultivating new agricultural business entities [15–17]. In terms of exploring the factors influencing high-quality agricultural development, most of the existing literature employs methods such as the Logarithmic Mean Divisia Index (LMDI) decomposition method and the Stochastic Impacts by Regression on Population, Affluence, and Technology (STIRPAT) model. These studies have concluded that technological innovation capability, industrial capital intensity, industrial structure, innovative human capital, and environmental regulations can effectively impact high-quality agricultural development, though the research conclusions revealed vary [18]. As exemplified by Li et al. (2021), the process of household registration urbanization and permanent urbanization has been found to enhance the quality of agricultural development by facilitating more efficient resource allocation, deeper capital investment, and wider technological dissemination [19]. Han et al. (2022) put forth the argument that challenges including the "small-scale, fragmented, and disorganized" nature of agricultural enterprises, the significant loss of rural labor, agricultural environmental pollution, and the low agricultural total factor productivity have posed considerable impediments to achieving high-quality agricultural development, thereby limiting the competitiveness of China's agricultural products and sector [20].

The third dimension entails innovative inquiries into the underlying mechanisms and pathways that digital inclusive finance exerts its influence on through the attainment of high-quality agricultural development. As the concept of digital inclusive finance is relatively new, and its connotations are constantly enriching and evolving, there are few studies that delve deeply into the relationship between digital inclusive finance and high-quality agricultural development. The existing literature mainly focuses on whether digital inclusive finance, relying on the Internet, 5G, and big data technology, can promote high-quality agricultural development by correcting factor misallocation, alleviating credit constraints, and exerting energy-saving and carbon reduction effects. However, there are still some differences in the research conclusions that have been formed. Guo et al. (2022) and Shen et al. (2023) both pointed out that digital inclusive finance can promote the decentralization of the financial industry through financial deepening, digitization, and intelligence, breaking through the spatiotemporal limitations of traditional financial services and the "financial exclusion" of agricultural business entities, thereby alleviating financing constraints faced by agricultural development [21,22]. Such advancements facilitate farmers' adoption of agricultural machinery and innovative agricultural technologies, thereby enhancing agricultural production efficiency, minimizing resource consumption and environmental degradation, and ultimately advancing the intensification, industrialization, and modernization of agricultural production. Moreover, Hong et al. (2022) noted that as a promoter of information dissemination and sharing, digital inclusive finance enriches and improves the new agricultural socialized service system, driving the development of agriculture-related industries and facilitating the transfer of rural labor in various stages of agricultural production, circulation, sales, and services [23]. This alleviates the phenomenon of "excessive concentration" of labor in agricultural production and supports the efficient and green transformation of agriculture.

In summary, while numerous scholars have conducted extensive and impactful research on the nexus between digital inclusive finance and high-quality agricultural development, offering numerous insightful conclusions that provide both conceptual provocation and empirical insights for this study, there remain certain research gaps. The existing studies rarely combine digital inclusive finance and high-quality agricultural development within the same analytical framework, and there is an even greater lack of exploration of the intrinsic mechanism from the perspective of land circulation. Therefore, this article explores the impact of digital inclusive finance on high-quality agricultural development and its mechanism of action based on provincial panel data from 2011 to 2021 in China, providing valuable practical references for the construction of an agricultural powerhouse and the implementation of the rural revitalization strategy.

3. Theoretical Analysis and Research Hypothesis

3.1. Direct Influence of Digital Inclusive Finance on High-Quality Development of Agriculture

High-quality agricultural development refers to the joint participation of basic production factors such as labor, capital, and land. By expanding agricultural production scales, enhancing production efficiency, and optimizing the allocation of production factors, high-quality agricultural development aims to mitigate the negative externalities of agricultural production and achieve high-quality, vibrant, and low-pollution agricultural development. However, substantial capital investment is imperative for such development. Obstacles like fragmented agricultural capital demand and challenges in gathering farmers' credit information hinder direct support from traditional financial institutions. Amidst such significant capital requirements and pressure for green transformation, digital inclusive finance emerges as an inevitable trend. Leveraging digital technology, digital inclusive finance can significantly enhance capital matching efficiency and financial service accessibility in the agricultural sector at a low cost, reduce information asymmetry, and implement targeted risk mitigation, thereby providing robust financial support for agricultural production and innovation activities, ultimately fostering high-quality agricultural development [24]. Specifically, digital inclusive finance harnesses information technologies encompassing the internet, mobile communication networks, and vast data analytics to transcend geographical limitations, broaden the scope of rural financial services, and establish agricultural big-data-driven credit systems. This facilitates financial institutions in identifying potential capital seekers efficiently and precisely, thereby mitigating financial exclusion in remote rural regions and augmenting farmers' access to productive credit support [25]. The reduction in agricultural loan thresholds favorably contributes to encouraging the adoption of agricultural machinery and equipment, improved crop varieties, and innovative agricultural technologies among smallholder farmers, large-scale growers, and agricultural enterprises. This augmentation not only elevates agricultural production efficiency but also mitigates environmental degradation stemming from over-cultivation and excessive fertilizer usage, thereby fostering the intensification, industrialization, and modernization of agricultural production [26,27]. At the same time, digital inclusive finance innovates agricultural financial models, such as agricultural crowdfunding, internet wealth management, Internet of Things insurance, and supply chain finance. These models can meet the capital needs of agricultural development while reducing agricultural production risks (such as crop failures, natural disasters, and sluggish sales of agricultural products), ensuring stable agricultural development [28]. Furthermore, due to its targeted marketing capabilities and inherent green attributes, digital inclusive finance can reduce moral hazards and adverse selection issues caused by information asymmetry. It possesses the capability to precisely and conditionally allocate adequate funds to environmentally sustainable business activities in domains such as plant cultivation, breeding, animal husbandry, and rural tourism. This facilitates the influx of high-quality production factors into green agriculture and other environmental protection projects, thus promoting the green transformation of traditional agricultural practices [29,30]. Furthermore, alongside the advancement of underlying financial technologies, new media platforms encompassing mobile payments, online lending, and mobile banking have gained widespread adoption in rural areas. Gradually integrating green and low-carbon concepts, environmental protection projects such as "Ant Forest" have been launched. Through online virtual energy collection and afforestation, these projects are ultimately transformed into real-world afforestation projects, directly promoting high-quality and green agricultural development [31].

Based on the above analysis, this paper puts forward research Hypothesis 1:

Hypothesis 1. *Digital inclusive finance can empower high-quality agricultural development by easing credit constraints, sharing agricultural production risks and promoting the development of green agriculture.*

3.2. Heterogeneity of Digital Inclusive Finance Affecting High-Quality Development of Agriculture

Due to differences in factors such as digital infrastructure construction, environmental regulation intensity, and financial development level, the efficiency of digital inclusive finance cannot truly benefit remote areas, resulting in significant heterogeneity in the impact of digital inclusive finance on the high-quality development of agriculture. Specifically, first, digital infrastructure construction, as a key material basis for breaking the problems of small-scale and extensive agricultural development, can effectively guarantee the conduct of agricultural production and business activities, enhance the resilience of the agricultural economy, and provide good support for accelerating the development of facility agriculture and the construction of high-standard farmland in rural areas. It is an important historical opportunity for achieving high-quality and green agricultural development. In regions where digital infrastructure is more comprehensive, the capabilities for information collection, transmission, sharing, and data processing are substantially bolstered. This aids digital inclusive finance in maximizing the long-tail effect of the digital economy, enabling real-time feedback on agricultural production data, refining crop production processes, and attaining precise allocation of pesticides, fertilizers, and water resources. This can promote the flow and sharing of factor resources throughout the agricultural industry chain, improve production models, and reduce agricultural carbon emissions [32–34]. At the same time, the development of digital platforms facilitates farmers' direct access to high-quality online education through internet channels, improving their professional abilities and overall quality, enabling them to adapt to employment situations at different stages and meet the needs of different industrial developments, thereby supporting the modernization of the agricultural industry. In addition, the "siphon effect" released by the improvement in digital infrastructure attracts more and more rural surplus labor to transfer to non-agricultural sectors in cities and towns, accelerating the alleviation of the problem of "excessive concentration" of agricultural labor, thereby driving the effective circulation of land resources and improving the efficiency of green agricultural production.

Moreover, in accordance with the "compliance cost" effect of environmental regulation, an escalation in the rigor of environmental regulations will subsequently increase pollution control costs for market participants. These costs include the recuperation expenses for agricultural waste, the procurement costs of clean production factors, and the acquisition or upgrade expenses for agricultural machinery and equipment, as well as the costs associated with environmental restoration. This crowds out the productive investments of agricultural producers, especially the research and development investments in clean production technologies that require large upfront investments and long cycles, leading to the preference of "two highs and one surplus" industries for areas with looser environmental regulations to achieve cost savings [18,35]. In areas with stronger environmental regulation, agricultural producers will reflect on issues such as low factor utilization and high pollution emissions in their production processes, prompting them to adopt green production technologies to optimize factor allocation, reduce pollution emissions, and increase product added value, thereby alleviating or offsetting the energy-saving hard constraints brought by environmental regulation policies. At the same time, as the government pays increasing attention to environmental protection issues, financial products such as green securities, green bonds, and green funds are constantly being enriched, guiding capital elements to flow towards rural new formats such as rural tourism, rural e-commerce, circular agriculture, and smart agriculture, meeting the funding needs of eco-friendly agricultural industries, promoting the green upgrading of the agricultural industrial structure, and thus driving the high-quality development of agriculture [36,37].

Drawing upon the aforementioned analysis, this paper advances research Hypothesis 2:

Hypothesis 2. *Due to differences in digital infrastructure construction, environmental regulation intensity, and location factors, there exists heterogeneity in the impact of digital inclusive finance on the high-quality development of agriculture.*

3.3. Channel Mechanism of Land Circulation

Land, being the focal point of agricultural development research, serves as an indispensable means of production in agricultural production activities. Given the gradually tightening constraints of resources and the environment, the key to promoting high-quality agricultural development lies in overcoming the obstacles of small-scale fragmented operations among smallholder farmers. With the deepening implementation of land transfer policies, farmers have expanded their business scale by transferring land, optimized factor allocation, and altered factor output elasticity, thereby promoting high-quality agricultural development to a certain extent [38,39]. Specifically, first, the expansion of land transfer scale helps overcome the long-standing practical challenges faced by the domestic agricultural production sector, such as land fragmentation, high cost depletion, and low economic efficiency. This transformation guides agricultural production from "survival ethics" to "profit maximization", providing the possibility for contiguous land management and large-scale planting and breeding [40,41]. Additionally, under stable land ownership, land transfer stimulates the integration of agricultural machinery technology applications and advanced management techniques, maximizing factor combination productivity, thereby reducing agricultural non-point source pollution and enhancing agricultural environmental efficiency. Furthermore, land transfer facilitates the redistribution of land property rights among entities with varying behavioral capabilities and preferences for operational decision-making, thereby enabling market participants to engage in industrial specialization activities grounded in their comparative advantages [42]. For instance, farmers possessing non-agricultural comparative advantages can enhance the congruency between their remaining land and other production factors by transferring surplus land, thereby facilitating more efficient planning and consolidation of fragmented land parcels. Moreover, land transfer is not a simple process of transferring in or out. It promotes the integration of land parcels and forms a policy "combination punch" with agricultural infrastructure construction projects, contributing to the rational planning and layout of various aspects such as fields, roads, mountains, water, and villages, ultimately forming a modern agricultural operation mode characterized by "interconnected fields", "integrated field and water systems", and "integrated rural villages". Lastly, as the country places increasing emphasis on farmer training and strongly advocates for agricultural professionals to engage in the agricultural sector, agricultural human capital continues to sink into rural areas, giving rise to a new breed of agricultural operators such as large-scale grain farmers and professional farmers. After rural land is transferred to these new agricultural operators, they are more inclined to adopt modern agricultural production techniques and scientific management models due to their more comprehensive agricultural knowledge structure [43]. This not only promotes agricultural specialization and professionalization but also contributes to the promotion of appropriate land-scale operations, enhancing agricultural "increment, quality, and efficiency", ultimately achieving high-quality agricultural development [44].

However, land transfer, as a contractual process between land transferors and transferees, faces challenges in rural remote areas due to limited information exchange and high transaction costs, significantly hindering the improvement in rural land transfer systems. Digital inclusive finance, as an information dissemination carrier, alleviates information asymmetry between land supply and demand subjects and provides effective solutions to credit constraints such as "difficult and expensive financing" and "where to acquire the money" that limit land transfer and agricultural expansion. This not only effectively supports the development of land transfer and promotes the rational allocation of land but also avoids resource waste caused by fragmented planting, thereby contributing to high-quality agricultural development.

Based on the above analysis, this paper puts forward research Hypothesis 3:

Hypothesis 3. *Land circulation serves as a pivotal intermediary in the process where digital inclusive finance fosters high-quality agricultural development.*

4. Research Design and Data Sources

4.1. Variable Selection

4.1.1. Explained Variable

High-quality agricultural development (HAD) pertains to the attainment of two significant objectives: "quality enhancement" in terms of environmental improvement and reduction in resource consumption, alongside "efficiency augmentation" through the application of novel technologies and seamless trade, all while fulfilling the "incremental" aspirations such as agricultural production growth and income augmentation. In other words, it aims to comprehensively achieve the triple goals of "increment, quality improvement, and efficiency enhancement" in agriculture. In light of this, adhering to the overarching principles of high-quality economic development and the blueprint of rural revitalization strategic planning, this article adopts the research framework of Tang et al. (2022) and Wen et al. (2023) to devise an evaluation index system for high-quality agricultural development encompassing five key dimensions: innovation, coordination, greenness, openness, and sharing (as detailed in Table 1) [45,46]. Specifically, innovation serves as the core driving force for high-quality agricultural development, able to accelerate the breakthrough of agricultural "bottleneck" technologies through innovative inputs such as agricultural fiscal investment and agricultural machinery applications, thereby driving the improvement in agricultural product quality. Coordination is a characteristic of high-quality agricultural development, which requires the coordinated development of the primary, secondary, and tertiary industries in rural areas. Greenness, as the form of high-quality agricultural development, demands that agriculture must follow a low-carbon, energy-saving, and environmentally friendly development path. Openness is the trend of achieving high-quality agricultural development. The long-term high-quality development of agriculture cannot be separated from the utilization and optimization of domestic and foreign resources and markets. It should actively participate in the international market of agricultural products, innovate agricultural trade strategies, and promote diversification of exports and imports. Sharing is the goal of high-quality agricultural development, which aims to satisfy the pursuit of farmers for increased income and a better life and achieve shared benefits.

Table 1. Evaluation index system of agricultural high-quality development.

Primary Index	Secondary Index	Three-Level Index	Specific Explanation	Attribute
Innovation	Innovation foundation	Agricultural mechanization level	Directly available data	+
		Proportion of agricultural financial investment	Proportion of fiscal expenditure of agriculture, forestry and water resources to total fiscal expenditure	+
		Proportion of demonstration counties of leisure agriculture	Proportion of leisure agriculture demonstration counties in the total number of counties in the local area	+
		Proportion of typical counties in rural entrepreneurial innovation	Proportion of typical counties of rural entrepreneurial innovation in the total number of counties in the local area	+
	Innovation benefit	Labor productivity	Proportion of total output value of agriculture, forestry, animal husbandry and fishery in the number of employees in the primary industry	+
		Land productivity	Proportion of total agricultural output value to total sown area of crops	+
		Number of green food enterprises	Number of certified units of green food in that year	+
		Grain yield per unit area	Proportion of grain output in the total grain planting area	+
		Effective irrigation area	Directly available data	+
Coordination	Industrial coordination	Agricultural industrial structure adjustment index	1—(Proportion of total agricultural output value to total agricultural output value)	+

Table 1. *Cont.*

Primary Index	Secondary Index	Three-Level Index	Specific Explanation	Attribute
Coordination	Urban–rural coordination	Binary contrast coefficient	Proportion of comparative labor productivity of primary industry to comparative labor productivity of secondary and tertiary industries	+
Green	Resource consumption	Usage of agricultural film per unit area	Proportion of agricultural film usage in sowing area	−
		Use intensity of agricultural diesel oil	Proportion of agricultural diesel oil in sowing area	−
		Per capita electricity consumption	Proportion of rural electricity consumption to employees in the primary industry	−
	Environmental pollution	Fertilizer application per unit area	Proportion of chemical fertilizer application rate to sowing area	−
		Application amount of pesticide per unit area	Proportion of pesticide application amount to sowing area	−
	Environmental protection	Forest coverage rate	Directly available data	+
Open	Resource optimization	Rural land circulation rate	Proportion of household contracted land transfer to agricultural land	+
		Proportion of investment in agricultural fixed assets	Proportion of fixed assets investment in agriculture, forestry, animal husbandry and fishery to total fixed assets investment	+
		Proportion of foreign direct investment in agricultural investment	Proportion of foreign direct investment in agricultural investment to total agricultural investment	+
	Market optimization	Market quantity of agricultural products	Directly available data	+
		Market turnover ratio of agricultural products	Proportion of agricultural products market turnover in the added value of the primary industry	+
		Dependence on import and export of agricultural products	Proportion of import and export trade volume of agricultural products in the added value of primary industry in China	+
Share	Standard of living	Income level of rural residents	Per capita net income of rural residents	+
		Overall affluence level of rural residents	Engel coefficient in rural areas	−
		Life richness of rural residents	Proportion of per capita education, culture and entertainment expenditure to per capita consumption expenditure	+
		Rural residents' attention to medical care	Proportion of per capita health care expenditure to per capita consumption expenditure	+
		Proportion of minimum living security for rural residents	Directly available data	−

Table 1. *Cont.*

Primary Index	Secondary Index	Three-Level Index	Specific Explanation	Attribute
Share	Benefit sharing	Income ratio of urban and rural residents	Proportion of disposable income of urban households in rural per capita disposable income	−
		Urban–rural consumption level ratio	Proportion of urban residents' per capita consumption expenditure to rural residents' per capita consumption expenditure	−
		Urban–rural consumption gap	Proportion of retail sales of consumer goods in towns and villages in the retail sales of consumer goods in the whole society	+

In the construction of the agricultural high-quality development index, the disparity in weight allocation among various sub-indicators critically impacts the reliability and precision of the evaluation outcomes. Therefore, this article employs the fixed-base range entropy weight method to assess and analyze the level of high-quality agricultural development. This approach integrates the entropy weight method with the fixed-base range method, which not only mitigates the subjectivity of weight assignment but also establishes a universally applicable reference framework through fixed-base methodology, thus reflecting the trends of change in both spatial and temporal dimensions. The specific computational procedures are outlined below:

Step 1: The above-mentioned positive and negative indicators in the agricultural high-quality development indicator system should be subject to non-dimensionalization. The treatment method is as follows:

$$Y_{ij}^t = \begin{cases} \dfrac{X_{ij}^t - \min\left(X_{ij}^t\right)}{\max\left(X_{ij}^t\right) - \min\left(X_{ij}^t\right)}, X_j^t \text{ is a positive indicator} \\[3mm] \dfrac{\max\left(X_{ij}^t\right) - X_{ij}^t}{\max\left(X_{ij}^t\right) - \min\left(X_{ij}^t\right)}, X_j^t \text{ is a negative indicator.} \end{cases} \tag{1}$$

Among them, X_{ij}^t represents the original data of the j index of the i province in the t year; and Y_{ij}^t is the data after dimensionless processing using the range method. For positive indicators, the greater the numerical value, the greater its contribution to the index and the better its performance, whereas for negative indicators, the smaller the numerical value, the greater its contribution to the index and the better its performance.

Step 2: Calculate the proportion of the indicator value in the t-th year under the j-th indicator.

$$P_{ij}^t = \frac{Y_{ij}^t}{\sum_{i=1}^n Y_{ij}^t}, i \in [1,n], j \in [1,m] \tag{2}$$

where P_{ij}^t represents the proportion of the i-th province under the j-th indicator in year t compared to that indicator. If the specific gravity value $P_{ij}^t = 0$, $lim_{Y_{ij}^t \to 0} P_{ij}^t \times \ln\left(P_{ij}^t\right) = 0$ is defined.

Step 3: Compute the index information entropy (E).

$$E_j^t = -[\ln(n)]^{-1} \times \sum_{i=1}^n \left[P_{ij}^t \times \ln\left(P_{ij}^t\right) \right] \tag{3}$$

where E_j^t denotes the information entropy of the (j)-th index in the (t)-th year. A smaller index information entropy signifies a higher degree of data dispersion, indicating a greater amount of information provided and thus a higher index weight. Conversely, a larger index information entropy leads to a smaller index weight.

Step 4: Calculate the weight of the j-th indicator.

$$W_j^t = \frac{\left(1 - E_j^t\right)}{\sum_{j=1}^{m}\left(1 - E_j^t\right)} \tag{4}$$

where W_j^t is the weight of the j-th index. The greater the weight of an indicator, the greater its contribution to the measurement result.

Step 5: Utilize the fixed-base range method to process the original data.

$$Z_j^t = \frac{X_j^t - \min\left(X_{j,min}^{2011}\right)}{\max\left(X_{j,max}^{2011}\right) - \min\left(X_{j,min}^{2011}\right)} \tag{5}$$

Here, Z_j^t represents the dimensionless index value of the (j)-th index in the (t)-th year after being processed using the fixed-base range method. X_j^t denotes the original data, while $X_{j,min}^t$ and $X_{j,max}^t$ signify the minimum and maximum values of the (j)-th index in the original data of all cities during the base year, respectively. This article designates 2011, the initial year of the sample, as the base year.

Step 6: Compute the comprehensive index. This paper utilizes the index weight determined by the entropy weight method to weight the dimensionless index value processed using the fixed-base range method, resulting in the derivation of the comprehensive index for agricultural high-quality development in each province:

$$S_j^t = \sum_{j=1}^{m}\left(W_j^t \times Z_j^t\right) \tag{6}$$

4.1.2. Explanatory Variable

The original intention of digital inclusive finance (DFI) is to improve the availability of financial-related services through the continuous popularization and strengthening of digital financial basic services, and to provide more convenient financial and credit services to all sectors of society, especially low-income groups such as rural underdeveloped areas and farmers, at low cost. In view of this, this paper refers to the practices of Ge et al. (2022) and Fu et al. (2024), and selects the digital inclusive finance index compiled by the Digital Finance Research Center of Peking University to measure the development level of digital finance [47,48]. The index is constructed based on extensive real-world transaction data from Ant Financial, comprehensively analyzing financial development in terms of coverage, usage depth, and the extent of digital support services. This approach aims to capture the convenience and inclusiveness fostered by the development of digital inclusive finance.

4.1.3. Mediating Variable

Land transfer (LT) refers to the process of activating the operation and use rights of rural land and promoting the orderly flow of rural land management rights. Driven by relevant policies of land transfer, the scale of domestic agricultural land transfer continues to expand. However, due to the widespread existence of problems such as "difficult and expensive loans" in rural areas, the scale of land transfer has not yet reached the ideal state. Most of the existing studies on the measurement of the probability of agricultural land transfer focus on micro data such as questionnaires, and there is no systematic research on the measurement of agricultural land transfer at the provincial level. In view of this, this paper refers to the practices of Wang et al. (2023) and Yang et al. (2023) and selects the total area of cultivated land transfer contracted by rural households (in ten thousand mu) as a measure of land transfer level [49,50]. This indicator includes the area of lease (subcontracting), transfer, exchange, joint-stock cooperation, and other forms of land transfer.

4.1.4. Control Variable

Given the multitude of macro and micro factors influencing inclusive growth, this article aims to minimize the bias stemming from omitted variables in model causal inference [25,51–53]. Consequently, it selects the following control variables based on the research perspectives found in existing literature: (1) rural education investment, quantified by the average years of education attained by rural residents; (2) internet popularization, measured by the number of rural broadband subscribers per 10,000 households in each province; (3) industrial structure upgrading, assessed by the ratio of the output value of the secondary industry to the tertiary industry; (4) urbanization level, represented by the proportion of urban to rural population in each province; and (5) fiscal environmental expenditure, calculated as the ratio of local fiscal expenditure on environmental protection to the local fiscal general budget expenditure.

4.2. Model Setting

To validate the direct influence of digital inclusive finance on the high-quality development of agriculture, combined with research Hypothesis 1 and the availability of data, this study constructs the following panel econometric model:

$$HAD_{it} = \alpha_0 + \alpha_1 DFI_{it} + \alpha_2 Control_{it} + v_t + \lambda_i + \varepsilon_{it} \tag{7}$$

where α_0 represents a constant term; α_1 and α_2, respectively, represent the regression coefficients to be fitted and calculated; subscripts i and t represent individuals and time, respectively; *Control* represents a series of control variables selected above; and λ_i and v_t represent individual fixation effect and time fixation effect, respectively. Individual-fixed effects can reflect the intrinsic differences among different individuals (which do not change over time), while time-fixed effects reflect the trend differences among cross-sections in different years (which do not change with individuals); and ε_{it} is a random disturbance term that obeys the white noise process.

Furthermore, to examine the channel through which digital inclusive finance influences the high-quality development of agriculture, particularly the mediating role of land circulation in this process, this paper, in line with research Hypothesis 3, employs the mediation effect model for fitting and computation. The specific model is detailed as follows:

$$LT_{it} = \beta_0 + \beta_1 DFI_{it} + \beta_2 Control_{it} + v_t + \lambda_i + \varepsilon_{it} \tag{8}$$

$$HAD_{it} = \gamma_0 + \gamma_1 DFI_{it} + \gamma_2 LT_{it} + \gamma_3 Control_{it} + v_t + \lambda_i + \varepsilon_{it} \tag{9}$$

Equations (1)–(3) together form the test equations of land circulation. According to the research ideas of Moon et al. (2015) and Ye et al. (2023), the precondition for verifying the mediation effect of land transfer is that parameter α_1 is significant [54,55]. Furthermore, if the sign of the indirect effect is consistent with that of the direct effect, it shows that the intermediary effect is established; otherwise, there is a masking effect.

4.3. Data Source

Adhering to the principles of data availability and consistency in statistical standards, this article selects panel data spanning from 2011 to 2021 for 30 provinces in China (excluding Tibet, Hong Kong, Macao, and Taiwan) as the research sample. The data sources and descriptive statistical analysis of each variable are shown in Table 2. For a few missing values, the Lagrange interpolation method is used to fill them in.

Table 2. Data sources and descriptive statistics of related variables.

Variable	Definition	Data Source	Obs	Mean	Std. Dev.	Min	Max
lnLT	Land transfer	China Rural Economic Management Statistics Yearbook	330	6.786	1.167	2.801	8.839
lnDFI	Digital financial inclusion	Peking University Digital Inclusive Finance Index (Third Edition)	330	5.556	0.682	2.026	6.136
lnHAD	High-quality agricultural development	China City Statistical Yearbook and various provincial statistical yearbooks	330	6.544	0.41	5.285	8.335
lnREI	Rural education investment		330	2.055	0.079	1.771	2.294
lnIP	The degree of internet popularization		330	3.963	0.272	3.186	4.521
lnISU	Industrial structure upgrading		330	0.126	0.415	−0.658	1.667
lnUR	Urbanization level		330	−0.537	0.198	−1.049	−0.110
lnFEE	Fiscal environmental expenditure		330	1.034	0.309	0.164	1.919

5. Empirical Result Analysis

5.1. Analysis of Benchmark Regression Results

The p values obtained from the likelihood ratio test and Hausman test both reject the null hypothesis at a significance level of 1%, thus prompting the adoption of the two-way fixed effects model as the benchmark for empirical testing in this paper. Additionally, to mitigate the potential influence of heteroscedasticity on the accuracy of cross-sectional data fitting, the Driscoll–Kraay (DK) estimation method is employed to process subsequent fitting results unless otherwise specified. The detailed benchmark regression results are presented in Table 3.

Table 3. Benchmark regression result.

Variable	(1)	(2)	(3)	(4)	(5)	(6)
lnDFI	0.238 *** (17.41)	0.181 *** (12.90)	0.108 *** (6.76)	0.105 *** (6.29)	0.956 *** (5.88)	0.098 *** (5.98)
lnREI		2.921 *** (8.43)	1.961 *** (5.75)	1.966 *** (5.75)	1.689 *** (5.04)	1.705 *** (5.08)
lnIP			0.422 *** (7.62)	0.427 *** (7.65)	0.016 (0.16)	0.013 (0.12)
lnISU				−0.014 (−0.73)	−0.028 (−1.45)	−0.029 (−1.50)
lnUR					1.241 *** (4.69)	1.239 *** (4.69)
lnFEE						−0.048 (−1.18)
Constant	5.221 *** (68.19)	−0.466 (−0.69)	0.237 (0.38)	0.230 (0.37)	3.148 *** (3.62)	3.166 *** (3.65)
N	330	330	330	330	330	330
R^2	0.6035	0.6992	0.7648	0.7654	0.7886	0.7901

Note: *** is significant at 1% significance levels.

For the sake of model robustness, a fitting approach was adopted by gradually incorporating control variables. As shown in Table 1, the results in column (5), which included all control variables, indicated that the estimated parameter for digital inclusive finance was 0.098 and significant at the 1% significance level. This suggests that digital inclusive finance has a significant promotional effect on high-quality agricultural development, initially confirming research Hypothesis 1. Potential explanations for this phenomenon lie in the utilization of digital inclusive finance's technological strengths, including artificial intelligence, blockchain, cloud computing, and big data. These technologies effectively diminish the reliance of traditional financial institutions on physical branches, thereby enhancing farmers' access to financial products such as internet insurance and investment management. This, in turn, facilitates the dispersal of financial resources and satisfies the productive capital requirements of agricultural development in less developed or "tail" regions. After obtaining sufficient funds, ordinary farmers and rural micro-enterprises will increase their application of agricultural machinery and equipment, agricultural varieties, and new agricultural technology equipment, reduce the use intensity of agricultural inputs such as fertilizer and pesticides, accelerate the realization of production scale and intensification, and thus provide new momentum for high-quality agricultural development. Concurrently, as green consumption gains prominence and the government tightens agricultural environmental regulations, digital inclusive finance not only mitigates financial exclusion in rural areas but also underscores the green credentials of agricultural output. To secure the backing of digital inclusive finance, agricultural production and operation entities are inclined to prioritize green investments in their operational activities. This approach aims to enhance pollution prevention and control capabilities in agricultural production processes, minimize agricultural carbon emissions and non-point source pollution, ultimately fostering the green transformation of traditional agriculture [56].

5.2. Robustness Test

The benchmark regression results confirm that digital inclusive finance can significantly promote high-quality agricultural development. To validate the robustness of the conclusions drawn, this paper employs three distinct methods for demonstration and interpretation, as evidenced in the specific estimation results presented in Table 4.

(1) Winsorization. Considering that agricultural production is extremely vulnerable to force majeure or major natural disasters, various agricultural production and operation entities will face significant survival risks, difficulties in cashing out agricultural products, extended financing periods for green investment projects, and market oversaturation, leading to outliers in the sample data and thus biasing the final regression results. Therefore, this paper performs a 1% winsorization on all continuous variables and then re-estimates using a two-way fixed effects model.

(2) Changing the sample size. Due to heterogeneity in factors such as digital infrastructure construction, innovation capabilities, resource endowments, and the scale of agricultural composite talents between China's municipalities (Beijing, Tianjin, Shanghai, and Chongqing) and other regions, as well as significant financial policy biases, this article re-estimates the model after excluding these samples.

(3) Replacing proxy indicators for explanatory variables. Referring to the research ideas of Lin et al. (2022), this paper replaces the overall digital inclusive finance index with the coverage breadth index and re-estimates using a two-way fixed effects model [57]. The coverage breadth embodies the financial broadening concept of digital finance, emphasizing the inclusion of more groups into the financial coverage and adjusting the flow of social funds through the rational allocation of financial resources. It is worth noting that according to the Peking University Digital Financial Inclusion Index, the coverage breadth index measures the usage of digital inclusive financial services from the perspective of account coverage, reflecting the outreach of digital financial services.

(4) Replace the model. When there are intra-group correlation, inter-group correlation, and contemporaneous correlation in the random disturbance term, the estimation results of the two-way fixed effects model may be biased. Moreover, in the baseline regression analysis mentioned above, to address potential issues of heteroscedasticity, autocorrelation, and cross-sectional dependence, the Driscoll–Kraay standard errors were utilized. However, apart from this standard error correction, the feasible generalized least squares (FGLS) method can also handle the three main threats posed by short panel data. Given the relatively small number of cross-sections, we allow each individual to have the same autoregressive coefficient during the estimation process, and employ the specific AR(1) autocorrelation structure that is characteristic of panel data.

Table 4. Robustness test and endogenous processing results.

Variable	Robustness				Endogenous Treatment	
	Tail Shrinking Treatment	Change the Sample Size	Replace Explanatory Variable	Replace the Model	IV2SLS	GMM
lnDFI	0.099 *** (7.83)	0.065 *** (4.35)	0.094 *** (5.19)	0.089 *** (4.71)	0.189 *** (2.83)	0.168 *** (3.30)
Control variable	Control	Control	Control	Control	Control	Control
Weak instrumental variable test					62.524	
Unidentifiable test					92.308 ***	
R^2	0.8030	0.6884	0.7616		0.7842	0.7917

Note: *** is significant at 1% significance levels.

Based on the empirical findings presented in Table 4, it is evident that the estimation coefficient associated with digital inclusive finance remains consistently positive in promoting agricultural high quality development, while the level of statistical significance remains largely unchanged. This consistency serves as a robust validation of the reliability and robustness of the aforementioned benchmark regression analysis.

5.3. Endogenous Treatment

Although this article controls for a large number of observable confounding factors when analyzing the relationship between digital inclusive finance and high-quality agricultural development, it still faces the endogenous problem of potential mutual causality between the two. For instance, high-quality agricultural development signifies a rationalized agricultural production paradigm and optimized resource allocation. This facilitates agricultural-oriented enterprises to actively raise and utilize capital, attaining intensification of operations and enhancing utilization efficiency, thereby bolstering the advancement of digital inclusive finance. Consequently, in adherence to the methodologies of causal inference within econometrics, this study employs a two-stage least squares approach to mitigate the endogeneity issue, effectively disentangling the underlying endogenous relationship between the two phenomena. Specifically, drawing on the research methods of Zhang et al. (2021) and Gao et al. (2022), we construct an interaction term between the number of fixed-line telephones per 100 people, the number of post offices per million people, and the number of internet users in the previous year as an instrumental variable for digital inclusive finance for model estimation [7,58]. The instrumental variable needs to be uncorrelated with the disturbance term while highly correlated with the explanatory variable. The number of fixed-line telephones and post offices can reflect the historical information transmission volume in a region to a certain extent. Digital inclusive finance transmits information through electronic computers and the internet, and digital inclusive finance in regions with high numbers of fixed-line telephones and post offices is more likely to thrive and develop. At the same time, considering that there are heteroscedasticity and series autocorrelation problems in macroeconomic variables, systematic GMM estimation has the advantages of solving the problems of unrecognized individual differences, the influence of variables that are not considered, and the correlation between variables and random terms. Therefore, referring to the practice of Teng et al. (2020), the systematic GMM model is introduced for estimation so as to obtain more accurate parameter estimation results [59].

As can be seen from Table 3, the models passed the tests of non-identifiability, weak instrumental variables, and over-identification. The test results indicate that the endogenous problem caused by missing important variables or bidirectional causality has no significant impact on the core conclusions, and even the estimation coefficients increased slightly.

5.4. Endogenous Treatment

5.4.1. Heterogeneity Analysis

(1) Location factor. Considering differences in agricultural policy support, resource endowment, and agricultural development status among regions, it is easy to cause mismatch of high-quality production factors, affecting the development level of digital inclusive finance and the agricultural ecological environment in various regions. Concurrently, amidst the incessant surge in urbanization, spatial interconnectedness emerges in agricultural carbon emissions and energy utilization among adjacent urban agglomerations, thereby exerting an influence on the agriculture's pursuit of high-quality green development [60]. In view of this, this paper divides the samples according to the basic characteristics of the three major economic zones of east, central, and west formed from China's coast to the interior, and further examines the differences in the impact of digital inclusive finance on the high-quality development of agriculture. As can be seen from Table 5, the marginal contribution of digital inclusive finance to the high-quality development of agriculture in the eastern region is the highest, followed by the western region, while the empirical results in the central region have not passed the significance test. Research Hypothesis 3 has been partially demonstrated. Meanwhile, this conclusion may be related to regional layout planning and the development stage it is in. Most cities in the eastern region have strong economic strength and abundant technological resources, providing good hardware support and growth environment for the development of digital inclusive finance. This facilitates the precise and swift delivery of financial services to the agricultural sector, thereby satisfying the diverse requirements for high-quality agricultural development. At the same time, financial institutions in the eastern region have high innovation capabilities and service awareness in the field of digital inclusive finance, and can actively use digital technology to develop financial products and services suitable for agricultural development, such as agricultural supply chain finance and rural e-commerce finance, effectively improving the coverage and penetration of financial services.

(2) Digital infrastructure construction. The improvement in digital infrastructure construction creates a favorable technical environment for the development of digital inclusive finance, which not only improves the network coverage and communication quality in rural and remote areas but also provides strong support for the availability and coverage of financial services. Concurrently, to execute the "Action Plan for Promoting Big Data Development" promulgated by the State Council, the National Development and Reform Commission initiated pilot policies for comprehensive big data experimentation in eight regions, encompassing Guangdong, Shanghai, and Beijing, in 2016. These policies aim to lead the flow of technology, materials, funds, and talents through data flow, and promote the sharing, integration, collaboration, and efficient utilization of social production factors. Therefore, these regions themselves have better digital infrastructure, and their digital financial systems and product

supply are also more developed and mature. Given this, this article takes the national comprehensive big data pilot policy as the sample classification standard to explore whether different levels of digital infrastructure will cause deviations in the practice of digital inclusive finance. As can be seen from Table 5, the regression coefficient of digital inclusive finance in the pilot cities of the national comprehensive big data pilot zone is 0.120, higher than that in non-pilot cities, and both have passed the 1% significance test. Research Hypothesis 3 has been partially demonstrated. This result indicates that digital inclusive finance has a more significant role in promoting the high-quality development of agriculture in areas with better digital infrastructure, which is in line with expectations. The reason for this phenomenon lies in the fact that improved digital infrastructure construction helps to create a favorable financial ecological environment, promote cooperation and competition among financial institutions, and promote the formation of a financial infrastructure system with reasonable layout, effective governance, advanced reliability, interconnectivity, and flexibility [61]. This helps agricultural operators to more conveniently manage funds, make payments, and obtain loans, effectively addressing the issue of traditional financial services being unable to fully cover rural areas. At the same time, with the implementation of the Broadband China and Digital Rural Strategies, the ability to collect, transmit, and process information in various stages of agricultural production has been significantly enhanced. Through mobile applications, farmers can obtain real-time information on market prices, weather forecasts, and advanced agricultural technologies, enabling them to make better decisions and plan agricultural production. This helps to improve the efficiency and quality of resource utilization in agricultural development, effectively breaking through resource and environmental bottlenecks, and achieving an intensive utilization of agricultural development resources [62].

(3) Green finance. Amidst the constraints of limited endogenous financing and an indirect financing-dominated financial structure in China, bank credit emerges as a crucial source of capital for corporate innovation activities. In 2017, the Chinese government implemented green finance pilot policies across ten regions, including Zhejiang, Guangxi, Guizhou, and Xinjiang. These policies seek to diversify financing avenues for green funds, establish robust mechanisms for disclosing corporate environmental responsibility information, steer social capital towards active participation in green project investments, and ultimately foster the green transformation of traditional industries and sustainable economic and societal development. This also implies that market entities are facing stronger environmental regulations, forcing them to focus more on proactive prevention rather than end-of-pipe emission reduction. Given this context, this article uses the green finance reform pilot policy as a sample classification standard to explore whether digital inclusive finance has a differentiated impact on the high-quality development of agriculture in regions with stronger environmental regulations and more sophisticated green financial systems. As shown in Table 5, the impact coefficient of digital inclusive finance on the high-quality development of agriculture is higher in the green finance policy pilot regions, with a value of 0.584, and it passed the 1% significance test. Research Hypothesis 3 has been partially demonstrated. This result indicates that digital inclusive finance plays a more significant role in promoting the high-quality development of agriculture in regions with stronger environmental regulations, which is consistent with expectations. Green finance, as a new financial strategy, focuses on financial institutions as the mainstay, actively leveraging various channels to incentivize market entities to transition towards cleaner and low-carbon production activities, thereby effectively achieving pollution control and environmentally friendly production [63]. Driven by the pilot policies, the establishment of a sound green financial standard system and the implementation of incentive policies, combined with digital technology and the core concepts of inclusive finance, provided agricultural operators with a series of green financial products and continuously innovated the financial service system.

At the same time, through the establishment of fiscal special funds, direct financial support can be provided to agricultural green and low-carbon projects, facilitating their rapid development. Furthermore, the synergistic interplay between the digital governance paradigm of digital inclusive finance and green finance synergistically drives the reconfiguration of low-carbon civilization, thereby augmenting farmers' low-carbon literacy. This, in turn, provides ample endogenous impetus for advancing digital governance in agricultural and rural regions and fostering a novel form of green civilization. Ultimately, these concerted efforts jointly expedite the green and low-carbon transformation of green rural areas, green agriculture, and small- and micro-enterprises, laying a solid foundation for the high-quality and sustainable development of agriculture [64].

Table 5. Heterogeneity analysis results.

| Variable | Heterogeneity Analysis | | | | | | |
| | Location Factor | | | Digital Infrastructure Construction | | Green Finance | |
	Eastern Region	Central Region	Western Region	Pilot Area	Non-Pilot Area	Pilot Area	Non-Pilot Area
lnDFI	0.101 ***	0.014	0.072 ***	0.120 ***	0.048 ***	0.584 ***	0.107 ***
	(5.27)	(0.29)	(4.39)	(6.30)	(2.06)	(3.08)	(5.17)
Control variable	Control	Control	Control	Control	Control	Control	Control
Regional effect	Control	Control	Control	Control	Control	Control	Control
Time effect	Control	Control	Control	Control	Control	Control	Control
R^2	0.7929	0.6361	0.9296	0.8217	0.6930	0.9117	0.6371

Note: *** is significant at 1% significance levels.

5.4.2. Mechanism Analysis

To validate the mechanism of land circulation in the facilitation of high-quality agricultural development through digital inclusive finance, according to the recursive equations of intermediary effect, the results in Table 6 are obtained by fitting.

Table 6. Test results of intermediary effect.

Variable	*lnHAD*	*lnLT*	*lnHAD*
lnDFI	0.137 ***	0.480 ***	0.103 **
	(3.12)	(3.23)	(2.03)
lnLT			0.223 ***
			(16.76)
Control variable	Control	Control	Control
Regional effect	Control	Control	Control
Time effect	Control	Control	Control

Note: ***, and ** are significant at 1% and 5% significance levels.

According to Table 6, the results of the third segment of the mediation effect equation group indicate that the fitting coefficient between digital inclusive finance and land circulation for the high-quality development of agriculture is significantly positive, indicating that both variables can promote the high-quality development of agriculture. The results of the second segment equation show that digital inclusive finance can significantly promote the development of land circulation. Based on the magnitude, significance, and signs of the coefficients in the benchmark regression and mediation equation group, it is evident that the partial mediation effect of land circulation is valid, accounting for 78.13% of the total effect, thus validating Hypothesis 3. Digital inclusive finance truly realized short-cycle digital financial services such as "land mortgage cloud loans", "agricultural credit cloud loans", and "borrow and repay anytime" by establishing shared credit data for farmers and focusing on data-based credit enhancement, forming a financing model of "agricultural big data + finance" to meet the financing needs of agricultural operators for scale operations and promote agricultural-related projects such as land contracting and farmland transfer [65]. Concurrently, digital inclusive finance also exerts a positive influence on land circulation in aspects of digital securities and digital insurance. The securitized transfer of farmland facilitated by digital technology not only fosters market-based pricing of land resources but also aids in steering social capital towards rural areas, thereby enhancing profitability in the rural factor market. Additionally, digital agricultural insurance effectively utilizes risk compensation mechanisms to safeguard the expedited transfer of rural land and revitalize land and other factor resources. When farmland is transferred to new agricultural operators, such as large-scale farmers, they are inclined to adopt modern agricultural production technologies and scientific management practices due to their more comprehensive agricultural knowledge base compared to individual farmers. This process not only promotes agricultural specialization and division of labor but also helps achieve a scientific allocation of agricultural factor inputs, thereby improving agricultural production efficiency, operational efficiency, and environmental efficiency [66].

6. Research Conclusions and Policy Recommendations

6.1. Research Conclusions

The evolution of digital inclusive finance, centered on data elements, has emerged as a pivotal catalyst for revolutionizing agricultural development methods and attaining sustainable green growth. Utilizing panel data spanning from 2011 to 2021 across 30 provinces in China (excluding Tibet, Hong Kong, Macao, and Taiwan), this study validates the direct influence of digital inclusive finance on the advancement of agricultural quality. Furthermore, employing a mediation effect model, we delve into the indirect impact mechanism mediated by land circulation. Our findings yield several insights: Primarily, digital inclusive finance is a significant enhancer of agricultural quality development, a conclusion upheld by rigorous robustness tests and endogeneity adjustments. Secondly, a heterogeneous analysis reveals that digital inclusive finance notably propels agricultural quality in China's eastern regions, areas with robust digital infrastructure, and financially advanced regions. Thirdly, through channel analysis, we discover that land circulation serves as a positive mediator in the process of digital inclusive finance bolstering agricultural quality development.

6.2. Policy Recommendations

The above research conclusions provide important theoretical and empirical evidence for China to deepen the application of digital inclusive finance, promote the transformation and upgrading of traditional agriculture, and cultivate new competitive advantages and new momentum for agricultural development. In view of this, this article proposes the following policy recommendations:

Firstly, the digital technology hardware infrastructure construction in the less developed regions of central and western China should be strengthened to empower the implementation of digital inclusive finance. Specifically, the central and local governments should establish special funds to support the construction of digital technology hardware infrastructure in less developed regions of central and western China, including but not limited to network coverage, data center construction, and cloud computing platform establishment. This will optimize the layout of digital technology hardware, ensure coverage of various financial service demand points in urban and rural areas, and improve the availability of financial services. Concurrently, government departments should actively encourage private capital to collaborate with the government, adopting a Public–Private Partnership (PPP) model to jointly invest in digital infrastructure construction, and establish reasonable mechanisms for revenue distribution and risk sharing. On the one hand, the investment and construction of the industrial internet system built by 5G, cloud computing, big data, and artificial intelligence should be strengthened to lay the foundation for the application of digital inclusive finance. On the other hand, the construction of rural digital inclusive finance sharing platforms should be strengthened to support the development of digital industries such as "Internet +" agriculture, rural e-commerce, and smart agriculture, and enhance the modernization level of the agricultural industry chain.

Secondly, rural land transfer policies should be improved and moderate scale operation in agriculture promoted. Specifically, the first priority is to improve the detailed rules and regulations governing the management of agricultural land management rights transfer, standardize the procedures and contract content of agricultural land transfer, including but not limited to critical elements such as the duration, area, price, payment method, and breach of contract responsibilities, thereby ensuring the transparency and simplicity of the agricultural land transfer process and safeguarding the legitimate rights and interests of both parties involved in the transfer. The second step is to increase subsidies for farmland transfer and promote the subsidy schemes for farmland transfer, encouraging farmers to learn from and imitate the experience of a small number of farmers who have benefited from farmland transfer, and urging farmers to voluntarily "wash their feet and go to the fields" to achieve real "de-involution" in agriculture and promote the scaled operation of agricultural production and management. The third step is to strengthen the connection between urban and rural employment departments during the process of farmers' professional non-agriculturalization, promote the transfer of rural redundant labor, and enable rural labor to better integrate into urban life. The fourth step is to rely on digital inclusive finance service platforms to achieve smooth financing for both the transferor and transferee of farmland, better leveraging the driving role of digital inclusive finance and farmland transfer in the high-quality development of agriculture. Meanwhile, enterprises are encouraged and supported to conduct recruitment activities in rural areas, providing rural laborers with more employment opportunities and job choices. Fourthly, on the existing digital inclusive financial service platforms, a module for agricultural land transfer financing services should be added to provide convenient and efficient financing services to both parties involved in the transfer. Additionally, cooperation with financial institutions should be strengthened to explore financial product innovations such as agricultural land transfer mortgage loans, aiming to reduce the financing costs and risks for farmers.

Thirdly, coordination and cooperation among multiple entities should be actively sought and the connection between government functional institutions and market financial institutions should be strengthened. On the one hand, environmental regulation policies must effectively curb environmental pollution during agricultural production to protect rural ecological environments while avoiding being overly stringent to prevent imposing excessive economic burdens on farmers and agricultural enterprises. The government needs to formulate reasonable environmental regulation intervals to ensure that policies achieve both environmental protection goals and promote sustainable development in agricultural production. On the other hand, the market must fully leverage digital inclusive finance as a financial tool to exert its growth, inclusiveness, and poverty reduction effects on agriculture, rural areas, and farmers. In practice, it is necessary to balance the relationship between economic growth and green development, embarking on a new path of agricultural green and high-quality development where government regulation and market regulation complement each other. The government should strengthen supervision and guidance for agricultural green development to ensure the effective implementation of policies and the orderly operation of the market. At the same time, the government should respect market rules, giving full play to the decisive role of market mechanisms in resource allocation, and promoting the marketization, industrialization, and socialization of agricultural green development.

6.3. Research Limitations

Although this article provides some empirical insights for the government's decision-making in the field of digital inclusive finance and promoting high-quality green agricultural development, there are still certain limitations. Firstly, regarding data granularity and data collection periodicity, this paper utilizes provincial-level data to discuss the impact of China's digital inclusive finance development on agricultural high-quality development from 2011 to 2021. The relatively short span of data years may fail to capture long-term trends or structural changes in the agricultural sector. Furthermore, future research could explore how digital inclusive finance influences the production processes of agricultural enterprises, subsequently affecting agricultural high-quality development, by adjusting research methods and perspectives. Secondly, from the perspective of research areas, major grain-producing areas are important regions for China's grain supply. However, due to the quasi-public nature of grain and the low industry profit rate, the economic strength and financial resources of these areas are relatively weak. Therefore, future research can include the high-quality development of agriculture in China's 13 major grain-producing areas in the research framework to draw more profound conclusions. Finally, this study belongs to empirical research in statistics and econometrics, and the research conclusions do not provide detailed operational plans. In future research, it would be beneficial to use case studies to deeply analyze the specific measures and experiences of financial institutions in promoting high-quality agricultural development through digital inclusive financial products from specific cases.

Author Contributions: Conceptualization, X.G. and J.Y.; Methodology, Q.X.; Software, Q.X.; Formal Analysis, Q.X.; Investigation, X.G. and J.Y.; Resources, Q.X.; Writing—Original Draft, Q.X.; Writing—Review and Editing, Q.X.; Supervision, Q.X.; Project Administration, X.G. and J.Y. All authors have read and agreed to the published version of the manuscript.

Funding: This research was funded by the Fujian Province Innovation Strategy Research Project (grant number: 2022R0043), the Philosophy and Social Sciences Young Scholars Development Program of Huaqiao University (grant number: 18SKGC-QG07) and the Natural Science Foundation of Fujian Province (grant number: 2022J01320).

Institutional Review Board Statement: Not applicable.

Informed Consent Statement: No description in this study involves humans.

Data Availability Statement: The datasets used during the current study are available from the corresponding author on reasonable request.

Conflicts of Interest: The authors declare no conflicts of interest.

References

1. Vinuesa, R.; Azizpour, H.; Leite, I.; Balaam, M.; Dignum, V.; Domisch, S.; Felländer, A.; Langhans, S.D.; Tegmark, M.; Nerini, F.F. The role of artificial intelligence in achieving the sustainable development goals. *Nat. Commun.* **2020**, *11*, 233. [CrossRef]
2. Federici, S.; Tubiello, F.N.; Salvatore, M.; Jacobs, H.; Schmidhuber, J. New estimates of CO_2 forest emissions and removals: 1990–2015. *For. Ecol. Manag.* **2015**, *352*, 89–98. [CrossRef]
3. Pu, G.L. Achieving agricultural revitalization: Performance of technical innovation inputs in farmland and water conservation facilities. *Alex. Eng. J.* **2022**, *61*, 2851–2858. [CrossRef]
4. Zhang, C.K.; Li, Y.; Yang, L.L.; Wang, Z. Does the development of digital inclusive finance promote the construction of digital villages?—An empirical study based on the Chinese experience. *Agriculture* **2023**, *13*, 1616. [CrossRef]
5. Zhang, W.; Huang, M.; Shen, P.C.; Liu, X.M. Can digital inclusive finance promote agricultural green development? *Environ. Sci. Pollut. Res.* **2023**, *11*, 29557. [CrossRef]
6. Guo, X.P.; Wang, L.T.; Meng, X.L.; Dong, X.T.; Gu, L.L. The impact of digital inclusive finance on farmers' income level: Evidence from China's major grain production regions. *Financ. Res. Lett.* **2023**, *58*, 104531. [CrossRef]

7.   Gao, Q.; Cheng, C.M.; Sun, G.L.; Li, J.F. The impact of digital inclusive finance on agricultural green total factor productivity: Evidence from China. *Front. Ecol. Evol.* **2022**, *10*, 905644. [CrossRef]
8.   Wang, Y.F.; Xie, L.; Zhang, Y.; Wang, C.Y.; Yu, K. Does FDI promote or inhibit the high-quality development of agriculture in China? An agricultural GTFP perspective. *Sustainability* **2019**, *11*, 4620. [CrossRef]
9.   Zheng, G.T.; Wang, W.W.; Jiang, C.; Jiang, F. Can Rural Industrial Convergence Improve the Total Factor Productivity of Agricultural Environments: Evidence from China. *Sustainability* **2023**, *15*, 16432. [CrossRef]
10. Wang, G.F.; Mi, L.C.; Hu, J.M.; Qian, Z.Y. Spatial analysis of agricultural eco-efficiency and high-quality development in China. *Front. Environ. Sci.* **2022**, *10*, 847719. [CrossRef]
11. Huang, J.; Duan, X.Y.; Li, Y.L.; Guo, H.T. Spatial-temporal evolution and driving factors of green high-quality agriculture development in China. *Front. Environ. Sci.* **2023**, *11*, 1320700. [CrossRef]
12. Yang, L.; Guan, Z.Y.; Chen, S.Y.; He, Z.H. Re-measurement and influencing factors of agricultural eco-efficiency under the 'dual carbon' target in China. *Heliyon* **2024**, *10*, 24944. [CrossRef] [PubMed]
13. Chen, Y.F.; Miao, J.F.; Zhu, Z.T. Measuring green total factor productivity of China's agricultural sector: A three-stage SBM-DEA model with non-point source pollution and CO_2 emissions. *J. Clean. Prod.* **2021**, *318*, 128543. [CrossRef]
14. Chen, Z.; Li, X.J.; Xia, X.L. Measurement and spatial convergence analysis of China's agricultural green development index. *Environ. Sci. Pollut. Res.* **2021**, *28*, 19694–19709. [CrossRef] [PubMed]
15. Huang, X.H.; Yang, F.; Fahad, S. The impact of digital technology use on farmers' low-carbon production behavior under the background of carbon emission peak and carbon neutrality goals. *Front. Environ. Sci.* **2022**, *10*, 100218. [CrossRef]
16. Wang, H.; Tang, Y.T. Spatiotemporal Distribution and Influencing Factors of Coupling Coordination between Digital Village and Green and High-Quality Agricultural Development—Evidence from China. *Sustainability* **2023**, *15*, 8079. [CrossRef]
17. Zou, Y.N.; Cheng, Q.P.; Jin, H.Y.; Pu, X.F. Evaluation of green agricultural development and its influencing factors under the framework of sustainable development goals: Case study of Lincang city, an underdeveloped mountainous region of China. *Sustainability* **2023**, *15*, 11918. [CrossRef]
18. Xu, L.Y.; Jiang, J.; Du, J.G. The dual effects of environmental regulation and financial support for agriculture on agricultural green development: Spatial spillover effects and spatial-temporal heterogeneity. *Appl. Sci.* **2022**, *12*, 11609. [CrossRef]
19. Li, J.K.; Chen, J.Y.; Liu, H.G. Sustainable agricultural total factor productivity and its spatial relationship with urbanization in China. *Sustainability* **2021**, *13*, 6773. [CrossRef]
20. Han, J.S.; Yang, Q.; Zhang, L. What are the priorities for improving the cleanliness of energy consumption in rural China? Urbanization advancement or agriculture development? *Energy Sustain. Dev.* **2022**, *70*, 106–114. [CrossRef]
21. Guo, H.; Gu, F.; Peng, Y.L.; Deng, X.; Guo, L.L. Does digital inclusive finance effectively promote agricultural green development?— A case study of China. *Int. J. Environ. Res. Public Health* **2022**, *19*, 6982. [CrossRef] [PubMed]
22. Shen, Y.; Guo, X.Y.; Zhang, X.W. Digital financial inclusion, land transfer, and agricultural green total factor productivity. *Sustainability* **2023**, *15*, 6436. [CrossRef]
23. Hong, M.Y.; Tian, M.J.; Wang, J. Digital inclusive finance, agricultural industrial structure optimization and agricultural green total factor productivity. *Sustainability* **2022**, *14*, 11450. [CrossRef]
24. Li, H.; Shi, Y.; Zhang, J.; Zhang, Z.; Zhang, Z.; Gong, M. Digital inclusive finance & the high-quality agricultural development: Prevalence of regional heterogeneity in rural China. *PLoS ONE* **2023**, *18*, e0281023. [CrossRef] [PubMed]
25. Xie, L.J.; Li, W.J. Can Digital Financial Inclusion Drive High-Quality Agricultural Development in China's Border Areas? *Acad. J. Bus. Manag.* **2023**, *5*, 33–41. [CrossRef]
26. Wang, Y.F.; Liu, J.; Huang, H.H.; Tan, Z.X.; Zhang, L.C. Does Digital Inclusive Finance Development Affect the Agricultural Multifunctionality Extension? Evidence from China. *Agriculture* **2023**, *13*, 804. [CrossRef]
27. Moahid, M.; Khan, G.D.; Bari MD, A.; Yoshida, Y. Does access to agricultural credit help disaster-affected farming households to invest more on agricultural input? *Agric. Financ. Rev.* **2023**, *83*, 96–106. [CrossRef]
28. Chi, M.J.; Guo, Q.Y.; Mi, L.C.; Wang, G.F.; Song, W.M. Spatial Distribution of Agricultural Eco-Efficiency and Agriculture High-Quality Development in China. *Land* **2022**, *11*, 722. [CrossRef]
29. Zhao, J.H.; Gong, Y.Y. Analysis on the Mechanism of Digital Economy Boosting the High-quality Development of Agriculture in China. *E3S Web Conf.* **2020**, *218*, 02009. [CrossRef]
30. Zhao, Y.; Zhu, J.C.; Zhang, Y. Measurement, Temporal-spatial Evolution Characteristics and Countermeasures of High-quality Agricultural Development in China. *E3S Web Conf.* **2021**, *228*, 02013. [CrossRef]
31. Zhou, L.; Zhang, S.; Zhou, C.; Yuan, S.; Jiang, H.; Wang, Y. The impact of the digital economy on high-quality agricultural development—Based on the regulatory effects of financial development. *PLoS ONE* **2024**, *19*, e0293538. [CrossRef] [PubMed]
32. Nie, S.S. Influence of Rural Infrastructure Construction on Agricultural Total Factor Productivity. *Agric. For. Econ. Manag.* **2021**, *4*, 65–68.
33. Tian, Y.; Du, W.; Hua, Y.W.; Fan, L.J. Will infrastructure construction drive Chinese rural families to develop their own enterprises? *Appl. Econ. Lett.* **2022**, *29*, 1665–1669. [CrossRef]
34. Zhang, H.; Wu, D.L. The Impact of Transport Infrastructure on Rural Industrial Integration: Spatial Spillover Effects and Spatio-Temporal Heterogeneity. *Land* **2022**, *11*, 1116. [CrossRef]

35. Chen, K.; Liu, Y. The impact of environmental regulation on farmers' income: An empirical examination based on different sources of income. *Environ. Sci. Pollut. Res. Int.* **2023**, *30*, 103244–103258. [CrossRef] [PubMed]

36. Sun, Y.C. Environmental regulation, agricultural green technology innovation, and agricultural green total factor productivity. *Front. Environ. Sci.* **2022**, *10*, 955954. [CrossRef]

37. Zhou, Z.Q.; Liu, W.Y.; Wang, H.L.; Yang, J.Y. The Impact of Environmental Regulation on Agricultural Productivity: From the Perspective of Digital Transformation. *Int. J. Environ. Res. Public Health* **2022**, *19*, 10794. [CrossRef] [PubMed]

38. Zang, D.G.; Yang, S.; Li, F.H. The Relationship between Land Transfer and Agricultural Green Production: A Collaborative Test Based on Theory and Data. *Agriculture* **2022**, *12*, 1824. [CrossRef]

39. Gao, J.; Zhao, R.R.; Lyu, X. Is There Herd Effect in Farmers' Land Transfer Behavior? *Land* **2022**, *11*, 2191. [CrossRef]

40. Wu, C.L. Research on the Predicament and Upgrading Path of Rural Land Transfer under the Background of Rural Revitalization—An Empirical Study Based on Sichuan Countryside. *Front. Soc. Sci. Technol.* **2022**, *4*, 17–20. [CrossRef]

41. Ma, G.Q.; Dai, X.P.; Luo, Y.X. The Effect of Farmland Transfer on Agricultural Green Total Factor Productivity: Evidence from Rural China. *Int. J. Environ. Res. Public Health* **2023**, *20*, 2130. [CrossRef]

42. Li, C.M.; He, G. Study on the Effects of Agricultural Land Transfer on Agricultural Economic Growth. *IOP Conf. Ser. Earth Environ. Sci.* **2020**, *615*, 012004. [CrossRef]

43. Fei, R.L.; Lin, Z.Y.; Chunga, J. How land transfer affects agricultural land use efficiency: Evidence from China's agricultural sector. *Land Use Policy* **2021**, *103*, 105300. [CrossRef]

44. Tan, J.; Cai, D.L.; Han, K.F.; Zhou, K. Understanding peasant household's land transfer decision-making: A perspective of financial literacy. *Land Use Policy* **2022**, *119*, 106189. [CrossRef]

45. Tang, Y.; Chen, M.H. The impact of agricultural digitization on the high-quality development of agriculture: An empirical test based on provincial panel data. *Land* **2022**, *11*, 2152. [CrossRef]

46. Wen, Y.; Zhuo, S. The impact of the digital economy on high-quality development of agriculture: A China case study. *Sustainability* **2023**, *15*, 5745. [CrossRef]

47. Ge, H.; Li, B.; Tang, D.; Xu, H.; Boamah, V. Research on digital inclusive finance promoting the integration of rural three-industry. *Int. J. Environ. Res. Public Health* **2022**, *19*, 3363. [CrossRef] [PubMed]

48. Fu, C.L.; Sun, X.Y.; Guo, M.T.; Yu, C.Y. Can digital inclusive finance facilitate productive investment in rural households?—An empirical study based on the China household finance survey. *Financ. Res. Lett.* **2024**, *61*, 105034. [CrossRef]

49. Wang, B.; Shao, X.Y.; Yang, X.; Xu, H.H. How does land transfer impact rural household income disparity? An empirical analysis based on the micro-perspective of farmers in China. *Front. Sustain. Food Syst.* **2023**, *7*, 1224152. [CrossRef]

50. Yang, H.; Huang, Z.; Fu, Z.; Dai, J.; Yang, Y.; Wang, W. Does land transfer enhance the sustainable livelihood of rural households? Evidence from China. *Agriculture* **2023**, *13*, 1667. [CrossRef]

51. Fang, L.; Hu, R.; Mao, H.; Chen, S. How crop insurance influences agricultural green total factor productivity: Evidence from Chinese farmers. *J. Clean. Prod.* **2021**, *321*, 128977. [CrossRef]

52. Li, H.; Tang, M.; Cao, A.; Guo, L. Assessing the relationship between air pollution, agricultural insurance, and agricultural green total factor productivity: Evidence from China. *Environ. Sci. Pollut. Res.* **2022**, *29*, 78381–78395. [CrossRef] [PubMed]

53. Zhang, H.; Li, Y.; Sun, H.X.X.; Wang, X.H. How Can Digital Financial Inclusion Promote High-Quality Agricultural Development? The Multiple-Mediation Model Research. *Int. J. Environ. Res. Public Health* **2023**, *20*, 3311. [CrossRef] [PubMed]

54. Moon, H.R.; Weidner, M. Dynamic linear panel regression models with interactive fixed effects. *Econom. Theory* **2015**, *33*, 158–195. [CrossRef]

55. Ye, F.; Yang, Z.; Yu, M.; Watson, S.; Lovell, A. Can market-oriented reform of agricultural subsidies promote the growth of agricultural green total factor productivity? Empirical evidence from maize in China. *Agriculture* **2023**, *13*, 251. [CrossRef]

56. He, Z.M.; Chen, H.C.; Hu, J.W.; Zhang, Y.T. The impact of digital inclusive finance on provincial green development efficiency: Empirical evidence from China. *Environ. Sci. Pollut. Res. Int.* **2022**, *29*, 90404–90418. [CrossRef]

57. Lin, Q.H.; Cheng, Q.W.; Zhong, J.F.; Lin, W.H. Can digital financial inclusion help reduce agricultural non-point source pollution?—An empirical analysis from China. *Front. Environ. Sci.* **2022**, *10*, 1074992. [CrossRef]

58. Zhang, J.P.; Chen, S.Y. Financial Development, Environmental Regulation and Economic Green Transformation. *J. Financ. Econ.* **2021**, *47*, 78–93. [CrossRef]

59. Teng, L.; Ma, D.G. Can digital finance promote high-quality development? *J. Stat. Study* **2020**, *37*, 80–92. [CrossRef]

60. Guo, X.Y.; Yang, J.Y.; Shen, Y.; Zhang, X.W. Prediction of Agricultural Carbon Emissions in China Based on GA-ELM Model. *Front. Energy Res.* **2023**, *11*, 1245820. [CrossRef]

61. Tang, Y.; Chen, M.H. The Impact Mechanism and Spillover Effect of Digital Rural Construction on the Efficiency of Green Transformation for Cultivated Land Use in China. *Int. J. Environ. Res. Public Health* **2022**, *19*, 16159. [CrossRef] [PubMed]

62. Zhao, W.; Liang, Z.Y.; Li, B.R. Realizing a Rural Sustainable Development through a Digital Village Construction: Experiences from China. *Sustainability* **2022**, *14*, 14199. [CrossRef]

63. Guo, X.Y.; Yang, J.Y.; Shen, Y.; Zhang, X.W. Impact on green finance and environmental regulation on carbon emissions: Evidence from China. *Front. Environ. Sci.* **2024**, *12*, 1307313. [CrossRef]

64. Han, X.; Wang, Y.; Yu, W.L.; Xia, X.L. Coupling and coordination between green finance and agricultural green development: Evidence from China. *Financ. Res. Lett.* **2023**, *58*, 104221. [CrossRef]

65. Peng, K.L.; Yang, C.; Chen, Y. Land transfer in rural China: Incentives, influencing factors and income effects. *Appl. Econ.* **2020**, *52*, 5477–5490. [CrossRef]
66. Cao, H.; Zhu, X.Q.; Heijman, W.; Zhao, K. The impact of land transfer and farmers' knowledge of farmland protection policy on pro-environmental agricultural practices: The case of straw return to fields in Ningxia, China. *J. Clean. Prod.* **2020**, *277*, 123701. [CrossRef]

sustainability

Article

Study on Sustainable Operation Mechanism of Green Agricultural Supply Chain Based on Uncertainty of Output and Demand

Qianyi Wang [1,2], Minghui Ni [3], Wei Wen [1], Ruijuan Qi [1] and Qiwen Zhang [1,*]

[1] College of Economics and Management, Northeast Agricultural University, Harbin 150030, China; wqy19900406@163.com (Q.W.); wenwei21@163.com (W.W.); ruijuan_qi@163.com (R.Q.)
[2] School of Management, Heilongjiang University of Science and Technology, Harbin 150022, China
[3] College of Economics and Management, Heilongjiang Institute of Technology, Harbin 150006, China; nmhfzk@sina.com
* Correspondence: qiwen_zhang2023@163.com

Abstract: A future trend in agricultural development is to promote the green transformation of agriculture and realize the transformation from extensive consumption to environmentally friendly consumption. However, in the process of circulating green agricultural products, the output and demand are uncertain, and the cooperation of various entities in the supply chain is unstable, which leads to the risk of interruptions to the supply chain, and then leads to ineffective supply chain operations for green agricultural products. Therefore, under the background of double uncertainty of output and demand, combined with CVaR theory and considering the risk avoidance degree of farmers, a Stackelberg game model of a "firm + farmer" two-level green agricultural product supply chain was constructed, and the supply coordination mechanism was studied. The results show that a benefit-sharing contract can effectively coordinate the supply chain of green agricultural products under the double uncertainty of output and demand and obtain optimal greenness, agricultural input, order quantity, and optimal inventory factors under centralized decision making. The optimal production decisions of farmers and the optimal pricing decisions of companies are obtained under decentralized decision making. The benefit-sharing contract is used to coordinate the supply chain, and the overall incomes of farmers, companies, and the supply chain improved after the coordination. The research results can enrich the relevant research on coordinating green agricultural products supply chains under the uncertainty of output and demand and provide a reference for ensuring the effective and stable operation of supply chains.

Keywords: random output and demand; green agricultural product supply chain; coordinating mechanism

Citation: Wang, Q.; Ni, M.; Wen, W.; Qi, R.; Zhang, Q. Study on Sustainable Operation Mechanism of Green Agricultural Supply Chain Based on Uncertainty of Output and Demand. *Sustainability* **2024**, *16*, 5460. https://doi.org/10.3390/su16135460

Academic Editors: Fotios Chatzitheodoridis, Efstratios Loizou and Achilleas Kontogeorgos

Received: 20 May 2024
Revised: 19 June 2024
Accepted: 21 June 2024
Published: 27 June 2024

1. Introduction

Sustainable agricultural development is an important topic, which first appeared in the text of Agenda 21 of the United Nations Conference on Environment and Development in 1992 [1,2]. In the past few decades, the world has faced the challenges of a growing population, food security, and environmental sustainability [3–5]. In order to increase agricultural output, traditional agriculture consumes agricultural resources excessively, and excessive use of fertilizers and pesticides leads to deterioration of the ecological environment and the continuous occurrence of food safety incidents. With the increasing attention of people to food safety, environmental protection, and sustainable development, the quality and safety of agricultural products have received more attention by consumers [6,7]. Green agricultural products have green, safety, nutritional, and other characteristics, so the market demand for green agricultural products is gradually growing [8,9]. And to a certain extent, this has driven the transformation of agricultural production methods [10,11].

In terms of output, green agricultural products have greater uncertainty risk and strong seasonality. Under the influence of natural factors such as weather, climate change leads to irregular changes in temperature, precipitation, and seasonality; natural disasters such as drought, flood, hail, and hurricanes; and problems of soil and water resources, which will all affect the growth of green agricultural products to varying degrees [12,13]. In terms of demand, there are differences in consumer demand and cognition in the market, and consumers in some regions may have higher demand for green agricultural products; meanwhile, in other regions, consumers' cognition and acceptance of green agricultural products may be low [14,15]. Therefore, under the influence of various factors, the output and demand of green agricultural products are often accompanied by uncertainties, leading to the risk of interruptions to the supply chain for green agricultural products [16–18]. The management of the supply chain of green agricultural products involves many links from farmers to consumers, including production, processing, transportation, storage, and sales. Both output and demand are uncertain, and the main body of each link makes decisions based on the principle of maximizing their own interests, which has a negative impact on the operational efficiency of the supply chain. Supply chain coordination is the main way to achieve supply chain management [19]. Given the double uncertainty of output and demand, how can resources be allocated and inventory management be optimized to improve supply chain returns? How can effective strategies and methods be developed to improve the efficiency and flexibility of the supply chain and avoid the risk of supply chain disruption? These problems need to be further studied and analyzed, in order to ensure the sustainable operation of the green agricultural product supply chain.

By studying the uncertain characteristics of output and demand of green agricultural products and their impact on the supply chain, it was found that when the output and demand of green agricultural products face double uncertainty, this uncertainty often leads to mismatches between the output and demand ends in the circulation process of green agricultural products, thus restricting the overall operation efficiency of the supply chain of green agricultural products. Based on game theory, this paper discusses how to coordinate the "company + farmer" type green agricultural product supply chain. Considering the risk aversion attitude of decision makers, this paper analyzes how to make optimal decisions between centralized and decentralized companies and farmers based on the CVaR theory, and puts forward a coordination mechanism for revenue-sharing contracts to enhance the stability and operation efficiency of the supply chain, making the whole supply chain reach the Pareto optimal. In particular, we want to answer and analyze the following questions:

1. How could a Stackelberg game model be used to analyze the optimal production decision of farmers and the optimal order decision of companies in the "company + farmer" green agricultural product supply chain under centralized and decentralized models?
2. Can revenue-sharing contracts improve the overall profits of the supply chain? What is the impact of farmers' risk aversion attitude on the supply chain?
3. What are the effects of revenue-sharing contracts on the greenness of agricultural products and the expected profits of companies and farmers by numerical simulation?

The rest of this article is divided into several sections. We review the literature in Section 2. In Section 3, we describe the problem definition and the basic model. In Section 4, we examine how farmers and firms make optimal decisions in centralized and decentralized models. We study the use of revenue-sharing contract coordination in Section 5. Section 6 provides numerical analysis to verify the validity of the model. In Section 7, we summarize the article and suggest possible future research topics.

2. Literature Review

In today's complex international environment, geopolitical conflicts, epidemics, inflation, and other uncertain factors increase the risk of supply chain disruption [20,21]. Under this uncertain environment, supply chain network design must be optimized, and optimal decisions must be made based on the risks faced by the supply chain [22]. Understanding agricultural product supply chain risk is based on the study of supply chain risk, and the research scope is limited to agricultural products. Scholars have defined the risk of agricultural product supply chain as shown in Table 1.

Table 1. Definitions and types of agricultural product supply chain risk.

Representative Scholar	Main Point
Imbiri S et al. [23]	Agricultural product supply chain risk is the uncertainty that affects the main body and the target of agricultural product supply chain.
Eremenko V et al. [24]	It is believed that agricultural activities directly depend on nature, climate, and geographical environment, which lead to uncertainties of agricultural enterprises in the supply chain of agricultural products.
Zhao G et al. [25]	The risk analysis of the agricultural product supply chain using TISM and MICMAC methods reveals that weather and politics are identified as the most critical factors disrupting the entire supply chain process.
Sharma R et al. [26]	Agricultural supply chain can be divided into supply risk, demand risk, financial risk, logistics and infrastructure risk, management and operational risk, policy and regulatory risk, and biological and environmental risk.
Bo Y et al. [27]	Agricultural product supply chain risk is divided into internal risk, external risk, and intermediate risk.
He Xuefeng et al. [28]	Agricultural supply chain risk is divided into environmental risk, cooperation risk, demand risk, supply risk, logistics risk, and information risk.
Wang W et al. [29]	The risk of green supply chain of agricultural products is divided into internal risk and external risk, and the internal risk is composed of organizational risk and quality safety risk.

Chauhan A et al., based on an ISM structural equation model and a DEMATEL decision and evaluation model, concluded that the decisive factors that can improve the overall income of an agricultural product supply chain are output and demand [30]. Therefore, on the basis of understanding the supply chain of agricultural products, coordinating the supply chain of agricultural products under the uncertainty of output and demand is deeply studied. Shi Y and Wang F studied the impact of uncertain output and market demand on the expected profit of each subject of the supply chain and the overall return of the supply chain, and coordinated the supply chain by adjusting the revenue-sharing contract [31]. Giri B C et al. studied a three-tier agricultural supply chain consisting of two suppliers and one buyer. In the face of uncertain output and demand, fixed-price contracts and a new revenue-sharing contract were adopted to coordinate the supply chain and avoid the risk of supply chain disruption [32]. Chen T et al. studied the two-level agricultural product supply chain composed of a manufacturer and a retailer under the dual uncertainty of output and demand, adopted the mixed contract of "quantity discount + revenue sharing" to coordinate the supply chain, and verified the coordination effectiveness through numerical analysis [33]. Liu X et al. built a two-level agricultural supply chain composed of companies and farmers under the uncertainty of agricultural production and market demand. In order to coordinate the income distribution between the two sides of the supply chain, an income distribution coordination mechanism based on income-sharing coefficient and margin was proposed [34].

Research on the supply chain of green agricultural products mainly focuses on the decision making and coordination of the supply chain. Su studied the model of "connecting agriculture to supermarkets" and analyzed the optimal decision making of supermarkets under different conditions based on Stackelberg game theory, and adopted "wholesale price + order subsidy" to coordinate the supply chain and improve the profits of supply chain subjects [35]. Tan constructed a secondary agricultural supply chain considering the freshness and greenness of agricultural products, analyzed the optimal decision making of supply chain members via game theory, coordinated the supply chain through cost-sharing contracts, and narrated the profit gap between producers and retailers [36]. Considering consumer preferences, Hu Q et al. studied how farmers' cooperatives, introducing online channels and green brand building, make decisions under the dual-channel supply chain and contractual cooperation model of green agricultural products [37]. Hu H et al. constructed a two-level green agricultural product supply chain of retailers and cooperatives, in which cooperatives provided green input and retailers provided fresh-keeping input. Based on a game analysis on whether fairness was considered, it was more beneficial to consider the fairness of agricultural cooperatives [38]. Cao Y et al. constructed a green agricultural product supply chain for cooperatives, enterprises, and environmentally sensitive consumers, and concluded that the degree of greenness is decided jointly by cooperatives and enterprises, and green efforts in the supply chain of cost-sharing contracts and buyback contracts are adopted for decision making [39]. Cui L et al. constructed a two-level green agricultural product supply chain composed of farmers and retailers, adopted a revenue-sharing contract for coordination, and found that its effectiveness was positively correlated with consumers' green cognitive sensitivity [40].

Based on the application of conditional value-at-risk theory in agricultural supply chains, some scholars use CVaR to describe the degree of risk avoidance to coordinate agricultural supply chain. YE output and demand uncertainty set the wholesale price of agricultural supply chain based on CVaR, and is affected by the interaction between farmers' risk aversion and yield uncertainty; meanwhile, the retail price is affected by the interaction between demand uncertainty and price elasticity. Agricultural supply chains with uncertain output and demand orders are coordinated under the mechanism of revenue sharing and cost sharing [41]. Peng (2023) constructed a second-order agricultural supply chain with risk-averse farmers and a risk-neutral company, used conditional value at risk (CVaR) to obtain the optimal output of agricultural products, and introduced option contracts to enable farmers to respond to the price fluctuation risks of agricultural products by exercising their rights, in order to achieve supply chain coordination [42]. Liao C studied a three-level agricultural supply chain composed of a risk-averse farmer, a risk-neutral supplier, and a risk-averse retailer, and adopted three options contracts, wholesale price contracts, and replenishment cost-sharing contracts under the CVaR criterion based on random demand [43]. C Liao et al. built a three-level fresh product supply chain composed of suppliers, distributors, and retailers, used CVaR to measure retailers' decision making in risk avoidance and preference, and adopted option contracts to coordinate the supply chain. It was concluded that the aversion and preference towards retailers will not affect the option strike price, but only the option purchase price of distributors and retailers [44].

To sum up, the main research object of agricultural supply chain coordination research under the double uncertainty of output and demand is fresh agricultural products, and there are few systematic studies on the green agricultural product supply chain under the double uncertainty of output and demand; the risk avoidance degree of farmers is not considered in the coordination analysis for green agricultural product supply chains. Therefore, this paper takes green agricultural products as the research object, considers their greenness, environmental protection, and other characteristics, adds greenness into the Stackelberg game model, uses CVaR theory to consider the risk avoidance degree of farmers, and deeply studies how to coordinate the supply chain to help supply chain members make optimal decisions and improve the income of each subject in the supply chain to achieve Pareto optimization.

3. The Basic Model

3.1. Problem Description

Considering the characteristics of green agricultural products themselves, the production and sales of green agricultural products generally take leading processing enterprises as the core supply chain model. Therefore, under the double uncertainty of output and demand, this paper analyzes the "company + farmer" type green agricultural product supply chain, and establishes a basic model of a two-level green agricultural product supply chain composed of a risk-averse farmer and risk-neutral company. The input of farmers, the greenness of agricultural products, and the order quantity and price of the company can be obtained under the maximum profit. At the same time, the supply chain coordination decision-making model under the revenue-sharing contract is introduced to consider the fixed costs of farmers' farm tools and daily consumption. The cost of inputs such as seeds, pesticides, fertilizers, and efforts to produce green agricultural products are all expressed as the product of agricultural input and the production cost per unit of agricultural input for farmers. Under the double uncertainty of output and demand, two variables of farmers' shortage loss and unit salvage value are added to the model. Farmers and companies sign sales contracts; since the output is random, the actual output is $Y = \mu I$, where μ is the random output rate of farmers, $\mu \in [0, A_1]$; A_1 represents the upper bound of the random variable; I is the agricultural inputs of farmers; and $G(\cdot)$ and $g(\cdot)$ are the cumulative distribution and probability density function of green agricultural product output. Thus, we put forward the following assumptions:

Assumption 1. *The company is assumed to be risk-neutral and perfectly rational, with the ultimate goal of maximizing expected profits. Farmers are risk avoiders, and the ultimate goal is utility maximization.*

Assumption 2. *The farmer and the company sign a mutually agreed order based on their expectations, which specifies the price of the purchased unit of green agricultural products at the end of production (the order price); ω_1, $R(\cdot)$ and $r(\cdot)$ are the cumulative distribution and probability density function of ω_1.*

In addition, the stock loss of farmers is greater than the unit salvage value, that is, $C_r > V_s$; the residual value of farmers' unit surplus agricultural products is lower than the unit production cost, that is, $C_s > V_s$, to ensure that farmers cannot obtain profits through excessive production.

Assumption 3. *The information of farmers and companies is symmetrical, and both farmers and companies know the cost, income, and sales strategy of each other.*

Assumption 4. *Demand uncertainty is the biggest risk factor facing the company. In order to consider the impact of greenness on market demand, the relationship of the market demand parameter is $x(p) = ap^{-b}g$, $(a > 0, b > 0)$; $F(\cdot)$ and $f(\cdot)$ are the cumulative distribution of demand and the probability density function. Market demand is considered as follows:*

$$D(x) = x(p)\varepsilon$$

Among them, ε represents the market variables and random demand, $\varepsilon \in [A, B]$. Consider the existence of the following inventory factors:

$$z = \frac{q}{x(p)}$$

Therefore, the corresponding order quantity can be rewritten as follows:

$$q = zx(p)$$

Therefore, the decision variable can be converted into the selling price p and the inventory factor z.

Assumption 5. *Market demand will decrease as the sales price increases, but if the greenness of green agricultural products increases, the market demand will also increase.*

3.2. Income Function of Farmers and Companies

Based on the above assumptions, if the company does not provide any income contract mechanism to coordinate farmers, the income function of farmers' green agricultural products is as follows:

$$\pi_F(p, z, g, \omega_1) = \omega_1 zx(p) - C_s I - \frac{1}{2}kg^2 \tag{1}$$

The revenue function of the company's green agricultural products is as follows:

$$\pi_E(p, z, g) = p\min\{q, D\} + V_r[q - D]^+ - C_r q - \omega_1 zx(p) \tag{2}$$

Therefore, when the company cooperates with farmers, the revenue function of the entire supply chain is the following:

$$\pi = p\min\{q, D\} + V_r[q - D]^+ - C_r q - C_p[q - \mu I]^+ - C_s I - \frac{1}{2}kg^2 \tag{3}$$

Among them, $p\min\{q, D\}$ represents the sales revenue of green agricultural products; $V_r[q - D]^+$ represents the company's residual income; $C_r q$ is the company cost; $C_p[q - \mu I]^+$ represents the backorder cost; $C_s I$ represents the input cost of farmers' agricultural capital; and $\frac{1}{2}kg^2$ represents the production cost of green products, where k is the cost coefficient of greenness and g is the greenness.

4. Decision-Making Model

4.1. Centralized Decision Model

In the case of centralized decision making, the whole green agricultural product supply chain is taken as the research object, and the profit maximization of the whole supply chain is taken as the decision-making goal. At this time, with the farmer and the company as a whole, all the green agricultural products produced by the farmer are responsible for sales by the company. According to the above model's assumptions, the profit function of the green agricultural product supply chain under the decision is described as follows:

$$\pi = p\min\{zx(p), D\} + V_r[zx(p) - D]^+ - C_r zx(p) - C_p[zx(p) - \mu I]^+ - C_s I - \frac{1}{2}kg^2 \tag{4}$$

The definition is as follows:

$$S(q) = E(\min(zx(p), D)) = \int_0^q Df(D)dD + zx(p)\int_q^\infty f(D)dD \tag{5}$$

Since D represents the market size at p = 0, it conforms to the uniform distribution of $[\alpha, \beta]$. Therefore, the following uniform distribution can be obtained:

$$\left[\alpha - \varepsilon a p^{-b}g, \beta - \varepsilon a p^{-b}g\right]$$

Further, Formula (2) can be rewritten as follows:

$$S(q) = \frac{1}{\beta - \alpha}\left[zx(p) - \left(\beta - \varepsilon a p^{-b}g\right) - \frac{(zx(p))^2 + (\alpha - \varepsilon a p^{-b}g)^2}{2}\right] \tag{6}$$

Since the output factors of farmers conform to the uniform distribution of (0, 1), we have the following:

$$S(q, I) = E(\min(zx(p), \mu I)) = zx(p) - I\int_0^{\frac{q}{I}} G(y)dy = zx(p) - \frac{1}{2}\frac{(zx(p))^2}{I} \tag{7}$$

In the centralized decision-making mode, the expected return of the whole supply chain is the following:

$$
\begin{aligned}
E(\pi) &= pS(zx(p)) + V_r[zx(p) - S(q)]^+ - C_r zx(p) - C_p[zx(p) - S(q,I)]^+ - C_s I - \tfrac{1}{2}kg^2 \\
&= (p - V_r)S(q) + (V_r - C_p S - C_r)zx(p) - C_p S(zx(p), I) - C_s I - \tfrac{1}{2}kg^2
\end{aligned}
\tag{8}
$$

Below, the optimal solution conditions are used to find the optimal input, the optimal retail price, the optimal order quantity, and the optimal greenness; thus, there are the following:

$$\frac{dE(\pi)}{dI} = 0$$

$$\frac{dE(\pi)}{dp} = 0$$

$$\frac{dE(\pi)}{dq} = 0$$

$$\frac{dE(\pi)}{dg} = 0$$

$$\frac{dE(\pi)}{dz} = 0$$

Then, we can obtain the following:

$$
I^* = \sqrt{\frac{1}{2}\frac{C_p}{C_s}}\, q^*
\tag{9}
$$

$$
g^* = \frac{d}{k}\left(p - C_r - \sqrt{2C_p C_s}\right)
\tag{10}
$$

The ratio of the company's order quantity to the producer's agricultural inputs is as follows:

$$
\delta^* = \frac{q^*}{I^*} = \sqrt{\frac{2C_s}{C_p}}
\tag{11}
$$

With respect to the partial derivative of q, the optimal order quantity under centralized decision making can be obtained as follows:

$$
q^* = \left(\beta - \varepsilon a p^{-b} g^*\right) - \frac{[(C_r - V_r) + 2\sqrt{2C_s C_p}]}{p - V_r}\sqrt{\frac{C_p}{2C_s}}
\tag{12}
$$

Therefore, we obtain the following:

$$
I^* = \sqrt{\frac{1}{2}\frac{C_p}{C_s}}\left(\left(\beta - \varepsilon a p^{-b} g^*\right) - \frac{[(C_r - V_r) + 2\sqrt{2C_s C_p}]}{p - V_r}\sqrt{\frac{C_p}{2C_s}}\right)
\tag{13}
$$

Plug q^* and g^* into the following equation to obtain the following:

$$
q^{*2} = 2q^*\left(\beta - 2\varepsilon a \left(p^*\right)^{-P} g^* + kV_r\right) - \left(\alpha - k\varepsilon a \left(p^*\right)^{-P} g^*\right)\left(\alpha - 3k\varepsilon a \left(p^*\right)^{-P} g^* + 2kV_r\right)
$$

Using the equation above, we obtain p^*. Therefore, when the distribution function of the stock factor z is an increasing generalized failure rate function, the optimal stock factor is z^*, as follows:

$$z^* = \frac{q^*}{ap^{*-b}g} \tag{14}$$

4.2. Decentralized Decision Model

4.2.1. Farmer Production Optimal Decision

When there is no coordination, the decision of the green agricultural supply chain is a typical wholesale price contract decision. For realism, here we consider the Stackelberg game in which farmers act first, and the information of the two is symmetric, namely, the wholesale price of agricultural products. It is determined by the farmer according to the company's sales price response function. Farmers can determine the agricultural inputs and greenness according to the market demand distribution function, and the company as a follower decides the sales price and order quantity.

The profit function of farmers is as follows:

$$\pi_s = \omega zx(p) + V_s[\mu I - zx(p)]^+ - C_p[zx(p) - \mu I]^+ - C_s I - \frac{1}{2}kg^2 \tag{15}$$

The profit function of farmers' sales expectation is the following:

$$E(\pi_s) = (\omega - C_p)zx(p) - (C_p - V_s)S(q, \mu I) + V_s\mu I - C_s I - \frac{1}{2}kg^2 \tag{16}$$

According to the generalized definition of CVaR, the decision objective function for farmers with risk avoidance characteristics to select production yield q can be described as follows:

$$CVaR(\pi_F(q,g)) = \max_{v \in R}\left\{v + \frac{1}{\eta}E(\min(\pi_s(q,g)) - v, 0)\right\} \tag{17}$$

where v represents the value at risk at a given confidence level η. $\eta \in (0,1]$ represents the degree of risk avoidance of farmers.

We define a function as follows:

$$f(q, g, v, \delta, I, z) = v + \frac{1}{\eta}E(\min(\pi(q,g)) - v, 0)$$

Therefore, we obtain the following:

$$I^* = \text{argmax}_{I>0}\max_v f\left(q^*, g^*, v, \delta^*, I, z^*\right)$$

The ratio of the company's order quantity to the producer's agricultural inputs is the following:

$$\delta^* = \text{argmax}_{I>0}\max_v f\left(q^*, g^*, v, \delta, I, z^*\right)$$

The optimal order quantity is the following:

$$q^* = \text{argmax}_{I>0}\max_v f\left(q, g^*, v, \delta^*, I, z^*\right)$$

The optimal inventory factor is as follows:

$$z^* = \text{argmax}_{z>0}\max_v f\left(q^*, g^*, v, \delta^*, I^*, z\right)$$

The optimal greenness is the following:

$$g^* = \text{argmax}_{g>0}\max_v f\left(q^*, g, v, \delta^*, I^*, z^*\right)$$

Substitute (1) into the above formula to obtain the following:

$$f(q, g, v, \delta, I, z) = v - \frac{1}{\eta} E(v - \pi_F(p, z, g, \omega_1))^+ = v - \frac{1}{\eta} \int_{\omega_1}^{P} \left(\omega z x(p) - C_s I - \frac{1}{2} k g^2 \right)^+ dR(\omega)$$

where $(z)^+ = \max\{z, 0\}$ represents the projection operator.

For fixed p, z, and g, $\max_v f(q, g, v, \delta, I, z)$ can be discussed in three cases:

When $v < \omega z x(p) - C_s I - \frac{1}{2} k g^2$, $f(q, g, v, \delta, I, z) = v$, the $\frac{\partial f(p, g, z, v)}{\partial v} = 1 > 0$

When $\omega z x(p) - C_s I - \frac{1}{2} k g^2 \leq v < p z x - C_s I - \frac{1}{2} k g^2$, the

$$f(q, g, v, \delta, I, z) = v - \frac{1}{\eta} \int_{\omega_1}^{P} \left(\omega z x(p) - C_s I - \frac{1}{2} k g^2 \right) dR(\omega)$$

Therefore, we have the following:

$$\frac{\partial f(q, g, v, \delta, I, z)}{\partial v} = 1 - \frac{1}{\eta} R(\omega_0) - \frac{1}{\eta} \left[R \left(\frac{v + C_s I + \frac{1}{2} k g^2}{z x} \right) - R(\omega_0) \right] = 1 - \frac{1}{\eta} R \left(\frac{C_s I + \frac{1}{2} k g^2}{z x} \right)$$

It is worth noting that in the following:

$$\left. \frac{\partial f(q, g, v, \delta, I, z))}{\partial v} \right|_{v = \omega z x(p) - C_s I - \frac{1}{2} k g^2} = 1 - \frac{1}{\eta} R(\omega_1)$$

$$\left. \frac{\partial f(q, g, v, \delta, I, z)}{\partial v} \right|_{v = \omega z x(p) - C_s I - \frac{1}{2} k g^2} = 1 - \frac{1}{\eta} R(\omega_h) = 1 - \frac{1}{\eta} < 0$$

When $v \geq p z x - C_s I - \frac{1}{2} k g^2$, we obtain the following:

$$f(q, g, v, \delta, I, z) = v - \frac{1}{\eta} \int_{\omega_1}^{P} \left(v - p z x + C_s I + \frac{1}{2} k g^2 \right) dR(\omega) - dR(\omega)$$

Therefore, we obtain the following:

$$\frac{\partial f(q, g, v, \delta, I, z)}{\partial v} = 1 - \frac{1}{\eta} \left[R \left(\frac{v + C_s I + \frac{1}{2} k g^2}{z x} \right) \right] = 1 - \frac{1}{\eta} < 0$$

Let v^* be the optimal yield because $f(p, g, z, v)$ is a concave function of v, which can be obtained according to the above three cases as follows:

$$v^* = \begin{cases} \omega z x(p) - C_s I - \frac{1}{2} k g^2 & \text{if } \omega_1 \geq R^{-1}(\eta) \\ R^{-1}(\eta) z x - C_s I - \frac{1}{2} k g^2 & \text{else} \end{cases}$$

By introducing v^*, the optimal retail price and greenness distribution of farmers can be obtained, and the optimal agricultural input volume, greenness, and optimal wholesale order price of farmers, respectively, are the following:

$$I^* = \sqrt{\frac{1}{2} \frac{C_p - V_s}{C_s - V_s \mu}} \left(\left(\beta - 2 \varepsilon a \left(p^* \right)^{-P} g^* \right) - \frac{(\omega + C_r - V_r)(\beta - \alpha)}{p - V_r} \right) \tag{18}$$

$$g^* = \frac{1}{k} \left((\omega - V_s) d - \sqrt{2(C_p - V_s)(C_s - V_s \mu)} \right) \tag{19}$$

The best order wholesale price is as follows:

$$\omega^* = C_s I^* + \frac{1}{2} k g^{*2} - V_s \mu I^* \tag{20}$$

4.2.2. Company Pricing Optimal Decision

The profit function of the firm is the following:

$$\pi_r = p\min\{q, D\} + V_r[q - D]^+ - \omega q - C_r q \tag{21}$$

where ωq represents the procurement cost.

The profit function of sales expectation is as follows:

$$E(\pi_r) = (p - V_r)S(q) - (\omega - V_r + C_r)q \tag{22}$$

Making use of the following:

$$\frac{dE(\pi)}{dp} = 0$$

$$\frac{dE(\pi)}{dq} = 0$$

The optimal order quantity of the company can be obtained as follows:

$$q^* = \left(\beta - 2\varepsilon a\left(p^*\right)^{-P}g^*\right) - \frac{(\omega + C_r - V_r)(\beta - \alpha)}{p - V_r} \tag{23}$$

Plug q^* and g^* into the following equation:

$$q^{*2} = 2q^*\left(\beta - 2\varepsilon a\left(p^*\right)^{-P}g^* + kV_r\right) - \left(\alpha - k\varepsilon a\left(p^*\right)^{-P}g^*\right)\left(\alpha - 3k\varepsilon a\left(p^*\right)^{-P}g^* + 2kV_r\right)$$

Using the above equation, we obtain p^*.

Therefore, the following results:

$$p^* = \frac{a\beta}{a - 1}\frac{(\omega + C_r - V_r)(\beta - \alpha)}{p - V_r} \tag{24}$$

5. Revenue-Sharing Coordination Mechanism

First of all, the company determines the sales price from the perspective of maximizing the overall benefits of the supply chain. The advantage of this is that it can stimulate the consumer market and satisfy consumers with green agricultural products at a lower price. After negotiation, farmers also sell green agricultural products to the company at a lower wholesale price. Secondly, the company needs to share the income from the sale of the product to the market with the farmers according to a certain proportion. λ represents the share proportion retained by the company from its sales revenue. The share proportion of farmers is $1 - \lambda$. Through the change in the profit-sharing coefficient, the agricultural input of farmers, the greenness of agricultural products, the order quantity of the company, and the retail price, agricultural products will change with the change in the profit-sharing coefficient, finally affecting changes in their respective expected profits.

The profit function of the firm after the introduction of the contract is the following:

$$\pi_r = \lambda p\min\{q, D\} + V_r[q - D]^+ - \omega q - C_r q \tag{25}$$

The profit function of sales expectation is as follows:

$$E(\pi_r) = (\lambda p - V_r)S(q) - (\omega - V_r + C_r)q \tag{26}$$

Making use of the following:

$$\frac{dE(\pi)}{dp} = 0$$

$$\frac{dE(\pi)}{dq} = 0$$

The optimal order quantity of the company can be obtained as follows:

$$q^* = \left(\beta - \varepsilon a \left(p^*\right)^{-P} g^*\right) - \frac{(\omega + C_r - V_r)(\beta - \alpha)}{\lambda p - V_r} \tag{27}$$

Plug q^* and g^* into the following equation to obtain the following:

$$\lambda q^{*2} = 2q^* \left(\lambda\left(\beta - 2k\varepsilon a \left(p^*\right)^{-P} g\right) + kV_r\right) - \left(\alpha - k\varepsilon a \left(p^*\right)^{-P} g\right)\left(\lambda\left(\alpha - 3k\varepsilon a \left(p^*\right)^{-P} g\right) + 2kV_r\right)$$

Using the equation above, we obtain p^*.

After introducing the contract, the profit function of farmers is the following:

$$\pi_s = (1 - \lambda)p\min\{q, D\} + \omega q + V_s[\mu I - q]^+ - C_p[q - \mu I]^+ - C_s I - \frac{1}{2}kg^2 \tag{28}$$

The profit function of sales expectation is as follows:

$$E(\pi_s) = (1 - \lambda)pS(q) + (\omega - C_p)q - (C_p - V_s)S(q, \mu I) + V_s\mu I - C_s I - \frac{1}{2}kg^2 \tag{29}$$

Making use of the following:

$$\frac{dE(\pi)}{dI} = 0$$

$$\frac{dE(\pi)}{dg} = 0$$

The optimal agricultural investment amount and greenness of the company can be obtained as follows:

$$I^* = \sqrt{\frac{1}{2}\frac{C_p - V_s}{C_s - V_s\mu}}\left(\left(\beta - 2\varepsilon a \left(p^*\right)^{-P} g^*\right) - \frac{(\omega + C_r - V_r)(\beta - \alpha)}{\lambda p - V_r}\right) \tag{30}$$

$$g^* = \frac{1}{k}\left(d(1 - \lambda) + (\omega - V_s)d - \sqrt{2(C_p - V_s)(C_s - V_s\mu)}\right) \tag{31}$$

Under the circumstance that the output and demand of green agricultural products are uncertain, the purpose of introducing a benefit-sharing contract in the above model is to make farmers and companies no longer make decisions with the goal of maximizing their own interests, but instead tend to follow a centralized decision-making mode. While improving the overall interests of the supply chain, it also improves the interests of farmers and companies. We make use of the following:

$$\left(\beta - \varepsilon ap^{-b}g\right) - \frac{(\omega + C_r - V_r)(\beta - \alpha)}{\lambda p - V_r} = \left(\beta - \varepsilon ap^{-b}g\right) - \frac{[(C_r - V_r) + 2\sqrt{2C_sC_p}]}{p - V_r}\sqrt{\frac{C_p}{2C_s}}$$

Thus, when the share proportion is the following:

$$\lambda = \frac{V_r}{p} + \frac{(\omega + C_r - V_r)(p - V_r)}{p[(C_r - V_r) + 2\sqrt{2C_sC_p}]} \tag{32}$$

Order volume has reached coordination.

When we have the following:

$$\sqrt{\frac{1}{2}\frac{C_p-V_s}{C_s-V_s\mu}}\left(\left(\beta-\varepsilon ap^{-b}g\right)-\frac{(\omega+C_r-V_r)(\beta-\alpha)}{\lambda p-V_r}\right)=\sqrt{\frac{1}{2}\frac{C_p}{C_s}}\left(\left(\beta-\varepsilon ap^{-b}g\right)-\frac{\left[(C_r-V_r)+2\sqrt{2C_sC_p}\right]}{p-V_r}\sqrt{\frac{C_p}{2C_s}}\right)$$

Then, the share proportion becomes the following:

$$\lambda=\frac{V_r}{p}+\frac{1}{p}\frac{(\omega+C_r-V_r)(\beta-\alpha)}{\left(\beta-\varepsilon ap^{-b}g\right)\left(1-\sqrt{\frac{1}{2}\frac{C_p(C_s-V_s\mu)}{C_s(C_p-V_s)}}\right)+\frac{\left[(C_r-V_r)+2\sqrt{2C_sC_p}\right]}{p-V_r}\sqrt{\frac{1}{2}\frac{C_p(C_s-V_s\mu)}{C_s(C_p-V_s)}}}$$

The amount of agricultural capital input of farmers has reached coordination. When we have the following:

$$\frac{1}{k}\left(d(1-\lambda)+(\omega-V_s)d-\sqrt{2(C_p-V_s)(C_s-V_s\mu)}\right)=\frac{1}{k}\left((\omega-V_s)d-\sqrt{2(C_p-V_s)(C_s-V_s\mu)}\right)$$

And then, the share proportion is as follows:

$$\lambda=\frac{C_r+\omega-V_s+\sqrt{2C_pC_s}}{p}-\frac{\sqrt{2(C_p-V_s)(C_s-V_s\mu)}}{pd}$$

The greenness is in harmony.

6. Numerical Analysis

In order to analyze the role of the overall coordination mechanism, consider the parameter assignments as shown in Table 2.

Table 2. Parameter assignments.

Argument	Numerical Value	Argument	Numerical Value
μ_1	0.4	C_s	4
k	4	V_s	2
C_p	20	ω	21
C_r	6	V_r	3
d	1	a	2
α	60	β	90

According to the parameter assignments in Table 2, they are substituted into the green agricultural product supply chain Jupiter under uncertain output and demand. Through calculation, it can be seen that the company's selling price is 27.2 in the decentralized decision-making state. The order quantity from farmers is 6.9; the amount of agricultural input of farmers is 14.2; and the greenness of agricultural products is 2.7. In the centralized decision-making state, the selling price of sellers is 29.2, and the order quantity of farmers is 24.4. The agricultural input of farmers is 35.2, and the greenness of agricultural products is 2.9.

Under centralized decision making, the expected profit of the agricultural supply chain is 92.3. In decentralized decision making, the expected profits of farmers and sellers are 43.1 and 9.3, respectively, and the expected profit of the agricultural product supply chain is 52.4. According to the calculated numerical results, when the green agricultural product supply chain is under centralized decision making, the greenness of the product, the selling price, the order quantity, and the agricultural input quantity are all better than those under decentralized decision making, which is also in line with the theory because the centralized decision making is the result under ideal conditions and can be used as a reference for the decision making of farmers and companies. In decentralized decision making, on the one hand, farmers are reluctant to increase the input of agricultural materials due to the uncertainty of output and demand; on the other hand, sellers face the same problem and order smaller quantities to reduce their own risks. After the introduction of the revenue-sharing contract, within a reasonable revenue-sharing coefficient range, set the range as $0.75 \leq \lambda \leq 0.8$ of λ and take 0.01 as the step to calculate the corresponding value of each decision variable and the profit of the supply chain.

6.1. The Influence of Income-Sharing Coefficient on the Profits of Farmers and Companies

Figure 1a,b show that with an increase in the income-sharing coefficient, the profits of both farmers and companies show an upward trend. We hypothesize that the wholesale price is determined by the farmer according to the company's price response function. After introducing the revenue-sharing contract, the selling price also decreased to a certain extent. Although the wholesale price of farmers is lower than that of decentralized decision making, the company's order quantity from farmers increased, and farmers were willing to increase their input of agricultural resources, so the profits of farmers and companies show an upward trend. This is also in line with the actual situation. Within a reasonable range of the income-sharing coefficient, the greenness of green agricultural products is higher than that of ordinary agricultural products, and the sales price is lower than before the introduction of the contract, which stimulates the consumption desire of consumers and promotes increases in the profits of farmers and companies.

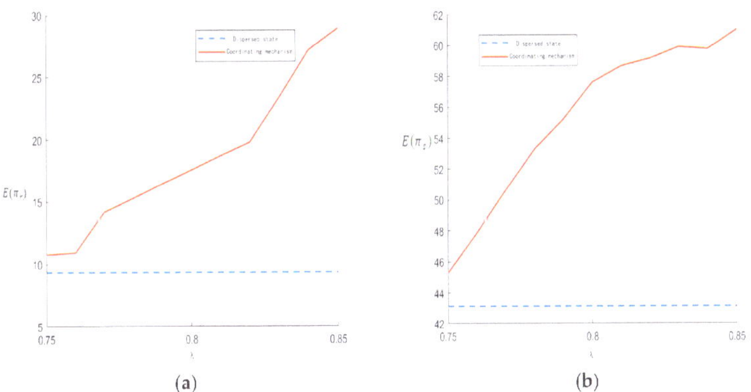

(a)

(b)

Figure 1. (**a**) The influence of income-sharing coefficient on farmers' expected profit. (**b**) The effect of revenue-sharing coefficient on the firm's expected profit.

6.2. Effect of Income-Sharing Coefficient on Greenness of Agricultural Products

As can be seen from Figure 2, the greenness of agricultural products in centralized decision making is in a higher state than that in decentralized decision making. This is because decentralized decision making requires more green costs to improve the greenness of agricultural products, so farmers make decisions based on the principle of maximizing their own interests and are reluctant to increase green production costs. After the introduction of the revenue-sharing contract, when the income of farmers from the company decreases with the increase, the greenness of agricultural products will be further affected. This is also in line with the reality. For farmers, in order to maintain a profitable state, farmers will choose to reduce the greenness of agricultural products to reduce costs.

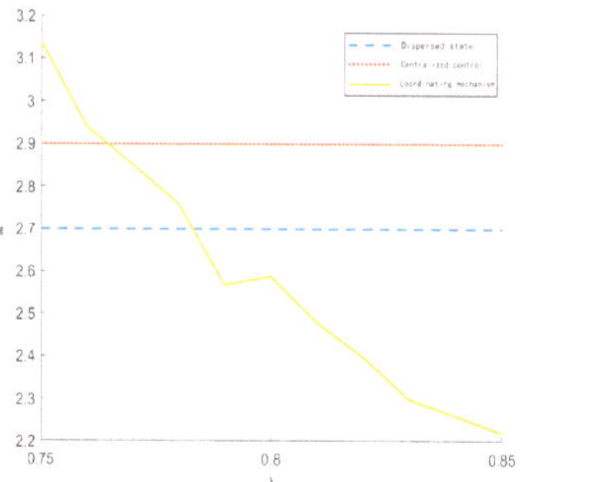

Figure 2. Effect of income-sharing coefficient on greenness of agricultural products.

6.3. Effect of Income-Sharing Coefficient on Order Quantity and Agricultural Input Quantity

It can be seen from Figure 3a,b that when farmers and companies agree to introduce the revenue-sharing contract, both the input of agricultural capital and the ordered quantity of green agricultural products show a rising trend. This result not only meets the demand of market consumers for green agricultural products, but also meets the necessary conditions for the profit growth of farmers and sellers, forming a good trend of farmers increasing their agricultural input and companies increasing their orders. In real life, when the two reach a sales agreement, the company receives a lower wholesale price, and inevitably increases the quantity of green agricultural products ordered; Farmers will also increase their agricultural inputs to meet the company's increased orders, forming a virtuous circle.

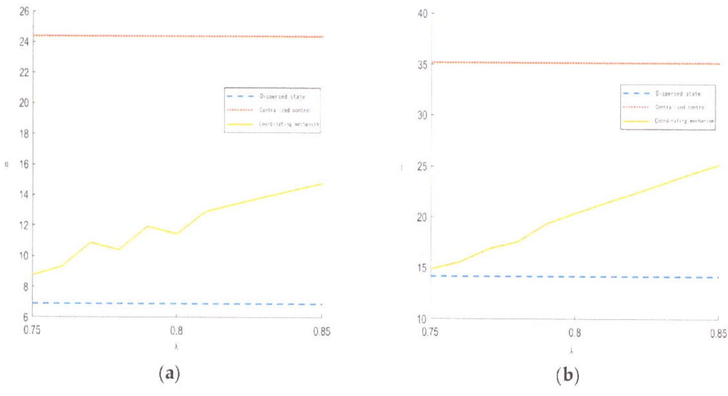

(a)　　　　　　　　　　　　　　　　(b)

Figure 3. (**a**) Effect of revenue-sharing coefficient on order quantity. (**b**) Effect of income-sharing coefficient on agricultural capital input.

7. Discussion and Conclusions

The internal and external environments of the supply chain for green agricultural products is complex, and its ability to resist natural risks and market risks is poor. In addition, the supply chain for green agricultural products contains a large number of subjects, and the chain is long. When signing and executing contracts, some subjects in the chain have opportunistic tendencies, and there are problems of low operational efficiency in the supply chain system. Under the double uncertainty of output and demand, the output end and demand end of the supply chain of green agricultural products face an imbalance phenomenon, which leads to the risk of interruptions to the supply chain of green agricultural products. On this basis, this paper constructed a Stackelberg game model of "firm + farmer" green agricultural product supply chain and introduced CVaR theory to study the coordination mechanism of a green agricultural product supply chain under the double uncertainty of output and demand, considering the risk avoidance degree of farmers. The benefit-sharing contract was used to coordinate the supply chain. After the coordination, the income of farmers, the income of companies, and the income of the whole supply chain improved to different degrees, so that the supply chain was able to reach the Pareto optimal. The effectiveness of the model was verified through numerical simulation.

In theory, this paper provides a new perspective for the study of green agricultural product supply chain coordination. Currently, the study of agricultural product supply chains mainly focuses on the supply chain of fresh agricultural products, with few studies on issues related to green agricultural product supply chains. Most research on supply chain coordination focuses on the coordination of multi-agent supply chains, price, and service competition under random demand. From the perspective of double uncertainty of output and demand, this paper combined theory with practice to analyze the sustainable operation mechanism of a green agricultural product supply chain under the uncertainty of production and demand, and to some extent filled gaps in terms of coordinating the green agricultural product supply chain under the uncertainty of production and demand.

The results show that the greenness, selling price, order quantity, and input quantity of green agricultural products are better than those parameters under decentralized decision making. In decentralized decision making, farmers are reluctant to increase their input of agricultural materials due to the double uncertainty of output and demand, and companies also order smaller quantities to reduce their own risks. After introducing a revenue-sharing contract, the selling price also decreases to a certain extent, and the company increases the quantity of its orders from farmers. Within a reasonable range of the revenue-sharing coefficient, the greenness of green agricultural products is higher than that of ordinary agricultural products, and the selling price is lower than before the introduction of the contract; this stimulates the consumption desire of consumers and can increase the profits of farmers and companies. The greenness of agricultural products in centralized decision making is higher than that in decentralized decision making. The introduction of revenue-sharing contracts can increase the agricultural capital input of farmers and the amount of green agricultural products ordered by companies. After coordination, farmers and companies no longer make decisions to maximize their own interests, but tend to follow a centralized decision-making mode. Through numerical simulation analysis, it was verified that the revenue-sharing contract improved the overall incomes of farmers, companies, and the supply chain, making the supply chain reach the Pareto optimal; thus, the overall operation efficiency and stability of the supply chain improved.

Under the double uncertainty of output and demand, there are still many problems in operating a green agricultural products supply chain, which lead to the risk of interruptions to the supply chain. This paper analyzed a supply chain of green agricultural products in the form of "company + farmer", but there are some limitations. In future research, we can consider how to coordinate the dual-channel supply chain with the double uncertainty of output and demand. In addition, on the basis of the existing research on the "bullwhip effect" and the "robustness model", the future focus will be on the study and measurement of supply chain stability and the stability of the supply chain of green agricultural products under the uncertainty of output and demand. In the future, we can further evaluate the performance of green agricultural products supply chains under the uncertainty of output and demand, in order to improve their overall efficiency.

Author Contributions: Conceptualization, Q.W. and Q.Z.; methodology, R.Q.; software, M.N.; validation, W.W., Q.W. and M.N.; formal analysis, Q.Z.; investigation, Q.W. and R.Q.; writing—original draft preparation, Q.W.; writing—review and editing, Q.W., W.W. and M.N.; supervision, Q.Z. All authors have read and agreed to the published version of the manuscript.

Funding: This research was supported by the Natural Science Foundation of Heilongjiang Province, China, under the project of "Research on Financial Risk warning, blocking and Recovery Mechanism of Online Farm product Supply chain", grant number LH2023G016.

Institutional Review Board Statement: Not applicable.

Informed Consent Statement: Not applicable.

Data Availability Statement: The data presented in this study are available on request from the corresponding author. The data are not publicly available due to personal privacy and non-open access to the research program.

Conflicts of Interest: The authors declare no conflicts of interest.

References

1. Thomas, C. The United Nations Conference on Environment and Development (UNCED) of 1992 in Context. *Environ. Politics* **1992**, *1*, 250–261. [CrossRef]
2. Cars, M.; West, E.E. Education for sustainable society: Attainments and good practices in Sweden during the United Nations Decade for Education for Sustainable Development (UNDESD). *Environ. Dev. Sustain.* **2015**, *17*, 1–21. [CrossRef]
3. Pachapur, P.K.; Pachapur, V.L.; Brar, S.K.; Galvez, R.; Le Bihan, Y.; Surampalli, R.Y. Food security and sustainability. *Sustain. Fundam. Appl.* **2020**, *17*, 357–374. [CrossRef]
4. Roy, P.; Mohanty, A.K.; Dick, P.; Misra, M. A Review on the Challenges and Choices for Food Waste Valorization: Environmental and Economic Impacts. *ACS Environ. Au* **2023**, *3*, 58–75. [CrossRef] [PubMed]
5. Tallard, G.; Adenuer, M.; Deconinck, K.; Gouarin, G. Potential impact of dietary changes on the triple challenge facing food systems: Three stylised scenarios. *OECD Food Agric. Fish. Pap.* **2022**, 1–47. [CrossRef]
6. Liu, R.; Gao, Z.; Snell, H.A.; Ma, H. Food safety concerns and consumer preferences for food safety attributes: Evidence from China. *Food Control* **2020**, *112*, 107157. [CrossRef]
7. Yang, W.; Fang, L. Consumer Willingness to pay for food safety attributes in China: A meta-analysis. *J. Int. Food Agribus. Mark.* **2021**, *33*, 152–169. [CrossRef]
8. Yu, X.; Guo, L.; Jiang, G.; Song, Y.; Muminov, M.A. Advances of organic products over conventional productions with respect to nutritional quality and food security. *Acta Ecol. Sin.* **2018**, *38*, 53–60. [CrossRef]
9. Xu, J.; Zhong, J.; Zhang, B.; Li, X. Green Labelled Rice Shows a Higher Nutritional and Physiochemical Quality Than Conventional Rice in China. *Foods* **2021**, *10*, 915. [CrossRef] [PubMed]
10. Ye, F.; Yang, Z.; Yu, M.; Watson, S.; Lovell, A. Can market-oriented reform of agricultural subsidies promote the growth of agricultural green total factor productivity? Empirical evidence from maize in China. *Agriculture* **2023**, *13*, 251. [CrossRef]
11. Huang, X.; Feng, C.; Qin, J.; Wang, X.; Zhang, T. Measuring China's agricultural green total factor productivity and its drivers during 1998–2019. *Sci. Total Environ.* **2022**, *829*, 154477. [CrossRef] [PubMed]
12. Wubie, A.A. Review on the impact of climate change on crop production in Ethiopia. *J. Biol. Agric. Healthc.* **2015**, *5*, 103–111.
13. Abubakar, A.; Ishak, M.Y.; Uddin, M.K.; Zangina, A.S.; Ahmad, M.H.; Danhassan, S.S. Impact of climate change and adaptations for cultivation of millets in Central Sahel. *Environ. Sustain.* **2023**, *6*, 441–454. [CrossRef]
14. Sánchez-Bravo, P.; Chambers, E.; Noguera-Artiaga, L.; Sendra, E.; Chambers, I.V.E.; Carbonell-Barrachina, Á.A. Consumer understanding of sustainability concept in agricultural products. *Food Qual. Prefer.* **2021**, *89*, 104136. [CrossRef]
15. Nguyen, H.; Le, H. The effect of agricultural product eco-labelling on green purchase intention. *Manag. Sci. Lett.* **2020**, *10*, 2813–2820. [CrossRef]
16. Deschênes, O.; Greenstone, M. The economic impacts of climate change: Evidence from agricultural output and random fluctuations in weather: Reply. *Am. Econ. Rev.* **2012**, *102*, 3761–3773. [CrossRef]
17. Cheng, S.; A López, M. Synergy analysis of agricultural economic cycle fluctuation based on ant colony algorithm. *Open Phys.* **2018**, *16*, 978–988. [CrossRef]
18. Chen, W.; Li, J.; Jin, X. The replenishment policy of agri-products with stochastic demand in integrated agricultural supply chains. *Expert Syst. Appl.* **2016**, *48*, 55–66. [CrossRef]
19. Zhao, Z.; Min, K.J. Blockchain Traceability Valuation for Perishable Agricultural Products Under Demand Uncertainty. *Int. J. Oper. Res. Inf. Syst.* **2020**, *11*, 1–32. [CrossRef]
20. Tamasiga, P.; Onyeaka, H.; Bakwena, M.; Happonen, A.; Molala, M. Forecasting disruptions in global food value chains to tackle food insecurity: The role of AI and big data analytics—A bibliometric and scientometric analysis. *J. Agric. Food Res.* **2023**, *14*, 100819. [CrossRef]
21. Maria, S.A.; Jesse, L. Global Supply Chain Disruptions and Inflation During the COVID-19 Pandemic. *Review* **2022**, *104*, 78–91.
22. Ivanov, D. Two views of supply chain resilience. *Int. J. Prod. Res.* **2024**, *62*, 4031–4045. [CrossRef]
23. Imbiri, S.; Rameezdeen, R.; Chileshe, N.; Statsenko, L. A novel taxonomy for risks in agribusiness supply chains a systematic literature review. *Sustainability* **2021**, *13*, 9217. [CrossRef]
24. Eremenko, V.; Shumilin, P.; Shamkina, E.; Shumilina, V.; Makartsova, T.; Kravchenko, A.; Evsyukova, O.; Filippova, A.; Yatsuk, D. Adaptation of accounting in agribusiness risk management system. *E3S Web Conf.* **2020**, *164*, 10006. [CrossRef]
25. Zhao, G.; Liu, S.; Lopez, C.; Chen, H.; Lu, H.; Mangla, S.K.; Elgueta, S. Risk analysis of the agri-food supply chain: A multi-method approach. *Int. J. Prod. Res.* **2020**, *58*, 4851–4876. [CrossRef]
26. Sharma, R.; Shishodia, A.; Kamble, S.; Gunasekaran, A.; Belhadi, A. Agriculture supply chain risks and COVID-19: Mitigation strategies and implications for the practitioners. *Int. J. Logist. Res. Appl.* **2020**, 1–27. [CrossRef]

27. Bo, Y.; Han-Meng, S.U. A Review of Risk Identification of Multinational Agricultural Products Supply Chain. In Proceedings of the 2018 5th International Conference on Management Science and Management Innovation (MSMI 2018), Wuhan, China, 20–22 April 2018; Atlantis Press: Amsterdam, The Netherlands, 2018; pp. 234–238.

28. He, X.; Zhang, C.; Jing, Y.; Ai, X. Risk Evaluation of Agricultural Product Supply Chain Based on BP Neural Network. In Proceedings of the 2019 16th International Conference on Service Systems and Service Management (ICSSSM), Shenzhen, China, 13 July 2019.

29. Wang, W.; Cao, Q.; Liu, Y.; Zhou, C.; Jiao, Q.; Mangla, S.K. Risk management of green supply chains for agricultural products based on social network evaluation framework. *Bus. Strategy Environ.* **2024**, *3*, 1–22. [CrossRef]

30. Chauhan, A.; Kaur, H.; Yadav, S.; Jakhar, S.K. A hybrid model for investigating and selecting a sustainable supply chain for agri-produce in India. *Ann. Oper. Res.* **2020**, *290*, 621–642. [CrossRef]

31. Shi, Y.; Wang, F. Agricultural supply chain coordination under weather-related uncertain yield. *Sustainability* **2022**, *14*, 5271. [CrossRef]

32. Giri, B.C.; Majhi, J.K.; Chaudhuri, K. Coordination mechanisms of a three-layer supply chain under demand and supply risk uncertainties. *RAIRO-Oper. Res.* **2021**, *55*, S2593–S2617. [CrossRef]

33. Chen, T.; Liu, C.; Xu, X. Coordination of Perishable Product Supply Chains with a Joint Contract under Yield and Demand Uncertainty. *Sustainability* **2022**, *14*, 12658. [CrossRef]

34. Liu, X.; Shen, X.; You, M. Study on coordination and optimization of contract farming supply chain based on uncertain conditions. *Sci. Program.* **2020**, *2020*, 8858812. [CrossRef]

35. Su, H.; Cui, Y.; Liu, B. Research on the supply chain model based on the differentiation of green agricultural products. In *MATEC Web of Conferences*; EDP Sciences: Les Ulis, France, 2018; Volume 232, p. 02012.

36. Tan, M.; Tu, M.; Wang, B.; Zou, T.; Cheng, H. A Two-Echelon Agricultural Product Supply Chain with Freshness and Greenness Concerns: A Cost-Sharing Contract Perspective. *Complexity* **2020**, *1*, 1–13. [CrossRef]

37. Hu, Q.; Xu, Q.; Xu, B. Introducing of Online Channel and Management Strategy for Green Agri-food Supply Chain based on Pick-Your-Own Operations. *Int. J. Environ. Res. Public Health* **2019**, *16*, 1990. [CrossRef] [PubMed]

38. Hu, H.; Li, Y.; Li, M.; Zhu, W.; Chan, F. Optimal decision-making of green agricultural product supply chain with fairness concerns. *J. Ind. Manag. Optim.* **2023**, *19*, 4926–4948. [CrossRef]

39. Cao, Y.; Tao, L.; Wu, K.; Wan, G. Coordinating joint greening efforts in an agri-food supply chain with environmentally sensitive demand. *J. Clean. Prod.* **2020**, *277*, 123883. [CrossRef]

40. Cui, L.; Guo, S.; Zhang, H. Coordinating a green agri-food supply chain with revenue-sharing contracts considering retailers' green marketing efforts. *Sustainability* **2020**, *12*, 1289. [CrossRef]

41. Ye, F.; Lin, Q.; Li, Y. Coordination for contract farming supply chain with stochastic yield and demand under CVaR criterion. *Oper. Res.* **2020**, *20*, 369–397. [CrossRef]

42. Hongjun, P.; Meng, Y. Coordination of Order-agriculture Supply Chain Based on CVaR under Option Contract. *Oper. Res. Manag. Sci.* **2023**, *32*, 131.

43. Liao, C.; Lu, Q.; Lin, L. Coordinating a three-level contract farming supply chain with option contracts considering risk-averse farmer and retailer. *PLoS ONE* **2023**, *18*, e0279115. [CrossRef]

44. Nan, C.; Cai, J.; Han, W. Coordination of Supply Chain of a Three-level Fresh Products Based on Conditional Value at Risk. In Proceedings of the 2021 IEEE International Conference on Industrial Engineering and Engineering Management (IEEM), Singapore, 13–16 December 2021; pp. 153–157.

Article

An Analysis of the Eco-Efficiency of the Agricultural Industry in the Brazilian Amazon Biome

Gabriela Mayumi Saiki [1,*], André Luiz Marques Serrano [1,*], Gabriel Arquelau Pimenta Rodrigues [1], Carlos Rosano-Peña [2], Fabiano Mezadre Pompermayer [3] and Pedro Henrique Melo Albuquerque [2]

[1] Professional Post-Graduate Program in Electrical Engineering (PPEE), Department of Electrical Engineering (ENE), Technology Faculty, University of Brasilia (UnB), Brasilia 70910-900, DF, Brazil; gabriel.arquelau@redes.unb.br

[2] Faculty of Economics, Administration and Accounting (FACE), Department Administration (ADM), University of Brasilia (UnB), Brasilia 70910-900, DF, Brazil; crosano@unb.br (C.R.-P.); pedro.albuquerque@bellevuecollege.edu (P.H.M.A.)

[3] Institute for Applied Economic Research (IPEA), Brasilia 70390-025, DF, Brazil; famepomper@hotmail.com

* Correspondence: mayumisaiki.gms@gmail.com (G.M.S.); andrelms@unb.br (A.L.M.S.)

Citation: Saiki, G.M.; Serrano, A.L.M.; Rodrigues, G.A.P.; Rosano-Peña, C.; Pompermayer, F.M.; Albuquerque, P.H.M. An Analysis of the Eco-Efficiency of the Agricultural Industry in the Brazilian Amazon Biome. *Sustainability* **2024**, *16*, 5731. https://doi.org/10.3390/su16135731

Academic Editors: Fotios Chatzitheodoridis, Efstratios Loizou and Achilleas Kontogeorgos

Received: 4 June 2024
Revised: 18 June 2024
Accepted: 25 June 2024
Published: 4 July 2024

Abstract: The exponential growth of the agricultural industry in the Amazon region has brought about notable economic advancements. However, this growth has substantially cost the region's ecosystems, manifesting in increased deforestation and biodiversity degradation within the Amazon forest. This article is dedicated to assessing the eco-efficiency of agricultural production in Amazon Biome municipalities. It places particular emphasis on identifying critical determinants through the utilization of the classic Data Envelopment Analysis (DEA) model for efficiency computation, super-efficiency models for distinctive characterization, bootstrap computational techniques for robust resampling, and the Malmquist index for calculating annual eco-efficiency indices of each Decision-Making Unit (DMU). An exploration of the correlation between efficiency and meteorological attributes of the municipalities is conducted. The findings of this study reveal the following significant points: Eco-efficient municipalities within the Amazon Biome can serve as benchmarks for other DMUs striving to attain optimal input–output levels, most municipalities in the Amazon Biome operate close to the productive frontier due to the prevalent technology employed in their agricultural activities, the nature of the technological frontier's return suggests that small and large DMUs possess eco-efficiency potential, and the current dataset does not yield conclusive evidence regarding a direct correlation between the variables. Leveraging this information, strategic pathways can be formulated to drive economic development in tandem with the sustainability of Amazon Biome municipalities. These strategies promise to foster social, economic, and environmental benefits for the populace while providing valuable insights to inform future research within this thematic domain.

Keywords: farming industry; ecosystem; bootstrap; sustainable development; productivity analysis

1. Introduction

The issue of sustainable development became significantly relevant after the publication of the Club of Rome's report, "The Limits to Growth" [1]. The report warned that the planet could not support continued economic growth without exceeding the Earth's carrying capacity due to the increasing pressure on energy and natural resources, pollution, and climate change. This prompted the United Nations to establish conferences and commissions worldwide to study and discuss sustainable development and develop commitments and solutions.

Thus, in the Brundtland Report published by the United Nations World Commission on Environment and Development entitled "Our Common Future" [2], the definition of sustainable development emerged based on the concept of the "triple bottom line" of sustainability. The three dimensions are outlined as follows. (i) The economic dimension refers to the traditional search for profit and financial value creation. It involves assessing an organization's ability to be financially healthy. (ii) According to the social dimension, a production unit must consider its impact on society and the communities in which it operates. This can include corporate social responsibility; the quality of relations with employees, customers, and suppliers; and contributions to social development. (iii) According to the environmental dimension, an organization assesses its environmental impact and contribution to long-term sustainability. This involves issues such as reducing greenhouse gas emissions, managing natural resources efficiently, minimizing waste, and adopting sustainable practices and products, meaning that an organization's success should not only be measured by its financial performance but also by how it contributes to people's well-being and the preservation of the environment.

Since its inception, the concept of sustainable development has been adopted in the theoretical literature. It is practiced by companies and countries seeking a balance between financial prosperity, social responsibility, and environmental preservation. From this perspective, we can highlight the study by Heyder and Theuvsen [3], where the authors developed a survey of 170 companies in the agricultural industry of small and large multinational corporations in Germany related to social and environmental dimensions. In addition, other studies have proposed the construction of multidimensional sustainability indicators to evaluate the performance of the agricultural industry's production system. Some of them can be highlighted, such as [4–7].

Moving in this direction, the 26th Conference of Nations on Climate Change, which took place in Glasgow, Scotland, referred to the conservation of forests. During this conference, according to the United Nations website (https://news.un.org/en/story/2021/11/1104642 (accessed on 24 June 2024), 110 countries that are home to 85% of the world's forests signed a declaration committing to halt and reverse deforestation by 2030. This declaration was an essential step towards the preservation of the environment. However, the most challenging task lies in achieving this regional and local commitment without compromising the sustainable development and resilience to climate change of families who live and work in forests. They are producers with limited resources to maintain and improve productivity and competitiveness. They are marginalized from primary social services and impacted by the independence of adult children who give up and migrate to the urban environment or are fragmenting family property [8].

As a result, introducing ever stricter environmental protection regulations has caused great concern due to their social effects. This situation is no different in the Brazilian Amazon rain forest, where decision-makers face two challenges. This is due to the need to make communities more productive and efficient to offer more products of higher quality and competitive prices and, thus, to be able to face direct competition with agriculture on an industrial scale [9]. The second challenge arises from the evidence that agricultural intensification and expansion in forest areas can generate irreversible environmental damage that economic benefits may not offset [6].

Removing native primary vegetation cover is known to compromise the ecosystem services of flora and fauna, which are also at risk of degradation. In addition, deforestation causes erosion, a reduction in soil nutrients, and an increase in greenhouse gas (GHG) emissions, which interfere with rainfall and the planet's temperature. This creates a vicious cycle, as changes in temperature and rainfall patterns can harm agricultural activity itself [8,10]. There is also the fact that increased territorial expansion and human exploitation of the environment are increasingly depleting native forests. Climate change is known to cause direct effects, such as changes in soil moisture and temperature regimes. However, these changes also cause what Butterbach-Bahl and Dannenmann [11] call indirect effects, such as an increase in carbon dioxide (CO_2) in the atmosphere and soil dethatching (N_2).

In the Amazon Biome, this phenomenon is no different, making the issue of deforestation one of the main problems facing Brazil [12]. There are several consequences of deforestation in the Amazon. Among these consequences, we can see decreases in rainfall (220 to 640 mm/year) and evaporation (164 to 500 mm/year) and a slight increase in temperature (0 °C to 3 °C) [6,13–16]. From these observations, it can be inferred that the impact on the biome is significant, causing changes in other regions and biomes across the planet, such as a decrease in precipitation, an increase in temperature, and even a decrease in the efficiency of local agriculture [17]. Therefore, a rapid change is needed in how these natural resources are used and preserved [18].

Another factor causing deforestation in the Amazon is mechanized farming, which promotes burning [19]. The typical fires that cause deforestation are called high-frequency fires. These fires have contributed to more than 40% of the fires detected in recent years, while on the Bolivian border and in the states of Mato Grosso, Pará, and Rondônia, this figure is as high as 84% [19]. However, it is interesting to review how the agricultural industry operates and how it would be possible to improve the use of resources without failing to produce what is needed more efficiently. For this reason, it is impossible to think of reducing production to protect the environment. Instead, it is necessary to think of ways to develop production sustainably by introducing new technologies and improving planting methods.

This means that production must be performed in a way that does not harm the environment and consumes as few resources as possible. For this reason, one of the keywords in this research is efficiency. By efficiency, we mean production that delivers the maximum output level while spending as few resources as possible [20], thereby reaching the optimum point between the ratio of input and output within the production frontier. In this way, it is expected to produce as much food as possible, consuming as few environmental resources and emitting as little CO_2 as possible. DEA is used to assess efficiency.

This method consists of non-parametric mathematics that, using linear programming algorithms, can compare entities that carry out similar processes of transforming inputs into outputs to define an efficiency index between them in which the most efficient compared entity receives an index value equal to 1. The others receive a score below one proportional to how much less efficient they are. As a traditional technique, DEA is being popularized and has gained a lot of relevance, especially in research and academic circles. DEA is the only way to evaluate the efficiency of a transformation process by considering multiple inputs and outputs without having to transform these components into a single value, such as financial value.

In this study, we intend to address issues related to agricultural production in the municipalities of the Brazilian Amazon Biome. Our aim is to measure the eco-efficiency of municipalities belonging to the Amazon Biome, taking into account agricultural production and environmental preservation variables. To describe eco-efficient municipalities and, thus, provide a strategic vision for critical (inefficient) municipalities of how agriculture in the Amazon Biome can achieve more efficient production with minimal environmental impact and use of resources, using bootstrap computer models. We mapped the Brazilian municipalities considered to be outliers given the latest Brazilian agricultural census of 2017 and evaluated the eco-efficiency trend of the municipalities during the period from 2006 to 2017 using the Malmquist Index technique.

The application of Data Envelopment Analysis (DEA) to evaluate agriculture has been a consolidated practice in the literature [21–25]. However, most studies focus on China and continents such as Europe [26], where they seek to propose alternatives to make agriculture more efficient [27]. Few studies have focused on Brazil, since comparative analysis is carried out considering more than one region or municipality [28] or other regions of the Brazilian territory. On the other hand, it is interesting to have a view on a national scale. However, there has been little focus on trying to understand the relationship between inputs and outputs in one of the world's main environmental reserves (the Amazon); conducting eco-efficiency analysis using bootstrap and computational methods, as well as the Malmquist index to calculate eco-efficiency over time; identifying the dynamics of efficiency in the agricultural sector; and considering data from the period from 2006 to 2017 [29].

The remainder of this article is organized as follows. In Section 2, we present a theoretical discussion about the main concepts involved in data envelopment analysis. Section 3 presents the materials and methods used to carry out the research, describes in detail the mathematical models for calculating eco-efficiency using bootstrap computational methods, as well as the approach using the super efficiency to detect anomalous data, and the application of the Malmquist Index. Then, Section 4 presents a discussion of the main results obtained in this research. Finally, the article ends with Section 5, concisely describing the main results found by the research, as well as its contributions to the literature.

2. Theoretical Framework

Efficiency is a concept based on economics, mathematics, and operational research to reach an optimum point between the ratio of inputs to be transformed (x) and the outputs generated (y), given a technological set (T) so that $T = \{(x, y)|x$ can produce $y\}$ in a concept of production possibilities (CPP), whose properties are described by Färe [30]. In this way, eco-efficiency represents the factors involving economic, social, and environmental issues that form a tripod to make up the inputs and outputs of the production process. The inputs, (x) represent the ecological costs necessary to generate the desirable products (y) [31]. In theory, there is a minimization of inputs (x) and a maximization of outputs (y). Nonetheless, this relationship is not generalized for cases of desirable inputs, such as preserved areas, and undesirable outputs, such as CO_2 emissions; this relationship is inverted, given a technological set $(T = \{(x, y)|x)$ can produce $1/y\}$.

The DMUs (Decision-Making Units)—Inefficient DMUs are inserted within the CPP, formed by linear segment combinations of efficient DMUs that include a convex figure. However, they do not present optimal performance between the ratio of inputs and outputs [32]. The upper part of the frontier segment represents the (T^δ) of the CPP, formed by a negative space (R^{p+q}) and structured by the x and y ε (R^q) of the DMUs. From this discussion, the concept of eco-efficiency emerges. It is achieved when the set of n DMUs (decision-making units) that form the production frontier, composed of x and y, present the lowest possible quantity of inputs (related to the environmental cost of the process) and the highest possible amount of outputs (related to the impact that activities can have on the environment).

This distance between the inefficient point and the benchmarks made up of the frontier of efficient DMUs is called the Euclidean distance. Therefore, to achieve technical efficiency, it can be input-oriented (when inputs are minimized, without changing the level of outputs) and output-oriented (when outputs are maximized, without changing the status of inputs) [32], where the inverse of technical efficiency is the radial efficiency described in [31]. Thus, the DEA model is a mathematical model that, by receiving the input and output variables of a process, can define how efficient the entity carrying it out is [33].

To do this, the model uses several entities that carry out the same transformation process to, in a comparative analysis, find the limit of efficiency and how each of these entities (DMUs) is classified with respect to this limit. With enough DMUs, creating a scatter plot with all these DMUs distributed along the Y-axis, representing the outputs that a given DMU produces, is possible. The X-axis is the axis that quantifies the inputs that this DMU needs to consume to reach this production level. Thus, Figure 1 is a graphic example in which DMUs A, B, C, D, and E are considered benchmarks of excellence, i.e., those that have achieved technical efficiency [32].

Still analyzing the graph in Figure 1, it can be seen how much more DMU_E would need to produce to achieve technical efficiency and how much less it should consume—or even a combination of the two. In this context, the efficiency frontier is then generated by drawing a line between the points where it is positioned on the graph. The DEA model can calculate an efficiency index given the distance from the end to the line, and it is up to the researcher to indicate in which direction the line size should be measured. Next, the model can be oriented towards input, i.e., how much less a given DMU_i could consume to maintain the output production level.

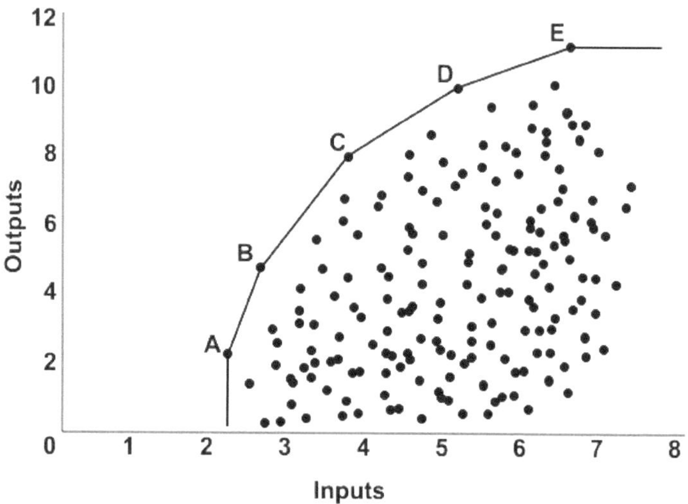

Figure 1. Scatter plot of DMU-DEA.

In this way, all the DMUs are compared, and each is given an efficiency index with the symbol θ. This efficiency index ranges from 0 to 1, with 1 being the maximum, corresponding to the achievement of technical efficiency. By multiplying θ by the inputs of a given (DMU_i), we obtain the ideal amount of inputs for it to achieve technical efficiency. To calculate θ, a set of n observations is needed, considering $S_n = \{x_i, y_i \, to \, i = \{1, \ldots n\}\}$. With this set, it is possible to estimate the efficiency frontier and the efficiency index $(\hat{\theta})$ for each DMU after solving a series of mathematical inequalities. The model determines the technological level (T) using an estimator $(\hat{T}_{CRS})$; with this estimator (X_i, Y_i), for each DMU_i, we can calculate the efficiency index by solving the following mathematical problem.

$$\hat{\theta}_i \, (x_i, y_i)_{OI-CRS} = max\{\hat{\theta}_i \mid \hat{\theta}_i(y_i \leq Y\lambda, \, x_i \geq X\lambda), \lambda \in \mathbb{R}^n_+\} \tag{1}$$

where $X = [X_i, \ldots, X_n]$ and $Y = [Y_i, \ldots, Y_n]$ are the matrices representing the sets of inputs and outputs of the n observed DMUs; X_i and Y_i represent the vectors of inputs consumed and outputs produced by a given DMU_i, respectively; and $\lambda = [\lambda_1, \ldots, \lambda_n]$ are the combinations of inputs and outputs that make it possible to achieve the most excellent efficiency.

On the other hand, the model can also be output-oriented. In this orientation, the focus is on understanding how much more a given DMU_i could produce. Therefore, the technical efficiency index, in this case, ranges from 1 for greater efficiency to infinity and is represented by the Greek letter ϕ. However, if we calculate $1/\phi$, we also obtain an index from 0 to 1. The same DMU_i should receive similar efficiency indices, regardless of whether the orientation is towards inputs or outputs. As with input-oriented DEA, product-oriented DEA also allows you to calculate how much DMU_i should produce to be technically efficient by multiplying ϕ by the DMU's outputs. Using the same principles for input-oriented DEA, we would have the following formula for output-oriented DEA:

$$\hat{\phi}_i \, (x_i, y_i)_{OO-CRS} = min\{\hat{\phi}_i \mid \hat{\phi}_i(y_i \leq Y\lambda, \, x_i \geq X\lambda), \lambda \in \mathbb{R}^n_+\} \tag{2}$$

In addition to orientation, the DEA model can vary in analyzing DMU efficiencies. Charnes et al. [34] developed the first model that did not consider the variation in efficiency that scale can provide. They then created a DEA model called CRS for constant return to scale.

However, there are cases where the return to scale is not constant. For example, producing the maximum batch of a machine in industry is more efficient than producing half that batch. This is because, simultaneously, people and machines can be used for both scenarios, while the former produces twice as much as the latter. In this case, it is unfeasible for the industry in the second scenario to improve its efficiency without increasing its production scale, since it cannot use half a machine to reduce its inputs.

To solve this problem, Banker et al. [33] created the variable return to scale (VRS) model, which calculates the maximum efficiency of each DMU_i, taking into account the DMU's level of production scale. Hence, in the previous example, the company could be technically efficient in both scenarios, since the first and second machines could operate on the efficiency frontiers of their scales. The concept of allocative efficiency was created for DMUs that are technically efficient and at the optimal scale. Allocatively efficient DMUs are DMUs that manage to produce more efficiently than all others DMUs, operating the ideal scale for the production process. They also tend to have the highest profit among all their peers [32,35–37].

Thus, eco-efficiency is achieved when a DMU obtains the highest possible level of desired outputs with a given level of inputs and environmental impact or requires the lowest possible amount of inputs and environmental costs to produce a given number of outputs. Its measurement results are obtained by calculating the Euclidean distance that separates each DMU from the border formed by the benchmarks. Thus, it is possible to define the following two measures of technical efficiency: (i) Farrell's technical efficiency oriented to maximize outputs with a given input level (θ_0) and (ii) Farrell's technical efficiency aimed at minimizing the inputs with a given level of products (θ_I). According to [32], technical efficiencies are inverse to [31] radial efficiency.

The evolution of the model is described in Figure 2.

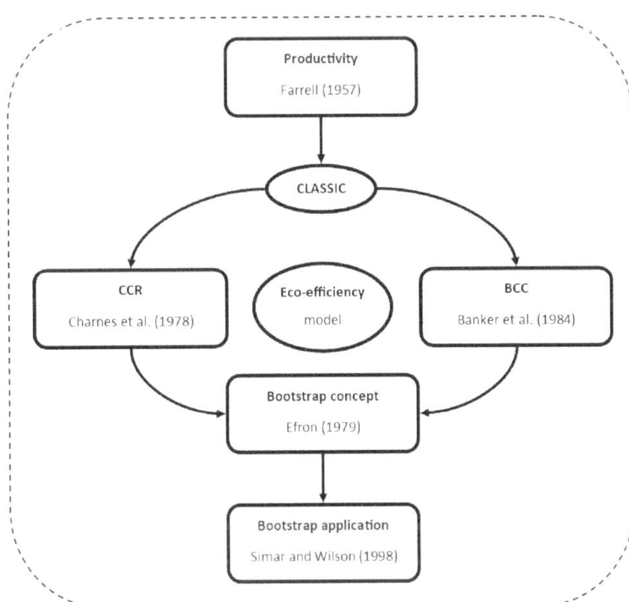

Figure 2. Framework for the evolution of the methodological theorist [32–34,38,39].

3. Materials and Methods

3.1. Detection and Outliers Using Super-Efficiency

The DEA model calculates its indices based on a comparative analysis between peers. However, it has some limitations when there is a considerable range of productive units to be analyzed, making it necessary to use additional models to support the CRS, configuring the case of the super-efficiency model [35,36]. Thus, *outliers* can generate super-efficient DMUs. This means that these DMUs are far removed from the other DMUs and cause a distortion in the efficiency frontier. This method calculates each DMU's impact on the model to determine whether it is an atypical case, and the greater the effect, the more atypical the unit of measurement—DMU.

To perform this calculation, the model removes a particular DMU and recalculates the efficiency of the other DMUs. The volume of the initial data set created by the distribution of the efficiencies generated by the model is calculated. After removing DMU_i, the volume is calculated and compared with the initial volume. Therefore, the closer to 0, the more significant the impact that DMU_i will have on the total volume in the data set, which means that it is an outlier [40]. This process is repeated for all DMUs and must be performed again after removing each outlier, since one outlier can mask the existence of another.

The researcher must decide how many DMUs should be drawn from the database they are working with. Despite that, studies such as [41] suggest approximately 10%.

3.2. Stochastic DEA Model—Bootstrap

A significant criticism of DEA models is their deterministic nature, since they consider past events and do not account for stochastic effects. It would, therefore, be relevant to use statistical models in conjunction with DEA to validate hypotheses, confidence intervals, and correlation analyses. With this improvement, we can have more confidence in the efficiency results and correlation with exogenous variables.

For these two reasons, the bootstrap method is used in this research. The concept of bootstrapping arises from the studies reported in [38], based on which Simar and Wilson [39] applied the DEA stochastic model to a data generation process (DMP). Bootstrapping is a statistical technique for manipulating sampling to create a probability distribution of a dependent variable from a single sample, which is necessary when the nature of the data in the original model is not exhaustively known. The data generation process is described in four steps as follows:

I.   The efficiency (θ_i) of the original sample is calculated for each DMU_i, where $(i = 1, 2, \ldots, n)$ considering (x_i, y_i), considering that it is a linear programming problem (LPP) as described in the Equations (1) and (2), depending on the type of orientation to be adopted.

II.   A new distribution is created, generated by simulated samples $(x^*_i = [x^*_1, x^*_2, \ldots, x^*_n])$ based on the original sample $(x_i = [x_1, x_2, \ldots, x_n])$ of the same size, so $x^*_i = x_i \frac{\theta_i}{\theta^*_i}$. Then, the efficiency scores are calculated for each generated DMU_i, and each value is stored, resulting in a set of estimates $(\hat{\theta}^*_{b,i})$.

$$\hat{\theta}^*_{b,i} = min\{\theta_i \mid \theta_i(y_i \leq Y\lambda, \ x_i \geq X\lambda), \lambda \in \mathbb{R}^n_+\} \tag{3}$$

III.   The observations are replaced B times (for the estimate to be significant, usually $B \geq 1000$) with simulated comments that respect the rules of the original sample with $(B = 1, 2, \ldots, b)$, and the calculations are repeated on top of each simulation. In this way, it is possible to understand how the values of the dependent variable or the estimator (θ^*) respond to variation in the sample. To estimate the standard error, the D_p error of the replications is used.

$$D_p(\theta^*) = \sqrt{\frac{\sum_{b=1}^{B}(\theta^*_b - \bar{\theta}^*)}{B - 1}} \tag{4}$$

IV.   With that, the confidence limits are calculated, having their intervals determined by the default mode with $\alpha = 95\%$ for each estimate, with $\{\hat{\theta}^*_{i,b}, b = 1, \ldots, B\}$, making the results even more robust and reliable [39].

3.3. Calculation of the Return to Scale Test

The choice of which model to use to calculate efficiency is not an arbitrary one because pre-adopting a model without investigating the nature of the behavior of the Constant Return on Scale or Variable Return on Scale (CRS or VRS) of the technological frontier (T) can cause the result to be biased. This is because, in the case of a model with constant returns to scale, it is not taken into account that certain DMUs that are close in terms of efficiency may have particularities, so total efficiency is calculated. It is analyzed that not every DMU considered efficient in the VRS model will be regarded as efficient in the CRS model [42].

The null hypothesis (H_0) considers the model to be constant (CRS), given consistent returns to scale, and the alternative hypothesis (H_1) believes the model to be variable (VRS), given variable returns to scale. By design, the efficiency calculated by the constant model is always lower than the efficiency calculated by the variable model, so $\theta_{i,CRS} \leq \theta_{i,VRS}$; therefore, the estimator (S) is calculated by the ratio of the sum of $\theta_{i,CRS}$ and $\theta_{i,VRS}$.

$$S = \frac{\sum_{i=1}^{n} \theta_{i,CRS}}{\sum_{i=1}^{n} \theta_{i,VRS}} \leq 1 \tag{5}$$

With an estimator value of $S = 1$, H_0 is considered to hold, although this value is hardly reached. Therefore, when the estimator S is close to 1, the model to be considered is CRS. In H_a, the estimator (S) is significantly less than 1, so VRS is the model considered. To state whether it is significantly smaller, the critical value (C_α) is obtained with a significance level of $\alpha = 5\%$, so $S \leq C_\alpha$. Given that the natural distribution of the original sample is not exhaustively known, bootstrapping is applied with the boot.sw98 package inside the R environment to calculate the efficiencies $(\theta_{i,CRS}$ and $\theta_{i,VRS})$ according to the resampling method to obtain the estimated value of S [43]. In this way, a decision is made between H_0 and H_a based on the result of Equation (5).

3.4. Malmquist Index Model

To clarify this question, the Malmquist Index was developed, named after Sten Malmquist [44]. To calculate the efficiency gain in period T compared to period $T - X$, the technical efficiency frontiers must first be calculated using DEA for the DMUs in both periods. When comparing DMU efficiencies over time, two variables can be measured, namely the slope of the efficiency frontier, which would mean that there is a new technology enabling DMUs to be more efficient—a phenomenon is known as the frontier shift effect; and the catch-up effect, or pairing, which is when a given DMU_i has decreased its distance to the efficient frontier. The Malmquist Index is the multiplication of these two variables.

The two scenarios are presented in Figure 3. A frontier shift can be seen in the efficiency frontier created by DMUs A, B, and C, representing the efficiency frontier in period $T + 1$, since it is closer to the Y-axis and further away from the X-axis. The decrease can be seen in the pairing effect in the distance from DMU_M to the efficiency frontier for the period it is in. In period T, DMU_M was at a distance of X from the border, while in period $T + 1$, it was at a distance of 0.7_X.

The matching effect, or catch-up, is the result of continuous improvements in production processes, using the same technology. Therefore, the comparison of technical efficiency between two periods can be defined as follows:

$$\frac{\theta_k^t \left(x_k^t, y_k^t \right)}{\theta_k^{t-1} \left(x_k^{t-1}, y_k^{t-1} \right)} \tag{6}$$

where:

$\theta_k^t\left(x_k^t, y_k^t\right) = DMU_K$ is technical efficiency in a given period (t);
$\theta_k^{t-1}\left(x_k^{t-1}, y_k^{t-1}\right) = DMU_K$ is technical efficiency in a given period of $t + 1$.
Frontier shift can be calculated according to the following formula:

$$\sqrt{\frac{\theta_k^{t-1}\left(x_k^{t-1}, y_k^{t-1}\right)}{\theta_k^{t}\left(x_k^{t-1}, y_k^{t-1}\right)} \frac{\theta_k^{t-1}\left(x_k^{t}, y_k^{t}\right)}{\theta_k^{t}\left(x_k^{t}, y_k^{t}\right)}} \tag{7}$$

Therefore, multiplying Equation (6) by Equation (7) and passing the second term of the expression (7) into the square root, the word is transformed into the product of the pairing by the frontier shift. Consequently, a result less than 1 means an improvement in the DMU's technical efficiency index, and an effect greater than 1 implies a worsening in the DMU's technical efficiency index. Therefore, the same can be said for the Malmquist Index, calculated with constant returns to scale (*Mo*) as follows:

$$Mo = \frac{\theta_k^t\left(x_k^t, y_k^t\right)}{\theta_k^{t-1}\left(x_k^{t-1}, y_k^{t-1}\right)} \cdot \sqrt{\frac{\theta_k^{t-1}\left(x_k^{t-1}, y_k^{t-1}\right)}{\theta_k^{t}\left(x_k^{t-1}, y_k^{t-1}\right)} \frac{\theta_k^{t-1}\left(x_k^{t}, y_k^{t}\right)}{\theta_k^{t}\left(x_k^{t}, y_k^{t}\right)}} \tag{8}$$

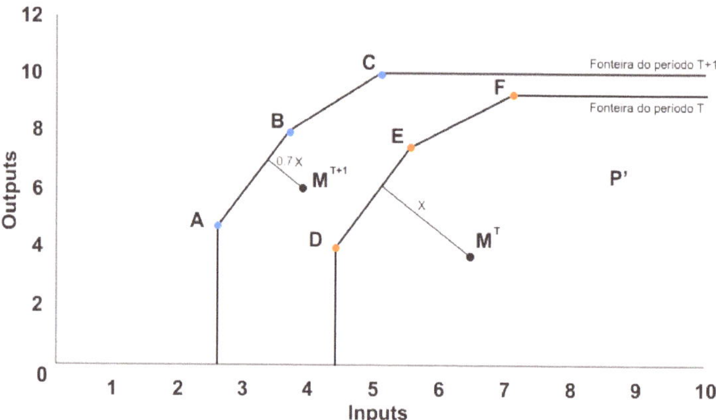

Figure 3. Frontier shift—Malmquist Index.

3.5. Database

To assess the agricultural eco-efficiency of the municipalities that make up the Amazon Biome, a set of variables available in the 2006 and 2016 Agricultural Censuses was adopted, and a time series was generated using linear regression to fill in all the missing years. The sector's classic inputs and outputs were considered, plus one positive and one negative externality.

To calculate the municipalities' operational efficiencies of agricultural production, the variables used as inputs were people engaged in agricultural work, hectares dedicated to rural production, and hectares preserved, the latter being a desirable input. The selected outputs were the value of agricultural production in thousands of BRL and CO_2 emissions as an undesirable output. All these variables were at the municipality level. As a general rule, the following classic inputs and outputs were used in the modeling:

X_1—People engaged in agricultural industry;
X_2—Hectares dedicated to agricultural production;
X_3—Preserved hectares;
Y_1—Value of agricultural production in thousands of BRL;
Y_2—CO_2 emissions with an undesirable output;
Z_1—Average temperature;

Z_2—Precipitation.

The area variables (hectares dedicated to production and hectares preserved), as well as the production value and people employed, come from the official Brazilian Institute of Geography and Statistics (IBGE) (https://www.ibge.gov.br/estatisticas/economicas/agricultura-e-pecuaria/9827-censo-agropecuario.html (accessed on 24 June 2024)) databases according to the censuses carried out in 2006 and 2017. For the years for which there is no official census information by the municipality, a simulation was performed using the estimated growth in annual production in the state according to CONAB (https://www.conab.gov.br/ (accessed on 24 June 2024)) (National Supply Company) year-on-year, combined with an estimate of how much each municipality should have as a share (in %) within each state, using an arithmetic progression calculation between the municipalities' claims in 2006 and the municipalities' claims in 2017. All the consolidated data and results found by this research are available and can be consulted on figshare (https://doi.org/10.6084/m9.figshare.25958827 (accessed on 24 June 2024)).

4. Results and Discussions

4.1. Analysis Variables

Figures 4 and 5 show the area dedicated to agricultural activities in each municipality as a percentage of total municipal area in 2006 and 2017, respectively. The municipalities with a more proportional area dedicated to agriculture are along the eastern and southern borders of the Amazon Forest. From 2006 to 2017, the increase in agricultural area occurred mostly in the municipalities that already had strong agricultural activities.

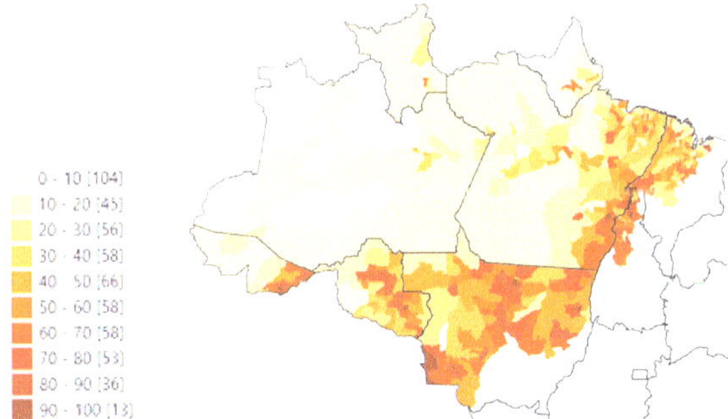

Figure 4. Area dedicated to agricultural production in 2006—% of municipality area (number of municipalities in each group is presented in brackets).

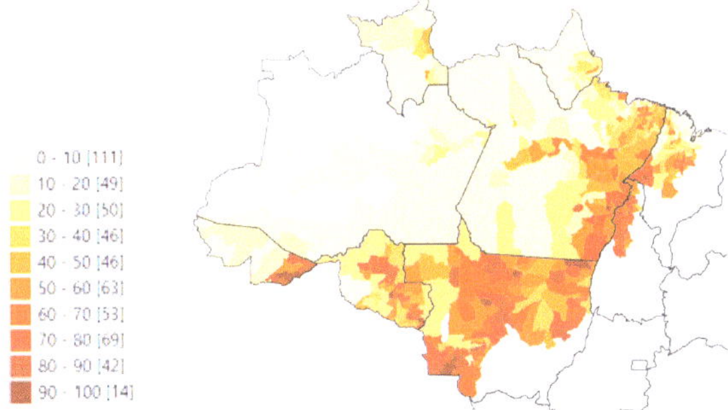

Figure 5. Area to agricultural production in 2017—% of municipality area—(number of municipalities in each group is presented in brackets).

Figures 6 and 7 present the protected area as a percentage of the total municipal area in 2006 and 2017, respectively. Two characteristics of these maps need attention. (i) The majority of the municipalities have a very low percentage of protected area (see the number of municipalities in each group—numbers in brackets in the map caption). (ii) The municipalities with more protected area are the same as those with high agricultural area. These contradictory results come from the definition of the protected area variable in the censuses, as it represents protected areas on farms, not including Indigenous reservation areas, (environment) conservation units, and other public areas (with no private owners). Hence, the municipalities with known intact forests (in the center-west of the Amazon Forest area) are shown with a low percentage of protected area because they have very small areas of private farms. For this study, however, the definition of the protected area variable considering only the area on farms is adequate, since it is derived from farmers' decisions on how much of an area to allocate for agriculture and environmental protection.

The data regarding CO_2 emissions were downloaded from the official site of Greenhouse Gas Emission and Removal Estimation Systems (SEEG) (https://plataforma.seeg.eco.br/total_emission (accessed on 24 June 2024)). This variable is measured in tons of carbon dioxide emitted by the agribusiness and property transformation but with an annual uninterrupted update since 2000. After that, an official database was collected from INMETRO using all the automatic measuring agencies in each state for every hour of every day to analyze the relation between the production efficiency index and the recorded weather. Then, the average temperature for each year in each state was calculated, and these values were used for each municipality that composes the Brazilian Biome.

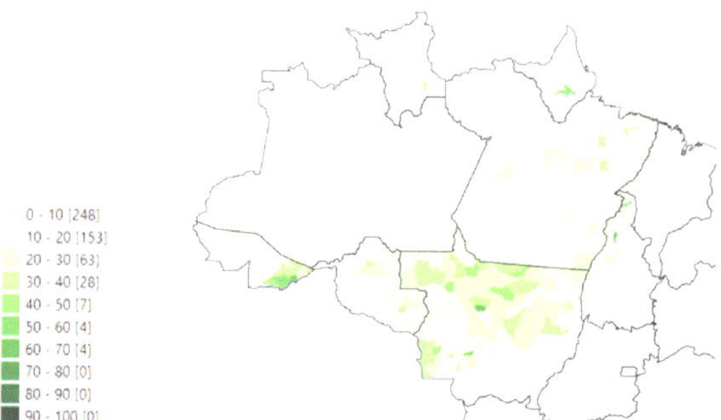

Figure 6. Protected area on farms in 2006—% of municipality area—(number of municipalities in each group is presented in brackets).

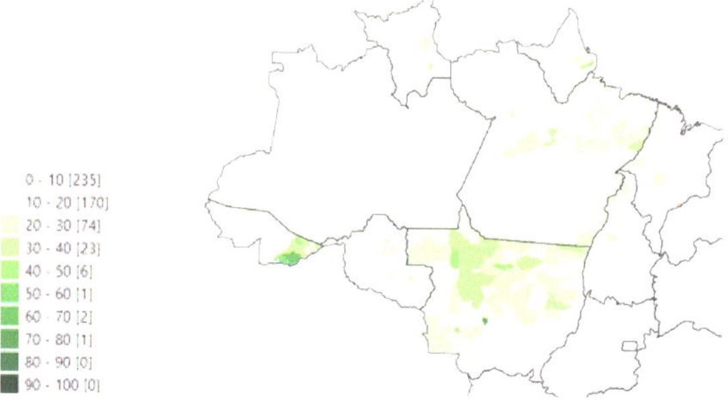

Figure 7. Protected area on farms in 2017—% of municipality area—(number of municipalities in each group is presented in brackets).

4.2. Removing the Outliers

Next, given super-efficient DMUs, the Bogetoft and Otto [40] method was applied. The model is based on the assumption that a super-efficient DMU, considered to be an outlier, masks an efficient DMU, so looping is applied to recalculate the efficiencies with an *input* orientation and remove the super-efficient DMUs until there are no more super-efficient DMUs. To achieve this, there is a pause criterion in the process of eliminating *outliers* given by $\frac{V_f}{V}$, with a value of less than 0.7, where a DMU could produce 70% of what it has and still be considered a technically efficient DMU. The *sdea()* function predefined in the "Benchmarking" package was used for the calculation. Given the deaR model, the efficiencies of DMUs with an output and input orientation were calculated using the processes in the RVE box.

The looping removal model is applied again each year. Each year should have a few DMUs considered super-efficient, creating a list of all the super-efficient DMUs to be removed from all the years, i.e., from the entire database. We used 0.7 for the efficiency index, since 150 DMUs would be removed from the sample with this parameter, which is no more than 30% of the database used in [41]. To compare the time windows, data processing is needed to give continuity to the efficiency gain, where the frontier shift index for year T is multiplied by the index for year $T-1$; then, the index for the year $T+1$ is multiplied by the product of the two previous indices in an accumulative manner. Using the efficiency index of the years, a cross base is created with the annual meteorological information, and the multivariate linear regression of the floor is analyzed, with the meteorological data being the independent variable and the efficiency of the year as the dependent variable.

Another technique is to create cut-off ranges by estimating DEA with super-efficiency. Figure 8 shows the municipalities considered outliers and super-efficient in 2017 that were excluded from the Malmquist Index analysis below.

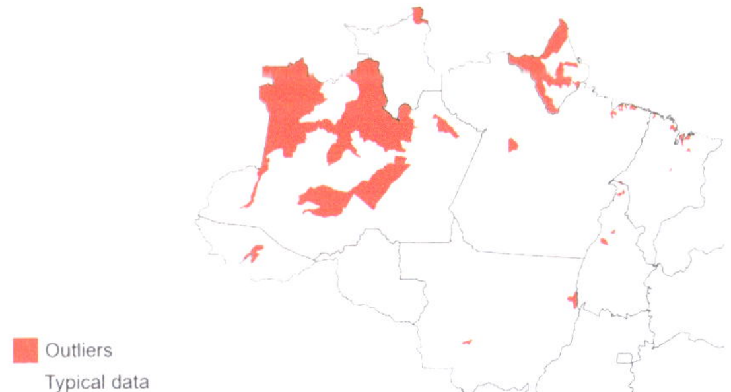

Figure 8. Municipalities considered outliers in 2017.

The data cloud technique was used because it is robust for identifying outliers. In summary, the volume of the combined matrix of inputs and outputs is observed, where a significant reduction in this volume following the removal of a DMU indicates that this unit is an outlier. As can be seen in Figure 8, most of these DMUs are in undeveloped areas in Amapá and Amazonas states and some very urbanized municipalities (but small in total area) in Pará, Maranhão, and Tocantins states (sast of the map).

Totaling the number of outliers found by the cloud technique, 95 DMUs were removed, which represents around 16.99% of the total. This shows that the model adopted based on the super-efficiency cloud technique was more effective at detecting outliers compared to that reported in [29], which identified around 4.5% of outliers, most of which were in Pará, while the cloud technique brought in many outliers from the states of Amazonas and Amapá.

4.3. Scale Return Test

The result found for the critical value was $C_\alpha = 0.3854$, and the estimated value was 0.9605, which is higher than the critical value (C_α). With these results, there was no statistical evidence to reject the null hypothesis, accepting that the best model for this problem is the CRS (constant return to scale). The conclusion, therefore, must be that scale has no direct influence on the efficiency of agricultural production in the municipalities of the biome included in this study, so small, medium, and large producers can all become eco-efficient, reaching the technological frontier.

What, at first, may be a controversial result considering the analysis of studies that used all Brazilian municipalities as a sample, compared to previous studies carried out in the same region (Amazon), they also pointed out that the result of the scale test is constant [29], confirming the null hypothesis. When analyzing the productivity of an agricultural unit, the main determining variables are soil, climate, and the technologies used. Whether the production unit is larger or smaller is not affected by soil and climate. What determine these conditions are geography and the location of these units. As for technology, a team with greater production power could have more resources to invest in machinery. However, as the object of this study is entire municipalities, this effect is diluted among the various production units within each municipality and is therefore regressed to an average.

4.4. Eco-Efficiency of Municipalities the Amazon Biome

First, it is essential to analyze the current scenario of the agricultural industry in the Amazon to measure potential gains and even map out action plans to capture these gains. The results shown in Table 1 were achieved using DEA for inputs with constant returns to scale. Thus, it is possible to see that, on average, DMUs should produce 9.9% more than they did in 2017 to become efficient. In addition, it is also possible to see that the worst DMU should have approximately twice as much, or 197.5% of what it produces.

These results show how current agricultural industry practice, on average, has become more sustainable, since most DMUs are close to technical efficiency, given that, in the first quartile, DMUs would only need to produce 1.5% more to achieve efficiency and, in the third quartile, this figure increases to 12.3%. Few municipalities would need to improve their production significantly to achieve technical efficiency, so only 25% of municipalities would need to increase production by more than 12.3%.

Table 1. Summary of data regarding the efficiency of municipalities.

Min.	First Q	Median	Mean	Third Q	Max.
1.000	1.015	1.051	1.099	1.123	1.975

Figure 9 shows the results of CRS efficiency by municipality in 2017. The following geographical concentrations of the most efficient DMUs can be seen: (i) one group along the southern border in the state of Mato Grosso, where highly technical soybean and corn plantations are the main agricultural activities; (ii) one group in the state of Amapá (north of the map), where farming activities are lacking; and (iii) one group in the west of the state of Amazonas (west of the map), where farming activities are scarce. On the other hand, less efficient DMUs are located in the southeast of Pará state, where farmers focus on raising beef cattle, and in Rondônia state (southwest of the map), where agricultural activities are spread across several crops. These results provide insights on how to achieve eco-efficiency for agriculture in the region, as discussed later.

Table 2 shows that the variables relating to inputs and outputs between the municipalities have heterogeneous values, coinciding with the type of return to scale identified in the CRS model, where small and large producers can both be considered eco-efficient, even though they have different levels of resources and results. This is the case of Juara, which has proportionally higher values for inputs and outputs compared to the municipality of Rosário, but both are considered eco-efficient.

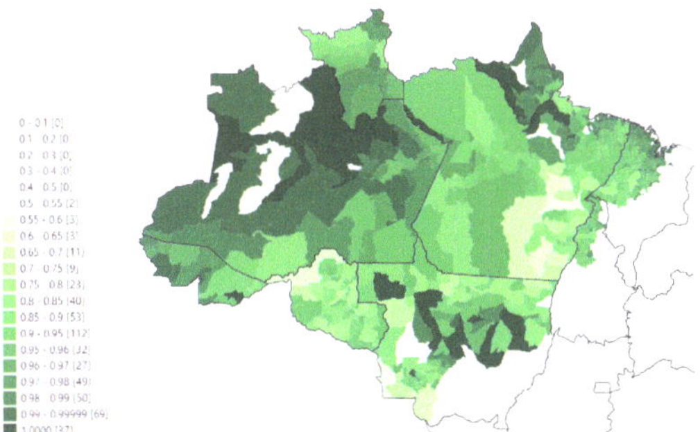

Figure 9. CRS efficiency by municipality in 2017—(number of municipalities in each group is presented in brackets).

Table 3 lists all the municipalities considered efficient, i.e., they have an eco-efficiency index equal to 1, totaling 37 eco-efficient municipalities. This represents 6.22% of the initial 595 municipalities, a low figure at first. Still, considering that most of the municipalities are close to the production frontier, it can be concluded that the DMUs considered efficient do have an optimum level of total efficiency.

Municipalities such as Novo Airão (AM) and Presidente Figueiredo (AM) have also been considered efficient in studies considering different variables [29], which demonstrates the excellent management of these municipalities. As DEA investigates direct relationships between the inputs and outputs of processes, the choice of variables has a strong impact on the DMUs that will be considered efficient, so there may be divergences, even if the analysis is carried out in the same region but considers different variables. For example, the DMUs of Marabá, Cáceres, Vila Bela da Santíssima Trindade, São Félix do Xingu, Novo Repartimento, Santa Maria das Barreiras, Porto Velho, Água Azul do Norte, and Cumaru do Norte had already been designated as inefficient, and Rosano-Peña et al. [29] pointed out that Santa L. do Paruá (MA), Garrafão do Norte (PA), Anajatuba (MA), Governador Nunes F. (MA), and were also Itapecuru Mirim (MA) were also inefficient DMUs.

Table 2. Variables of efficient municipalities, inputs, and outputs.

		Input		Input	Output	Output
		Input		**Desirable Input**	**Output**	**Undesirable Output**
ID	**Municipality**	**Production Hectares 10^3**	**Employed People (10^3)**	**Preserved Hectares 10^2**	**Revenue BRL 10^3**	**CO_2 Emissions Tons 10^3**
2109601	Rosário	1	16.46	3	3	7
1300060	Amaturá	6	25.15	50	9	3
1600154	Pedra Branca do Amapari	33	20.43	289	16	6
1600279	Laranjal do Jari	30	11.99	259	14	13
5105101	Juara	1522	54.69	8461	226	2016
2101350	Bacurituba	1	7.69	3	0	22
5107248	Santa Carmem	234	9.30	1136	464	190
1301308	Codajás	70	38.78	610	22	9
1302108	Japurá	3	12.56	20	3	4
1302801	Maraã	4	26.26	15	17	5
1303205	Novo Airão	8	17.40	65	5	3
1303700	Santo Antônio do Içá	3	33.29	24	6	4
1304062	Tabatinga	3	87.79	12	13	2
1304203	Tefé	21	117.49	155	65	8
1500305	Afuá	134	121.28	1201	92	37

Table 2. *Cont.*

		Input		Desirable Input	Output	Undesirable Output
		Input				
ID	Municipality	Production Hectares 10^3	Employed People (10^3)	Preserved Hectares 10^2	Revenue BRL 10^3	CO_2 Emissions Tons 10^3
1503002	Faro	6	3.33	43	1	30
1503101	Gurupá	57	52.83	523	27	18
1507961	Terra Alta	2	3.45	9	2	6
2106755	Miranda do Norte	20	3.04	18	1	18
5106224	Nova Mutum	766	47.29	2.630	1957	233
5106307	Paranatinga	1300	46.75	5.268	827	263
5107156	Reserva do Cabaçal	105	6.08	657	0	34
5107925	Sorriso	828	49.37	2011	2812	432
5108907	Nova Maringá	612	14.74	3966	515	297
1200708	Xapuri	494	55.20	3758	19	435
1300409	Barcelos	8	27.47	52	23	4
1300839	Caapiranga	4	6.20	26	5	6
1303536	Presidente Figueiredo	197	76.92	1642	64	33
1500701	Anajás	163	29.00	1323	20	19
1504000	Limoeiro do Ajuru	37	141.21	317	51	13
2104909	Guimarães	1	24.50	2	2	7
2111201	São José de Ribamar	1	27.06	2	1	4
5101407	Aripuanã	1112	55.18	7085	33	1012
5102702	Canarana	790	26.23	4107	756	246
5107065	Querência	834	57.70	3813	1546	551
1300102	Anori	22	29.31	186	15	3
5105259	Lucas do Rio Verde	316	26.05	828	1111	55

The results of eco-efficiency also indicate that the global average was 0.920, but on average, when compared between the states, the values tend to diverge. The state with the highest average was Amazonas, unlike the results reported in [29], where the state with the best performance was Amapá, with a score of 0.720, in contrast to the results found in this study of 0.984, followed by Acre (0.937), Roraima (0.939), Mato Grosso (0.884), Amazonas (0.978), Pará (0.905), Tocantins (0.945), Maranhão (0.962), and Rondônia (0.835). There was a 26.83% difference between the scores, demonstrating that the detection of a greater number of outliers had a significant impact on the results.

Table 3. Efficient municipalities.

ID IBGE	Municipality	DEA Index CRS Output	ID IBGE	Municipality	DEA Index CRS Output	ID IBGE	Municipality	DEA Index CRS Output
2109601	Rosário	1	1304203	Tefé	1	1300839	Caapiranga	1
1300060	Amaturá	1	1500305	Afuá	1	1303536	Presidente Figueiredo	1
1600154	Pedra Branca do Amapari	1	1503002	Faro	1	1500701	Anajás	1
1600279	Laranjal do Jari	1	1503101	Gurupá	1	1504000	Limoeiro do Ajuru	1
5105101	Juara	1	1507961	Terra Alta	1	2104909	Guimarães	1
2101350	Bacurituba	1	2106755	Miranda do Norte	1	2111201	São José de Ribamar	1
5107248	Santa Carmem	1	5106224	Nova Mutum	1	5101407	Aripuanã	1
1301308	Codajás	1	5106307	Paranatinga	1	5102702	Canarana	1
1302108	Japurá	1	5107156	Reserva do Cabaçal	1	5107065	Querência	1
1302801	Maraã	1	5107925	Sorriso	1	1300102	Anori	1
1303205	Novo Airão	1	5108907	Nova Maringá	1	5105259	Lucas do Rio Verde	1
1303700	Santo Antônio do Içá	1	1200708	Xapuri	1			
1304062	Tabatinga	1	1300409	Barcelos	1			

4.5. Evolution of Eco-Efficiency Using the Malmquist Index

The Malmquist index was calculated for series from 2006 to 2017 (link: https://doi.org/10.6084/m9.figshare.25958827 (accessed on 24 June 2024)). Using the Malmquist function, the frontier shift, pairing effect, and Malmquist index are calculated for each municipality. To calculate the frontier effect for the year, a geometric mean of all the frontier shift indices for all the DMUs in the corresponding year is used. The first year is discarded, since the calculation is made by comparing efficiency in years T and $T - 1$. To solve the problem, the data for 2006 is repeated, simulating 2005, where 2006 has a frontier displacement index equal to 1. Figure 10 shows the aggregated index (2006–2017).

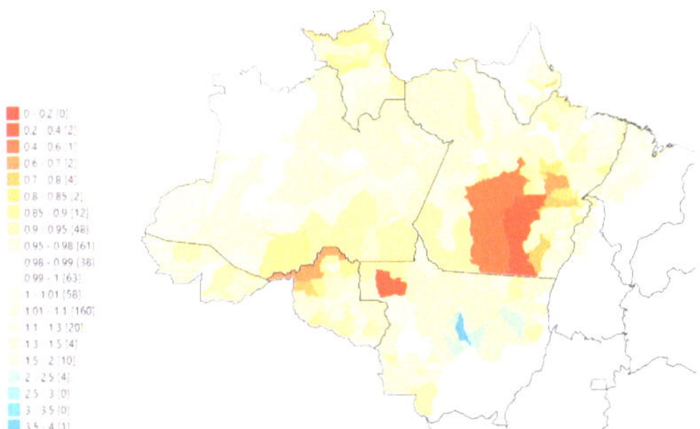

Figure 10. Malmquist Index from 2006 to 2017 by municipality (number of municipalities in each group presented in brackets).

The municipalities that improved most in the period are in the state of Mato Grosso (South of the map)m where highly technical soybean and corn plantations became the main activity. Municipalities where eco-efficiency worsened are in the southeast of Pará state, a region where the raising of beef cattle increased considerably, and in Rondônia state (west of Mato Grosso state), where agriculture with several crops increased. Although a great increase in the value of agricultural production was observed in these DMUs, it was accompanied by a large increase in emissions. Another important result from the Malmquist analysis is that DMUs that are very eco-efficient but have very small agricultural production achieved minimal progress in the period.

4.6. Optimizing Variables to Identify Potential Improvement

In absolute terms, as the chosen orientation is towards output, the possible impact on each variable (Y_1 and Y_2) is analyzed if all the municipalities converge towards efficiency. The values are described in Table 4, where the CO_2 emission variable represents reduction or savings, generating ecological gains, while the production value represents the increase needed to turn inefficient DMUs into efficient ones. BRL is the Brazilian currency; USD 1.00 = BRL 4.90 in August 2023.

Table 4. Potential production gains.

Variable	Value
Revenue	3,195,002,000 (R$)
CO_2 emission	25,849,560,000 (ton)

Due to the way data envelopment analysis is calculated, the efficiency index is applied equally to all variables, so if the index is 1.5, the municipality should produce 50% more of all products. Therefore, the ranking of the cities in relative values will remain static, regardless of the variable being analyzed. However, this ranking can change in nominal values. With this in mind, Table 5 lists the least efficient municipalities.

Table 5. Worst municipalities analyzed by simulation.

IBGE Index	Municipality	DEA Index CRS Output	IBGE Index	Municipality	DEA Index CRS Output
1504208	Marabá	1.975094	5102504	Cáceres	1.626327
5105507	Vila Bela da Santíssima Trindade	1.883153	1507300	São Félix do Xingu	1.601048
1505064	Novo Repartimento	1.781707	1506583	Santa Maria das Barreiras	1.573696
1100205	Porto Velho	1.703062	1500347	Água Azul do Norte	1.537795
1502764	Cumaru do Norte	1.694544			

Although it does not have the worst efficiency index, the municipality with the most significant potential for gains in production value is São Félix do Araguai, totaling of BRL 182 million, and the municipality with the most potential for savings in CO_2 emissions is São Félix do Xingu, which could save approximately 1.7 billion tons, as can be seen in Table 6.

Table 6. Municipalities with the greatest potential in terms of production value and CO_2 emission savings.

Variable	Municipality	Potential
Revenue	São Félix do Araguai	183,353,000.00 (BRL)
CO_2 emissions	São Félix do Xingu	−1,683,912,000.00 (tons)

It is also possible to measure the impact of efficiency gains on input savings. However, it is important to remember that you shouldn't look at outputs and inputs simultaneously, since the profits from these two perspectives are mutually exclusive. The results for savings in production hectares and people employed, as well as the potential for increasing the area preserved, can be found in Table 7.

Table 7. Economic potential.

Variable	Value
Employed people	242,438 (un)
Production area	13,916,750 (ha)
Preserved area	6,248,745 (ha)

The municipality with the most significant potential for gain, São Félix do Xingu, is presented in Table 8, along with the values needed to improve each variable to achieve efficiency.

Table 8. São Félix do Xingu' economic opportunity.

Variable	Quantity
Employed people	8455 (un)
Production area	924,292 (ha)
Preserved area	578,568 (ha)

For the statistical modeling part, a linear regression study was carried out. The efficiency index of the municipalities was chosen as a dependent variable. The independent variables, or predictors, were the meteorological characteristics, volume of precipitation, average temperature, average humidity, and standard deviation of temperature over the months. The study sought to simulate the different seasons and capture the impact of this natural phenomenon and the various crops that can be planted depending on the season. It is also interesting to understand production and efficiency by state.

Table 9 presents the values achieved for each of the variables analyzed in 2017, as well as the state's total efficiency index, calculated as the geometric mean of the efficiencies of the municipalities in each state.

Still analyzing Table 9, it can be seen that the state with the highest efficiency is Amapá, with an index of 1.017, followed by Amazonas, which obtained a result of 1.02. The state with the worst recorded efficiency was Rondônia. Despite being the state with the third-highest financial value of production, the amount of CO_2 emitted and the number of hectares used for production were much higher than ideal. It is also interesting to understand production and efficiency by state. Table 9 presents the values achieved for each of the variables analyzed in 2017, as well as the states' total efficiency indices, calculated as the geometric mean of the efficiencies of the municipalities in each state.

Table 9. Summary of states.

State	Efficiency Index	Input		Desirable Input	Output	Undesirable Output
		Input		Desirable Input	Output	Undesirable Output
		Production Hectares 10^3	Employed People (10^3)	Preserved Hectares 10^2	Revenue BRL 10^3	CO_2 Emission Tons 10^3
AC	1.069	4233	126,514	2544	442	5870
AM	1.023	3733	307,201	2150	1294	3102
AP	1.017	1501	30,732	871	226	822
MA	1.041	4652	293,574	862	865	7794
MT	1.140	38,164	283,069	15,349	26,914	45,209
PA	1.111	27,637	951,857	10,338	6036	42,360
RO	1.203	9220	270,812	2319	1582	28,552
RR	1.067	2636	67,070	1181	395	1792
TO	1.059	2867	51,596	669	211	4170

The ideal consumption quantities per state are presented in Table 10. If the states reach these values, production remains constant, with the values in the "Revenue" and "CO_2 Emissions" columns shown in Table 9.

Therefore, Table 10 represents the ideal level of consumption while maintaining the quantity produced. By comparing these tables, it is possible to calculate the number of inputs that could be saved or boosted in the case of the desirable input of preserved area. The state of Rondônia, for example, could use approximately two million hectares and fifty-four thousand fewer people to maintain its production level and still increase the preserved area by about seventy thousand hectares.

On the other hand, the states could increase production. In this scenario, the states' consumption would maintain the same values in the "Production Hectares", "Employed People", and "Preserved Hectares" columns of Table 10.

Table 10. Ideal consumption by state.

State	Production Hectares 10^3	Employed People (10^3)	Preserved Hectares 10^2
AC	3903	117,526	2738
AM	3521	297,925	2263
AP	1467	29,940	888
MA	4232	280,095	964
MT	32,929	241,189	17,856
PA	22,309	845,996	12,830
RO	7313	216,768	3007
RR	2402	62,525	1261
TO	2651	48,023	725

The ideal production values, maintaining current consumption, can be found in Table 11. Rondônia, which is the state with the worst eco-efficiency index in the Amazon Biome, would need to increase the value of production by BRL 341 thousand and reduce CO_2 emissions by approximately6000 to reach the production frontier given by the DMUs in the northern region.

Table 11. Ideal production by state.

State	Revenue BRL 10^3	CO_2 Emissions Tons 10^3
AC	470.2961	5292.594
AM	1334.37	2834.646
AP	231.9256	795.1486
MA	950.0968	7111.449
MT	28,795	37,305.74
PA	6792.038	32,472.07
RO	1923.489	22,531.48
RR	435.6717	1644.752
TO	226.3446	3834.731

Table 12 shows an analysis of the number of inputs that could be saved. In the fourth quartile, since it is the quartile with the worst DMUs, there would have to be a reduction of eleven million hectares and one hundred and seventy-nine thousand people, as well as an increase of five hundred and eighty-four thousand hectares preserved to reach the level of efficiency, provided that the level of production remains constant.

Table 12. Consumption reduction potential per quartile.

Quartile	Production Hectares 10^3	Employed People (10^3)	Preserved Hectares 10^2
1	−26	−2374	14
2	−428	−15,480	200
3	−1584	−45,162	651
4	−11,879	−179,423	5384

As for Table 13, how much more could the municipalities produce to reach the maximum possible production for each quartile? Again, in quartile four, the value of production would have to increase by approximately BRL two million, CO_2 emissions would have ot be reduced by about twenty-three thousand tons, assuming that inputs remain constant, i.e., do not change.

Table 13. Ideal production per quartile.

Quartile	Revenue BRL 10^3	CO_2 Emissions Tons 10^3
1	20.00	−19.06
2	251.65	−388.72
3	627.22	−2305.30
4	2296.13	−23,136.49

By summarizing the efficiency quartiles, it is also possible to measure the results divided into efficiency groups. This summary can be found in Table 14. Between quartile 1 and quartiles 2 and 3, there is no significant difference between the determined eco-efficiency indices of 1.005, 1.031, and 1.085, respectively, at which point the hypothesis that, on average, there is no considerable distance between the eco-efficient and inefficient DMUs is, once again, confirmed. This can also be seen in the difference between the first quartile and the fourth quartile, where the eco-efficiency index values are 1.005 and 1.265, respectively, and there is no considerable discrepancy between these indices.

Table 14. Quartile summary.

		Input			Output	
		Input		Desirable Input	Output	Undesirable Output
Quartil	Efficiency Index	Production Hectares 10^3	Employed People (10^3)	Preserved Hectares 10^2	Revenue BRL 10^3	CO_2 Emissions Tons 10^3
1	1.005	12,763	463,090	6704	13,174	8163
2	1.031	13,374	539,044	6133	8420	11,606
3	1.085	19,919	578,013	7546	7686	27,829
4	1.265	48,586	802,278	15,900	8684	92,074

Various combinations of independent variables were investigated to find the best predictive model for the average annual efficiency index. The most assertive model with static significance, defined by the p-value test, used average dry bulb temperature, average dew-point temperature, average relative humidity, annual rainfall volume, and the standard deviation of temperature over the months.

The model calculated using these variables to predict the efficiency index achieved a p value of 0.0007 and an R2 of 0.9721. The figures presented by the model are encouraging. However, a closer look at the statistics reveals that this relationship may not be one of cause and effect. Efficiency has increased steadily over the years without showing much variation or volatility in this growth, which tends to be explained by advancing technology crop growth linked to the agricultural industry. At the same time, with respect to the analyzed climate phenomena, it can be said that the world is undergoing a gradual process of global warming. As such, the correlation may result from two phenomena showing the same trend.

4.7. Impact of Climate Variables on Eco-Efficiency Indices

It is essential to understand the influences of climate variables (Z_1, Z_2) on the eco-efficiency of municipalities, as they can represent gains or losses in efficiency over time. All regions of the planet are subject to climate variations caused by factors such as global warming and increases in periods of rain and drought, as well as natural phenomena such as *El Niño*, which, when they occur, can significantly reduced the productive capacity of areas dedicated to the agricultural sector, generating reductions in essential inputs for the world economy. Therefore, it is understood that correlations between meteorological characteristics and productive efficiency can enable the development of alternative technologies to better prepare producers to face the above phenomena. However, it was not possible to prove a direct correlation between the variables. Even if the numbers were good and passed statistical tests, it was impossible to rule out the hypothesis that made them a coincidence. Although this hypothesis cannot be ruled out, there are studies that have been able to prove this relationship between eco-efficiency indices and climate factors, providing vital information for the environmental decision-making process, such as the work of Rosano-Peña et al, who concluded that the changes caused by the decrease in rainfall and the increase in temperature would have a positive impact on the mountainous region of the TMCF in Mexico, specifically in an area located in the Sierra Madre Oriental between 2016 and 2017.

Finally, we emphasize challenges related to the availability of data to obtain longer periods to determine relevant inputs and outputs, as well as the lack of open access to data for other non-agricultural contexts, such as industry and business. The research fulfilled its guidelines and achieved results relevant to its data structure, providing indicators of social, environmental, and economic development. According to [45], this balance between the search for economic and social development, aligned with the appropriate use of natural resources, arouses interest in studies focused on the area to promote a set of actions that have social and environmental impacts due to the updates that have occurred in recent years in agribusiness, which have led the country in the search for sustainable development.

4.8. Implications

The contributions of this study regarding theoretical implications are represented by the use of the DEA super-efficiency, bootstrap, and Malmquist Index methods to measure the efficiency of the Amazon Biome in the period from 2006 to 2017. The super-efficiency model was used to remove outliers. Many studies do not consider the removal of outliers when calculating the efficiency of models that use DEA [21,24,25,36,46]. Failing to address outliers can lead to biased results, as the DEA model is highly sensitive to this type of variable. An approach based on bootstrap computational models was used to calculate the efficiencies of municipalities in 2017, which also makes the results more robust. This is because the DEA model does not naturally make use of parametric models, resulting in the resampling provided by bootstrapping to create confidence intervals. It uses the Malmquist model to provide a systematic analysis of the development of agricultural eco-efficiency in the Amazon Biome.

The study's practical implications show that measuring efficiency in the agricultural sector of municipalities in the Amazon Biome in Brazil can help in proposing public policies and strategic planning for decision makers such as farmers, business people, governors, mayors, and the federal government. This can contribute to greater economic and sustainable development in the region, improve residents' quality of life, and respect the biodiversity and environmental wealth of this important area. This study also identifies potential improvements for inefficient municipalities and highlights which municipalities are efficient, providing models that can be used to improve the management of these areas.

5. Conclusions

The results point to an essential interpretation that most municipalities are already operating at a satisfactory level of efficiency, given the technological level available; around 7.17% of the municipalities are already working at an eco-efficient scale.

The production frontier of the municipalities results in a technological behavior of constant returns to scale (CRS), which is a relevant result for understanding factors that involve inequality between producers in municipalities because, given the results of the model, small, medium, and large producers can all be eco-efficient (total efficiency), given the CRS frontier within the Amazon Biome. Hence, these differences occur regardless of the production level, so there will be inefficient and efficient small, medium, and large producers.

We also recommend redoing the DEA modeling with different groups of municipalities, separating them by state or region rather than comparing all the municipalities with each other. In addition, comparing the results with those obtained in other areas of the country and even other countries makes sense. Adding input or output variables to the survey would also enrich the analysis. This would allow a larger volume of data and different situations to be tested.

Author Contributions: Conceptualization, G.M.S. and A.L.M.S.; Data curation, G.M.S., A.L.M.S. and C.R.-P.; Formal analysis, F.M.P.; Investigation, G.M.S., A.L.M.S. and C.R.-P.; Methodology, G.M.S., A.L.M.S. and C.R.-P.; Supervision, P.H.M.A.; Validation, G.M.S., A.L.M.S., G.A.P.R., C.R.-P. and F.M.P.; Writing—original draft, G.M.S., A.L.M.S. and G.A.P.R.; Writing—review and editing, G.M.S., A.L.M.S., G.A.P.R. and F.M.P. All authors have read and agreed to the published version of the manuscript.

Funding: This research was funded by the University of Brasilia (UnB).

Institutional Review Board Statement: Not applicable.

Informed Consent Statement: Not applicable.

Data Availability Statement: The data presented in this study are openly available in Figshare. [Gabriela Mayumi Saiki] [https://doi.org/10.6084/m9.figshare.25958827] [25958827].

Acknowledgments: The authors would like to thank the Brazilian National Confederation of Industry (CNI) for partially supporting this project and for their support and collaboration throughout this research project.

Conflicts of Interest: The authors declare no conflict of interest.

References

1. Meadows, D.H.; Meadows, D.L.; Randers, J.; Behrens, W.W. *Limites do Crescimento: Um relatório Para o Projeto do Clube de Roma sobre o dilema da Humanidade*; Perspectiva São Paulo: São Paulo, Brazil, 1972.
2. Keeble, B.R. The Brundtland Report: 'Our Common Future'. *Med. War.* **1988**, *4*, 17–25. [CrossRef]
3. Heyder, M.; Theuvsen, L. Determinants and effects of corporate social responsibility in German agribusiness: A PLS model. *Agribusiness* **2012**, *28*, 400–420. [CrossRef]
4. Bockstaller, C.; Girardin, P. How to validate environmental indicators. *Agric. Syst.* **2003**, *76*, 639–653. [CrossRef]
5. Morse, S.; McNamara, N.; Acholo, M.; Okwoli, B. Sustainability indicators: The problem of integration. *Sustain. Dev.* **2001**, *9*, 1–15. [CrossRef]
6. Ângelo, H.; De Almeida, A.N.; Serrano, A.L.M. Determinants of Brazil's demand of pulpwood. *Sci. For. Sci.* **2009**, *1*, 491–498.
7. Bauler, T. An analytical framework to discuss the usability of (environmental) indicators for policy. *Ecol. Indic.* **2012**, *17*, 38–45. [CrossRef]
8. Donatti, C.I.; Harvey, C.A.; Martinez-Rodriguez, M.R.; Vignola, R.; Rodriguez, C.M. Vulnerability of smallholder farmers to climate change in Central America and Mexico: Current knowledge and research gaps. *Clim. Dev.* **2019**, *11*, 264–286. [CrossRef]
9. Morton, J.F. The impact of climate change on smallholder and subsistence agriculture. *Proc. Natl. Acad. Sci. USA* **2007**, *104*, 19680–19685. [CrossRef] [PubMed]
10. Kummu, M.; Heino, M.; Taka, M.; Varis, O.; Viviroli, D. Climate change risks pushing one-third of global food production outside the safe climatic space. *ONE Earth* **2021**, *4*, 720–729. [CrossRef] [PubMed]
11. Butterbach-Bahl, K.; Dannenmann, M. Denitrification and associated soil N2O emissions due to agricultural activities in a changing climate. *Curr. Opin. Environ. Sustain.* **2011**, *3*, 389–395. [CrossRef]
12. Trigueiro, W.R.; Nabout, J.C.; Tessarolo, G. Uncovering the spatial variability of recent deforestation drivers in the Brazilian Cerrado. *J. Environ. Manag.* **2020**, *275*, 111243. [CrossRef] [PubMed]
13. Dickinson, R.E.; Henderson-Sellers, A. Modelling tropical deforestation: A study of GCM land-surface parametrizations. *Q. J. R. Meteorol. Soc.* **1988**, *114*, 439–462. [CrossRef]
14. Henderson-Sellers, A.; Gornitz, V. Possible climatic impacts of land cover transformations, with particular emphasis on tropical deforestation. *Clim. Change* **1984**, *6*, 231–257. [CrossRef]
15. Lean, J.; Warrilow, D. Simulation of the regional climatic impact of Amazon deforestation. *Nature* **1989**, *342*, 411–413. [CrossRef]
16. Shukla, J.; Nobre, C.; Sellers, P. Amazon deforestation and climate change. *Science* **1990**, *247*, 1322–1325. [CrossRef] [PubMed]
17. Werth, D. The local and global effects of Amazon deforestation. *J. Geophys. Res.* **2002**, *107*, LBA 55-1–LBA 55-8. [CrossRef]
18. Simberloff, D. The role of science in the preservation of forest biodiversity. *For. Ecol. Manag.* **1999**, *115*, 101–111. [CrossRef]
19. Morton, D.; Defries, R.; Randerson, J.; Giglio, L.; Schroeder, W.; van der Werf, G. Agricultural intensification increases deforestation fire activity in Amazonia. *Glob. Change Biol.* **2008**, *14*, 2262–2275. [CrossRef]
20. Aigner, D.; Lovell, C.; Schmidt, P. Formulation and estimation of stochastic frontier production function models. *J. Econom.* **1977**, *6*, 21–37. [CrossRef]
21. Angulo-Meza, L.; GonzálezAraya, M.; Iriarte, A.I.; RebolledoLeiva, R.; Mello, J.C.S. A multiobjective DEA model to assess the eco-efficiency of agricultural practices within the CF+DEA method. *Comput. Electron. Agric.* **2019**, *161*, 151–161. [CrossRef]
22. Lachaud, M.A.; Bravo-Ureta, B.E.; Ludena, C.E. Agricultural productivity in Latin America and the Caribbean in the presence of unobserved heterogeneity and climatic effects. *Clim. Change* **2017**, *143*, 445–460. [CrossRef]
23. Yang, L.; Zhou, Y.; Meng, B.; Li, H.; Zhan, J.; Xiong, H.; Zhao, H.; Cong, W.; Wang, X.; Zhang, W.; et al. Reconciling productivity, profitability and sustainability of small-holder sugarcane farms: A combined life cycle and data envelopment analysis. *Agric. Syst.* **2022**, *199*, 103392. [CrossRef]
24. Lu, L.C.; Chiu, S.Y.; Chiu, Y.h.; Chang, T.H. Sustainability efficiency of climate change and global disasters based on greenhouse gas emissions from the parallel production sectors—A modified dynamic parallel three-stage network DEA model. *J. Environ. Manag.* **2022**, *317*, 115401. [CrossRef] [PubMed]
25. Shang, M.; Pei, Y.; Chen, C.; Shin, Y.; Zhu, M. Assessing Manufacturing Efficiency in Central Plains Cities: A Three-Stage DEA and Malmquist Index Approach. *J. Urban Dev. Manag.* **2023**, *2*, 196–210. [CrossRef]
26. Taoumi, H.; Lahrech, K. Economic, environmental and social efficiency and effectiveness development in the sustainable crop agricultural sector: A systematic in-depth analysis review. *Sci. Total Environ.* **2023**, *901*, 165761. [CrossRef] [PubMed]
27. Li, W.; Han, X.; Lin, Z.; Rahman, A.U. Enhanced Pest and Disease Detection in Agriculture Using Deep Learning-Enabled Drones. *Acadlore Trans. AI Mach. Learn.* **2024**, *3*, 1–10. [CrossRef]
28. Carauta, M.; Grovermann, C.; Heidenreich, A.; Berger, T. How eco-efficient are crop farms in the Southern Amazon region? Insights from combining agent-based simulations with robust order-m eco-efficiency estimation. *Sci. Total Environ.* **2022**, *819*, 153072. [CrossRef] [PubMed]

29. Rosano-Peña, C.; Silva, J.V.B.; Serrano, A.L.M.; Vieira Filho, J.E.R.; Kimura, H. Eco-Efficiency of Agriculture in the Amazon Biome: Robust Indices and Determinants. *World* **2022**, *3*, 753–771. [CrossRef]
30. Färe, R. *Fundamentals of Production Theory*; Springer: Berlin/Heidelberg, Germany, 1988; pp. IX, 163. [CrossRef]
31. SHEPHARD, R.W. *Theory of Cost and Production Functions*; Princeton University Press: Princeton, NJ, USA, 1970.
32. Farrell, M.J. The Measurement of Productive Efficiency. *J. R. Stat. Society. Ser. A (Gen.)* **1957**, *120*, 253–290. [CrossRef]
33. Banker, R.; Charnes, A.; Cooper, W. Some Models for Estimating Technical and Scale Inefficiencies in Data Envelopment Analysis. *Manag. Sci.* **1984**, *30*, 1078–1092. [CrossRef]
34. Charnes, A.; Cooper, W.; Rhodes, E. Measuring the efficiency of decision making units. *Eur. J. Oper. Res.* **1978**, *2*, 429–444. [CrossRef]
35. Wilson, P.W. Detecting Outliers in Deterministic Nonparametric Frontier Models with Multiple Outputs. *J. Bus. Econ. Stat.* **1993**, *11*, 319–323. [CrossRef]
36. Yang, Q.; Wan, X.; Ma, H. Assessing Green Development Efficiency of Municipalities and Provinces in China Integrating Models of Super-Efficiency DEA and Malmquist Index. *Sustainability* **2015**, *7*, 4492–4510. [CrossRef]
37. Rosano-Pena, C.; De Almeida, C.A.R.; Rodrigues, E.C.C.; Serrano, A.L.M. Spatial dependency of eco-efficiency of agriculture in São Paulo. *Braz. Bus. Rev.* **2020**, *17*, 328–343. [CrossRef]
38. Efron, B. Bootstrap Methods: Another Look at the Jackknife. *Ann. Stat.* **1979**, *7*, 1–26. [CrossRef]
39. Simar, L.; Wilson, P.W. Sensitivity analysis of efficiency scores: How to bootstrap in nonparametric frontier models. *Manag. Sci.* **1998**, *44*, 49–61. [CrossRef]
40. Bogetoft, P.; Otto, L. *Benchmarking with DEA, SFA, and R*; Springer: New York, NY, USA, 2011. [CrossRef]
41. Banker, R.D.; Chang, H. The super-efficiency procedure for outlier identification, not for ranking efficient units. *Eur. J. Oper. Res.* **2006**, *175*, 1311–1320. [CrossRef]
42. Simar, L.; Wilson, P.W. Non-parametric tests of returns to scale. *Eur. J. Oper. Res.* **2002**, *139*, 115–132. [CrossRef]
43. Simar, L. Estimating efficiencies from frontier models with panel data: A comparison of parametric, non-parametric and semi-parametric methods with bootstrapping. *J. Product. Anal.* **1992**, *3*, 171–203. [CrossRef]
44. Malmquist, S. Index numbers and indifference surfaces. *Trab. Estad.* **1953**, *4*, 209–242. [CrossRef]
45. Rosano-Peña, C.; Guarnieri, P.; Sobreiro, V.A.; Serrano, A.L.M.; Kimura, H. A measure of sustainability of Brazilian agribusiness using directional distance functions and data envelopment analysis. *Int. J. Sustain. Dev. World Ecol.* **2014**, *21*, 210–222. [CrossRef]
46. Mardani, A.; Zavadskas, E.K.; Streimikiene, D.; Jusoh, A.; Khoshnoudi, M. A comprehensive review of data envelopment analysis (DEA) approach in energy efficiency. *Renew. Sustain. Energy Rev.* **2017**, *70*, 1298–1322. [CrossRef]

 *sustainability*

Article

Agricultural Production Efficiency and Differentiation of City Clusters along the Middle Reaches of Yangtze River under Environmental Constraints

Lei Wang [1], Yi Zhang [1], Jingyi Xia [1], Zilei Wang [1] and Wenjing Zhang [2,*]

[1] School of Business, Hohai University, Nanjing 320100, China; 20141924@hhu.edu.cn (L.W.); 2163810203@hhu.edu.cn (Y.Z.); 2263810204@hhu.edu.cn (J.X.); hhuwzl@163.com (Z.W.)

[2] School of Management, Xi'an University of Finance and Economics, Xi'an 710000, China

* Correspondence: 521laynehewei@163.com

Citation: Wang, L.; Zhang, Y.; Xia, J.; Wang, Z.; Zhang, W. Agricultural Production Efficiency and Differentiation of City Clusters along the Middle Reaches of Yangtze River under Environmental Constraints. *Sustainability* **2024**, *16*, 6126. https://doi.org/10.3390/su16146126

Academic Editors: Fotios Chatzitheodoridis, Efstratios Loizou and Achilleas Kontogeorgos

Received: 29 May 2024
Revised: 7 July 2024
Accepted: 12 July 2024
Published: 18 July 2024

Abstract: The improvement of overall agricultural efficiency in the city clusters along the middle reaches of the Yangtze River is crucial for promoting stable regional agricultural production and ensuring food security. This study employs the SBM (slack-based measure) model with the unexpected environmental outputoutputs, including agricultural surface pollution and agricultural carbon emissions, and the SFA (stochastic frontier approach) model to investigate the overall agricultural efficiency and its influencing factors in 31 prefecture-level cities in the middle reaches of the Yangtze River urban agglomeration from 2008 to 2021. The research findings indicate the following: (1) Without eliminating the impact of environmental variables, the overall agricultural efficiency in the middle reaches of the Yangtze River city clusters shows a rise–fall–stability trend and limited level. The scale of production input is relatively reasonable, but there is inefficiency in the utilization of factor resources. (2) The SFA model reveals that economic development, urbanization construction, industrial structure, and government influence have significant but different impacts on agricultural production factor input. Accelerating economic development is helpful for reducing excessive inputs of agricultural capital, labor, planting area, agricultural film, and irrigation. Increasing the level of urbanization can promote the efficient allocation of planting area and effective irrigation area. The improvement of industrialization level pushes the rational input of planting area and agricultural film, but it may also lead to excessive input of agricultural capital, labor, pesticides, and effective irrigation area. Expanding government influence can restrain the excessive use of pesticides. (3) After eliminating environmental variables, there is a low and slow declining trend of the overall agricultural efficiency over time. Neither production scale efficiency nor pure technical efficiency reached optimal levels; the former one is significantly lower than the latter. In terms of spatial distribution, there exists a "higher in the west and lower in the east" feature, with obvious and expanding regional efficiency differences and high-efficiency areas gradually concentrating in the Wuhan urban circle. In summary, this article puts forward the following suggestions: optimize the structure of the government's support for agriculture, focusing on the construction of agricultural infrastructure and the support for green production in agriculture; improve the research and development and promotion of green production technology and encourage the establishment of the use of resources and recycling; and absorb the population of farmers who have been transferred to urban areas reasonably and orderly under the adjustment of industrial structure.

Keywords: city clusters along the middle reaches of Yangtze River; environmental constraints; heterogeneous analyses; overall agricultural efficiency; unexpected output SBM model; SFA model

1. Introduction

The city clusters along the middle reaches of Yangtze River, connecting east and west as well as north and south, play an indispensable role in the pattern of regional development in China. In April 2015, the State Council approved the implementation of the

"Development Plan for City Clusters along the Middle Reaches of Yangtze River" [1]. The city clusters along the middle reaches of Yangtze River have become a new growth pole for China's economic development. The urban agglomeration in the middle reaches of the Yangtze River is also one of China's main grain-producing areas, and it plays a strategic role in China's food security, but there exists an increasing gap between the region's agricultural and economic development. The urban—rural dual structure has led to serious phenomena such as the massive loss of rural labor and the encroachment of agricultural resources by urban development [2]. Taking the secondary city clusters along the middle reaches of the Yangtze River—the urban agglomeration around Changsha-Zhuzhou-Xiangtan—as an example, in 2018, the rural labor force in the region decreased by about 0.19 million people, and the arable land decreased by about 4.3 thousand hectares. The massive loss of agricultural production factors is a serious threat to the region's stable agricultural production and food security. Under the rigid constraints of production factors, it is necessary to study the agricultural production efficiency in the region and explore how to promote the effective allocation of production factors and enhance agricultural production efficiency. The mismatch of agricultural capital, labor, and land hinder the improvement of agricultural production efficiency to a large extent [3]. Therefore, it is of great significance to study the overall efficiency of agricultural production and analyze the driving factors in the city clusters along the middle reaches of Yangtze River under the decreasing production factors. For example, agricultural labor helps promote the improvement of the regional agricultural production efficiency and guarantee food security.

Agricultural production efficiency has always been a hot topic for scholars. Some scholars have tried to analyze the level of agricultural production efficiency through the construction of evaluation index system qualitatively and quantitatively. Previous studies evaluated the level of China's agricultural development in terms of resource-saving utilization, environmental protection, and production supply capacity [4]. Some scholars constructed an evaluation index system to analyze the level of China's agricultural development [5]; to set evaluation standards for the level of low-carbon agricultural development [6]; and to evaluate the level of sustainable agricultural development in China [7]. Although scholars try to construct a comprehensive index system of agricultural production efficiency as best they can and fully consider the factors of ecological environment, the conventional evaluation index system can only be suitable in the overall situation to evaluate the level of agricultural production, while it cannot facilitate a good response to the efficiency of regional agricultural production. Regarding the measurement of efficiency, scholars generally use the data envelopment model (DEA), and this model is also widely used in the study of agricultural production efficiency. The DEA model was formally proposed by Charnes in 1978 [8], followed by scholars such as Restuccia and Vollrath, who used this model to analyze the problem of agribusiness efficiency in the world [9,10]. Ruttan used this model to explore the regional agricultural production efficiency under the factor constraints of environment, resources, and scientific and technological innovation [11]. In recent years, Chinese scholars have also used the DEA model to measure agricultural production efficiency. Scholars utilized the method to measure interprovincial agricultural total factor productivity in China [12]. Xing [13] discussed the influencing factors based on the efficiency measure using the Tobit model. DEA models are divided into radial and non-radial models. The radial DEA model assumes that input and output vary proportionally, while the non-radial model relaxes this assumption, allowing them to vary disproportionally. Moreover, the traditional radial model usually underestimates the directional distance function of the model when evaluating the input–output efficiency of agricultural production, which is unfavorable for reflecting the actual efficiency of each decision unit [14]. In order to overcome this problem, the non-radial data envelopment model has been gradually applied in the evaluation of agricultural production efficiency. It can be used to study the comprehensive efficiency of agricultural production in counties of Hebei Province [15]. The efficiency of agricultural production in Heilongjiang Reclamation Area was evaluated, with which scholars further analyzed the influencing factors by using the FGLS (feasible generalized least squares) model [16]. The traditional DEA model cannot eliminate the interference of management and stochastic factors on the efficiency, so some scholars have introduced the stochastic frontier method (SFA) to propose the three-phase DEA model, taking environmental variables into account, and applied the three-stage DEA model to study the efficiency of agricultural production in 2008 [17]. Urbanization level, industrial development level, transportation status, financial support, and other factors have been taken as environmental variables in the studies [18–20].

Ecological environmental protection has raised increasing attention in society, and more and more scholars have begun to explore the problem of environmental pollution generated in agricultural production. Environmental pollution is a non-desired output of agricultural production, and ignoring this output in the measurement of agricultural production efficiency will inevitably lead to biased results [21]. The previous DEA model could not analyze the efficiency measurement including non-expected outputs, so some scholars proposed the SBM model with environmental pollution non-expected outputs [22]. Pan and Lv incorporated the agricultural non-point source pollution into non-desired outputs to measure the level of agricultural production efficiency [23,24]. A few studies have measured the agricultural environmental efficiency and agricultural green total factor productivity by applying the SBM model with carbon emission as the non-expected output [25–27]. In addition, some scholars adopted the nitrogen surplus of agricultural land and agricultural non-point source pollution as the non-desired outputs of agricultural production [28,29]. Some scholars have taken the Yellow River Basin and the Yangtze River Economic Belt as objectives to explore the problem of agricultural production efficiency with the negative externalities of production [30,31].

Nowadays, there are many studies focusing on agricultural production efficiency, but they still have some shortcomings. Firstly, the existing studies are mostly focused on the macro level of the country or the province, and the practical guiding significance of the conclusions and recommendations is limited. Secondly, there is a relative lack of studies based on the micro level of the municipal area; most of the studies are limited to a single province, so there is a lack of inter-regional coordinated analysis. In addition, the existing studies mainly take capital, labor, and land into consideration. They lack consideration of non-desired outputs such as environmental pollution of agricultural production, which results in an inaccurate result of real overall agricultural production efficiency of the region.

The middle reaches of the Yangtze River, as an important grain-producing area in China, are endowed with unique agricultural resources. According to statistics, in 2021, the grain output of the middle reaches of the Yangtze River accounted for 11.76% of the national total production, making a great contribution to China's food security. However, due to the long history of agriculture and dense population in the region, there is a trend of shrinking proportion and increasing fragmentation of arable land. This helps increase application of agrochemicals, surface source pollution, and agricultural carbon emission, and other problems are becoming more and more prominent. The middle reaches of the Yangtze River are now facing the challenge of sustainable development of agriculture. Based on this, this paper takes 31 prefecture-level cites in the middle reach of the Yangtze River from 2008 to 2021 as an example, using the SBM model and the SFA model with the undesired outputs of the environment to analyze the overall agricultural production efficiency with the hope of finding the influencing factors from both macro and micro perspective and paths of agricultural efficiency under environmental constraints. Moreover, the study aims to provide new ideas for the city clusters along the middle reaches of Yangtze River, improve the overall efficiency of agriculture, and guide the transformation and upgrading of regional agricultural production. This article clarifies the correlation between environmental factors, management factors, and agricultural production efficiency, enriching the theoretical foundation of agricultural environmental governance. At the same time, the applicability of SBM model and SFA model is expanded, which provides useful supplements to the governance factors in current agricultural production.

2. Model

2.1. Stage 1: SBM Modeling of Undesired Outputs

This paper constructs the most optimal production frontier for each year for the 31 prefectural cities in the middle reaches of the Yangtze River as a decision unit. Assume that each decision unit has m input terms and s_1 desired output terms with s_2 non-desired output terms; the matrix form can be expressed, respectively, as $X = [x_1 \cdots x_{31}] \in R^{m \times 31}$; $Y^g = \left[y_1^g \cdots y_{31}^g \right] \in R^{s_1 \times 31}$; $Y^b = \left[y_1^b \cdots y_{31}^b \right] \in R^{s_2 \times 31}$; $X > 0$, $Y^g > 0$, $Y^b > 0$; the production possibility frontier of the resulting decision unit P can be expressed as follows:

$$P = \left\{ \left(x, y^g, y^b \right) \middle| x \geq X\lambda, y^g \leq Y^g\lambda, y^b \geq Y^b\lambda, \lambda \geq 0 \right\} \tag{1}$$

In Equation (1), $\lambda \in R^{31}$ is the intensity vector that assigns weights to each observation when constructing the production set, assuming constant returns to scale since no constraints are imposed on their sum. In the case of constant returns to scale, it can improve the stability of the research in this paper, and assuming constant returns to scale is a common situation in applying the SBM model, so this paper assumes constant returns to scale. To bring the decision unit under the condition of non-desired outputs, $DMU_1 \left(x_1, y_1^g y_1^b \right)$ is efficient to reach the optimal frontier; then, its input–output term needs to satisfy the following conditions: $x_1 \leq x$, $y_1^g \geq y^g$, $y_1^b \leq y^b$, and $\left(x, y^g, y^b \right) \in P$. Therefore, the non-expected output SBM model constructed in this paper is specified as follows:

$$\rho = min \frac{1 - \frac{1}{m}\sum_{i=1}^{m} \frac{s_i^-}{x_{io}}}{1 + \frac{1}{s_1+s_2} \left(\sum_{r=1}^{s_1} \frac{s_r^g}{y_{ro}^g} + \sum_{r=1}^{s_2} \frac{s_r^b}{y_{ro}^b} \right)} \tag{2}$$

$$\begin{cases} x_o = X\lambda + s^- \\ y_0^g = Y^g\lambda - s^g \\ y_o^b = Y^b\lambda + s^b \\ s^- \geq 0, s^g \geq 0, s^b \geq 0, \lambda \geq 0 \end{cases} \tag{3}$$

where the vector $s^- \in R^m$ and $s^b \in R^{s_2}$ denote the excess of the decision unit input term and the undesired output term, respectively. $s^g \in R^{s_1}$ denotes the amount of shortfall in the desired output of the decision unit, while $0 < \rho \leq 1$. Under the non-desired output condition, when $\rho = 1, s^- = 0, s^g = 0$, and $s^b = 0$, the production efficiency of the decision unit reaches the optimal frontier; when $0 < \rho < 1$, the optimal frontier is not reached, which means that the production efficiency can be improved by optimizing the input–output structure of the decision unit.

2.2. Stage 2: Similarity SFA Analytical Model

The efficiency values obtained from the non-expected SBM model in the first stage are affected by the external environment, random errors, and internal management factors. The influence of these three factors can be eliminated by using the similar SFA analysis model as follows:

$$S_{mn} = f(z_n; \beta^m) + v_{mn} + u_{mn}a \tag{4}$$

where $m = 1, 2, \ldots$ denotes the first m input; $n = 1, 2, \ldots$ indicates the first n input; S_{mn} indicates the first n decision unit and m the slack variables for the inputs; z_i and β^m denote the environment variables and parameters to be estimated, respectively; $f(z_n; \beta^m)$ denotes the effect of environmental variables on the redundant variables S_{mn}. As scholars usually do, in this study, we chose to use $f(z_n; \beta^m) = z_n * \beta^m$. Equation (4) $v_{mn} + u_{mn}$ is the composite error term. $v_{mn} \sim N\left(0, \delta_{vm}^2\right)$ reflects the effect of statistical noise, and $u_{mn} > 0$ reflects managerial inefficiencies. We assume that u_{mn} obeys a half-normal distribution $\mu_{mn} \sim N^+\left(0, \delta_{um}^2\right)$, and v_{mn} and u_{mn} are independent of each other and of the environment variables. It is a good idea to make $\gamma = \delta_{um}^2 / \delta_{um}^2 + \delta_{vm}^2$; when γ tends to 1, managerial factors have the greatest influence in the inefficient decision-making unit; when γ tends to 0, random interference factors have the greatest influence.

Using the measurements of Equation (3), the input and output data of each decision-making unit were adjusted as shown in Equation (5):

$$x_{mn}^A = x_{mn} + \left[max_n\{z_n\hat{\beta}^m\} - z_n\hat{\beta}^m \right] + \left[max_n\{\hat{v}_{mn}\} - \hat{v}_{mn} \right], n = 1, 2, \ldots ; m = 1, 2, \ldots,$$

$$y_{mn}^{gA} = y_{mn}^g + \left[max_n\left\{ z_n\hat{\beta}_{mn}^g - z_n\hat{\beta}_{mn}^g \right\} \right] + \left[max_n\left\{ \hat{v}_{mn}^g \right\} - \hat{v}_{mn}^g \right] \tag{5}$$

$$y_{mn}^{bA} = y_{mn}^b + \left[max_n\left\{ z_n\hat{\beta}_{mn}^b - z_n\hat{\beta}_{mn}^b \right\} \right] + \left[max_n\left\{ \hat{v}_{mn}^g \right\} - \hat{v}_{mn}^b \right]$$

where x_{mn}^A denotes the input-adjusted values, and y_{mn}^{gA} and y_{mn}^{bA} denote the adjusted desired and non-desired output values, respectively.

2.3. Stage 3

The input–output efficiencies were estimated again by applying the non-desired output SBM model in conjunction with the adjusted input and output volumes. The input–output efficiencies were estimated to eliminate the effects of environmental variables and random errors this time.

3. Data and Indicators

3.1. Data Sources

Based on the "Development Plan for City Clusters along the Middle Reaches of the Yangtze River", this paper takes 31 prefecture level cites in the Middle Reach of the Yangtze River as objective and divides them into three sub-city clusters: (1) Wuhan city group (urban agglomeration around Wuhan in Hubei Province): Wuhan City, Huangshi City, Jingzhou City, Yichang City, Xiangyang City, Ezhou City, Jingmen City, Xiaogan City, Huanggang City, Xianning City, Xiantao City, Tianmen City, and Qianjiang City; (2) Chang-Zhu Tan city group (the Changsha-Zhuzhou-Xiangtan city group in Hunan Province): Changsha City, Zhuzhou City, Xiangtan City, Hengyang City, Yueyang City, Changde City, Yiyang City, and Loudi City; and (3) Poyang Lake city group (clusters around Poyang Lake in Jiangxi Province): Nanchang City, Jingdezhen City, Pingxiang City, Jiujiang City, Xinyu City, Yingtan City, Ji'an City, Yichun City, Fuzhou City, and Shangrao City. Among these, Huanggang City, Xianning City, Xiantao City, and Ezhou City were excluded due to serious missing data. The geographical location of the city clusters along the middle reaches of the Yangtze River in China is shown in Figure 1. The data of input–output variables and external influencing factors of agricultural production in this paper come from "Statistical Yearbook of Hubei Province", "Statistical Yearbook of Hunan Province", "Statistical Yearbook of Jiangxi Province", and "Statistical Yearbook and Statistical Bulletin of Prefecture-level Municipalities" in 2009–2022 [32–35]. For the missing data, this paper uses quadratic exponential smoothing to supplement them and prevent model-run failures caused by 0 values in the raw data.

3.2. Selection of Indicators

3.2.1. Input–Output Indicators

Based on the data availability and consistency of statistical caliber, the total output value of agriculture, forestry, animal husbandry, and fishery in municipalities was selected as the desired output variable, and the agricultural carbon emissions and agricultural non-point source pollution were selected as the non-desired output variables [36]. Considering the inflation and taking 2008 as the base period, the gross output value of agriculture, forestry, animal husbandry, and fishery was converted by the index of their gross output value, and the converted gross output value was used as the desired output variable.

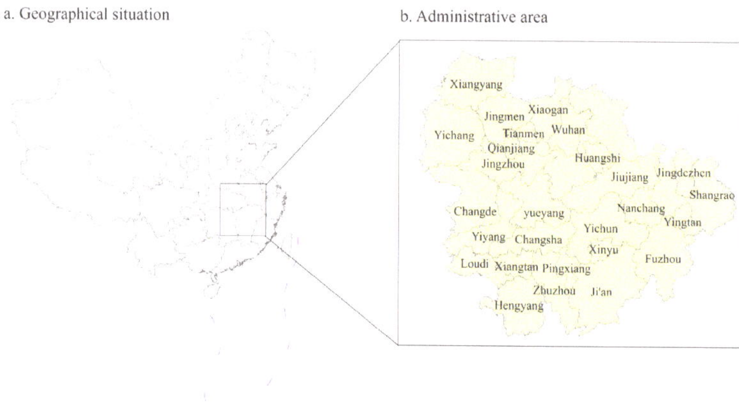

Figure 1. The geographical location of the city clusters along the middle reaches of the Yangtze River in China.

Agricultural production cannot exist without factors such as agricultural capital, labor, land, fertilizer, pesticide, agricultural film, and irrigation. Due to the lack of statistical data on agricultural capital, the degree of agricultural electrification, i.e., the total power of agricultural machinery, was selected as a proxy variable for capital input. Labor force was selected as the labor input of agriculture, forestry, animal husbandry, and fishery in each municipality. Since the shifting and abandonment of cultivation in different regions, the cultivated area cannot represent the land input situation, so the sown area of crops in different regions was adopted as the land input. Effective irrigation area can reflect the degree of hydration of agricultural production and the drought-resistance ability of arable land, so effective irrigation area was taken as one of the input factors. Agricultural fertilizer can improve soil fertility and is also an important measure to increase the yield of crops per unit area, so it was selected as an input indicator of agricultural production. Pesticides can prevent diseases, insects, grass, and other hazards to crops and are indispensable agricultural production materials. Agricultural film is conducive to crop germination and seedling emergence, which improves the soil's ability to retain water and fertilizer. As an important factor in stable and high agricultural production, it was used as an input element.

With reference to the earlier studies [15,37,38], when measuring the total agricultural carbon emissions in each city combined with the selection of input indicators, the study estimated the carbon emissions from agricultural farming and irrigation, pesticides, agricultural films, agricultural fertilizer use, agricultural electrification, etc. Due to the diversity and complexity of agricultural carbon emission sources, the formula constructed for calculating the total agricultural carbon emissions is as follows: .

$$E_t = \sum E_{ti} = \sum S_{ti}\theta_i \tag{6}$$

where E_t is the total carbon emissions from agricultural production; E_{ti} denotes t city and i carbon emissions from the carbon source category; S_{ti} denotes t city i amount of inputs from the class of carbon sources; θ_i denotes the number of i carbon emission coefficients for each type of carbon source. The carbon emission coefficients for each type of carbon source are shown in Table 1. During analysis, CO_2 and N_2O were uniformly converted to standard carbon.

Agricultural non-point pollution was calculated with reference to the previous studies [39], which use the equal weight assignment method (each weight was 0.25), and measured by the weighted value of the average input of fertilizer, pesticide, and agricultural film.

Table 1. Carbon emission factors for major carbon sources.

Carbon Source	Emission Factor	Reference Source
Agricultural tillage and irrigation	579.0800 kg/hm^2	ORNL (Oak Ridge National Laboratory, USA)
Pesticide	4.9341 kg/kg	ORNL (Oak Ridge National Laboratory, USA)
Agricultural film	5.1800 kg/kg	IREEA (Institute of Agricultural Resources and Ecology, Nanjing Agricultural University)
Fertilizer use	0.8956 kg/kg	[40]
Agricultural electrification	Average carbon emission factor of electricity kg/kWh	[41]

3.2.2. Description of Indicators of External Environmental Factors of Overall Agricultural Efficiency

Based on the availability of data and the related literature [42–45], this study hypothesized that the external drivers of overall agricultural efficiency in the city clusters along the middle reaches of the Yangtze River are mainly divided on the following four aspects:

(1) Economic development: This paper uses per capita GDP to measure the level of regional economic development. The level of economic development is a key factor affecting a region's technology level, environmental protection inputs, etc., which will impose greater impact on the overall efficiency of regional agriculture;

(2) Urbanization development: The proportion of the urban population to the total population stands for the level of urbanization development. In the process of urbanization expansion, the agricultural population will migrate to cities, making it easier to consolidate the fragmented arable land, thus promoting the large-scale operation of agricultural production, which will finally affect the overall agriculture efficiency within the region;

(3) Industrial structure: The proportion of the secondary industry is used to indicate the factors of regional industrial structure. Most of the existing studies found that industrial structure plays an important role in the resource utilization, such as the stage and mode of regional industrial development. Whether the type of industry is highly water-consuming or high-polluting or not, it will affect the overall efficiency of regional agriculture;

(4) Government influence: This paper uses the government's expenditure budget on agriculture, forestry, and water affairs to measure the influence of regional governments. The government's investment in the construction of water infrastructure, guidance for industrial transformation, and related environmental policies all play an important role in the overall efficiency of regional agriculture.

The selection and interpretation of indicators for measuring agricultural production efficiency in the city clusters along the middle reaches of the Yangtze River are shown in Table 2.

Table 2. Categories of Indicator Selection and Interpretation.

Indicator	Variables	Variables Interpretation
Inputs	Agricultural capital X_1	Total power of agricultural machinery (billion watts)
	Agricultural labor force X_2	Number of people employed in agriculture, forestry, animal husbandry, and fishery (10,000 people)
	Land X_3	Area sown with crops (thousands of hectares)
	Fertilizer X_4	Fertilizer use (million tons)
	Pesticides X_5	Pesticide use (tons)
	Agricultural film X_6	Agricultural film use (hundred tons)
	Irrigation X_7	Effective irrigated area (thousand hectares)
Output	Expected output Y_1	Agriculture forestry, livestock, and fisheries GDP (billion yuan)
	Undesired output Y_2	Total carbon emissions from agriculture (tons) Agricultural surface source pollution (tons)

Table 2. *Cont.*

Indicator	Variables	Variables Interpretation
External Environment	Economic development E_1 Urbanization development E_2 Industrial structure E_3 Government influence E_4	Per capita GDP (million yuan) Share of urban population in total population Share of secondary industry Government expenditure on agriculture, forestry, and water resources (ten thousand yuan)

Based on the previous analysis, the theoretical relationship between factor allocation, external environment, and agricultural production efficiency can be seen in Figure 2.

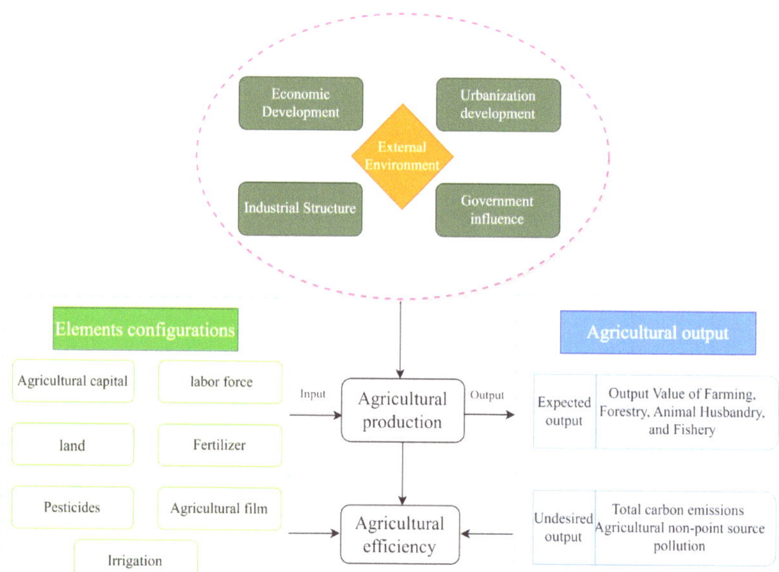

Figure 2. Theoretical framework of factor allocation, external environment, and agricultural production efficiency.

4. Empirical Analysis

4.1. Phase 1: SBM Model Analysis of Non-Expected Outputs

Based on the input–output indicator system of agricultural production and the SBM model of non-expected output, the measurement results of the comprehensive agricultural efficiency value and the characteristics of time variation of the city clusters along the middle reaches of the Yangtze River from 2009–2021 are shown in Figure 3.

Overall agricultural efficiency shows an upward–declining–steady trend, with obvious stage changes in efficiency. The comprehensive efficiency of agriculture sharply declined from 0.744 in 2008 to 0.486 in 2021. The first turning point appeared in 2009, which reached the highest point of 0.825, then showed a decreasing trend year by year after 2009; the second turning point appeared in 2013, and the comprehensive efficiency of agricultural production declined to the relative low point of 0.466. The scale effects of agricultural green production are good; the level of pure technical efficiency is lower than scale efficiency. As a whole, the average value of comprehensive efficiency in the city clusters along the middle reaches of the Yangtze River in 2008–2021 is 0.609, which is not a high level. The average level of scale efficiency is 0.820, and the average level of pure technical efficiency is 0.736.

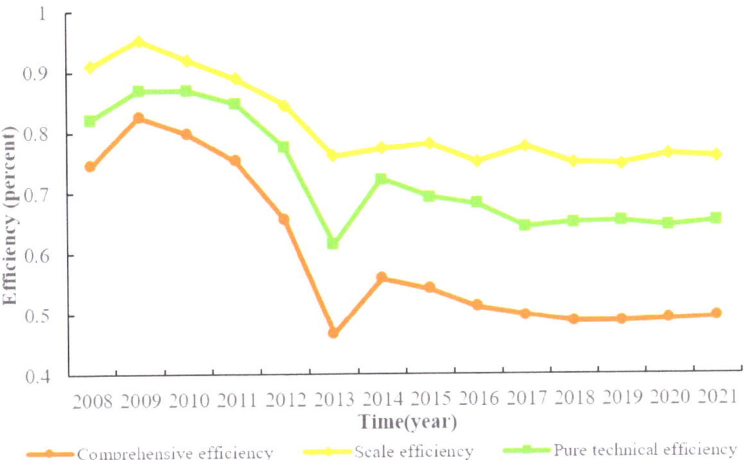

Figure 3. Decomposition of Agricultural Comprehensive Efficiency and Temporal Characteristics in the City Clusters along the Middle Reaches of the Yangtze River from 2008 to 2021.

The first stage of efficiency measurement contains the influence of environmental variables and random factors, which cannot truly reveal the actual situation of agricultural production efficiency in the city clusters along the middle reaches of the Yangtze River; it was therefore necessary to further eliminate the interference of exogenous variables.

4.2. Stage 2: SFA Model Analysis

The redundancy value of each input variable in the first stage SBM model was used as an explanatory variable, and the external environmental factors of agribusiness efficiency were used as explanatory variables. The paper uses Frontier 4.1 software for SFA regression, whose results are shown in Table 3.

Table 3. Stage 2: SFA regression results.

Slack Variables	Capital X_1	Labor X_2	Land X_3	Fertilizer X_4	Pesticides X_5	Agricultural Film X_6	Irrigation X_7
The constant term	−7041.21 ***	−139.722 ***	6133.37 *	−10.454	−8.921	−228.61 *	304.701 ***
Economy E_1	−0.316 ***	−0.0001 **	−0.1251 **	0	0.323	−0.17435 *	−0.0006 ***
Town E_2	1515.33 ***	1.710 ***	−1368.23 ***	−0.037	5.257 ***	−0.8291	−2.318 ***
Industry E_3	52,660.75 ***	67.183 **	−2735.98 *	15.108	47.543 **	−218.258 **	101.266 **
Government E_4	0.016 *	0.0001 ***	161.13	0.000017 ***	−0.011 **	149.24	0.00015 ***
σ^2	1.34×10^9 ***	3273.3 ***	484,373.1 ***	136.555 ***	2765.43 ***	4543.251 **	17,433.86 ***
γ	0.751 ***	0.893 ***	0.81 ***	0.534 ***	0.732 ***	0.825 ***	0.951 ***
Log likelihood function	3432.010	−1634.953	−3370.60	−1105.50	−448.59	−1438.36	−1482.714
LR test of the one-sided error	179.820	168.75	151.80	12.132	61.37	66.37	254.261
Mean	2.939	89.244	481.559	15.84	0.665	75.489	217.548
Variance	417.639	3200.977	104,447.353	95.478	2710.607	1,581,073.334	17,826.701

Note: *, **, and *** indicate significant at 10%, 5%, and 1% significance levels, respectively; numbers in parentheses are the corresponding estimated *t*-statistic variables.

The log likelihood function value from SFA regression results (log likelihood) and likelihood ratio test (LR test) show a good estimation effect. Most coefficients of each environmental variable on the input variables can pass the test of significance, which indicates that the external environmental variables have a significant influence on redundant value for each input variable. It can also be seen that each of the agricultural input slack variable's

γ values pass the test at a 1% level of significance, indicating that management factors play a dominant role in agricultural inputs, and it is necessary to strip out environmental and stochastic factors.

The impact of the level of economic development, urbanization level, industrial structure, and government influence on the redundancy value of each agricultural input was examined through the SFA model. When the coefficient of SFA regression is positive, it indicates a positive impact of environmental variables on inputs, which will increase the input redundancy value and the waste of resources; when the coefficient is negative, the environmental variables will reduce the input redundancy value, which is conducive to the saving of agricultural inputs. The SFA regression results are shown in Table 3.

(1) Level of economic development: The level of economic development has a significant negative impact on the slack variables of agricultural capital, agricultural labor, sown area, agricultural film, and irrigation, which means as the economy develops, the redundancy leading to the above five agricultural inputs will decrease. The level of economic development reflects, to a certain extent, the service management capability of the region. Agricultural trusteeship services were increasingly recognized and accepted by farmers; the mechanized operation of agricultural trusteeship occupied the form of farmers' self-purchased mechanical operations, which promoted the effective allocation of agricultural capital; the outsourcing of nursery services also reduced the overuse of agricultural film to a certain extent. With the development of the economy, there are more employment opportunities to absorb surplus agricultural workers. This promotes the rational allocation of land in different industries, making inefficient and even idle arable land being utilized. However, the effect of the level of economic development on the slack variable of agricultural fertilizer use is not significant, indicating that there is no significant relationship between the economic development and agricultural fertilizer use;

(2) Level of urbanization: The level of urbanization has a significant positive effect on the slack variables of agricultural capital, agricultural labor, and pesticides and a significant negative effect on the slack variables of sown area and effective irrigated area. With the increase of urbanization, the above five factors show an over-input problem, and resources are not effectively allocated. The urbanization expansion has promoted the outflow of rural labor. The women and children left in the rural area have pushed for the popularization of mechanized operations. It was found in the surveys that, although farmers household their own agricultural machinery and equipment, the usage rate is generally low. At present, the traditional fine and intensive operation of agriculture is still an important mode of production, coupled with the rural "home-based care" model for elderly people. Agricultural workers are "locked up", preventing the orderly flow of rural labor. Over all, the rural labor force is on the outflow trend, and the reduction of a young and strong labor force promotes the use of chemicals to eliminate pests and diseases, resulting in excessive pesticide inputs. Urbanization has promoted the construction of high-standard farmland and better allocation of land resources, improving the effective use of arable land;

(3) Industrial structure: The level of industrialization development has a significant positive effect on the slack variables of agricultural capital, agricultural labor, pesticides, and effective irrigation area and a significant negative effect on the slack variables of sown area and agricultural film. Accompanied by the development of industry, the feeding effect of industry on agriculture is more obvious, promoting the application of science and technology in agricultural production. Machinery and equipment have been popularized among farmers so as to liberate the labor force gradually, but there are many obstacles to the rational flow of excess labor. Scientific and technological progress has led to the development of water conservancy, improving the irrigation capacity of farmland water conservancy facilities, but some land did not bring considerable output with limited condition. Pesticides have become an important means of preventing and controlling pests and diseases; agricultural science and technology have enhanced the use effectiveness of pesticide, and coupled with increasing consumers' demand on agricultural products' quality, this has led to excessive use of pesticides. The development of industrialization

is inseparable from the demand for land, which, to a certain extent, has optimized the allocation of land between industry and agriculture. The recycling of agricultural films also benefits from the development of industrialization, thus reducing the excessive use of agricultural films;

(4) Government influence: Government influence mainly reflects the support for agriculture, which has a significant positive effect on the slack variables of agricultural capital, agricultural labor, fertilizer and effective irrigated area and a significant negative effect on the pesticide slack variable. The government's agricultural machinery subsidy policy is one of the key factors contributing to the redundancy of agricultural capital, while the possession of further complete agricultural machinery and equipment by farmers leads to the redundancy of labor. Expenditures on agriculture, forestry, and water affairs are also used for water conservancy expenditures and fertilizer subsidies, resulting in overuse of fertilizers and over-irrigation. Pesticide subsidies are now mainly for large grain farmers or those who purchase green pesticides. Farmers can purchase pesticides in accordance with the relevant guidelines, reducing the blindness of purchases and promoting the rational use of pesticides.

It can be seen that environmental variables have an important impact on agricultural factor inputs, which further affects agricultural production efficiency. Therefore, it is necessary to adjust the original input variables and then eliminate the impact of environmental variables to examine the level of agricultural production efficiency.

4.3. Stage 3: Adjusted Unexpected Output SBM Empirical Results

According to Equation (5), the original input variables were adjusted, and the adjusted input variables and the original output variables were introduced into the SBM model for efficiency measurement to obtain the actual situation of agricultural production efficiency in the city clusters along the middle reaches of the Yangtze River in the third stage.

4.3.1. Characteristics of Time Evolution of Comprehensive Agricultural Efficiency

Figure 4 shows the value of overall agricultural efficiency and the evolutionary trend over time in the city clusters along the middle reaches of the Yangtze River.

(1) The comprehensive agricultural efficiency is low, with a slowly decreasing trend. From 2008 to 2021, the average value of the comprehensive efficiency of agricultural production in the city clusters along the middle reaches of the Yangtze River was 0.51. It was reduced from 0.530 in 2008 to 0.436 in 2021, and the decline was not obvious. One of the reasons for the low comprehensive efficiency of agricultural production in the city clusters along the middle reaches of the Yangtze River is the low scale efficiency, which has been below 0.6 for a long period of time, while the continuous decline in pure technical efficiency has led to a slow decline in comprehensive efficiency. In 2013, there was the H7N9 avian influenza epidemic; poultry farming suffered an unprecedented damage, and the expected output of agriculture declined, resulting in a value of agricultural production efficiency below 0.5. In recent years, the growth of food production in the middle reaches of the Yangtze River has been obviously weak. The construction of the urban agglomeration occupied a large amount high-quality arable land, which is also one of the reasons for the continued decline in comprehensive efficiency, with the 2021 comprehensive efficiency falling to 0.436. Despite the rising agricultural GDP in the city clusters along the middle reaches of the Yangtze River, excessive inputs of production materials such as fertilizers, pesticides, and agricultural machinery have led to an increase in non-desired outputs. This accelerated the continued decline in the comprehensive efficiency of agricultural production. The agricultural production efficiency after the adjustment of input variables changed more gently than before the adjustment, indicating that environmental variables have a non-negligible impact on agricultural production efficiency. Under such low agricultural comprehensive efficiency conditions, due to the relatively developed industry in the urban agglomeration of the middle reaches of the Yangtze River, the loss of rural labor force will be more severe;

(2) The scale efficiency of agricultural production basically maintains a stable trend, and the level of pure technical efficiency shows a slow downward trend. The average level of scale efficiency is only 0.567, and the average level of pure technical efficiency is 0.912. Scale efficiency reflects the impact of the expansion of production scale on productivity. The scale efficiency of agricultural production in the city clusters along the middle and lower reaches of the Yangtze River is low; the scale effect is not obvious, which is also an important factor contributing to the low comprehensive efficiency of agriculture in the region. Most of the urban agglomerations in the middle reaches of the Yangtze River are in developed areas. The development of cities devours agricultural land and, coupled with China's small per capita arable land area, makes it difficult to realize the scale of land in the region, and the inputs of other factors of production make it difficult to realize economies of scale. Pure technical efficiency reflects the comprehensive management level of agricultural production and the impact of technological upgrading on productivity. The level of pure technical efficiency in this region is high compared to the level of scale efficiency, both above 0.8, but showing a slow downward trend in 2008–2021.The high level of pure technical efficiency in the city clusters along the middle and lower reaches of the Yangtze River is due to the advantages of technological innovation and promotion; the developed industrial development has provided technological support for agricultural production, and the effect of industry feeding agriculture has gradually appeared [46]. However, the enthusiasm of farmers for technology acceptance is not high. Correspondingly, research and development institutions lack the motivation for research and development. The promotion of the role of technology in agricultural production has not been fully realized, and the inhibitory effect on agricultural production has become more and more obvious.

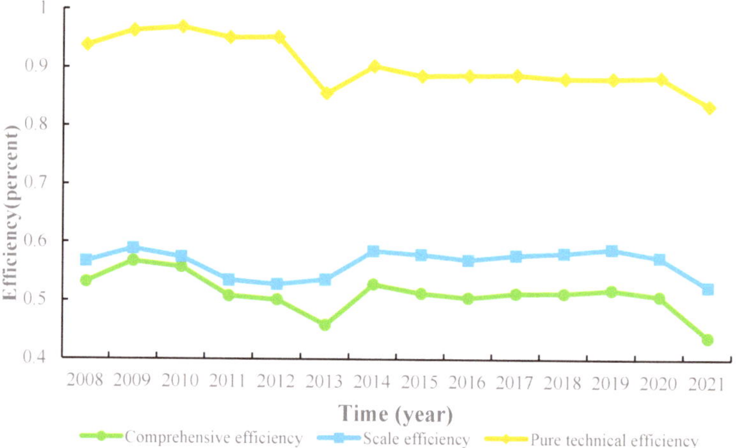

Figure 4. Comprehensive Agricultural Comprehensive Efficiency and Its Decomposition in the City Clusters of the Middle Reaches of the Yangtze River from 2008 to 2021.

4.3.2. Characteristics of Spatial Distribution of Comprehensive Agricultural Efficiency

(1) The comprehensive agricultural efficiency of the city clusters along the middle reaches of the Yangtze River shows a distribution of "high in the west and low in the east". During the period of 2008–2021, the comprehensive agricultural efficiency of the urban agglomeration in the middle reaches of the Yangtze River has decreased from the west to the east generally, and in the dynamic change period of the comprehensive agricultural efficiency of the three major sub-city clusters, the regional agricultural production efficiency was relatively stable in general, but the difference between them increased. The average comprehensive efficiency level of agricultural production in the Wuhan city group is the highest, reaching 0.636. The average comprehensive efficiency level of the city cluster around Changsha city, Zhuzhou city, and Xiangtan city (Chang-Zhu-Tan city group) is

the second highest at 0.599; and the average comprehensive efficiency of agricultural production in the city cluster around Poyang Lake is the lowest at only 0.343. As can be seen from Figure 5, the Wuhan city group has the fastest growth rate in comprehensive agricultural production efficiency and has been overtaking the agricultural production efficiency of the Chang-Zhu-Tan city group since 2011. This may stem from the leading role of the Wuhan city group in the region, whose high-quality technological resources provide vitality to agricultural production. Figure 6 also shows that the pure technological efficiency of the Wuhan city group is significantly higher than that of the urban agglomeration around Poyang Lake and the urban agglomeration around Chang-Zhu-Tan, but the pure technological efficiency of the three regions has not significantly improved over time. The reason for this result may be the insufficient promotion of agricultural technology and the low transformation rate of agricultural science and technology. The difference in the pure technology efficiency of the three regions has gradually increased. In 2008, the pure technical efficiency was highest in the Chang-Zhu-Tan city group (0.97), followed by the Wuhan city group (0.938), and the city cluster around Poyang Lake is the lowest (0.909). Although none of the three reaches the pure technical efficiency optimum, they are all closer to 1. After 2008, the pure technical efficiency of the city cluster around Chang-Zhu-Tan continued to decline, reaching its lowest value in 2013, and then stayed at 0.8–0.9. The pure technical efficiency of the urban agglomeration around Poyang Lake is in a declining trend, but the rate of decline is slower than that of the urban agglomeration around Chang-Zhu-Tan, causing it to surpass the urban agglomeration around Chang-Zhu-Tan in 2013. The scale efficiency of all three regions is low, and the scale efficiency value does not even reach 0.8; the scale efficiency of the Poyang Lake city group is much lower than that of the other two regions, while there is not much difference between the scale efficiency of the Wuhan city group and the Chang-Zhu-Tan city group. Due to the low scale efficiency of the city cluster around Poyang Lake, the comprehensive efficiency of agricultural production in this region is the lowest compared with the other two regions. See Figure 7 for details. The average comprehensive agricultural productivity and its decomposition in all these cities above can be found in Figure 8. It can be seen that the non-scale of land is one of the important factors constraining the improvement of agricultural productivity;

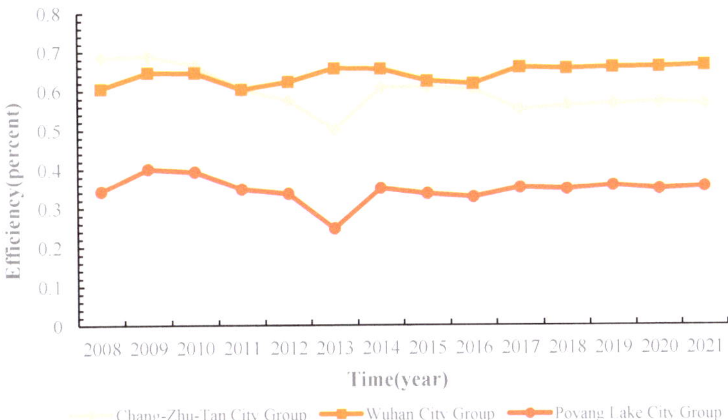

Figure 5. Evolutionary trend of comprehensive agricultural efficiency in urban agglomerations in the middle reaches of the Yangtze River, 2008–2021.

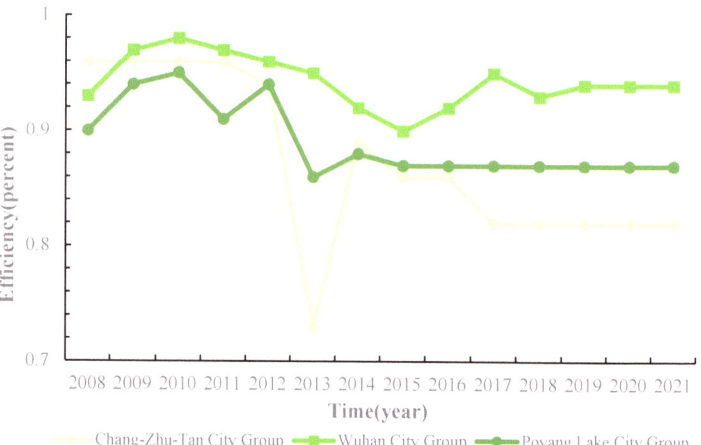

Figure 6. Evolutionary trend of pure technical efficiency in agriculture in the city cluster in the middle reaches of the Yangtze River, 2008–2021.

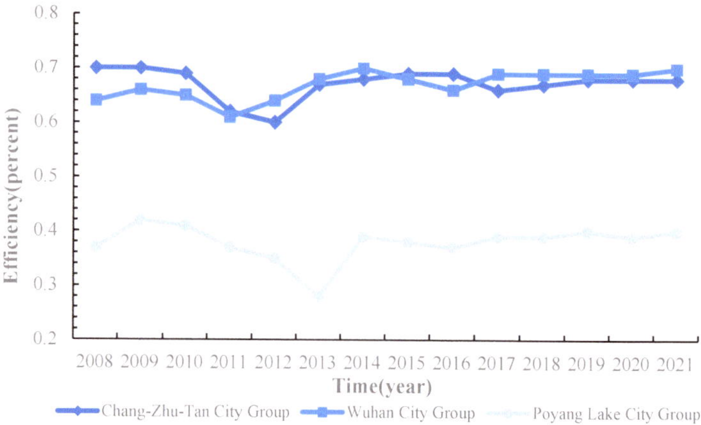

Figure 7. Evolutionary trend of agricultural scale efficiency in urban agglomerations in the middle reaches of the Yangtze River, 2008–2021.

(2) The inter-regional efficiency differences in comprehensive agricultural efficiency in the middle reaches of the Yangtze River sub-city cluster are obvious, and the gap is gradually increasing. Referring to the related study [47], the comprehensive efficiency of agricultural green production in the city clusters along the middle reaches of the Yangtze River is classified into four types, that is, the high-efficiency level (0.8E $\leq$ 1), the medium–high-efficiency level (0.6E $\leq$ 0.8), medium–low-efficiency level (0.4E $\leq$ 0.6), and the low-efficiency level (0.4 $\leq$ E) (see Figure 9). From the micro level, the cities with higher comprehensive efficiency of agricultural production are mainly in the Wuhan city group, in which Wuhan, Yichang, Xiangyang, and other places have higher comprehensive efficiency of agriculture, belong to the high-efficiency type, and have the "continuation effect" of high efficiency of agricultural green production. The cities with low efficiency of agricultural production are mainly in the Chang-Zhu-Tan city group and the Poyang Lake city group. The low-efficiency cities of agricultural production are mainly gathered in the city group around Chang-Zhu-Tan and the city group around Poyang Lake, in which the high-efficiency areas in the city group around Chang-Zhu-Tan are gradually reduced to

medium–high-efficiency areas, and there is a tendency of increase in the low-efficiency areas. Overall, with the passage of time, the high-efficiency area of agricultural green production gradually concentrated to the Wuhan city group, and the high-efficiency area gradually decreased, showing the trend of sinking to low efficiency.

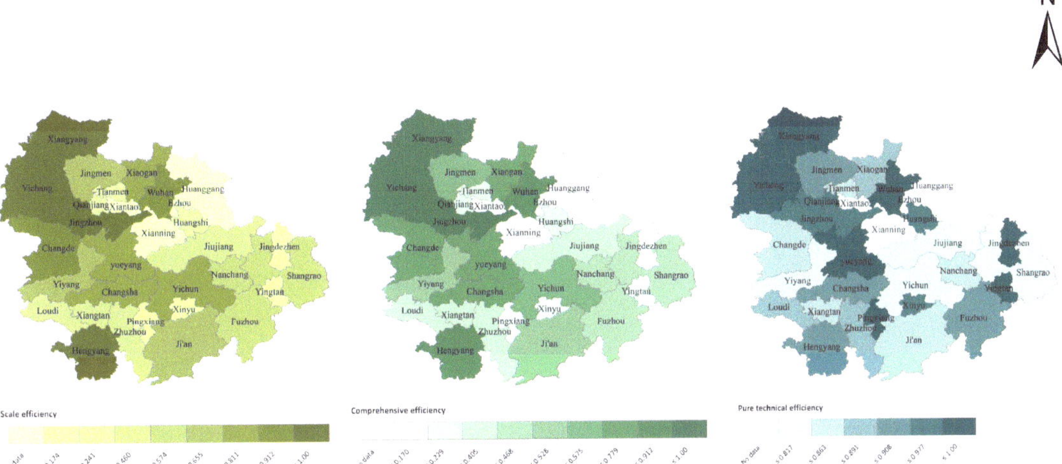

Figure 8. Average Comprehensive Agricultural Productivity and its Decomposition in Cities of the City Cluster in the Middle Reaches of the Yangtze River, 2008–2021.

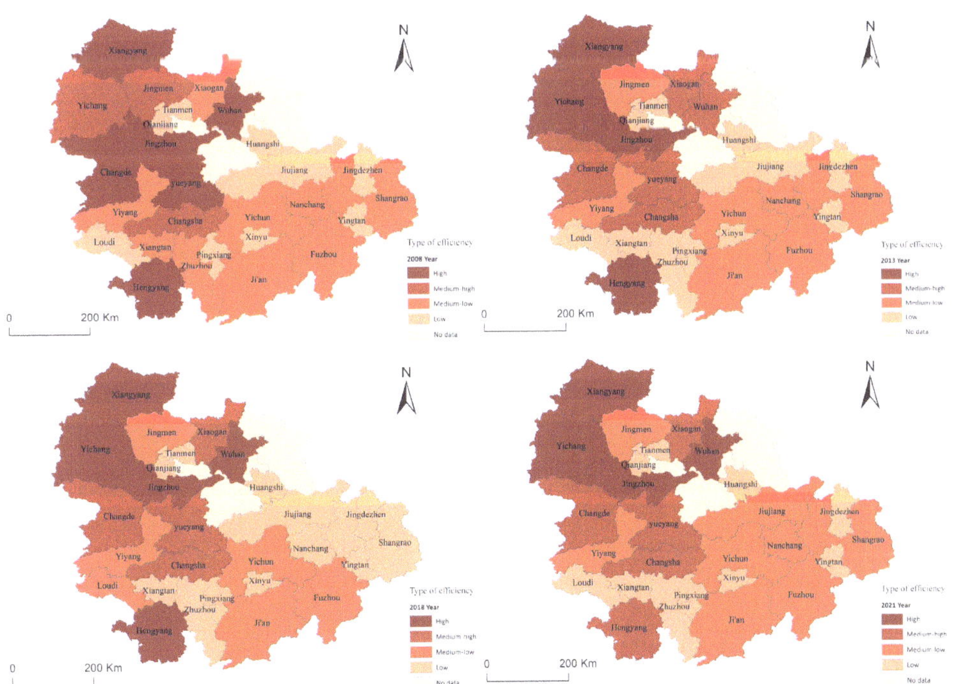

Figure 9. Clustering of comprehensive agricultural efficiency in cities of the city cluster in the middle reaches of the Yangtze River in 2008–2021.

5. Conclusions and Recommendations

5.1. Conclusions

Based on the input–output index system of agricultural green production, this paper uses the non-expected output SBM model and SFA model to measure the comprehensive agricultural efficiency and the external factors of the 31 prefectural-level cities in the middle reaches of the Yangtze River city clusters in the years 2008–2021. The paper draws the overall trend and distribution characteristics of agricultural comprehensive efficiency in the region and identifies important factors affecting agricultural production factor inputs. The specific conclusions are as follows:

(1) Without eliminating the influence of environmental variables, the comprehensive efficiency of agriculture in the city clusters along the middle reaches of the Yangtze River shows an upward–declining–steady trend, with obvious stage changes. The scale effect of green agricultural production is good, and the level of pure technical efficiency is lower than the scale efficiency. As a whole, the comprehensive efficiency level of the city cluster in the middle reaches of the Yangtze River is not high. The scale of production factor inputs such as agricultural land is more reasonable, but its pure technical efficiency level is low, and there exists inefficient utilization of factor resources;

(2) The results of the SFA model show that economic development, urbanization construction, industrial structure, and government influence have important impacts on agricultural production factor inputs. Accelerating economic development is conducive to reducing excessive inputs of agricultural capital, agricultural labor, sown area, agricultural film, irrigation, etc. Increasing the level of urbanization can promote the effective allocation of sown area and effective irrigated area but results in the redundancy of inputs of agricultural capital, agricultural labor, and pesticides. Increasing the level of industrial development can promote the reasonable inputs of sown area and agricultural film while also leading to the inputs of agricultural capital, agricultural labor, pesticides, pesticides, and effective irrigation area. Expanding the influence of the government can inhibit the excessive use of pesticides, but government subsidies make the input redundant in agricultural capital, labor, fertilizers, and effective irrigation area. It can be seen that environmental variables have an impact on the inputs of agricultural production factors and further affect the agricultural production efficiency. Therefore, it is necessary to eliminate environmental variables to examine the true level of production efficiency;

(3) The results after eliminating environmental variables show that, from the perspective of time evolution, the comprehensive efficiency of agriculture is low and presents an overall slow decline trend. The production scale efficiency and pure technical efficiency have not reached a better level, with the former significantly lower than the latter, and the low scale efficiency hinders the improvement of the comprehensive efficiency of agriculture. For the spatial distribution, the comprehensive efficiency of agriculture in the city cluster in the middle reaches of the Yangtze River shows the distribution of "high in the west and low in the east". With the passage of time, there are obvious differences in the efficiency of the comprehensive efficiency of agriculture, and the gap between the regions is gradually increasing. The high-efficiency area is gradually gathering to the Wuhan city group and shows the trend of increasing to the low-efficiency area;

(4) The above research conclusions clarify the impact of various factors on the comprehensive efficiency of agriculture in the Yangtze River Basin, pointing out the direction for the sustainable development of agriculture in the region in the future. In the future, attention should be paid to the reasonable investment of agricultural labor, pesticides, telephone bills, etc., while meeting the needs of agricultural development and avoiding excessive investment.

5.2. Policy Recommendations

Based on the above conclusions, the following recommendations are put forward to improve the comprehensive efficiency of agriculture in the city clusters along the middle reaches of the Yangtze River and guide the transformation and upgrading of regional agricultural production:

Firstly, decision makers must optimize the structure of government support for agriculture, focusing on agricultural infrastructure construction and agricultural green production support. The policy of financial support for agriculture can improve the conditions of farmland, water conservancy, production roads, and other infrastructures as well as strengthen the construction of agricultural production service teams, integrate agricultural production resources, improve the efficiency of resource utilization, and provide a strong guarantee for agricultural production. Financial support for agriculture should be appropriately tilted to the farmland fertility protection subsidies; decision makers should optimize the purchase of agricultural machinery subsidies, carry out training for agricultural professionals, etc., and subsidies should be concentrated to the advantageous production areas. We suggest increasing investments in scientific and technological research and development, supporting the construction of agricultural science and technology innovation centers, guiding farmers to gradually adopt new technologies, and improving the level of sustainable agricultural development. At the same time, it is necessary to strengthen the protection of agricultural ecological environment, promote the coordinated development of agriculture and ecological environment, implement agricultural ecological restoration projects, protect arable land resources, and reduce the negative impact of agricultural production on the environment.

Secondly, we suggest improving the research and development and popularization of green production technology and encouraging the establishment of resource use and recycling. Decision makers should vigorously develop water-saving irrigation techniques and green pest control techniques to reduce resource consumption and environmental pollution, improve the phenomenon of "heavy use, light recycling", and reduce the cumulative effects of "reverse ecology". Strengthening the construction of a multi-level agriculture ecological cycle linking the planting, animal husbandry, breeding, production and processing, and tourism industries and constructing an ecological cycle industry chain can help to realize a closed-loop flow of resources. We recommend strengthening the construction of ecological agriculture demonstration zones and promoting sustainable agricultural production models such as organic agriculture, ecological planting, and the utilization of agricultural waste resources. At the same time, decision makers should establish a sound ecological agriculture evaluation system, monitor and evaluate the ecological benefits of demonstration areas, promote successful experiences, form replicable and promotable models, and promote the development of ecological agriculture throughout the country.

Thirdly, under industrial restructuring, the population of farmers moving to urban areas should be absorbed in a rational and orderly manner. Through the establishment of a social security system and the formulation of appropriate supportive and preferential policies, problems such as social security for landless peasants will be solved, thereby promoting the flow of urban and rural factors, raising the level of intensification of agricultural production, promoting the industrialization of agriculture, and raising the level of agricultural production efficiency. But we cannot ignore the issue of rural development. We should strengthen the construction of rural infrastructure including roads, water supply, electricity, etc., to improve the living conditions of rural residents as well as strengthen rural education and training, improve the skill level of farmers, and enhance their ability to adapt to urban life.

5.3. Prospect

In future research, we will make improvements in variable selection to enhance the accuracy and rigor of the article. For example, currently, the calculation method for non-point source pollution is based on equal weight assignment, and suitable methods have not yet been found to measure the varying degrees of far-reaching effects caused by pesticides, agricultural films, and fertilizers. We will look for better ways to solve this problem in the future.

Next, we will continue to study the relevant issues of agricultural comprehensive efficiency and add cities in the Yellow River Basin as new research objects. Compared to the Yangtze River Basin, the Yellow River Basin itself has poor ecological conditions, high sediment content, and a poor ecological environment. One of the biggest obstacles to agricultural development is the shortage of water resources [48]. In addition, the urban agglomeration in the Yangtze River Basin is economically developed but has uneven development, with the phenomenon of industry feeding agriculture. However, the agricultural development in the Yellow River Basin lacks this advantage [49], and due to a series of factors such as terrain, cultural customs, policies, etc., the comprehensive efficiency of agriculture in the Yellow River region may present a completely different picture. Based on this, we can further explore the impact of mutual support between industry and agriculture on the comprehensive efficiency of agriculture.

Author Contributions: Conceptualization, L.W.; data curation, Y.Z.; formal analysis, L.W.; investigation, J.X. and W.Z.; methodology, L.W.; resources, L.W.; software, Y.Z. and Z.W.; validation, J.X. and W.Z; visualization, Y.Z. and Z.W.; writing—original draft preparation, L.W.; writing—review and editing, Y.Z. and J.X. All authors have read and agreed to the published version of the manuscript.

Funding: This research was supported by the Fundamental Research Funds for the Central Universities (B230207006), General project of the National Social Science Foundation (19BGL176), and Shaanxi Provincial Social Science Foundation Commissioned Project (2021WT09).

Institutional Review Board Statement: Not applicable.

Data Availability Statement: The data presented in this study are available on request from the corresponding author.

References

1. Wang, J.N.; Luo, C.P. The Outline of China's Regional Coordinated Development Strategy in New Era. *Reform* **2017**, *30*, 52–67.
2. Wu, Q.; Ma, H.L.; Wu, L.; Lu, J. Spatial-temporal pattern evolution of agricultural economy of urban agglomeration in the middle reaches of the Yangtze River. *Guangdong Agric. Sci.* **2016**, *43*, 152–158.
3. Lei, S.H.; Tian, X.; Wang, C. Decomposition of the spatial effect of agricultural resource mismatch with agricultural green production efficiency. *Areal Res. Dev.* **2023**, *4*, 124–129. [CrossRef]
4. Sun, W.L.; Wang, R.B.; Jiang, Q.; Huang, S.N. Study on connotation and evaluation of the agricultural green development. *J. China Agric. Resour. Reg. Plan.* **2019**, *40*, 14–21.
5. Gong, Q.W.; LI, X.M. Construction and Measurement of Agricultural Green Development Index: 2005–2018. *Reform* **2020**, *311*, 133–145.
6. Xie, S.; Kuang, Y.; Huang, N.; Zhao, X. Study on construction and application of evaluation index system of low-carbon agriculture in Guangdong. *Ecol. Environ. Sci.* **2013**, *22*, 916–923.
7. Xin, L.; Hu, Z. Evaluation of Agriculture Sustainable Development Level in China. *J. Agric. Sci. Tech.* **2015**, *17*, 135–142.
8. Charnes, A.; Cooper, W.W.; Rhodes, E. Measuring the efficiency of decision making units. *Eur. J. Oper. Res.* **1978**, *2*, 429–444. [CrossRef]
9. Restuccia, D.; Yang, D.T.; Zhu, X.D. Agriculture and aggregate productivity: A quantitative cross-country analysis. *J. Monet. Econ.* **2008**, *55*, 234–250. [CrossRef]
10. Vollrath, D. Land Distribution and International Agricultural Productivity. *Am. J. Agric. Econ.* **2007**, *89*, 202–216. [CrossRef]
11. Ruttan, V.W. Productivity Growth in World Agriculture: Sources and Constraints. *J. Econ. Perspect.* **2002**, *16*, 161–184. [CrossRef]
12. Li, Q.N. A study of the driving factors of agricultural production efficiency in China. *J. Chongqing Univ. (Soci. Sci. Edit.)* **2015**, *21*, 37–46.
13. Xing, H.R.; Zhang, X.J.; Deng, Y. Empirical analysis on the relationship between agricultural production efficiency and its influencing factors—Based on the data of hubei province. *Chin. J. Agric. Res. Reg. Plan.* **2016**, *37*, 198–203.
14. Zhang, S.H.; Jiang, W. Energy Efficiency Measures: Comparative Analysis. *J. Quan. Technol.* **2016**, *33*, 3–24.
15. Li, Q.; Li, G.; Yin, C.; Liu, F. Spatial Characteristics of Agricultural Green Total Factor Productivity at County Level in Hebei Province. *J. Ecol. Rural Environ.* **2019**, *35*, 845–852.
16. Guo, S.Y.; Zhang, M.; Wu, G. Study on agricultural production efficiency and its influencing factors in Heilongjiang rec-lamation area. *J. Northeast Agric. Univ.* **2019**, *50*, 68–75.
17. Guo, J.H.; Ni, M.; Li, B. Research on Agricultural Production Efficiency Based on Three-Stage DEA Model. *J. Quan. Techl.* **2017**, *27*, 27–38.

18. Lu, Q.; Meng, X. Research on the Agricultural Production Efficiency of Jiangsu Province Based on the Three-stage Data Envelopment Analysis (DEA) Model. *J. Northeast Agric. Sci.* **2021**, *46*, 94–99.
19. Wang, L.; Yu, C.; Wang, M.; Du, D. Study on Policy Effect and Spatial-temporal Difference of Agricultural Production Efficiency in China—Empirical Analysis Based on Three-stage DEA Model. *Soft Sci.* **2019**, *33*, 33–39.
20. Xv, Y.J. The Study on Environmental Efficiency and Environmental Total Factor Productivity in China's Industry. Hunan University: Changsha, China, 2012.
21. Seiford, L.M.; Zhu, J. Modeling undesirable factors in efficiency evaluation. *Eur. J. Oper. Res.* **2002**, *142*, 16–20. [CrossRef]
22. Kaoru, T. A Slacks-Based Measure of Super-Efficiency in Data Envelopment Analysis. *Eur. J. Oper. Res.* **2002**, *143*, 32–41. [CrossRef]
23. Lv, N.; Zhu, L. Study on China's Agricultural Environmental Technical Efficiency and Green Total Factor Productivity Growth. *J. Agrotech. Econ.* **2019**, *04*, 95–103.
24. Pan, D.; Ying, R. Agricultural eco-efficiency evaluation in China based on SBM model. *Acta Ecol. Sin.* **2013**, *33*, 3837–3845. [CrossRef]
25. Tian, W.; Yang, L.; Jiang, J. Measurement and Analysis of China's Agricultural Environmental Efficiency from a Low-Carbon Perspective: An SBM Model Based on Undesired Output. *Chin. Rural Surv.* **2014**, *5*, 59–71+95.
26. Xiao, Q.; Luo, Q.; Zhou, Z. Dynamic Evolution and Spatial Differentiation of Agricultural Green Production Efficiency in China: An Analysis Based on DDF-Global Malmquist-Luenberger Index. *J. Agric. For. Econ. Manag.* **2020**, *19*, 537–547.
27. Ge, P.; Wang, S.J.; Huang, X.l. Measurement for China's agricultural green TFP. *Chin. Popul. Resour. Environ.* **2018**, *28*, 66–74.
28. Meng, X.; Zhou, H.; Du, L.; Shen, G. The Change of Agricultural Environmental Technology Efficiency and Green Total Factor Productivity Growth in China: Re-examination Based on the Perspective of Combination of Planting and Breeding. *Issues Agric. Econ.* **2019**, *6*, 9–22.
29. Liu, S.; Zhang, H.Y.; Cai, W.J. Spatial Pattern and Dynamic Evolution of Agricultural Green Total Factor Productivity in the Yellow River Basin. *J. Ecol. Rural Environ.* **2022**, *38*, 1557–1566.
30. Cao, L.J. Analysis of Measurement of Agricultural Green Total Factor Productivity and Regional Heterogeneity in the Yangtze River Economic Belt. *Ecol. Econ.* **2024**, *40*, 95–102.
31. Xiao, Q.; Zhou, Z.; Luo, Q. Study on Agricultural Green Production Efficiency and Its Spatial-Temporal Differentiation Characteristics in the Yangtze River Economic Belt. *J. Chin. Agric. Resour. Region. Plan.* **2020**, *41*, 15–24.
32. Statistical Yearbook of Jiangxi Province. Available online: https://www.jiangxi.gov.cn/col/col424/?uid=45663&pageNum=1 (accessed on 2 February 2024).
33. Statistical Yearbook of Hunan Province. Available online: https://www.hunan.gov.cn/zfsj/tjnj/tygl.html (accessed on 2 February 2024).
34. China Statistical Yearbook. Available online: https://www.stats.gov.cn/sj/ndsj/ (accessed on 22 February 2024).
35. Statistical Yearbook of Hunan Province. Available online: https://tjj.hubei.gov.cn/tjsj/sjkscx/tjnj/qstjnj/index.shtml (accessed on 2 February 2024).
36. Qian, Z.; Liu, S. On the characteristics and drivers of spatial correlation network of green low-carbon agricultural production efficiency in the Yellow River Basin. *J. Arid Land Resour. Environ.* **2024**, *38*, 27–38.
37. Wu, X.R.; Zhang, J.B.; Tian, Y.; Li, P. Provincial Agricultural Carbon Emissions in China: Calculation, Performance Change and Influencing Factors. *Resour. Sci.* **2014**, *36*, 129–138.
38. Su, Y. Research on Carbon Emissions and Driving Factors of Farmland Use in Xinjiang. Master's Thesis, Xinjiang Agricultural University, Xinjiang, China, 2013.
39. Jiang, S.; Zhou, J.; Qiu, S. Whether Moderate Scale Operation Can Inhibit Agricultural Non-point Source Pollution: An Empirical Experiment Based on the Dynamic Threshold Panel Model. *J. Agrotech. Econ.* **2021**, *7*, 33–48.
40. Duan, H.P.; Zhang, Y.; Zhao, J.B.; Bian, X.M. Carbon Footprint Analysis of Farml and Ecosystem in China. *J. Soil Water Conserv.* **2011**, *25*, 203–208.
41. Wang, Y.; Gu, S. Analysis of electricity carbon footprint and its ecological stress in China from 2006 to 2015. *Acta. Sci. Circumstantiae* **2018**, *38*, 4873–4878.
42. Liu, N.; Han, Y.; Wang, P. Has FDI Increased the Productivity of Agricultural Firms in China? Evidence from Panel Data on 99,801 Agricultural Firms. *Chin. Rural Econ.* **2018**, *4*, 90–105.
43. Zheng, Y. The reasons for the strong dependence of agricultural transformation and upgrading on the government and its countermeasures: And the optimization and transformation of the form of agricultural organization. *Issues Agric. Econ.* **2016**, *37*, 4–8.
44. He, y.; Qi, Y.B. Study on the influence of urbanization development on the technical efficiency of grain production---based on the panel data of 13 major grain-producing areas in china. *J. Chin. Agric. Resour. Reg. Plan.* **2019**, *40*, 101–110.
45. Ding, X.; He, J.; Wang, L. Inter-provincial water resources utilization efficiency and its driving factors considering un-desirable outputs: Based on SE-SBM and Tobit model. *Chin. Popul. Resour. Environ.* **2018**, *28*, 157–164.
46. Yang, L.; Yang, F. "Nurturing" or "Syphoning": The Relationship between Industry and Agriculture in China's Re-source-Based Cities. *Contem. Econo. Manag.* **2019**, *41*, 1–9.
47. Zhu, L.X.; He, R.; Zheng, W.S.; Wang, S.; Han, L. Study on Spatial-Temporal Pattern and Driving Factors of Urban Innovation Efficiency of Urban Agglomeration in the Middle Reaches of Yangtze River. *Environ. Yangtze Val.* **2019**, *28*, 2279–2288.

48. Outline of the Plan for Ecological Protection and Quality Development of the Yellow River Basin. Available online: www.gov.cn/zhengce/2021-10/08/content_5641438.htm (accessed on 8 February 2024).
49. Zhang, B.; Miao, C. Spatiotemporal changes and driving forces of land use in the Yellow River Basin. *Resour. Sci.* **2020**, *42*, 460–473. [CrossRef]

Article

Evaluating Sustainable Alternatives for Cocoa Waste Utilization Using the Analytic Hierarchy Process

Natalia Andrea Salazar-Camacho [1], Liliana Delgadillo-Mirquez [2], Luz Adriana Sanchez-Echeverri [1,*] and Nelson Javier Tovar-Perilla [2]

[1] Facultad de Ciencias Naturales y Matemáticas, Universidad de Ibagué, Ibagué 730003, Colombia; natalia.salazar@unibague.edu.co
[2] Facultad de Ingeniería, Universidad de Ibagué, Ibagué 730003, Colombia; liliana.delgadillo@unibague.edu.co (L.D.-M.); nelson.tovar@unibague.edu.co (N.J.T.-P.)
* Correspondence: luz.sanchez@unibague.edu.co; Tel.: +57-3157419058

Abstract: Cocoa production has emerged as an effective agricultural strategy to reduce conflict in Colombia, transitioning from coca to cocoa cultivation. While this shift has provided economic benefits, it has also resulted in the generation of substantial cocoa by-products. Although there are various alternative methods of utilizing these by-products, many farmers are unaware of them, and others lack the necessary tools to determine which alternative is the best to pursue. This study sought to explore sustainable options for cocoa waste utilization through the application of the Analytic Hierarchy Process (AHP). By employing technological surveillance, viable options for reusing cocoa residues were identified. The AHP results indicate that pellet production is a promising alternative for rural communities. It is also a potential source of energy that could address the community's need for alternative energy sources. Initially, other energy production alternatives were not explored. However, in response to the AHP findings, this study also explored the use of cocoa waste combined with animal manure for energy generation through anaerobic digestion.

Keywords: cocoa waste; renewable energy; Analytic Hierarchy Process; rural development

Citation: Salazar-Camacho, N.A.; Delgadillo-Mirquez, L.; Sanchez-Echeverri, L.A.; Tovar-Perilla, N.J. Evaluating Sustainable Alternatives for Cocoa Waste Utilization Using the Analytic Hierarchy Process. *Sustainability* **2024**, *16*, 7817. https://doi.org/10.3390/su16177817

Academic Editors: Fotios Chatzitheodoridis, Efstratios Loizou and Achilleas Kontogeorgos

Received: 30 July 2024
Revised: 28 August 2024
Accepted: 30 August 2024
Published: 8 September 2024

1. Introduction

Cocoa production stands out as an agricultural strategy for reducing conflict in Colombia. This has become especially clear since 2010, when the process of replacing coca crops with cocoa began [1]. Cocoa is profitable due to its high global demand, making it a suitable crop to replace illicit crops. This strategy has enabled many Colombian families to generate employment in rural areas and obtain legal economic income. In 2022, cocoa production reached 62,158 tons, with the department of Santander being the largest producer nationally with 36.8%, followed by Arauca with 16.9%, Antioquia with 8.3%, Tolima with 5.8%, Huila with 5.7%, and Nariño with 5.4% [2]. Tolima ranks fourth due to its favorable agroclimatic conditions, which give its crops a competitive advantage, facilitating their access to international markets [3].

Cocoa bean exports decreased from 11,309 tons in 2021 to 5721 tons by November 2022, a drop of 50.5%. However, exports of semi-processed and processed products increased by 21%. These figures reflect how, in recent years, the transformation of cocoa in Colombia has been promoted, generating greater income opportunities for entrepreneurs and the country. Nevertheless, this transformation has also increased the generation of cocoa fruit residues, which require proper disposal or comprehensive utilization for their valorization and the development of new intermediate or final products.

In rural areas, it is common for cocoa pod husks to be used as fertilizer for crops. However, excess husks in the soil can have negative effects, causing diseases in cocoa trees and phytosanitary problems, which can result in significant crop losses due to improper waste management and the proliferation of pests like the cocoa pod borer (CPB) [4]. Inadequate disposal of these residues raises environmental concerns due to the possible spread of diseases and bad odors [5,6]. Despite the existence of valuable applications for agricultural waste that can add significant value, many communities lack sufficient information on how to implement these applications effectively.

In contexts where multiple options are available, the decision-making process becomes intricate due to the diverse criteria that can affect the selection of the best alternative. Social, economic, technological, and environmental considerations often conflict, making it important to balance these competing factors when making decisions [7]. Multi-Criteria Analysis, also known as Multi-Criteria Decision Making (MCDM) or Multi-Criteria Decision Aid (MCDA) methods, addresses the challenge of making decisions that involve multiple criteria. These approaches can accommodate both quantitative and qualitative criteria and are characterized by conflicts among the criteria and challenges in designing or selecting alternatives [8]. In general, Multi-Criteria Decision Making (MCDM) methods are categorized into two main types: Multi-Objective Decision Making (MODM) and Multi-Attribute Decision Making (MADM). The key difference between these two categories lies in how alternatives are determined within the decision-making process. In MODM, alternatives are not predetermined, while in MADM, alternatives are predetermined and evaluated against a set of attributes [9]. In general, Multi-Criteria Analysis (MCA) is a good tool, particularly in areas and sectors where methods based on a single criterion prove inadequate, and where some impacts cannot be accurately quantified in monetary terms [10,11].

In the field of multi-criteria decision-making (MCDM), various methods are commonly employed, each offering unique advantages and facing specific limitations. Among the most widely used is the Analytic Hierarchy Process (AHP), valued for its ability to decompose complex decisions into a hierarchical structure, facilitating the integration of both qualitative and quantitative criteria [12]. Another popular method is the Technique for Order Preference by Similarity to Ideal Solution (TOPSIS), which has a straightforward approach to ranking alternatives based on their proximity to an ideal solution [13]. Additionally, PROMETHEE (Preference Ranking Organization Method for Enrichment Evaluation) stands out for its effectiveness in managing outranking problems, making it particularly useful in situations with conflicting criteria. Its visual tools, such as GAIA planes, provide decision-makers with an intuitive means of analyzing alternatives, enhancing the overall decision-making process [14].

Among multi-criteria decision-making (MCDM) methods, the Analytic Hierarchy Process (AHP) has been widely applied in the fields of agroindustry and agriculture, proving to be an important tool for decision-making in complex scenarios involving multiple criteria. In the agroindustrial sector, AHP has been used to optimize supply chain management and to select the most suitable technologies for processing agricultural products. For example, AHP has been employed to prioritize alternative production technologies, considering factors such as cost, efficiency, and environmental impact [15].

Similarly, in the agricultural sector, AHP has been used to support decisions related to crop selection and resource allocation. It has also been applied to determine the most appropriate crop rotation systems, considering soil conditions, water availability, and market demand [16]. Additionally, in the area of sustainable agriculture, AHP has been used to evaluate and prioritize water management strategies that balance economic viability with environmental sustainability [17].

The widespread use of AHP is grounded in its greater flexibility in handling subjective judgments compared to other methods, such as TOPSIS or PROMETHEE. This is a significant feature when working with criteria defined by local communities.

The utilization of cocoa waste can be approached using the AHP methodology, which can help optimize decision-making to maximize environmental, economic, and social benefits. The Analytic Hierarchy Process (AHP) is a structured technique used for organizing and analyzing decisions. AHP helps decision-makers set priorities and make optimal choices by breaking down complex decisions into a series of pairwise comparisons, then synthesizing the results to include both qualitative and quantitative criteria [18]. The utilization of cocoa waste can be addressed using the AHP methodology, which aids in optimizing decision-making to maximize environmental, economic, and social benefits. This approach not only enhances waste management efficiency but also promotes sustainable practices and improves the quality of life for the communities involved.

Recent studies have highlighted the utilization of cocoa waste in the production of biofuels in solid, liquid, and gaseous forms. According to Lu et al. (2018), the cocoa pod husk constitutes 75% of the total weight of the cocoa fruit. It is the main byproduct generated from the cocoa process and, due to its lignocellulose content, can be used as a biofuel source [19]. For example, the use of cocoa pod husks in biochar fuel production at relatively low pyrolysis temperatures resulted in a bioresource with a calorific value of 17.8 MJ/kg and a high potassium content, resembling lignite [20]. Furthermore, the pyrolytic conversion of cocoa pod husks at different temperatures (300 to 600 °C) produced a bio-oil with a calorific value of 36.23 MJ/kg, similar to diesel [21]. On the other hand, cocoa residues are a good source for bioethanol production. Fermentation using *Zymomonas mobilis* with an initial microorganism dose of 14% v/v achieved an alcohol content of 10.62% on the 8th day of the process [22] (Billah et al., 2020). Additionally, some research has shown that anaerobic digestion could be an effective process for treating cocoa waste despite its low biodegradability (0.41) due to its high lignin content [23] (Antwi et al., 2019). In these cases, chemical and hydrothermal pretreatments have been used to improve biogas performance [23,24]. Moreover, Rojas et al. (2020) demonstrated at a laboratory scale the production of hydrogen from cocoa mucilage using dark fermentation processes [25]. The maximum hydrogen production was achieved under mesophilic conditions because the hydrolysis stage was longer. In this context, cocoa waste has the potential to be a biomass source for biofuels due to its chemical composition. However, the small quantity of cocoa waste produced in agricultural units may pose a limitation for these technologies.

The use of waste in the cocoa supply chain to produce energy production not only adds value to the waste but also enhances the quality of life for community members. Some technologies can add value to these solid residues by transforming them into renewable energy through thermochemical processes such as combustion [26] and biochemical processes such as anaerobic digestion [27]. Combustion, which transforms biomass into heat in the presence of oxygen, may be limited by economic and technical factors at the local level [28], as it requires physical conditioning of the residue, such as drying and pelletizing. However, there are other processes that can produce energy with minimal costs. Anaerobic digestion (AD) is a biochemical process in which a microbial consortium, in the absence of oxygen, promotes the transformation of available organic matter into by-products such as biogas and nutrient-rich fluids [29]. By using low-cost technologies like biodigesters [30], the organic load and environmental impact of the residues can be reduced, generating valuable by-products that can improve the socioeconomic conditions of rural communities in low-income countries [31]. Therefore, the proper use of cocoa fruit residues and other residues generated in production units not only benefits soil health but also boosts the economy of producers.

This study conducts an analysis in a cocoa-producing community in Tolima, examining their harvesting and post-harvesting techniques. Subsequently, a hierarchical analysis is applied to identify alternative uses for the waste. Based on the results of the AHP, a thermochemical energy production alternative was prioritized. However, in this initial analysis, other types of energy generation were not considered relevant and, therefore, were not included in the AHP evaluation. Nevertheless, in response to the AHP findings, this study also explored, in parallel, the potential of cocoa husks and other residues produced in the production units as sources for biochemical energy generation, aiming to utilize organic waste. The study demonstrates an application for utilizing cocoa waste in energy generation through biodigesters.

2. Methodology

2.1. Characterization of the Production Process in the Region

A survey was designed and administered to farmers from the productive units in a community in Tolima with the aim of investigating agricultural practices throughout the cocoa production process. The survey focused on elements related to the cultivation, management, and handling of the products and by-products generated, as well as the stages and conditions of the cocoa transformation process carried out by the surveyed farmers.

For the survey design, the categorical analysis technique was employed, evaluating three specific categories: general information on cocoa cultivation, general practices and conditions of cocoa processing, and the identification of weaknesses in cocoa cultivation practices among farmers.

2.2. Analytical Hierarchy Process (AHP) Method

The Analytical Hierarchy Process (AHP) is a multi-criteria decision analysis (MCDA) method that uses ratio scales to measure performance on the considered criteria and the importance of these criteria [25]. The process can be divided into three main stages:

Hierarchical structure: In this stage, the criteria for evaluating the alternative solutions to the problem are defined.

Pair criteria comparison: In this stage, systematic comparisons between criteria and alternatives are made. Each comparison sets a relative value for each one of the criteria based on its importance.

Priority calculation: In this stage, a mathematical process is used to calculate the relative priorities of the criteria and alternatives based on the scores obtained in the comparison matrices. The result is a hierarchy of priorities that allows you to make informed and objective decisions.

In this study, the AHP method was used to determine which alternative methods of reutilizing cocoa wastes could have the most impact on the San Bernardo community. To identify alternatives for the use of residues of different cocoa wastes, technological surveillance was applied. The technological surveillance (TV) methodology developed for the exploration of alternatives was carried out using the technique of [32], which includes implementing a comprehensive network approach supported by the UNE 166006:2018 standard [33,34]. This standard considers communication to be one of the most important aspects of the management system and its use for the creation, storage, distribution, and utilization of information [35,36].

The Technological Surveillance model was designed to be divided into five sections (Figure 1).

Figure 1. Phases used for Technological Surveillance.

The first phase, called Identify, is used to determine the technologies to monitor, the information needs, and the key monitoring factors of interest to carry out a study that will help us design an effective strategy for each case. The second step, called Searching, consists of defining the search equations and the specialized databases to be used for the collection of information. This step allows the capture of relevant information and determines keywords and the different search parameters such as the period. After that, the step called Analysis corroborates that the information obtained in previous steps corresponds to the main focus of the search. In this case, the required information must respond to the requirements to identify physical and chemical procedures implemented to utilize cocoa by-products.

From downloaded information, the step called Valorization takes care of carrying out a depuration process that consists of exclusively extracting papers related to the products obtained from cocoa wastes. Some papers focus on the cocoa process and others on products from cocoa liquor. According to their content, papers were grouped into categories according to their stated alternative uses for cocoa wastes. In this way, analysis of the literature of leading authors and their contributions of research to produce valorization of cocoa wastes was conducted.

Finally, within the fourth phase, called Results Intelligence, the information about the articles was stored in a structured, categorized, and classified way. All of this allows the definition of alternatives to be analyzed in the Analytical Hierarchy Process (AHP) method.

2.3. Experimental Set-Up for Biogas Production of Cocoa Pond by Anaerobic Co-Digestion

The anaerobic co-digestion of cocoa pond shells with raw cow manure was evaluated in two concentration ratios: 50% C–50% M and 70% C–30% M. The assays were performed in amber batch reactors of 250 mL, with a load volume of 160 mL, at room temperature ($25 \pm 2\,^{\circ}\text{C}$, mesophilic range). A blank (cow manure content alone) was carried out to correct endogenous methane production. Biogas production was measured daily for 77 days. In all cases, the batch reactors were run in triplicate. The results were expressed as mean values and standard deviations. Standard statistical procedures were used, including a standard deviation, mean averaging, and absolute differences. The significance of the variance test was determined using variance analysis, with a significant level of 0.05 ($p \leq 0.05$) considered significantly different.

Cocoa pond shells and cow manure were collected from the communities of rural areas of Ibagué-Tolima (Colombia). Cocoa pond shells were chopped with a food processor into sections of 1 to 5 mm. These were stored in plastic bags at $4\,^{\circ}\text{C}$ for about 10 days. Cow manure was stored in a plastic bottle of 2 L for about 3 days at room temperature and in the absence of oxygen.

Biogas production was performed using water displacement equipment, with a measure range of 1 to 100 mL of biogas, and reported in a standard temperature and pressure (STP) condition. This equipment was built taking into account the principles of fluid pressure and the model suggested by Ojikutu and Osokoya (2014) and Esposito et al. (2012) [37,38].

Composite samples were analyzed for various physicochemical parameters, as Total Solids (TS), Volatile Solids (VS), moisture, and ash, according to Standard Methods for the Examination of Water and Wastewater [39].

For the kinetic model, the modified Gompertz model was applied to adjust the experimental biogas production (Equation (1)). This model has been applied in modeling methane and biogas production in the anaerobic digestion process of lignocellulose material [40,41]

$$\beta(t) = \beta_0 \cdot exp\left\{ -exp\left[\frac{\mu_m e}{\beta_0}(\lambda - 1) + 1 \right] \right\} \tag{1}$$

where $\beta(t)$ is the cumulative biogas yield (mL/gSV), β_0 is the final biogas production (mL/gSV), μ_m refers to the maximum biogas production rate (mL/gSV.d), λ is the lag phase time (d), e is equal to 2.72, and t means the anaerobic digestion time (d). Furthermore, the parameters (μ_m and λ) were estimated using a non-linear square method of Matlab R2019A (lsqnonlin function) for all tests.

3. Results and Discusion

3.1. Characterization of the Production Process in the Region

Colombian farmers follow the necessary steps to ensure high-quality cocoa pods, relying on experience and observation rather than standardized methods. After the harvest, the pods are opened, mucilage is extracted, and the fermentation process begins, typically taking several days. Once the beans reach the desired moisture level, they are dried under indirect sunlight and stored in sacks. Among the farmers surveyed, 60% are directly involved in farming, while 40% hold administrative roles. Cocoa trees range in age from 7 to 30 years, with newer trees under two years old. Cocoa is cultivated on plots ranging from 3.25 to 15 hectares, with productive areas between 0.5 to 4.5 hectares. The most common cocoa varieties are Criollo, Clonados, CCN51, and Hybrid/Trinitario. Peak harvests occur in June and November, while August and September are less favorable. The lack of irrigation systems leaves 80% of participants vulnerable to drought-related losses, and land slippage during the rainy months is also a significant challenge.

Cocoa farmers follow traditional practices for harvesting, fermentation, and processing. They prioritize pod maturity during harvest, with only 20% classifying beans by size. Fermentation completion is judged by appearance, duration, texture, temperature, and smell, without pH measurement. Drying takes 3–15 days depending on climate, with wood being the most common container material. About 40% of farmers outsource roasting, shelling, and grinding, while the rest rely on traditional methods using wood stoves and steel containers, guided by visual cues. Grinding usually requires two to three passes through manual or electric mills. Key challenges include a lack of potable water, reliance on low-tech post-harvest methods that hinder traceability, and the absence of a gas pipeline, which complicates some post-harvest processes.

3.2. Technological Surveillance and AHP Method

3.2.1. Search for Alternatives for the Use of Cocoa Waste

Table 1 shows the searching equations we used in databases. After carrying out the search in different databases with each of the equations described, a total of 1310 articles was obtained, of which only those whose publication date fell between the years 2015 and 2020 were included. It was also filtered by the type of document (scientific articles) and the type of access, resulting in 425 documents, of which those that were adapted to the case study valorization of cocoa residues in rural populations were selected. It is important to highlight that Latin American countries lead publications on the utilization of cocoa waste. In countries like Colombia, research on this topic has increased as many illicit crops have been replaced with cocoa crops.

Table 1. Equation used in databases for technological surveillance.

Equation (1)	Cocoa OR mucilage AND uses
Equation (2)	Cocoa And agribusiness OR harness OR pod husk
Equation (3)	Wastes And cocoa AND uses
Equation (4)	Cocoa OR cacaota AND Uses
Equation (5)	Cocoa OR Pod husk AND uses

3.2.2. Alternatives from Cocoa Waste

Once the documents were obtained, a list of uses for waste in the cocoa production process was compiled, presenting alternatives to add value to the by-products. This information was then filtered based on the alternatives that could be implemented in the community, focusing on cocoa pod husks and mucilage, which according to community interviews are the most problematic waste products. Five alternatives were selected that use mucilage or cocoa pod husks and are suitable for the community's conditions: pellets, flour, pectin, vinegar, and alcoholic beverages.

3.2.3. Selection Criteria

The selection criteria were determined according to the context and needs of the community. The selection criteria are detailed below:

- Availability: This criterion evaluates the availability of raw material, referring to the amount of waste generated. Since the evaluated alternatives are only produced from two types of waste: mucilage and cocoa pod husks, the highest score was given to products with a high content of cocoa pod husks, as this is the most generated waste in the studied community.
- Implementation: This criterion evaluates the equipment needed to produce each product, aiming for low investment requirements. To evaluate this, we identified the necessary equipment for each alternative through previously conducted technology surveillance. The alternatives were rated on a scale from 1 to 5, with 1 being the lowest investment and 5 the highest. The ranking of these alternatives was done with consideration of the limited resources available to the community involved in the study (Table 2).
- Commercialization: This criterion was evaluated through two sub-criteria: price and demand. For the price, products that generate the most profit upon sale were considered. This profit is assumed based on the market price value. The prices of the products were obtained through a search in various physical and online markets for similar or identical products available in the market (Table 3). For the evaluation of this criterion, the prices of each product are organized from highest to lowest, with 1 being the highest price and 5 being the lowest price.

Table 2. Equipment used in the manufacturing of each selected alternative and their respective score on the Saaty scale.

Alternative	Equipment	Value
Pellets	Mill, sieve, hydraulic press, electronic caliper, digital scale, compression machine	2
Pectin	Dehydrator, mill	5
Flour	1.5 mm sheet, dehydrator	1
Wine	Alcohol meter, bottles, chemical supplies for fermentation	3
Vinegar	Yeast, bottles, chemical supplies for fermentation	4

Table 3. Marketing Price in (COP).

Alternative	Commercialization Price (COP)	Value
Pellets	5000	3.5
Pectin	36,000	1
Flour	25,000	2
Wine	5000	3.5
Vinegar	1700	5

To determine demand, companies that commercialize products identical or similar to the proposed alternatives were contacted to obtain information about the average national consumption (Table 4). The alternatives were rated on a scale from 1 to 5, based on the average national consumption, with 1 being the lowest consumption and 5 the highest.

- Environmental Impact: This criterion evaluates the degree of environmental impact generated by the studied residues on the soil. Farmers do not adequately dispose of cocoa pod husks, placing them in areas near the cultivation sites [44]. Due to their high lignin content, the decomposition rates of the matter are low [45], leading to the accumulation of residual biomass on the plantations. This accumulation can cause the spread of vectors [46], diseases in the crops, and a reduction in soil quality [47].

Mucilage and the fermentation process generate leachates that cause the leaching of bases such as calcium, magnesium, aluminum, potassium, and sodium, reducing soil pH and soil stability due to increased porosity [48]. Additionally, the interviewed cocoa farmers stated that "mucilage is a harmful liquid for the plants where this residue falls, to the point of causing the vegetation to disappear if not treated". For this reason, higher scores were given to alternatives that utilized a greater proportion of mucilage.

Table 4. Annual consumption of the selected alternatives.

Alternative	Annual Consumption	Reference	Value
Pellets	2,486,122.92 T/year	Castañeda & Arciniegas, 2019 [42]	5
Pectin	450–550 T/year	Pectinas de Colombia SAS	2
Flour	0.02 T/year	Harinas del Valle, 2015 [43]	1
Wine	12,888,000 L/year	Empresa Productora de Vino	4
Vinegar	10,000,000 L/year	Sierra Morena Zapatca	3

Once the criteria were defined, a comparison matrix was created. This matrix compares the criteria with each other based on Saaty's scale [31]. The comparison was performed by experts in value addition processes. Table 5 shows the comparison matrix of the criteria. After obtaining this comparison, the matrix was normalized, and the priority vector was determined.

Table 5. Comparison Matrix of Established Criteria.

Criteria	Availability	Investment	Price	Demand	Environmental Impact
Availability	1	0.11	0.14	0.14	1
Investment	9	1	3	3	5
Price	7	0.33	1	0.14	3
Demand	7	0.33	7	1	3
Environmental Impact	1	0.2	0.33	0.33	1
Total	25	1.98	11.48	4.62	13

After this comparison, the selection criteria for each alternative are evaluated. Using the results from these matrices, the weight for each criterion and the average weight value/priority vector are determined (Table 6).

Table 6. Weight by criterion and priority vector of the AHP model.

Criteria	Weight Criterion	Weight Criterion/Priority Vector
Availability	0.22	5.10
Investment	2.52	5.84
Price	0.84	5.27
Demand	2.06	6.83
Environmental Impact	0.35	5.44
Average		5.70

With the average obtained from Table 6, the consistency ratio (CR) is determined.

$$CR = \frac{CI}{RI}$$

where CI is the consistency index and RI is the randomness index, which are calculated using the following equations:

$$CI = \frac{P - n}{n - 1}$$

where P is the average of the weights with the priority vector and n number of criteria.

$$RI = \frac{1.98(n-2)}{n}$$

where n is the number of criteria.

The consistency index allows measuring the degree of consistency between the paired opinions provided by the evaluators. The degree of consistency should be <0.1 to proceed with the AHP process. Table 7 shows the calculation of CI, RI, and CR.

Table 7. Consistency Index, Randomness Index, and Consistency Ratio.

$CI = \frac{5.70-5}{5}$ $CI = 0.17$	$RI = \frac{1.98(5-2)}{5}$ $RI = 1.19$	$CR = \frac{0.17}{1.19}$ $CR = 0.1$

Since the consistency ratio is less than 0.1, it is possible to calculate the weighted score of the alternatives with the selection criteria. Table 8 shows the overall rating of the AHP process for the selected alternatives evaluated against the criteria.

Table 8. Prioritization matrix of the proposed alternatives for the utilization of cocoa waste.

	Availability	Investment	Price	Demand	Environmental Impact	Result
Pellets	0.30	0.26	0.50	0.50	0.05	0.36
Pectin	0.30	0.50	0.03	0.13	0.05	0.28
Flour	0.04	0.07	0.07	0.03	0.43	0.08
Wine	0.06	0.13	0.26	0.07	0.43	0.15
Vinegar	0.30	0.03	0.13	0.26	0.05	0.13
Priority Vector	0.04	0.43	0.16	0.30	0.06	

From the application of the AHP method, it can be observed that the Pellets alternative has the highest-ranking score, meaning it is the most beneficial alternative according to the established selection criteria.

3.3. Potential Use of Cocoa Waste as an Energy Source

Pellet production is a promising alternative due to its potential use in energy production, which would address the community's need for alternative energy sources. Initially, other energy production alternatives were not explored. Nevertheless, in response to the AHP findings, this study evaluated the possibility of producing biofuel that utilizes not only cocoa waste but also farm animal manure (pigs, cows, rabbits, etc.). This option proves to be doubly beneficial for the community, as using this type of fuel for basic needs such as cooking, which is still done with firewood today, would positively impact the health and well-being of the inhabitants. Therefore, an experimental design has been developed for the anaerobic co-digestion of cocoa waste and farm animal manure to determine their potential for biogas production.

Two different ratios of cocoa pond shell (C) with cow manure raw (M) (50/50 and 70/30, respectively) were monitored daily to determine their biogas production. The biogas curves are shown in Figure 2. In addition, statistical analyses of the data showed a significant difference ($p < 0.05$) between the ratios tested. Table 9 shows the physical characteristics of substrates utilized. The volatile solids fraction is high in both substrates, indicating a significant amount of organic matter potentially available for conversion into biogas by anaerobic microorganisms.

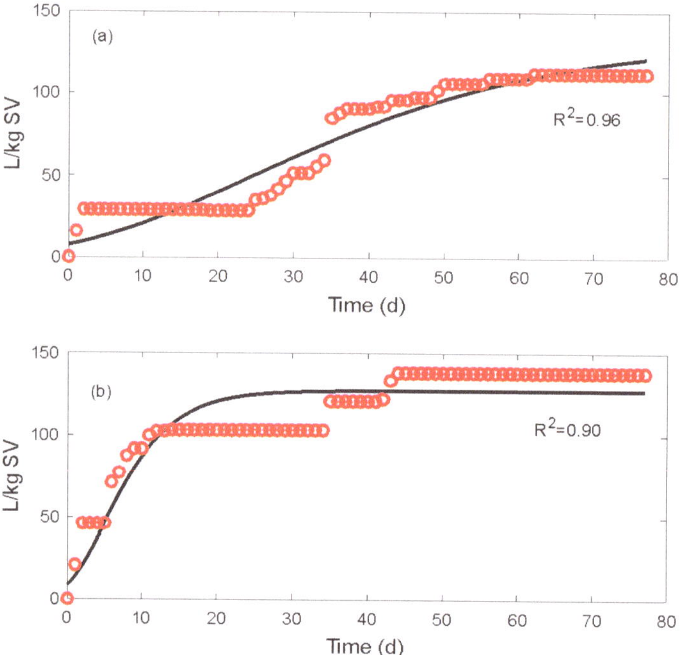

Figure 2. Cumulative biogas yield for different C/M ratios. Circles are experimental data and the solid line is the Gompertz model. (**a**) 50%/50% and (**b**) 70%/30%.

Table 9. Physical characteristics of substrates utilized. SV and Ash values in relation with dry matter.

Substrate	Moisture (%)	SV (g/100 g)	Ash (g/100 g)
Cocoa pond shell (Co)	87.09 ± 0.96	88.87 ± 0.82	11.13 ± 0.62
Cow manure (M)	82.42 ± 1.73	77.02 ± 6.93	22.98 ± 5.73

The results of the anaerobic co-digestion of the cocoa pond showed that the 70/30 concentration ratio produced slightly more biogas than the other one. However, both tests were positive for gas production. In general, the constituents originating from cocoa pond shells presented favorable characteristics for anaerobic biodegradation, with volatile solids between 63.2 and 92.6%. Similar results were obtained by Suhartini, et al. (2021) in their research on the co-digestion of cocoa pods with cocoa leaves for enhanced biogas production [48].

Overall, according to different authors, the key factors for the anaerobic digestion of lignocellulosic biomass could include the material composition, inoculum, temperature, TS content, carbon/nitrogen (C/N) ratio, particle size of the feedstock, concentration of inhibitors (e.g., ammonia), alkalinity of the system, and pretreatment methods [23,27,29,49].

These cocoa residues present an elevated lignocellulosic content. For example, Antwi et al. (2019) observed values of 26% cellulose and 21% lignin [23], and Acosta et al. (2021) detected 42.7% cellulose and 6.14% lignin [27]. These characteristics of lignocellulosic biomass are correlated with biogas yield; the correlation is positive with cellulose content and negative with lignin content [50]. In fact, these assays presented a major increase in biogas production when the cocoa pond was increased. This may be due to the higher substrate content and the choice of microorganisms used in the assays. In addition, the co-digestion strategy helps us to achieve a nutrient balance, an ideal carbon/nitrogen (C/N) ratio, which provides a synergistic effect to improve biogas yields [48,51]. Moreover,

Acosta et al., (2021) observed that the key contribution of manure co-digestion with cocoa seems to be the provision of buffering capacity, a source of nutrients, and potential key microorganisms [27]. According to Hagos et al. (2017), anaerobic co-digestion is the most used strategy to increase biogas production due to the lignocellulosic composition of cocoa waste, which has a complex chemical composition with recalcitrant characteristics, unfavorable to microbial activity [52].

According to Candia-García et al. (2018), the Gompertz model is useful to explain the lag time and sigmoidal growth curve of biogas production from the anaerobic digestion of lignocellulosic biomass [40]. Solid lines (Figure 2) show the model's results with a squared correlation coefficient (R2) close to unity, indicating a good fit in describing the biogas production in our assays. Table 10 shows the Gompertz model's parameters as obtained in the simulations. The lag phase time (λ) was short, with values of 0 and 1.4 d for cocoa pond shell/cow manure raw ratios of 50/50 and 70/30, respectively. Thus, microorganisms adapted quickly, synthesizing the necessary enzymes in anaerobic conditions. According to Kim et al. (2020) [53], in the anaerobic digestion of organic waste, a long lag phase can be caused by acidification due to the initial accumulation of volatile fatty acids (VFA) and a high VFA to volatile solids ratio. Our results indicate that co-digestion of cocoa pond shell and cow manure achieves a positive equilibrium with the substrate mixture, as evidenced by the short lag phase observed. The difference in the lag phase between the two treatments could be attributed to the amount of carbohydrates, which is higher in the 70/30 treatment.

Table 10. Gompertz model parameters.

Co/M Ratio	Final Biogas Production (β_0), mL/gSV	Maximum Biogas Production Rate (μ_m), mL/gSV.d	Lag Phase Time (λ), d	Squared Correlation Coefficient (R^2)
50/50	127.7	9.12	0	0.96
70/30	134.7	2.14	1.4	0.90

The maximum biogas production rate (m) was 9.12 and 2.14 mL/gSV.d and the final biogas production (0) was 127.7 and 134.7 mL/gSV, for 50/50 and 70/30, respectively. These results revealed that the 70/30 ratio could achieve better biogas production than the 50/50 ratio. Where cocoa pond shell was the primary substrate, biogas production was higher and faster, possibly because the anaerobic process was less sensitive to inhibition caused by intermediate compounds such as volatile fatty acids. Indeed, the 50/50 curve exhibited a stepwise pattern, indicating the presence of distinct and sequential hydrolysis processes among the substrates utilized. Generally, anaerobic co-digestion enhances biogas production when the components are combined in optimal ratios.

Indeed, a lower lag phase time and a higher maximum biogas production rate can mean a faster startup and a higher efficiency of anaerobic degradation [40]. Moreover, the co-digestion of cocoa waste and cow manure could significantly enhance biogas production rate when a suitable pretreatment is applied [23].

The results present the potential of cocoa residues for energy conversion through anaerobic co-digestion with raw cow manure. Anaerobic digestion (AD) is one of the most promising biotechnologies to degrade and convert lignocellulosic waste into bioenergy [29]. These findings are of key importance in the design and operation of anaerobic digestion reactors that have one agricultural residue as feedstock. Moreover, it can reduce the organic load and residue impact on the environment [49]. In fact, the digestate, a byproduct of anaerobic digestion, acts as a natural soil amendment, reducing the need for synthetic fertilizers [40]. Futhermore, AD technology represents an alternative method of reducing greenhouse gas emissions and improving the socioeconomic conditions of rural communities in low-income countries [31,54].

4. Conclusions

The application of the AHP methodology effectively identified the most promising alternatives for cocoa waste utilization, such as pellet production for energy generation. By prioritizing alternatives based on criteria relevant to the community, the method facilitates decision-making that aligns with local needs and resource availability.

The incorporation of not only economic criteria but also social and environmental factors into the AHP methodology ensured that the prioritization not only could optimize resource use and reduce environmental impact but also address the socioeconomic needs of the community.

Obtaining energy through anaerobic co-digestion using cocoa residues is possible and presents reasonable biogas yields. Indeed, raw cow manure is a suitable co-substrate in the co-digestion of cocoa waste, and both residues are common on agricultural farms.

The incorporation of anaerobic digestion into cocoa waste management strategies not only optimizes waste reuse for renewable energy but also offers socioeconomic benefits to rural communities, making it a viable and sustainable solution for agricultural waste valorization.

Author Contributions: Conceptualization, all authors; methodology, L.D.-M., N.A.S.-C. and L.A.S.-E.; validation, N.J.T.-P., L.D.-M. and L.A.S.-E.; writing—original draft preparation, all authors; writing—review and editing, N.J.T.-P. and L.A.S.-E.; project administration, L.D.-M. and N.A.S.-C. All authors have read and agreed to the published version of the manuscript.

Funding: This research was funded by Universidad de Ibagué under grant number 20_007_INT, which supported the project "Influence of the Handling and Operating Conditions of Artisanal Cocoa Beans on the Quality Attributes of Cocoa Liquor".

Data Availability Statement: Data is contained within the article.

Conflicts of Interest: The authors declare no conflicts of interest.

References

1. DNP Función Pública. *Documento CONPES 3654 de 2010*; Gestor Normativo; DNP Función Pública: Bogotá, Colombia, 2010.
2. Fedecacao. *Producción Cacaotera Presentó una Reducción del 10% en 2022 por Lluvias*; Fedecacao: Bogotá, Colombia, 2023.
3. Consejo Privado de Competitividad. *Informe Nacional de Competitividad 2019–2020*; Consejo Privado de Competitividad: Bogotá, Colombia, 2019.
4. Fidelis, C.; Rajashekhar Rao, B.K. Enriched Cocoa Pod Composts and Their Fertilizing Effects on Hybrid Cocoa Seedlings. *Int. J. Recycl. Org. Waste Agric.* **2017**, *6*, 99–106. [CrossRef]
5. Porto de Souza Vandenberghe, L.; Kley Valladares-Diestra, K.; Amaro Bittencourt, G.; Fátima Murawski de Mello, A.; Sarmiento Vásquez, Z.; Zwiercheczewski de Oliveira, P.; Vinícius de Melo Pereira, G.; Ricardo Soccol, C. Added-Value Biomolecules' Production from Cocoa Pod Husks: A Review. *Bioresour. Technol* **2022**, *344*, 126252. [CrossRef]
6. Mansur, D.; Tago, T.; Masuda, T.; Abimanyu, H. Conversion of Cacao Pod Husks by Pyrolysis and Catalytic Reaction to Produce Useful Chemicals. *Biomass Bioenergy* **2014**, *66*, 275–285. [CrossRef]
7. San Cristóbal, J.R. Multi-Criteria Decision-Making in the Selection of a Renewable Energy Project in Spain: The VIKOR Method. *Renew. Energy* **2011**, *36*, 498–502. [CrossRef]
8. Pohekar, S.D.; Ramachandran, M. Application of multi-criteria decision making to sustainable energy planning—A review. *Renew. Sustain. Energy Rev.* **2004**, *8*, 365–381. [CrossRef]
9. Caravaggio, N.; Caravella, S.; Ishizaka, A.; Resce, G. Beyond CO_2: A multi-criteria analysis of air pollution in Europe. *J. Clean. Prod.* **2019**, *219*, 576–586. [CrossRef]
10. Nautiyal, H.; Goel, V. Sustainability Assessment: Metrics and Methods. In *Methods in Sustainability Science*; Ren, J., Ed.; Elsevier: Amsterdam, The Netherlands, 2024; pp. 27–45.
11. Tovar-Perilla, N.J.; Bermeo-Andrade, H.P.; Torres-Delgado, J.F.; Gómez, M.I. Methodology to Support Decision-Making in Prioritization Improvement Plans Aimed at Agricultural Sector: Case Study. *DYNA* **2018**, *85*, 356–363. [CrossRef]
12. Saaty, T.L. *The Analytic Hierarchy Process*; McGraw-Hill: New York, NY, USA, 1980.
13. Hwang, C.L.; Yoon, K. *Multiple Attribute Decision Making: Methods and Applications*; Springer: Berlin, Germany, 1981.
14. Brans, J.P.; Vincke, P. A Preference Ranking Organization Method: The PROMETHEE Method for MCDM. *Manag. Sci.* **1985**, *31*, 647–656. [CrossRef]
15. Din, G.Y.; Yunusova, A.B. Using AHP for evaluation of criteria for agro-industrial projects. *Int. J. Hort. Agric.* **2016**, *1*, 6. [CrossRef]
16. Akinci, H.; Özalp, A.Y.; Turgut, B. Agricultural land use suitability analysis using GIS and AHP technique. *Comput. Electron. Agric.* **2013**, *97*, 71–82. [CrossRef]

17. Veisi, H.; Deihimfard, R.; Shahmohammadi, A.; Hydarzadeh, Y. Application of the Analytic Hierarchy Process (AHP) in a Multi-Criteria Selection of Agricultural Irrigation Systems. *Agric. Water Manag.* **2022**, *267*, 107619. [CrossRef]
18. Saaty, R.W. The Analytic Hierarchy Process—What It Is and How It Is Used. *Math. Model.* **1987**, *9*, 161–176. [CrossRef]
19. Lu, F.; Rodriguez-Garcia, J.; Van Damme, I.; Westwood, N.J.; Shaw, L.; Robinson, J.S.; Warren, G.; Chatzifragkou, A.; McQueen Mason, S.; Gomez, L.; et al. Valorisation Strategies for Cocoa Pod Husk and Its Fractions. *Curr. Opin. Green Sustain. Chem.* **2018**, *14*, 80–88. [CrossRef]
20. Tsai, C.H.; Tsai, W.T.; Liu, S.C.; Lin, Y.Q. Thermochemical Characterization of Biochar from Cocoa Pod Husk Prepared at Low Pyrolysis Temperature. *Biomass Convers. Biorefinery* **2018**, *8*, 237–243. [CrossRef]
21. Akinola, A.O.; Eiche, J.F.; Owolabi, P.O.; Elegbeleye, A.P. Pyrolytic Analysis of Cocoa Pod for Biofuel Production. *Niger. J. Technol.* **2018**, *37*, 1026–1031. [CrossRef]
22. Billah, M.; Agratiyan, T.D.; Ayu, D.; Erliyanti, N.K.; Saputro, E.A.; Yogaswara, R.R. Synthesis of Bioethanol from Cocoa Pod Husk Using *Zymomonas mobilis*. *Int. J. Environ. Innov. Sustain. Energy* **2020**, *1*, 31–34. [CrossRef]
23. Antwi, E.; Engler, N.; Nelles, M.; Schüch, A. Anaerobic Digestion and the Effect of Hydrothermal Pretreatment on the Biogas Yield of Cocoa Pod Residues. *Waste Manag.* **2019**, *88*, 131–140. [CrossRef]
24. Widjaja, T.; Nurkhamidah, S.; Altway, A.; Rohmah, A.A.Z.; Saepulah, F. Chemical Pre-Treatments Effect for Reducing Lignin on Cocoa Pulp Waste for Biogas Production. *AIP Conf. Proc.* **2021**, *2349*, 020058. [CrossRef]
25. Rojas, J.; Ramirez, K.; Velasquez, P.; Acevedo, P.; Santis, A. Evaluation of Bio-Hydrogen Production by Dark Fermentation from Cocoa Waste Mucilage. *Chem. Eng. Trans.* **2020**, *79*, 283–288. [CrossRef]
26. Syamsiro, M.; Saptoadi, H.; Tambunan, B.H. Experimental Investigation on Combustion of Bio-Pellets from Indonesian Cocoa Pod Husk. *Asian J. Appl. Sci.* **2011**, *4*, 712–719. [CrossRef]
27. Acosta, N.; Kang, I.D.; Rabaey, K.; De Vrieze, J. Cow Manure Stabilizes Anaerobic Digestion of Cocoa Waste. *Waste Manag.* **2021**, *126*, 508–516. [CrossRef] [PubMed]
28. Leckner, B.; Lind, F. Combustion of Municipal Solid Waste in Fluidized Bed or on Grate—A Comparison. *Waste Manag.* **2020**, *109*, 94–108. [CrossRef]
29. Dahunsi, S.O.; Osueke, C.O.; Olayanju, T.M.A.; Lawal, A.I. Co-Digestion of *Theobroma cacao* (Cocoa) Pod Husk and Poultry Manure for Energy Generation: Effects of Pretreatment Methods. *Bioresour. Technol.* **2019**, *283*, 229–241. [CrossRef] [PubMed]
30. Hernández-Sarabia, M.; Sierra-Silva, J.; Delgadillo-Mirquez, L.; Ávila-Navarro, J.; Carranza, L. The Potential of the Biodigester as a Useful Tool in Coffee Farms. *Appl. Sci.* **2021**, *11*, 6884. [CrossRef]
31. Kumar Khanal, S.; Lü, F.; Wong, J.W.C.; Wu, D.; Oechsner, H. Anaerobic Digestion beyond Biogas. *Bioresour. Technol.* **2021**, *337*, 125378. [CrossRef] [PubMed]
32. Sánchez-Torres, J.; Palop-Marro, F. *Herramientas de Software Para La Práctica En La Empresa de La Vigilancia Tecnológica e Inteligencia Competitiva: Evaluación Comparativa*; TRIZ: Valencia, Spain, 2002; Volume 1.
33. Arango Alzate, B.; Tamayo Giraldo, L.; Fadul Barbosa, A. Vigilancia Tecnológica: Metodologías y Aplicaciones. *Gestión Pers. Tecnol.* **2012**, *13*, 2–8.
34. *UNE 166006:2018*; Technology Watch System. Guidelines for Implementation. Asociación Española de Normalización: Madrid, Spain, 2018.
35. Carrillo-Zambrano, E.; Páez-Leal, M.C.; Suárez, J.M.; Luna-González, M.L. Modelo de Vigilancia Tecnológica Para la Gestión de Un Grupo de Investigación en Salud. *MedUNAB* **2018**, *21*, 84–99. [CrossRef]
36. Vega-Almeida, R.L.; Iglesias-Alfonso, C.; Moura-Delgado, M.; Cossio-Cárdenas, G. Plan de Comunicación Del Sistema de Inteligencia Colaborativa Para el Empresarial BioCubaFarma. *Rev. Cubana Inf. Cienc. Salud* **2020**, *31*, 1–24.
37. Ojikutu, A.O.; Osokoya, O.O. Evaluation of Biogas Production from Food Waste. *Int. J. Eng. Sci.* **2014**, *3*, 1–7.
38. Esposito, G.; Frunzo, L.; Liotta, F.; Panico, A.; Pirozzi, F. Bio-Methane Potential Tests to Measure the Biogas Production from the Digestion and Co-Digestion of Complex Organic Substrates. *Chem. Eng. Trans.* **2012**, *5*, 39–44. [CrossRef]
39. APHA/AWWA/WEF; American Public Health Association; American Water Works Association; Water Environment Federtion (APHA-AWWA-WEF). *Standard Methods for the Examination of Water and Wastewater*, 22nd ed.; American Water Works Association: Denver, CO, USA, 2001.
40. Candia-García, C.; Delgadillo-Mirquez, L.; Hernandez, M. Biodegradation of Rice Straw under Anaerobic Digestion. *Environ. Technol. Innov.* **2018**, *10*, 215–222. [CrossRef]
41. Li, D.; Liu, S.; Mi, L.; Li, Z.; Yuan, Y.; Yan, Z.; Liu, X. Effects of Feedstock Ratio and Organic Loading Rate on the Anaerobic Mesophilic Co-Digestion of Rice Straw and Cow Manure. *Bioresour. Technol.* **2015**, *189*, 319–326. [CrossRef]
42. Castañeda-Morales, Y.; Arciniegas-Benavides, N.C. *Diseño de una Planta Productora de Pellets de Madera a Partir Del Aprovechamiento de Residuos Forestales de Un Aserrío Ubicado en la Ciudad de Bogotá*; Universidad Distrital Francisco José de Caldas: Bogotá, Colombia, 2019.
43. Harinera del Valle. Harinera Del Valle. Available online: https://www.hv.com.co/ (accessed on 29 July 2024).
44. Rojas González, L.M. *Aprovechamiento de La Cáscara de Cacao Para La Elaboración de Un Biocomposito Con Aplicación En La Construcción Sostenible*; Universidad El Bosque: Bogotá, Colombia, 2019.
45. Navarro-García, G.; Navarro-García, S. *Fertilizantes Química y Acción*; Mundi-Prensa; Ediciones Mundi-Prensa: Madrid, Spain, 2014; Volume 1.

46. Villamizar-Jaimes, Y.; Rodriguez-Guerrero, J.S.; Leon-Castrillo, L.C. Characterization Physicochemical, Microbiological and Functional of Cacao Shell Flour (*Theobroma cacao* L.) Variety CCN-51. *Cuad. Act.* **2017**, *9*, 65–75.
47. Ramírez, A.Q.; González, Y.V.; Valencia, L.A.L. Effect of Solid Wastes Leachates on a Tropical Soil. *DYNA* **2017**, *84*, 283–290. [CrossRef]
48. Suhartini, S.; Hidayat, N.; Hadi, M.W.R. Co-Digestion of Cocoa Pods and Cocoa Leaves: Effect of C/N Ratio to Biogas and Energy Potential. *IOP Conf. Ser. Earth Environ. Sci.* **2021**, *733*, 012139. [CrossRef]
49. Messineo, A.; Maniscalco, M.P.; Volpe, R. Biomethane Recovery from Olive Mill Residues through Anaerobic Digestion: A Review of the State of the Art Technology. *Sci. Total Environ.* **2020**, *703*, 135508. [CrossRef]
50. Xu, F.; Wang, Z.-W.; Li, Y. Predicting the Methane Yield of Lignocellulosic Biomass in Mesophilic Solid-State Anaerobic Digestion Based on Feedstock Characteristics and Process Parameters. *Bioresour. Technol.* **2014**, *173*, 168–176. [CrossRef]
51. Mosquera, J.; Varela, L.; Santis, A.; Villamizar, S.; Acevedo, P.; Cabeza, I. Improving Anaerobic Co-Digestion of Different Residual Biomass Sources Readily Available in Colombia by Process Parameters Optimization. *Biomass Bioenergy* **2020**, *142*, 105790. [CrossRef]
52. Hagos, K.; Zong, J.; Li, D.; Liu, C.; Lu, X. Anaerobic Co-Digestion Process for Biogas Production: Progress, Challenges and Perspectives. *Renew. Sustain. Energy Rev.* **2017**, *76*, 1485–1496. [CrossRef]
53. Kim, M.-J.; Kim, S.-H. Conditions of Lag-Phase Reduction during Anaerobic Digestion of Protein for High-Efficiency Biogas Production. *Biomass Bioenergy* **2020**, *143*, 105813. [CrossRef]
54. Dinkler, K.; Li, B.; Guo, J.; Hülsemann, B.; Becker, G.C.; Müller, J.; Oechsner, H. Adapted Hedley Fractionation for the Analysis of Inorganic Phosphate in Biogas Digestate. *Bioresour. Technol.* **2021**, *331*, 125038. [CrossRef] [PubMed]

 *sustainability*

What Type of Public Library Best Supports Agricultural Economic Development? An Empirical Study Based on Rural China

Dimeng Zhang [1], Jiayao Li [2], Yingchi Ye [3], Rong Zhang [3] and Yuntao Zou [3,4,*]

[1] Hangzhou City University Library, Hangzhou City University, Hangzhou 310015, China; zhangdm@hzcu.edu.cn
[2] College of Art and Communication, China Jiliang University, Hangzhou 314423, China; 24a1305047@cjlu.edu.cn
[3] Future Front Interdisciplinary Research Institute, Huazhong University of Science and Technology, Wuhan 430074, China; arthur0713113@gmail.com (Y.Y.); d202187024@hust.edu.cn (R.Z.)
[4] School of Computer of Science and Technology, Huazhong University of Science and Technology, Wuhan 430074, China
* Correspondence: zouyuntao@hust.edu.cn

Citation: Zhang, D.; Li, J.; Ye, Y.; Zhang, R.; Zou, Y. What Type of Public Library Best Supports Agricultural Economic Development? An Empirical Study Based on Rural China. *Sustainability* **2024**, *16*, 8343. https://doi.org/10.3390/su16198343

Academic Editors: Fotios Chatzitheodoridis, Efstratios Loizou and Achilleas Kontogeorgos

Received: 4 September 2024
Revised: 22 September 2024
Accepted: 23 September 2024
Published: 25 September 2024

Abstract: Modernizing agricultural economies requires the infusion of knowledge and industrialization, necessitating the bridging of the "digital divide" and "talent gap" between urban and rural areas. Public libraries, as key knowledge dissemination institutions, play a crucial role in this process. This study aimed to explore how the development of such institutions can align with agricultural economic growth. Using China as a case study, where the Rural Revitalization Strategy has been implemented in recent years, including the extensive construction of rural public libraries and other infrastructure, we empirically analyzed the correlation between county-level public libraries and agricultural economic development from 2012 to 2019. The results show that the number of county-level public libraries and their assets, collection sizes, e-books, and professional staff have a significant positive impact on agricultural economics, while non-professional staff and facilities have a negative impact. It is recommended that future rural public library development should focus on enhancing professional standards and advancing digitalization and mobile internet integration, while being cautious about expanding the physical scale and staffing. This study fills a gap in the research on the correlation between rural public libraries and agricultural economics, and the methodology employed has a certain degree of general applicability. However, the applicability of the conclusions may be limited by China's unique national conditions.

Keywords: public libraries; rural; agricultural economic; DEA; Tobit

1. Introduction

1.1. The Value of Rural Public Libraries and Related Research

The construction of rural public libraries plays a significant role in promoting social, economic, and cultural development in rural areas, with key contributions in the following areas:

1. Knowledge Dissemination, Education Enhancement, and Skills Training: Public libraries provide access to a wide range of books and educational resources, which are particularly valuable in rural areas where educational facilities and resources are often lacking. Libraries help residents acquire knowledge and improve their cultural literacy, and they promote lifelong learning. Studies by Eve et al. [1] and Nielsen and Borlund [2] highlight the irreplaceable role of public libraries in lifelong learning, informal education, and continuous vocational skills education;

2. Information Access: Modern libraries offer more than just traditional borrowing services; they also provide e-books, online resources, and multimedia content. Many libraries host various training courses and workshops, such as those on agricultural technology, small business management, and computer skills, which are crucial for increasing farmers' income and employment opportunities. Research by Kinney [3], Abumandour [4], and Manžuch and Macevičiūtė [5] demonstrates that modern digital and networked public libraries are key measures in bridging the digital divide;

3. Catalyst for Economic Development: By providing educational resources and skills training, public libraries can enhance the quality of the rural workforce and promote diversified local economic development, helping to mitigate the effects of economic downturns. Moreover, the construction and operation of libraries create job opportunities, both directly and indirectly, contributing to economic growth. Taylor et al. [6] suggest that public libraries help people withstand the effects of economic recessions. Fairbairn and Lipeikaite [7] report that Macedonia's Braka Miladinovci Public Library provides services for unemployed young women, while South Africa's Masiphumelele Public Library serves impoverished suburban slums, aiming to help disadvantaged youth embark on positive career paths. In rural areas, Nikcevic-supported services targeting farmers, such as Serbia's Radislav Nikcevic Public Library, have been instrumental. This library's AgroLib online market allows farmers to share information on agricultural methods and sell their products. Macedonia's Goce Delchev-Stip Regional Public and University Library assists farmers in finding subsidy information and applying for grants online, while Lithuania's Pasvalys Marius Katiliskis Public Library offers ICT training, an online portal, and a desktop publishing center for farmers. Mehra et al. [8] found that rural public libraries have significant potential in driving economic development and sustainable economic vitality in the Appalachian region of the United States;

4. Community Centers and Cultural Exchange: Libraries often serve as community hubs, providing a public space for people to gather, interact, and organize events. This contributes positively to community cohesion, cultural preservation, and social integration. Audunson [9] found that public libraries have great potential to become low-intensity meeting places, which play a significant role in promoting tolerance and community integration. Flaherty and Miller [10] found that rural public libraries in the United States have a significant impact on community activities in remote areas, significantly promoting community health work and potentially leading to community transformation in the future;

5. Bridging the Digital Divide and Promoting Sustainable Development: As global attention to the Sustainable Development Goals increases, libraries can become platforms for disseminating environmental knowledge and sustainable agricultural practices, helping rural communities address challenges such as climate change and environmental degradation. Audunson et al. [11] demonstrate that public libraries serve as infrastructure for sustainable development, playing a significant role in bridging the digital divide and enhancing cultural inclusiveness. Mansour [12] argues that despite challenges such as weak infrastructure and inadequate collections and related facilities, high illiteracy rates among service recipients, a lack of funding and cooperation between relevant institutions, the improper training of library and information professionals, and a lack of research and surveys, Egypt's public libraries still contribute positively to sustainable development, although the number is far from sufficient. Panda and Das [13] suggest that public libraries play seven roles in sustainable development: informing, promoting, educating, creating resources, empowering, healing, and advocating. Wang et al. [14] also show that rural public cultural services, represented by public libraries, are crucial for narrowing the rural digital divide, including the physical and capacity divides. Digital public libraries also had unique advantages during the COVID-19 pandemic, with Xin [15] finding that they quickly transitioned their services online in response to the outbreak, ac-

tively providing reading therapy, quality digital resources, and reliable pandemic information, offering academic literature support to researchers, and alleviating the psychological pressure of isolated patients and medical staff.

In summary, the construction of public libraries not only enhances the knowledge level and quality of life for rural residents but also provides high-quality talent for rural areas and supports the sustainable development of entire communities. However, not all rural public libraries can achieve significant positive effects. Omeluzor et al. [16] found that rural libraries in Nigeria failed to fulfill their roles. However, scholars also argue that the failure stems from unmet information needs, a lack of up-to-date materials, a lack of awareness, illiteracy, language barriers, insufficient technical staff, and inadequate infrastructure and facilities. It is not just Third-World countries; studies by Bishop et al. [17] and Real and Rose [18] also show that rural public library construction in the United States has many shortcomings.

Given the above, the construction of rural public libraries should be carefully studied to determine how they can best integrate into and adapt to the surrounding social environment based on common economic, political, and social factors and conditions, and how they can be developed to meet the needs of agricultural economic development.

1.2. The Development of Rural Public Libraries in China

In recent years, the Chinese government has placed significant emphasis on rural development, introducing the Rural Revitalization Strategy. The implementation of this strategy relies heavily on the influx of individuals with modern higher education into rural areas. Sun et al. [19] emphasized that, in the context of rural revitalization, the primary focus should be on talent development and a people-centered approach. Wang et al. [20] found in their study on the coupling of the digital economy and rural economy that high-level technological innovation and rural management talent are of paramount importance. Given the significant role of public libraries in talent cultivation, the Chinese government has also placed particular emphasis on the construction of rural public libraries.

China's public library system has a long history, and, after decades of effort, it has developed into a nationwide network characterized by the following features:

1. Tiered Management System: The Chinese government established a public library system quite early, categorized by administrative levels into national, provincial, city, and county libraries, with each level responsible for library services and management within its respective jurisdiction, forming a multi-tiered management and service system [21]. Further down, cultural stations at the township and village levels also provide book-reading services. This system has been gradually perfected over the years and is still in use today [22].

In this tiered system, national-, provincial-, and city-level public libraries are located in the national capital of Beijing, provincial capitals, and prefecture-level cities, respectively, primarily serving urban residents. Although currently open to rural residents, their influence is limited due to geographic constraints. As for the township- and village-level cultural stations, due to uncertain and unstable funding sources, the number of books is very limited, and their impact is also restricted. Only county-level public libraries within the administrative system are closest to rural residents, and a significant proportion of their collections are specialized agricultural books.

A typical county-level public library, such as the one in Dongyang City (a county-level city) under Jinhua Municipality in Zhejiang Province, occupies a building area of 5200 square meters, holds a collection of 200,000 volumes, and is equipped with a book repository, information room, electronic reading room, and children's reading room, among other facilities. The library staff includes 18 employees, 40% of whom have an associate degree or higher. It accommodates 100,000 to 200,000 visitor instances annually [23]. For context, Dongyang City covers an area of 1747 square kilometers with a permanent population of 1.093 million;

2. Government Support and Investment: Within the administrative system, national-, provincial-, municipal-, and county-level public libraries receive comprehensive financial support from the respective levels of government, with management personnel incorporated into the civil service. China has also promoted the development of library services through the formulation of relevant policies and regulations. For example, the "Public Library Law of the People's Republic of China", promulgated in 2017 [24], clearly outlines the government's responsibilities and obligations in public library construction and services. In recent years, as China's economy has grown, the government has increasingly emphasized the construction of a public cultural service system, including the public library system [25].

Note that China is a state-owned country, and the funding for public libraries, including their construction, collection acquisition, operation, and the various activities organized by them, is of a public welfare nature and is supported by state finance. This differs significantly from public libraries in other countries and regions. Especially at the county level, public libraries, mainly serving rural areas and with limited visibility, are almost entirely unable to secure additional funding through private donations.

The advantage of this state-managed mechanism is that the basic funding is guaranteed, eliminating concerns about operational losses for libraries, which allows them to focus on library services. However, the disadvantages are also clear: the system is too rigid and lacks flexibility, making it extremely difficult to scale up or down. Personnel within this system also tend to lose their initiative and settle for the status quo;

3. Growing Emphasis on Digitalization and Informatization: With the advancement of information technology, China's public libraries have actively promoted digital transformation. Many libraries now offer e-books, online databases, and digital reading resources while leveraging information technology to enhance their service efficiency and quality. Additionally, due to China's centralized socialist system, the digital information construction of public libraries is uniformly advanced through administrative directives. For instance, in 2011, the Ministry of Culture of China launched the National Digital Library (NDL) project nationwide [26], significantly enriching the digital resources in the collections of public libraries at all levels within a few years, and making them available for external sharing.

Overall, due to the characteristics of China's state ownership, the construction of China's public library system is entirely undertaken by the government, with guaranteed funding sources, allowing for a greater focus on building library infrastructure and enhancing the service quality. However, this government-led public library system can also lead to issues such as rigid management and severe bureaucracy, resulting in inefficiencies. These problems may become more apparent in the development and promotion of rural public libraries and digital libraries.

Shao and colleagues, in their comparative study of public library evaluation systems in China and the United States [27], found that U.S. public libraries place greater emphasis on library services, while Chinese public libraries focus more on infrastructure and basic construction. This is likely because China, as a developing country, still has an underdeveloped public library system. Zhao and colleagues, through a survey of 31 senior library experts across 10 provinces in China, discovered that the development of public cultural services in Chinese libraries is unbalanced and inadequate [28]. They identified external factors such as economic development, government investment in cultural activities, and educational development, along with internal issues like weak management and insufficient investment in cultural services, as the primary influences on the development of public cultural services in libraries. This suggests that for Chinese public libraries to improve their services, they should seek financial support and enhance their internal management. Sun and colleagues [29], through factor and cluster analysis, found that there is a significant disparity in the development levels of public libraries between economically disadvantaged regions and more advanced areas in China, with even economically developed regions like Guangdong facing severe infrastructure deficiencies.

Regarding the construction of rural public libraries and digital public libraries in China, Bin and Miao, in their comparative study on Chinese public libraries and university libraries [30], identified several shortcomings in public libraries, such as the diversification of reader needs, a lack of general electronic publications, low usage skills for electronic publications, the adoption of fee-based services for electronic publications, and weak bargaining and pricing power. Yao and Zhao's research [31] highlighted deficiencies in some public libraries regarding homepage terminology standardization, unified database search platform selection, and the construction of navigation systems. Wanyan and Hu [32] pointed out that, despite years of development, digital cultural service expenditures remain low among the Chinese population, with significant disparities in digital cultural service participation between different population groups, particularly between urban and rural residents. Although there is strong interest in public digital cultural services, the actual utilization of these services in public libraries is low and fails to meet user quality expectations. Li and Xu [33] argued that while Chinese public libraries play a crucial role in digital resource development and public cultural services, they face contradictions and inconsistencies in areas such as the supply–demand relationship for resources, resource integration, service effectiveness and efficiency, marketing and promotion, and social participation.

1.3. Research Objective

In summary, the current consensus among scholars is that public knowledge and cultural dissemination institutions, such as public libraries, can support sustainable development in rural areas. However, there is still no clear conclusion on how public libraries in rural areas should be constructed to align with the future development of these regions. In recent years, China has vigorously promoted its Rural Revitalization Strategy, which includes the improvement in public cultural service systems and the promotion of public digital libraries. Nonetheless, various concerns remain, as highlighted by scholars' research.

Moreover, a search using the keywords "public library" and "agriculture" on academic platforms revealed that there is currently very little research that combines public libraries with agricultural economics. In China, studies of this kind are even rarer. As indicated in the Introduction of this paper, research on the impact of rural public libraries on agricultural economics in China is virtually nonexistent. This study aims to fill this gap by exploring the construction direction of rural public libraries that can support sustainable development in rural areas. Given the significant development of rural infrastructure under China's Rural Revitalization Strategy, this study uses the construction of public libraries in rural China as a case study. By analyzing the correlation between public libraries and agricultural economics across different regions of China, this research seeks to identify the experiences and lessons learned in the construction of rural public libraries, provide policy recommendations for future development, and contribute to the sustainable development of rural areas.

2. Materials and Methods

The research process for this study is outlined in Figure 1 and was as follows:

1. Data related to rural public libraries in China from 2012 to 2019 were collected;
2. The level and trends of rural economic development in China during the same period were assessed;
3. A correlation analysis was conducted based on the aforementioned results;
4. The characteristics of rural public libraries that are most beneficial for rural economic development were discussed and summarized.

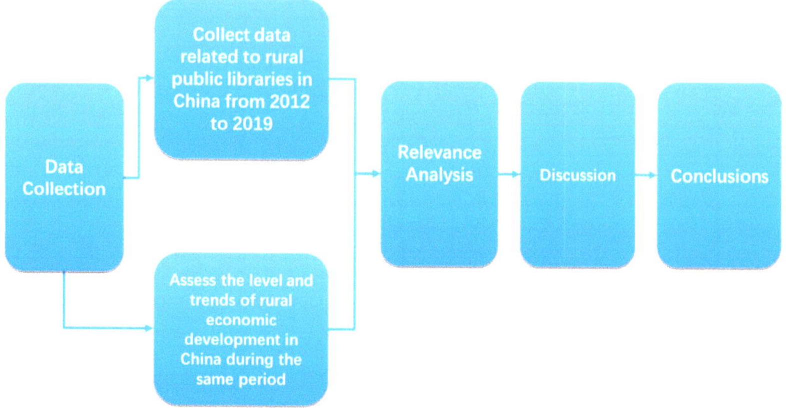

Figure 1. Flowchart of the study.

China comprises 34 provincial-level regions, among which Taiwan, Hong Kong, and Macau were not included in this study due to the significant differences in political systems and statistical standards. Beijing, Shanghai, Tianjin, and Chongqing, the four municipalities directly under the Central Government, predominantly focus on non-agricultural economies. The administrative setup and division of their subordinate units also differ greatly from other provinces and autonomous regions. The setup, collections, and target audiences of the county-level public libraries under their jurisdiction also differ significantly from those in other provinces and autonomous regions and thus were also excluded from this study. The remaining 27 provinces and autonomous regions represent the most significant part of China's agricultural economy and were sufficiently representative for this analysis.

Data sources: China Statistical Yearbook; China Rural Statistical Yearbook; China Library Yearbook; China Cultural and Heritage Statistics Yearbook; China Culture and Tourism Statistics Yearbook; China Education Statistics Yearbook; among others. In this study, all monetary values in Renminbi were adjusted using historical price indices to reflect constant prices, standardized to the 2019 RMB value, ensuring that the results are more in line with the actual conditions.

This study evaluated the level and trends of China's agricultural economic development by assessing the production efficiency. It then examined the impact of various data from rural public libraries on the agricultural economic efficiency to explore the future direction of rural public library development. Efficiency evaluation effectively mitigates the disparities in agricultural economies across different provinces and autonomous regions caused by factors such as geography, geology, climate, and historical development. Moreover, efficiency evaluation reflects the sustainability of agricultural economies: a higher efficiency indicates greater output with fewer resource inputs, signifying more advanced agricultural technology, more efficient management, and more rationalized resource allocation. This rationalization is associated with reduced carbon emissions, lower pollutant discharge, and improved sustainability.

Efficiency has long been used as an evaluation metric in economic studies. Schmookler and others used efficiency to evaluate the U.S. economy as early as 1952 [34]. Mundaca et al. employed efficiency evaluation in their study of energy economics [35]. Bukarica and Tomšić used efficiency to assess energy policy and sustainability [36]. In agricultural economics, Paul et al. analyzed U.S. farms and economies of scale using efficiency evaluation [37]. Reza et al. also used efficiency studies to reflect changes in agricultural productivity over 50 years worldwide [38].

This study employed Data Envelopment Analysis (DEA) to assess the production efficiency of China's agricultural economy. DEA is particularly well suited for evaluating the relative efficiency of decision-making units (DMUs) that involve multiple inputs and outputs. One of its key advantages is that it does not require the allocation of a priori weights to the inputs and outputs, making the analysis more objective and flexible, especially in complex economic and production systems. In this study, the 27 provinces and autonomous regions of China were treated as DMUs, and an input–output-oriented Slack-Based Measure (SBM) model was constructed for the analysis. The non-oriented SBM model can identify and measure the slack in individual input and output variables, providing a more comprehensive and detailed assessment than conventional DEA models.

The formula for the non-oriented SBM model is as follows [39]:

Let the set of DMUs be $j = \{1, 2, \ldots, n\}$, with each DMU having m inputs and s outputs. We denote the vectors of the inputs and outputs for DMU_j by $x_j = \left(x_{1j}, x_{2j}, \ldots, x_{mj}\right)^T$ and $y_j = \left(y_{1j}, y_{2j}, \ldots, y_{sj}\right)^T$, respectively. We define the input and output matrices (X and Y) as follows:

$$X = (x_1, x_2, \ldots, x_n) \in R^{m \times n} \text{ and } Y = (y_1, y_2, \ldots y_n) \in R^{s \times n}. \tag{1}$$

We assume that all data are positive (i.e., $X > 0$ and $Y > 0$).

In order to evaluate the relative efficiency of $DMU_o = (x_o, y_o)$, we solve the following linear program. This process is repeated n times for $o = (1, \ldots, n)$.

The non-oriented or both-oriented SBM efficiency (ρ_{IO}^*) is defined by [SBM-C]:

$$\rho_{IO}^* = \min_{\lambda, s^-, s^+} \frac{1 - (1/m)\sum_{i=1}^{m}\left(s_i^- / x_{io}\right)}{1 + (1/s)\sum_{r=1}^{s}(s_r^+ / y_{ro})}$$

Subjectto

$$x_{io} = \sum_{j=1}^{n} x_{ij}\lambda_j + s_i^- \ (i = 1, \ldots, m), \tag{2}$$

$$y_{ro} = \sum_{j=1}^{n} y_{rj}\lambda_j - s_r^+ \ (r = 1, \ldots, s),$$

$$\lambda_j \geq 0(\forall j), \ s_j^- \geq 0(\forall i), \ s_r^+ \geq (\forall r).$$

It is particularly important to note that this study employed the global SBM model [40], which treats all DMUs across different periods—namely, the input and output variables of the 27 provinces in China from 2012 to 2019—as a single reference set for model construction. This approach ensures that the efficiency scores obtained account for the influence of different periods. As a result, the efficiency scores and averages of each DMU in each period, as well as the mean efficiency scores of all DMUs in any given period, can be compared longitudinally. The efficiency scores derived from the global SBM model, which incorporates temporal variations, offer a more accurate and realistic fit when performing correlation analysis with rural public library-related data that also vary over time. Additionally, by aggregating the average efficiency scores of all DMUs across different periods, the global SBM model can be used to observe the development trends in the overall economic efficiency of China's agriculture.

In the correlation analysis, this study employed a regression model. It is important to note that the data related to rural public libraries used as independent variables in this study can only partially explain the dependent variable, which is the agricultural economic efficiency score, indicating a limitation of the model. However, the primary objective of this study was to identify the characteristics of public libraries that are most conducive to agricultural economic development and to clarify the future direction of rural library development. Therefore, the focus was on establishing the correlation between rural public libraries and agricultural economics rather than on determining causality, making the regression model a suitable method for this research.

This study specifically utilized a Tobit regression model because the dependent variable was the DEA efficiency score of agricultural economics, which is greater than 0 and less than or equal to 1. The Tobit regression model is designed to account for such censored dependent variables, using the maximum likelihood estimation method to effectively prevent bias in the model's estimates. It separates the effects of the censored and uncensored portions of the dependent variable, ultimately converting it into a probability model, making it more suitable for models with complex data characteristics. Moreover, the maximum likelihood method also offers a higher tolerance for multicollinearity among the variables compared to other methods.

3. Results

The input and output variables used in constructing the global SBM model for the agricultural economic efficiency across 27 provinces in China from 2012 to 2019 are presented in Table 1. The panel data for these input and output variables across the 27 provinces from 2012 to 2019 can be found in Supplementary Table S1.

Table 1. Input and output variables of the global SBM model for China's agricultural circular economy.

Indicator Categories	Indicators
Input Indicators	Rural Population
	Consumption of Chemical Fertilizers
	Consumption of Pesticides
	Consumption of Diesel Fuel
Output Indicators	Agriculture, Forestry, Animal Husbandry, and Fishery Total Output Value
	Primary Industry Added Value

It is important to note that certain input variables commonly considered significant, such as land area and irrigated area, were not included in this study. This decision was made due to the significant variations across the 27 provinces. For instance, in the northern regions, crops are harvested once a year, whereas in the southern regions, they may be harvested three times annually, leading to substantial differences in the output per unit area. Additionally, provinces like Inner Mongolia and Ningxia focus primarily on animal husbandry, while southern provinces have a larger proportion of their agricultural activities devoted to crop cultivation. Agricultural production in the northern and southern regions was specifically adjusted to account for these differences.

The input variables selected for this study are less affected by such regional disparities. For example, the "Consumption of Diesel Fuel" is used in both crop cultivation and animal husbandry, and its consumption is independent of the number of harvest cycles per year. Furthermore, these variables constitute a significant portion of the overall costs. Other inputs, such as seeds and plastic film, which have a smaller cost share and limited usage, were not included in this study, as their impact on the model results would have been minimal and might have introduced inaccuracies.

Regarding the output variables, the "Agriculture, Forestry, Animal Husbandry, and Fishery Total Output Value" represents the annual total output value of the agricultural economy across the 27 provinces, while the "Primary Industry Added Value" can be considered the gross profit of the agricultural economy in these provinces.

The DEA analysis in this study was conducted using DEARUN software, version V3.2.0.2. This software features a user-friendly graphical interface with built-in DEA model algorithms, eliminating the need for users to write custom code. The average agricultural economic efficiency scores for the 27 provinces and autonomous regions from 2012 to 2019 are shown in Table 2. TE represents the average technical efficiency score under the assumption of constant returns to scale, PTE represents the average pure technical efficiency under variable returns to scale, and SE represents the average scale efficiency score. Notably, the TE is the product of the PTE and SE. Detailed data from the model can be found in Supplementary Table S2.

Table 2. Mean scores of the global SBM model for the agricultural circular economy in 27 provinces and autonomous regions of China, 2012–2019.

	TE	PTE	SE
2012	0.500286	0.599746	0.863904
2013	0.521391	0.635033	0.852163
2014	0.52549	0.636519	0.852553
2015	0.534844	0.659064	0.843958
2016	0.566399	0.690096	0.84505
2017	0.5541	0.653512	0.867062
2018	0.594693	0.714161	0.852686
2019	0.68224	0.83108	0.833445

In selecting the independent variables for the Tobit model, this study first excluded national, provincial, and municipal public libraries in China, as these libraries are located in the capital, provincial capitals, or major regional cities and primarily serve urban residents. This study focused on county-level public libraries, as these libraries primarily serve the rural population.

The statistical data for China's county-level libraries, varying by year, include over 40 items. Based on the importance and completeness of the data, this study ultimately selected 13 variables related to county-level public libraries in 27 provinces and autonomous regions in China as the independent variables for the Tobit regression model:

1. The number of county-level public libraries, number of staff, number of professional staff, and total assets of county-level libraries, representing the actual scale of the county-level libraries in each province and autonomous region;
2. The revenue of county-level public libraries, representing government support, as the vast majority of county-level library revenue comes from government appropriations;
3. The number of books, the number of library cards issued, the building area, and the number of reading room seats, representing the professional development of county-level libraries;
4. The number of e-books and the number of public computers, representing the level of digitization of county level libraries and the role of public libraries in bridging the "digital divide";
5. The circulation number of library visits and the turnover rate of the collection, representing the operation and activity level of county-level public libraries.

Considering the differences between the provinces and autonomous regions, these values were normalized by dividing by the rural population of each province and autonomous region, resulting in 13 independent variables. Details of these variables are provided in Supplementary Table S3.

This study used the SPSSAU online analysis software to construct the Tobit model, with version number SPSSAU24.0. This software does not require downloading and comes with a variety of built-in data analysis model algorithms, allowing users to directly input data and construct models without the need to write model code. Additionally, it provides some basic analytical functions. Table 3 presents the likelihood ratio test results for the Tobit model (hereafter referred to as Model 1), with the TE scores as the dependent variable and 13 county-level public library data variables as the independent variables. The p-value is less than 0.05, indicating that Model 1 rejects the null hypothesis. The Akaike Information Criterion (AIC) and Bayesian Information Criterion (BIC) were used to assess the goodness of fit for the model's data, with smaller values indicating a better fit. Based on these values, Model 1 demonstrates a good fit.

Table 3. Likelihood ratio test for Tobit Model 1 with constant returns to scale.

Model	$-2\times$ Log-Likelihood	Cardinality	*df*	*p*	AIC	BIC
Intercept Distance	-122.651				-18.686	
Final Model	-240.775	118.123	13	0.000	-212.775	-165.521

Table 4 presents the results of Model 1, which was constructed using TE scores as the dependent variable and 13 county-level public library data variables as the independent variables.

Table 4. Results of Model 1 with constant returns to scale.

	Regression Coefficient	Standard Error	*z*-Value	*p*-Value	95% CI
Intercept	0.493	0.031	15.803	0.000	0.432~0.554
County Libraries per 10,000 Rural Residents	1.363	0.242	5.640	0.000	0.889~1.837
County Library Staff per 10,000 Rural Residents	-0.750	0.124	-6.069	0.000	-0.992~-0.508
County Library Professionals per 10,000 Rural Residents	0.618	0.157	3.948	0.000	0.311~0.925
County Library Collections per 10,000 Rural Residents	0.000	0.000	2.954	0.003	0.000~0.000
E-Books in County Libraries per 10,000 Rural Residents	0.000	0.000	4.475	0.000	0.000~0.000
Public Computers in County Libraries per 10,000 Rural Residents	0.016	0.027	0.588	0.557	-0.037~0.069
Library Cards Issued by County Libraries per 10,000 Rural Residents	0.000	0.000	1.721	0.085	-0.000~0.000
Circulation of County Libraries per 10,000 Rural Residents	-0.000	0.000	-2.497	0.013	-0.000~-0.000
Books Borrowed from County Libraries per 10,000 Rural Residents	-0.000	0.000	-0.792	0.428	-0.000~0.000
County Library Revenue per 10,000 Rural Residents	0.005	0.004	1.507	0.132	-0.002~0.013
County Library Assets per 10,000 Rural Residents	0.000	0.000	2.344	0.019	0.000~0.001
County Library Area per 10,000 Rural Residents	-0.002	0.001	-3.348	0.001	-0.003~-0.001
County Library Seats per 10,000 Rural Residents	0.016	0.006	2.510	0.012	0.003~0.028
Log (Sigma)	-1.976	0.048	-41.076	0.000	-2.071~-1.882

Dependent variable: TE; McFadden R^2: -0.963.

Table 5 presents the likelihood ratio test results for the Tobit model (hereafter referred to as Model 2), constructed by replacing the dependent variable with PTE scores. The *p*-value is also less than 0.05, indicating that Model 2 similarly rejects the null hypothesis. The AIC and BIC values are slightly higher compared to those of Model 1.

Table 5. Likelihood ratio test for Tobit Model 2 with variable returns to scale.

Model	$-2\times$ Log-Likelihood	Cardinality	*df*	*p*	AIC	BIC
Intercept Distance	-24.800				-18.686	
Final Model	-101.696	76.895	13	0.000	-73.696	-24.442

Table 6 presents the results of Model 2, which was constructed by replacing the dependent variable TE scores with PTE scores.

Table 6. Results of Model 2 with variable returns to scale.

	Regression Coefficient	Standard Error	*z*-Value	*p*-Value	95% CI
Intercept	0.635	0.043	14.758	0.000	0.550~0.719
County Libraries per 10,000 Rural Residents	1.782	0.334	5.344	0.000	1.128~2.436
County Library Staff per 10,000 Rural Residents	-0.884	0.170	-5.187	0.000	-1.218~-0.550

Table 6. *Cont.*

	Regression Coefficient	Standard Error	z-Value	p-Value	95% CI
County Library Professionals per 10,000 Rural Residents	0.975	0.216	4.511	0.000	0.551~1.398
County Library Collections per 10,000 Rural Residents	0.000	0.000	4.119	0.000	0.000~0.000
E-Books in County Libraries per 10,000 Rural Residents	0.000	0.000	1.439	0.150	−0.000~0.000
Public Computers in County Libraries per 10,000 Rural Residents	−0.099	0.037	−2.648	0.008	−0.172~−0.026
Library Cards Issued by County Libraries per 10,000 Rural Residents	0.000	0.000	2.197	0.028	0.000~0.000
Circulation of County Libraries per 10,000 Rural Residents	−0.000	0.000	−1.842	0.065	−0.000~0.000
Books Borrowed from County Libraries per 10,000 Rural Residents	−0.000	0.000	−1.321	0.186	−0.000~0.000
County Library Revenue per 10,000 Rural Residents	−0.005	0.005	−0.970	0.332	−0.015~0.005
County Library Assets per 10,000 Rural Residents	0.000	0.000	1.575	0.115	−0.000~0.001
County Library Area per 10,000 Rural Residents	−0.003	0.001	−3.191	0.001	−0.005~−0.001
County Library Seats per 10,000 Rural Residents	0.026	0.009	3.016	0.003	0.009~0.044
Log (Sigma)	−1.654	0.048	−34.385	0.000	−1.749~−1.560

Dependent variable: PTE; McFadden R^2: −3.101.

4. Discussion

The efficiency of China's agricultural economy during the period covered by this study is presented in Table 2. This study employed a global SBM model, which allowed for cross-period comparisons, and the data from Table 2 are visualized in Figure 2, which includes a linear trendline for a clearer interpretation. As shown in Figure 2, the TE scores of China's agricultural economy steadily improved from 2012 to 2019, with only a slight decline observed between 2016 and 2017, followed by a recovery that not only compensated for the previous dip but also surpassed the prior levels. Notably, 2018 marks the initiation of China's Rural Revitalization Strategy, suggesting that the comprehensive policies under this strategy, including the substantial development of rural public libraries, have been effective at enhancing the efficiency of China's agricultural economy

As previously mentioned in the Results Section, the TE is the product of the PTE and SE. Figure 2 also illustrates that the primary factor contributing to China's agricultural economic efficiency, as indicated by the TE score, is the SE, while the PTE generally remains lower. Scale efficiency evaluates whether increases in the input and output variables yield proportional returns, whereas the PTE encompasses factors beyond the scale, such as agricultural technology levels, production management, financing capabilities, and marketing. The results shown in Figure 2 suggest that relative to the scale of the industry, China's agricultural economy is weaker in areas related to pure technical efficiency.

However, the trend lines for the three scores reveal that the upward trend in the agricultural economic efficiency is driven by continuous improvements in the PTE. For the study period, the trend for the SE is declining, while the trend for the PTE shows a steep upward slope, closely aligning with the changes in the TE score. This indicates that the improvement in the pure technical efficiency has become the key factor driving the overall increase in the agricultural economic efficiency. The focus of this study on the construction of rural public libraries—aiming to disseminate agricultural knowledge in rural areas and bridge the "talent gap"—clearly falls within the scope of factors contributing to the pure technical efficiency in the agricultural economy.

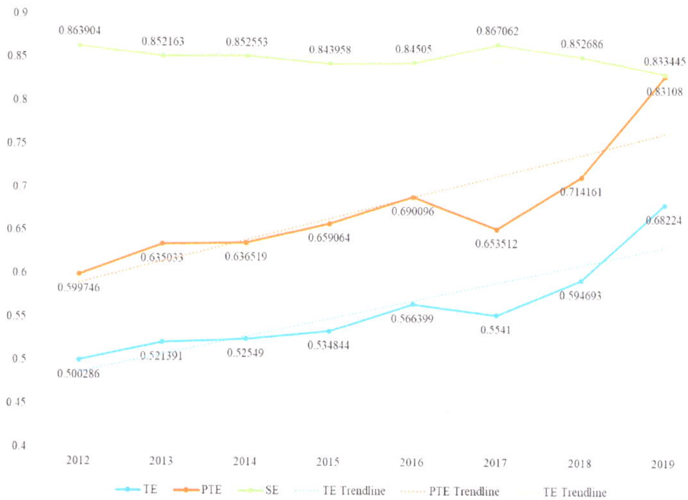

Figure 2. Agricultural economic efficiency scores and trends in China.

This study constructed two Tobit models with different dependent variables, a common approach to testing model robustness. Xu et al. [41] in their study on the impact of urbanization on rural land transfers, Su et al. [42] in their research on the digital economy, and Lin et al. [43] in their investigation of the impact of air pollution on technological innovation all employed the strategy of replacing the dependent variable to test the robustness of their models. The high similarity in the results of the two Tobit models in this study thus confirms the robustness of the models.

As shown in Table 4, Model 1, which uses TE scores as the dependent variable, six out of the thirteen independent variables related to county libraries have a significant positive impact on the agricultural economic efficiency TE scores under the assumption of the CRS. These variables are as follows: County Libraries per 10,000 Rural Residents; County Library Professionals per 10,000 Rural Residents; County Library Collections per 10,000 Rural Residents; E-Books in County Libraries per 10,000 Rural Residents; County Library Assets per 10,000 Rural Residents; and County Library Seats per 10,000 Rural Residents. Three variables have a significant negative impact on the TE scores: County Library Staff per 10,000 Rural Residents, Circulation of County Libraries per 10,000 Rural Residents, and County Library Area per 10,000 Rural Residents.

As shown in Table 6, Model 2, which uses PTE scores as the dependent variable, five out of the thirteen independent variables related to county libraries have a significant positive impact on the agricultural economic efficiency PTE scores under the assumption of the VRS. These variables are as follows: County Libraries per 10,000 Rural Residents; County Library Professionals per 10,000 Rural Residents; County Library Collections per 10,000 Rural Residents; Library Cards Issued by County Libraries per 10,000 Rural Residents; and County Library Seats per 10,000 Rural Residents. Three variables have a significant negative impact on the PTE scores: County Library Staff per 10,000 Rural Residents; Public Computers in County Libraries per 10,000 Rural Residents; and County Library Area per 10,000 Rural Residents.

Based on the combined results of the two Tobit models, the number of county-level public library-related variables that have a positive impact on the agricultural economic efficiency is significantly greater than those with a negative impact. This indicates that the construction and development of county-level public libraries in China have an overall positive effect on the agricultural economic efficiency. Numerous studies support the positive influence of rural public libraries on agricultural development, such as Bini Okiy's research on rural public libraries in Nigeria [44], Sultana et al.'s study on rural public libraries in West Bengal [45], as well as studies in developed countries like Svendsen et al.'s research on rural public libraries in Denmark [46] and Sikes' study on rural public libraries in Washington County [47].

County Libraries per 10,000 Rural Residents, County Library Assets per 10,000 Rural Residents, County Library Collections per 10,000 Rural Residents, and County Library Seats per 10,000 Rural Residents all demonstrate a significant positive impact on China's agricultural economic efficiency in both Tobit models. In contrast, County Library Area per 10,000 Rural Residents shows a significant negative impact on the efficiency of the cultural industry in both models. These findings suggest that while the number and total assets of county-level public libraries should continue to increase, resources should not be allocated to expanding physical space but rather to enhancing the collection of books and providing more opportunities for on-site reading. Sultana's research on rural public libraries in West Bengal [45] similarly found that library collections and related metrics are more crucial. Mansour et al.'s study of public libraries in Egypt [12] also indicates that infrastructure and collections are critical to the quality of public library services.

It is important to note that the independent variables used in the Tobit model of this study represent rural public library resources per 10,000 rural residents in China. However, due to ongoing urbanization, the rural population in China has been steadily declining, with the proportion of rural residents decreasing from 47.43% in 2012 to 39.40% in 2019, and the total rural population decreasing from 642.22 million in 2012 to 551.62 million in 2019 [48]. This trend implies that even without directly increasing the number of rural library resources, merely maintaining the current scale would lead to an annual increase in public library resources per 10,000 residents. Therefore, this further emphasizes that when the government strengthens the construction of rural public libraries, the focus should be on optimizing the structure of existing investments rather than on simply increasing resources such as building space.

The number of county library staff per 10,000 rural residents has a significant negative impact on the agricultural efficiency in China in both models. Conversely, the number of county library professionals per 10,000 rural residents has a significant positive impact in both models. This suggests that the number of non-professional staff and even the total number of staff in rural public libraries should be reduced, while the number of professional staff should be increased, indicating a need for greater professionalization in rural public libraries. As China is a state-owned country, the primary goal of government-managed public libraries is not profit making, funding sources are very stable, and staff management is within a fixed establishment. This model is likely to lead to rigid management and over-staffing, especially as non-professional staff may become a burden in the daily operation of public libraries. An excessive number of such staff can directly consume library resources and may lead to decreased work efficiency and quality. Future construction of rural libraries should particularly focus on improving the efficiency of human resources. De Jager et al. [49] also found in their study on South African rural public libraries that the quality of the staff significantly impacts the quality of library services. Rehman et al. [50], in their study on human resource management in public organizations in Pakistan, mentioned the same issue: non-professional staff within the establishment can severely affect the service quality. Research by Real and Rose [18] on remote rural public libraries in the United States found that the level of service, programming, and full-time equivalence of staff are important for effectively enhancing the library service quality. Mansour [12] also shows that the professionalization and informatization levels of library staff are very important.

The variable "E-Books in County Libraries" has a significant positive impact in Model 1, where the TE efficiency scores are the dependent variable, but shows no significant impact in Model 2, where the PTE efficiency scores are the dependent variable. Since the TE is the product of the PTE and SE, this indicates that the influence of "E-Books in County Libraries" on the agricultural economic efficiency is primarily reflected in the improvement in the scale efficiency (SE). In contrast, "Public Computers in County

Libraries" has a significant negative impact on the PTE efficiency scores, which may seem contradictory to the performance of the "E-Books in County Libraries" variable. However, it is important to note that e-books represent software, which can be accessed and read on various hardware devices, such as computers and mobile phones, while "Public Computers in County Libraries" represents hardware. Thus, considering these factors together, the future development of rural libraries should emphasize digitalization and mobile internet capabilities, focusing on increasing the availability of software resources, such as e-books, while appropriately reducing hardware facilities like computers to conserve resources.

It should be noted that the Chinese Copyright Law lists 12 scenarios for the fair use of intellectual property, of which the following are applicable to library and information institutions: (1) for individual learning, research, or appreciation, use of another's published works; (2) for the introduction, review of a work, or explanation of an issue, appropriately quoting another's published works within a work; (3) for the translation or limited reproduction of published works for classroom use or scientific research in schools, for use by teaching or research staff, but not for publication or distribution; (4) for libraries, archives, memorials, museums, art galleries, etc., to copy works in their collections for display or preservation purposes; (5) to sketch, paint, photograph, or video art works displayed in outdoor public places.

Based on this, in China, various types of libraries have achieved the online sharing of e-books to some extent. These libraries typically obtain e-books through legal channels, including purchasing licenses from publishers or using platforms that provide authorized access to e-books to manage copyright issues. These licenses allow libraries to lend digital copies to their patrons under specific conditions that respect copyright law or to offer online reading. Additionally, some libraries may participate in digital consortia, sharing resources among several institutions to further expand access while adhering to copyright agreements. Specifically for county-level public libraries, online shared e-books are of a public welfare nature, and the related copyright fees are paid by the libraries themselves, funded by the "Special Construction Funds for Digital Libraries" supported by finance. How these circulation methods impact the copyright of e-books is beyond the scope of this study.

E-books, characterized by rapid dissemination and appeal to younger audiences, are crucial for quickly bridging the "digital divide" and "talent gap" between urban and rural areas and addressing educational inequities between them. Real and Rose [18] found in their study on U.S. rural public libraries that wireless public internet access and the broadband capacity significantly impact the quality of rural library services, underscoring the importance of electronic information dissemination via the internet as a critical responsibility of contemporary public libraries, especially those in remote areas. Santoso et al.'s research suggests that e-books are more effective than physical books and are vital for improving education in rural schools in Indonesia [51]. Zhang and Zhou [52] argue that Education Informatization 2.0, which includes e-books, places a greater emphasis on educational equity. Additionally, Prasetianto et al.'s research indicates that digital reading can significantly enhance agronomy students' reading comprehension in an English-language environment [53]. In conclusion, the increase in e-books within rural public libraries plays a significant role in disseminating specialized knowledge to rural residents, thereby contributing to the development of more skilled professionals. These newly developed talents are crucial for restructuring the industry and enhancing the scale efficiency.

It is also noteworthy that the Circulation of County Libraries per 10,000 Rural Residents has a significant negative impact in Model 1, contrary to intuition, which generally suggests that the circulation of collections truly represents their value. It should be explained that, according to the China Library Yearbook, the circulation statistics only include printed materials and physical audiovisual documents (such as CDs), excluding the circulation of e-books. In today's era of the rapid development of mobile internet, printed materials and physical documents carried by CDs are being largely replaced by online-distributed e-books, which explains the annual decline in circulation. Correspondingly, there is an annual improvement in the agricultural economic efficiency, so the negative impact on the agricultural economic efficiency reflected in the model is normal. This also inadvertently underscores the importance of e-books and the construction of digital libraries from another perspective. This simultaneously highlights a regret: due to the lack of circulation data for e-books, the study lacks direct evidence of their impact.

5. Conclusions

In this empirical analysis using China as a case study, the following conclusions were drawn:

Between 2012 and 2019, the efficiency of China's agricultural economy showed a continuous improvement. Correlation analysis indicates that the number of rural libraries per capita has a positive impact on China's agricultural economy. Therefore, it is recommended to increase the number of rural public libraries per capita in rural areas.

During the period from 2012 to 2019, the agricultural economic efficiency in China progressively increased. Correlation analysis indicates that the number of rural libraries per capita among rural residents has a positive impact on China's agricultural economy. Currently, it is advisable to increase the number of rural public libraries per capita. Given the current situation in China, where urbanization is leading to a rapid decline in the rural population, in most cases, maintaining the current number of rural public libraries is sufficient to effectively increase the per capita availability.

The construction of rural public libraries should focus on enhancing their software capabilities and improving their service levels for readers. Specific measures include increasing the number of books, expanding the collection of e-books, and providing more reading seats for patrons. This implies that, initially, it is essential to enhance the management efficiency of libraries. Subsequently, for the same reasons, maintaining or slightly increasing the number of professional staff can achieve the same effect.

The hardware facilities of rural public libraries should be appropriately reduced, such as discontinuing the further construction of buildings and reducing the number of computers, among other measures. The funds saved should be allocated to the construction and maintenance of digital libraries. Special attention should be given to new public library constructions, particularly in underdeveloped areas with insufficient infrastructure.

In summary, this study suggests that future rural public libraries should become more professional, digital, and mobile internet-based. The future direction of development should focus on providing more professional services to rural residents through more robust information technology.

It is important to note that China's unique characteristics, such as its highly centralized government, may limit the applicability of these findings to other regions. However, the combined research methodology of using a global DEA model with a Tobit model, as applied in this study, possesses a degree of generalizability.

Regarding future developments of this study, plans are in place to address its current limitations. Initially, this research merely touched upon the correlation between rural public libraries and agricultural economic efficiency. Future efforts will attempt to delve into the causality between the two. Furthermore, further detailed investigations are planned, such as examining the impacts of various factors, including the types of collections in public libraries, their specialties, and publication years, as well as the influence of e-book circulation volumes and distribution methods.

Supplementary Materials: The following supporting information can be downloaded at https:// www.mdpi.com/article/10.3390/su16198343/s1, Table S1: Panel Data on Input and Output Variables of Agricultural Circular Economy in 27 Provinces. Table S2: Global SBM Model Scores for Agricultural Circular Economy in 27 Provinces and Autonomous Regions of China, 2012–2019. Table S3: Data on County-Level Public Libraries in 27 Provinces and Autonomous Regions of China, 2012–2019.

Author Contributions: D.Z.: conceptualization; formal analysis; validation; writing—original draft preparation; project administration. J.L.: software; investigation; validation; resources; supervision. Y.Y.: software; investigation; data curation; visualization. R.Z.: conceptualization; formal analysis; validation. Y.Z.: methodology; supervision; resources; writing—review and editing. All authors have read and agreed to the published version of the manuscript.

Funding: This research received no external funding.

Institutional Review Board Statement: Not applicable.

Informed Consent Statement: Not applicable

Data Availability Statement: Data will be provided upon request.

Conflicts of Interest: The authors declare no conflicts of interest.

References

1. Eve, J.; de Groot, M.; Schmidt, A.M. Supporting lifelong learning in public libraries across Europe. *Libr. Rev.* **2007**, *56*, 393–406. [CrossRef]
2. Nielsen, B.G.; Borlund, P. Libraries and Lifelong Learning. *Perspect. Innov. Econ. Bus.* **2014**, *14*, 94–102. [CrossRef]
3. Kinney, B. The internet, public libraries, and the digital divide. *Public Libr. Q.* **2010**, *29*, 104–161. [CrossRef]
4. Abumandour, E.-S.T. Public libraries' role in supporting e-learning and spreading lifelong education: A case study. *J. Res. Innov. Teach. Learn.* **2020**, *14*, 178–217. [CrossRef]
5. Manžuch, Z.; Macevičiūtė, E. Getting ready to reduce the digital divide: Scenarios of Lithuanian public libraries. *J. Assoc. Inf. Sci. Technol.* **2020**, *71*, 1205–1217. [CrossRef]
6. Taylor, N.G.; Jaeger, P.T.; McDermott, A.J.; Kodama, C.M.; Bertot, J.C. Public libraries in the new economy: Twenty-first-century skills, the internet, and community needs. *Public Libr. Q.* **2012**, *31*, 191–219. [CrossRef]
7. Fairbairn, J.; Lipeikaite, U. Small services big impact: Public libraries' contribution to urban and rural development. *Retrieved January* **2014**, *20*, 2015.
8. Mehra, B.; Bishop, B.W.; Partee, R.P. Small business perspectives on the role of rural libraries in economic development. *Libr. Q.* **2017**, *87*, 17–35. [CrossRef]
9. Audunson, R. The public library as a meeting-place in a multicultural and digital context: The necessity of low-intensive meeting-places. *J. Doc.* **2005**, *61*, 429–441. [CrossRef]
10. Flaherty, M.G.; Miller, D. Rural Public Libraries as Community Change Agents: Opportunities for Health Promotion. *J. Educ. Libr. Inf. Sci.* **2016**, *57*, 143–150. [CrossRef]
11. Audunson, R.; Aabø, S.; Blomgren, R.; Evjen, S.; Jochumsen, H.; Larsen, H.; Rasmussen, C.H.; Vårheim, A.; Johnston, J.; Koizumi, M. Public libraries as an infrastructure for a sustainable public sphere: A comprehensive review of research. *J. Doc.* **2019**, *75*, 773–790. [CrossRef]
12. Mansour, E. Libraries as agents for development: The potential role of Egyptian rural public libraries towards the attainment of Sustainable Development Goals based on the UN 2030 Agenda. *J. Librariansh. Inf. Sci.* **2020**, *52*, 121–136. [CrossRef]
13. Panda, S.; Das, S.K. Role of public libraries in promoting social sustainability for the united nations sustainable development goals (SDG): An exploratory study. *Libr. Waves* **2022**, *8*, 129–138.
14. Wang, M.; Hua, Y.; Sun, H.L.; Chen, Y. Bridging the rural digital divide: Avoiding the user churn of rural public digital cultural services. *Aslib J. Inf. Manag.* **2023**, *75*, 730–751. [CrossRef]
15. Xin, Z. Practices and thinking of public libraries in China during COVID-19. *IFLA J.* **2021**, *48*, 161–173. [CrossRef]
16. Omeluzor, S.U.; Oyovwe-Tinuoye, G.O.; Emeka-Ukwu, U. An assessment of rural libraries and information services for rural development. *Electron. Libr.* **2017**, *35*, 445–471. [CrossRef]
17. Bishop, B.W.; Mehra, B.; Partee Ii, R.P. The Role of Rural Public Libraries in Small Business Development. *Public Libr. Q.* **2016**, *35*, 37–48. [CrossRef]
18. Real, B.; Rose, R.N. *Rural Libraries in the United States: Recent Strides, Future Possibilities, and Meeting Community Needs*; American Library Association: Chicago, IL, USA, 2017.
19. Sun, R. Rural Education Revitalisation in the Context of National Cultural Revival. *J. Sociol. Ethnol.* **2022**, *4*, 38–48.
20. Wang, Y.; Huang, Y.; Zhang, Y. Coupling and Coordinated Development of Digital Economy and Rural Revitalisation and Analysis of Influencing Factors. *Sustainability* **2023**, *15*, 3779. [CrossRef]
21. Cheng, C. Libraries in China today. *Libri* **1959**, *9*, 105–110. [CrossRef]
22. Yi, Z. History of library developments in China. In Proceedings of the IFLA WLIC 2013—Future Libraries: Infinite Possibilities, Singapore, 21 August 2013.
23. Library, J. Dongyang Library. 2014. Available online: https://www.jhlib.com/jhhdxy/1903.htm (accessed on 1 August 2024).

24. Ministry of Culture and Tourism of the People's Republic of China. Public Libraries Law of the People's Republic of China. 2017/11/04. Available online: https://zwgk.mct.gov.cn/zfxxgkml/zcfg/fl/202012/t20201204_905426.html (accessed on 1 June 2024).
25. Central Government of the People's Republic of China. China. Opinions on Accelerating the Construction of a Modern Public Cultural Service System. 2015. Available online: https://www.gov.cn/gongbao/content/2015/content_2809127.htm (accessed on 1 June 2024).
26. Ministry of Culture and Tourism of the People's Republic of China. Notification from the Ministry of Culture Office on Issuing "Digital Library Promotion Project" Standards for Provincial and Municipal Digital Library Hardware Configuration. 2011. Available online: https://zwgk.mct.gov.cn/zfxxgkml/ggfw/202012/t20201205_916537.html (accessed on 1 June 2024).
27. Shao, H.; He, Q.; Cha, G.; Xi, Q. Comparison of the Assessment Systems of Public Libraries in the United States and China. *J. Aust. Libr. Inf. Assoc.* **2019**, *68*, 164–179. [CrossRef]
28. Zhao, Y.; Wan, Y.; Chun, J. An Unbalanced and Inadequate Development of the Chinese Public Libraries' Public Culture Services: An Investigation of 31 Senior Library Specialists. *Libri* **2021**, *71*, 293–306. [CrossRef]
29. Sun, Y. Analysis of the Development Level of Chinese Public Libraries. In Proceedings of the 2022 2nd International Conference on Business Administration and Data Science 2022, Kashi, China, 28–30 October 2022; pp. 1424–1432.
30. Bin, F.; Miao, Q. Electronic publications for Chinese public libraries: Challenges and opportunities. *Electron. Libr.* **2005**, *23*, 181–188. [CrossRef]
31. Yao, L.; Zhao, P. Digital libraries in China: Progress and prospects. *Electron. Libr.* **2009**, *27*, 308–318. [CrossRef]
32. Wanyan, D.; Hu, J. How to provide public digital cultural services in China? *Libr. Hi Tech* **2020**, *38*, 504–521. [CrossRef]
33. Li, Y.; Xu, T. From Digital Resource Construction to Public Cultural Services: The Innovation Approaches of Digital Culture Construction of Public Libraries. In Proceedings of the World Congress on Services, Chicago, IL, USA, 5–10 September 2021; Springer: Cham, Switzerland, 2022; pp. 106–112.
34. Schmookler, J. The changing efficiency of the American economy, 1869–1938. *Rev. Econ. Stat.* **1952**, 214–231. [CrossRef]
35. Mundaca, L.; Neij, L.; Worrell, E.; McNeil, M. Evaluating energy efficiency policies with energy-economy models. *Annu. Rev. Environ. Resour.* **2010**, *35*, 305–344. [CrossRef]
36. Bukarica, V.; Tomšić, Ž. Energy efficiency policy evaluation by moving from techno-economic towards whole society perspective on energy efficiency market. *Renew. Sustain. Energy Rev.* **2017**, *70*, 968–975. [CrossRef]
37. Paul, C.; Nehring, R.; Banker, D.; Somwaru, A. Scale economies and efficiency in US agriculture: Are traditional farms history? *J. Product. Anal.* **2004**, *22*, 185–205. [CrossRef]
38. Reza Anik, A.; Rahman, S.; Sarker, J.R. Five decades of productivity and efficiency changes in world agriculture (1969–2013). *Agriculture* **2020**, *10*, 200. [CrossRef]
39. Tone, K. Slacks-based measure of efficiency. In *Handbook on Data Envelopment Analysis*; Springer Science & Business Media: New York, NY, USA, 2011; pp. 195–209.
40. Golany, B.; Roll, Y. An application procedure for DEA. *Omega* **1989**, *17*, 237–250. [CrossRef]
41. Xu, D.; Yong, Z.; Deng, X.; Zhuang, L.; Qing, C. Rural-urban migration and its effect on land transfer in rural China. *Land* **2020**, *9*, 81. [CrossRef]
42. Su, J.; Su, K.; Wang, S. Does the digital economy promote industrial structural upgrading?—A test of mediating effects based on heterogeneous technological innovation. *Sustainability* **2021**, *13*, 10105. [CrossRef]
43. Lin, S.; Xiao, L.; Wang, X. Does air pollution hinder technological innovation in China? A perspective of innovation value chain. *J. Clean. Prod.* **2021**, *278*, 123326. [CrossRef]
44. Bini Okiy, R. Information for rural development: Challenge for Nigerian rural public libraries. *Libr. Rev.* **2003**, *52*, 126–131. [CrossRef]
45. Sultana, R. Rural library services: Lessons from five rural public libraries in West Bengal. *Int. J. Humanit. Soc. Sci. Invent.* **2014**, *3*, 27–30.
46. Svendsen, G.L.H. Public Libraries as Breeding Grounds for Bonding, Bridging and Institutional Social Capital: The Case of Branch Libraries in Rural Denmark. *Sociol. Rural.* **2013**, *53*, 52–73. [CrossRef]
47. Sikes, S. Rural public library outreach services and elder users: A case study of the Washington County (VA) Public Library. *Public Libr. Q.* **2020**, *39*, 363–388. [CrossRef]
48. National Bureau of Statistics of China. China Statistical Yearbook 2020. 2020. Available online: https://www.stats.gov.cn/sj/ndsj/2020/indexeh.htm (accessed on 1 March 2024).
49. De Jager, K.; Nassimbeni, M. Information literacy in practice: Engaging public library workers in rural South Africa. *IFLA J.* **2007**, *33*, 313–322. [CrossRef]
50. Rehman, M.S. Exploring the impact of human resources management on organizational performance: A study of public sector organizations. *J. Bus. Stud. Q.* **2011**, *2*, 1.
51. Santoso, T.N.B.; Siswandari, S.; Sawiji, H. The effectiveness of eBook versus printed books in the rural schools in Indonesia at the modern learning era. *Int. J. Educ. Res. Rev.* **2018**, *3*, 77–84. [CrossRef]

52. Zhang, L.; Zhou, Y. Education Informatization: An Effective Way to Promote Educational Equity. In Proceedings of the 2020 International Conference on Data Processing Techniques and Applications for Cyber-Physical Systems, Laibin, China, 11–12 December 2020; Springer: Singapore, 2021; pp. 837–845.
53. Prasetianto, M.; Maharddhika, R.; Trimus, S.E.P.L. The digital-mediated extensive reading on English Language learning of agriculture students. *J. Educ. Learn. (EduLearn)* **2024**, *18*, 107–115. [CrossRef]

Article

Economic Sustainability Foraging Scenarios for Ruminant Meat Production—A Climate Change Adapting Alternative

Rodica Chetroiu, Steliana Rodino *, Vili Dragomir, Petruța Antoneta Turek-Rahoveanu and Alexandra Marina Manolache

Research Institute for Agriculture Economy and Rural Development, 011464 Bucharest, Romania; rodica.chetroiu@iceadr.ro (R.C.); dragomir.vili@iceadr.ro (V.D.)
* Correspondence: steliana.rodino@yahoo.com

Abstract: Climate changes affect all agricultural production systems, directly or indirectly, including that of ruminant meat, through the limitation of forage resources sensitive to reduced water regimes and drought. The present paper assessed the economic sustainability of ruminant meat production in the context of climate change, with a particular focus on integrating bioeconomy principles through the use of drought-resistant crops such as sorghum and millet in livestock feed. This study included scenarios for two farm-level models, a sheep fattening farm and a cattle fattening farm, to determine the economic benefit and impact of integrating resilient crops in the total feed ration. The findings showed that the dry scenario system could offer economic and environmental advantages over traditional water-intensive crops like maize. The results demonstrated that replacing maize with sorghum or millet could result in a reduction in feed costs and enhanced economic benefit over the traditional feed system.

Keywords: meat; sheep; cattle; climate change; economic sustainability

Citation: Chetroiu, R.; Rodino, S.; Dragomir, V.; Turek-Rahoveanu, P.A.; Manolache, A.M. Economic Sustainability Foraging Scenarios for Ruminant Meat Production—A Climate Change Adapting Alternative. *Sustainability* **2024**, *16*, 9858. https://doi.org/10.3390/su16229858

Academic Editors: Efstratios Loizou, Achilleas Kontogeorgos and Fotios Chatzitheodoridis

Received: 20 September 2024
Revised: 31 October 2024
Accepted: 9 November 2024
Published: 12 November 2024

1. Introduction

Climate change is exerting substantial pressure on global agricultural systems, with increasing temperatures, changing precipitation patterns, and more frequent extreme weather events [1,2]. The consequences are impacting both crop yields and livestock production. In the European Union and Romania in particular, the agricultural sector is a fundamental sector of the economy, being a major contributor to the gross domestic product [2]. Agriculture is an extremely vulnerable sector in Romania since it is highly dependent on climatic conditions. As a result, farmers' livelihoods are seriously endangered, especially those managing small subsistence farms [3,4].

Regarding the livestock sector, the reduced availability and quality of forages have made the maintenance of productivity and profitability a difficult task [3]. Public expenditure and policy changes are being discussed all over the EU, aiming to seek solutions for adaptation of the agricultural production methods in order to overcome the challenges and provide resilience and sustainability [5]. As the sheep and beef sectors are experiencing pressures due to climate change, such as influences on physiological processes, on production and welfare, as well as through changes in the availability and quality of forages [6], new approaches, such as integrating drought-tolerant crops and applying resource efficiency principles integrated within the bioeconomy, are expected to make a significant contribution to agricultural viability and reducing the environmental footprint [6].

Climate change could affect both meat production and meat quality by reducing nutritional intake. The impact varies across regions due to differences in climate and agricultural conditions [1]. Adequate nutrition is essential for weight gain in ruminants, and forage is an essential component of ruminant nutrition. As forage quality varies greatly from one forage crop to another and nutritional needs vary between animal species, providing adequate feed for ruminants requires a balance [5]. Adaptation to current environmental and climate conditions in animal husbandry varies from technological solutions to changes in the management or structure of agricultural holdings, based on analysis of local or regional conditions. The adaptation of agricultural holdings aims to increase productivity, based on the current knowledge and experience of farmers. It will be necessary to pay attention to the stability and resilience of agricultural production and the income of farmers in vulnerable areas. Diversification of agricultural activities and sources of income, changes in the structure of agricultural holdings, and additional investments could also become necessary [4].

Meat is one of the main sources of protein for human consumption, being composed of amino acids, vitamins, minerals, etc. with nutritional properties [7]. At the same time, beef and sheep meat are of particular importance in traditional European and Romanian cuisine. Extreme weather conditions can lead to a reduction in the number of animals, affecting the level of production costs and the availability of meat. The European sheepfarming sector is facing market difficulties that affect both economic performance and long-term sustainability [8]. Small ruminant farming on the European continent is also facing competitiveness challenges caused by the partial decoupling of direct production payments and changes in the quality of life in rural areas [9].

Cereal productivity can be affected by climate change by lowering biomass, reducing plant quality, changing sowing or harvesting periods, etc. [10]. Thus, the provision of reduced feed both quantitatively and qualitatively determines both the failure to achieve the planned meat production indicators and the decrease in meat quality, but it even affects the vital functions of the animals.

Sorghum is considered the world's fifth most important cereal crop, judging by volume of production as well as area cultivated. Sorghum is used as a source of energy and protein, both in the feed of ruminants and nonruminants [11,12]. It is the cereal crop recording the highest levels of drought tolerance and is considered the 'camel of the crops' [13,14]. It has an energy content similar to that of corn, successfully replacing it in the feeding rations of ruminants that have an increased tolerance to moderate concentrations of food tannin [11]. However, considering the experiments in the case of cattle indicate a feed conversion of approximately 10% lower compared to those diets based on corn, in the case of sheep, the conversion was only 5%, and the replacement of 100–400 g/kg of corn with the same amount of sorghum protected the animals from parasite infection (*H. contortus*), and the color and quality of the meat were improved compared to the same diets based on corn [15–17]. It has also been shown that the introduction of tannins present in several forage plants (*L. corniculatus* and *sulla*) in ruminant diets [17], has beneficial effects on wool quality, milk production, reproductive performance, and body weight [18].

Moreover, the sorghum crop is ideal for the sustainability of the agro-food system. Considering the growth of the population leading to greater demand for food, the impoverishment of the soil in nutrients, and the loss of biodiversity, the sorghum crop could represent one of the insufficiently used and often neglected food resources that might be reconsidered [19,20].

The benefits of using sorghum and millet as fodder alternatives to green corn in raising cattle and sheep are multiple and have sustainability characteristics, such as being resistant to extreme climates, being sustainable for arid and semi-arid areas and for soils with a low degree of fertility, providing nutrients with a high protein and energy content, having a low need for inputs, having lower costs for fertilizers or pesticides, showing resistance to pests, etc. [21]

Previous research has shown that sorghum silage as a substitute for corn silage is a successful alternative in areas where corn cultivation is insecure. Sorghum grains can replace those of corn, barley, or wheat. In climate and soil conditions considered optimal for sorghum and millet, the maize suffers and is poorly productive.

Millet is a highly nutritious crop for health benefits in livestock and possesses the capacity for high yields with low input. Millet, especially pearl and finger millet varieties, has gained favor for its animal feed potential on account of nutrient composition and hardiness against adverse conditions. Further, finger millet has the potential for bioethanol production to be used as nutritious food with antioxidant properties, hence qualifying it for the title 'crop of the future' [22,23]. According to Hassan et al. (2021), millets are proximately rich in protein, dietary fiber, and micronutrients, hence promising from the perspective of nutritional relevance as alternatives to maize, particularly under semi-arid conditions with restricted water and poor fertility [24]. This nutritional profile makes it particularly beneficial for improving livestock health and productivity, especially in resource-limited settings [25].

Experiments in the tropical areas of India have indicated that the residues of sorghum and millet crops meet the requirements of cattle feed in most animal breeding systems [20]. The studies carried out to compare millet grains with corn and sorghum indicated that millet can be introduced into the diet of cattle for meat, but the diets must be formulated in such a way that a larger amount of protein from this cereal can be efficiently used as a replacement for additional protein [26]. Another study that investigated the feeding behavior of sheep and goats, fed randomly with corn, sorghum, and millet, highlights that the digestibility of crude protein in millet feed was higher compared to other feeds [27].

For example, it can be used in broiler diets at inclusion rates of at least 50% without compromising performance and egg laying. More importantly, in Brazil, it is being increasingly adopted as a cover crop in no-till soybean production systems due to its positive effect on soil health and nematode infestation [28]. Other studies also emphasized the value of millet for enhancing feed efficiency towards improved productivity in livestock. Indeed, inclusion of millet into animal feed may result in various positive impacts on nutrient utilization and general health status. According to Ganapathy et al. [29], including millet in animal feed might be very beneficial in terms of improvement in nutrient utilization and overall health. Therefore, millet is promising as a crop for enhancing productivity and sustainability in ruminant production in climate change-vulnerable areas, generally due to its hardiness under environmental stresses coupled with nutritional benefits.

Sustainability scenarios for ruminant meat production presented in this paper especially emphasize the economic aspects of the sustainability of the field, which ensures the continuity of meat production in conditions of climate change by applying nutritional interventions.

This research investigates how integrating bioeconomy principles into agricultural practices can mitigate the impacts of climate change on ruminant meat production by promoting the use of resilient, resource-efficient crops like sorghum and millet. The aim was to emphasize the benefits of using these two crops in a transition towards bioeconomy, surpassing their statute of underutilized crops.

For achieving this, several key research questions (RQ) were explored to address the economic sustainability of ruminant meat production under climate change within a bioeconomy framework as follows:

RQ 1: How can the integration of drought-resistant crops such as sorghum into animal feed impact the profitability of ruminant farms?

RQ2: How does the inclusion of millet in livestock diets affect the overall productivity and economic sustainability of ruminant farms under drought-prone conditions?

RQ3: What are the overall economic benefits of replacing water-intensive crops like maize with resilient alternatives in feed rations?

RQ4: How can forage alternatives contribute to a reduction in meat production decline in conditions of climate change?

Addressing these questions will provide a good understanding of the role of resilient agricultural practices in enhancing both economic and environmental sustainability in the livestock sector. Moreover, optimizing feed inputs will enhance the adoption of circular economy principles in resource management and improve the long-term sustainability of animal husbandry operations in the face of climate-induced challenges.

2. Materials and Methods

Considering the need to reduce the economic vulnerability of ruminant farms facing climate change, as well as the sensitivity of Romanian fodder systems to drought, economic calculations of the profitability of the use in feeding for young sheep and cattle for fattening were developed, with the use in rations of some crops less dependent on the summer water regime, such as millet and sorghum.

Scenarios were built and calculations were elaborated for 2 farms framed into medium size classes, as follows: (i) young sheep for a fattening farm with 1000 heads, with the initial weight of the animals of 15 kg/head, a final weight of 40 kg/head, and an average daily gain of 200 g/day/head; (ii) a farm of 100 young cattle for fattening, with an initial weight of 100 kg/head, a final weight of 450 kg/head, and an average daily gain of 1000 g.

For the establishment of the fodder rations, the energy norms nutritive unit meat (UNC) and digestible protein (PDI), developed by specialists from the Institute of Animal Biology and Nutrition of Romania, were considered for each species, depending on the weight of the animals and the average daily gain. In the case of young sheep for meat, for an average daily gain of 200 g and the life weight at sale of 40 kg/head, animal nutritionists recommended norms of 1.23 nutritive unit meat (UNC) and 100 g digestible protein (PDI) per day. For young cattle for fattening, for an average daily gain of 1000 g and life weight at sale of 450 kg/head, the norms were 7.60 nutritive unit meat (UNC) and 562 g digestible protein (PDI) per day.

The fodder rations used for young sheep for fattening contain sorghum silage, maize silage, alfalfa hay, corn grains, barley grains, and sunflower meal, the fattening period being 140 days, from 12 to 40 kg live weight, generally during May–September. For young fattening bulls, the fattening period lasts one year; in the warm season, the feed is made with alfalfa hay, green millet, sunflower meal, and barley grains; and in the cold season with alfalfa hay, millet silage, sunflower meal, and corn grains.

The summer ration is that in the warm seasons, and the winter ration refers to the cold period of the year. The two types of fodder rations cover the entire period of the year.

Apart from sorghum and millet, as resilient fodder, administered in different forms in fodder rations, the other types of fodder also used, such as alfalfa hay, sunflower meal, and corn and barley grains, as classic fodder, are included to cover the nutritive requirements of the rations. The role of millet and sorghum is to replace the lack of corn when it is supplied in reduced quantities due to periods of drought.

Elaboration of estimates and budgets of revenues and expenses was based on the specific average technological allocations of inputs (feed, energy, medicines, other material expenses, the cost of labor, other fixed expenses, etc.) for each species and on average free market input prices. Optimizing production factors was part of the strategy to increase economic efficiency in conditions of competitiveness and environmental limitations [30].

The income and expenditure budgets were based on determining different categories of costs and incomes and estimating profitability. The value of production was determined by multiplying the unit price by the average production yield. Total costs were calculated by summing variable and fixed costs. Variable costs were determined by summing the costs with forages, biologic material costs, energy costs, medicines, other materials costs, supply costs, and insurance costs. Fixed costs include labor costs, general costs, interest rates, and amortization. Taxable income was determined by the difference between the value of total production and total production costs. In total, 10% of taxes have been deducted from taxable income, resulting in net income. Rate of return (%) was determined by dividing taxable income by total costs for the main production.

Based on the relations known from the economic literature, technical–economic and risk indicators of the activity were calculated according to the following formulas:

$$TC = VC + FC,$$

where

TC—total costs, VC—variable costs, FC—fixed costs

$$VC = Fc + bc + ec + mc + oc + sc + inc,$$

where
Fc—forage costs, bc—biologic material cost, ec—energy cost, mc—medicines cost, oc—other materials costs, sc—supply cost, and inc—insurance cost

$$FC = lc + gc + ic + ac,$$

where
lc—labor cost, gc—general costs, ic—interest rate, and ac—amortization cost

$$VP = VM + VS,$$

where
VP—value of production, VM—value of main production, and VS—value of secondary production

$$TI = VP - TE,$$

where
TI—taxable income, VP—value of production, and TE—total expenses (total costs)

$$NI = TI - t$$

where
NI—net income, TI—taxable income, and t—taxes

$$RR = TI/CM$$

where
RR—rate of return, TI—taxable income, and CM—costs for main production

$$MC = Fc + bc + mc + oc,$$

where
MC—material costs, Fc—forage costs, bc—biologic material cost, mc—medicines cost, and oc—other materials costs

$$P = Up - Pc,$$

where
P—profit, Up—unitary price, and Pc—production cost

$$MVC = VP - VC,$$

where
MVC—margin on variable costs and VP—value of production

$$VC = Fc + bc + ec + mc + oc + sc + inc,$$

where
Fc—forage costs, bc—biologic material cost, ec—energy cost, mc—medicines cost, oc—other materials costs, sc—supply cost, and inc—insurance cost

$$MVC\% = MVC/VP \times 100$$

where

MVC—margin on variable costs and VP—value of production

$$BeV = FC \text{ per head}/MVC\% \times 100,$$

where

BeV—break-even point in value units and FC—fixed costs,

$$MVC\% - MVC/VP \times 100$$

where

MVC—margin on variable costs and VP—value of production

$$BeP = BeV/Pc,$$

where

BeP—break-even point in physical units, BeV—break-even point in value units, and Pc—production cost

$$ORR = BeV/VM \text{ per head} \times 100,$$

where

ORR—operating risk rate, BeV—break-even point in value units, and VM—value of main production

$$SI = (VM \text{ per head} - BeV)/VM \text{ per head, in which SI–security index,}$$

where

VM—value of main production and BeV—break-even point in value units.

3. Results

3.1. Young Sheep for Fattening Farm

In the fodder ration for young fattening sheep, sorghum silage replaces half of the amount of maize silage (as a crop affected by drought), the amounts being 1.5 kg of each, along with 0.4 kg of alfalfa hay and 0.1 kg of concentrates such as corn and barley grain and sunflower meal; these quantities cover the feed norms recommended by animal nutrition specialists (Table 1).

Table 1. Sorghum silage ratio/head.

Forages	kg/Head/Day	UNC	PDI (g)	Quantity/Head/Period	Price, RON/kg	Value/Head/Period
Sorghum silage	1.50	0.41	17.10	210	0.30	63
Maize silage	1.50	0.3	19.5	210	0.15	32
Alfalfa hay	0.40	0.20	30.00	56	0.9	50
Corn grains	0.10	0.14	7.30	14	1.08	15
Barley grains	0.10	0.12	6.5	14	1.09	15
Sunflower meal	0.10	0.07	22.6	14	1.17	16
Total		1.23	103.00			191.7
Norm		1.23	100.00			

UNC—nutritive unit meat; PDI—digestible protein; RON—Romanian national currency. Source: Authors' own elaboration.

The ration structure includes 40% each corn silage and sorghum silage, 11% alfalfa hay, and 3% each corn, barley, and sunflower meal (Figure 1).

In the meat production activity on the farm, several limiting production factors are involved (land, forages, biological material, labor force, etc.); therefore, the economic evaluation must take place in relation to the resources used [31].

In the budget of income and expenditures, feed and biological material occupy approximately equal shares within the variable expenses, which represent 86.3% (429,639 RON) of the total expenses (640,000 RON). The other cost elements are energy, medicine, other material costs, supplies, and insurance. Within fixed expenses, labor costs have the largest share, representing 59%, the rest being general costs, interest, and depreciation (Table 2).

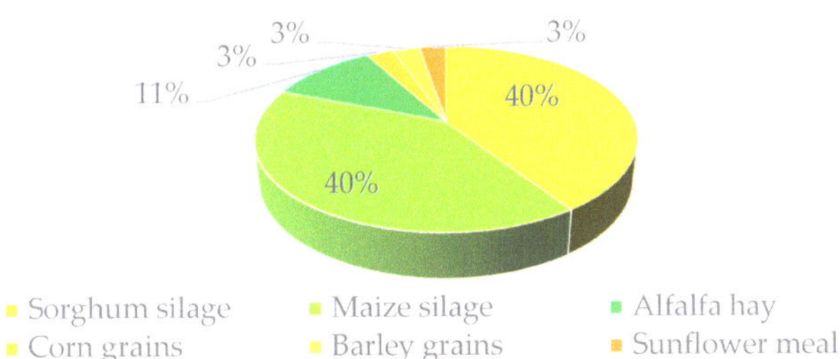

Figure 1. The structure of the fodder ration.

Table 2. Budget of income and expenditures on farm.

Indicators	Average Daily Gain 200 g, Life Weight at Sale 40 kg/Head		
	RON/Head	**RON/kg**	**Value/Farm**
Value of production	640.0	16.00	640,000
Value of main production	640.0	16.00	640,000
Subsidies	0.0	0.00	0.00
Gross product	640.0	16.00	640,000
Total costs	497.7	12.44	497,656
Costs for main production	497.7	12.44	497,656
Variable costs	429.6	10.74	429,639
Forage costs	191.7	4.79	191,660
Biologic material cost	195.6	4.89	195,600
Energy cost	10.0	0.25	10,000
Medicines	14.0	0.35	14,000
Other material costs	6.0	0.15	6000
Supply	10.0	0.25	10,032
Insurance cost	2.3	0.06	2347
Fixed costs	68.0	1.70	68,017
Labor cost	40.0	1.00	40,000
General expenses	10.4	0.26	10,432
Interest rates	12.6	0.31	12,586
Amortization costs	5.0	0.13	5000
Taxable income	142.3	3.56	142,344
Taxes and fees	14.2	0.36	14,234
Net income + subsidies	128.1	3.20	128,109
Rate of return%	28.6	28.6	29
Net income rate%	25.7	25.7	26
Production cost	497.7	12.44	497,656
Price	640.0	16.00	640,000

Source: Authors' own elaboration.

The calculations indicated that the activity of the young sheep for the fattening farm is a profitable activity, with the production cost (12.44 RON/kg) being lower than the average prices (16.00 RON/kg) recorded on the free market, so the profitability rate is 28.6%.

3.2. Young Cattle for Fattening Farm

For the cattle for the fattening farm, summer and winter rations were calculated, using millet as fodder adapted to drought conditions, in summer completely replacing the green fodder, and in winter as silage (Table 3).

Table 3. Green millet summer ratio/head.

Forages	kg/Head/Day	UNC	PDI (g)	Quantity/Head /Period	Price, RON/kg	Value/Head/Period
Alfalfa hay	0.50	0.25	37.50	91	0.90	82
Green millet	5.00	1.15	97.00	910	0.18	164
Sunflower meal	0.48	0.34	108.48	87	1.17	102
Barley grains	4.90	5.88	318.50	892	1.09	972
Total		7.62	561.48			1320
Norm		*7.60*	*562.00*			

UNC—nutritive unit meat; PDI—digestible protein; RON—Romanian national currency. Source: Authors' own elaboration.

In the summer ration structure, 46% is green millet, 45% is barley, 5% is alfalfa hay, and 4% is sunflower meal (Figure 2).

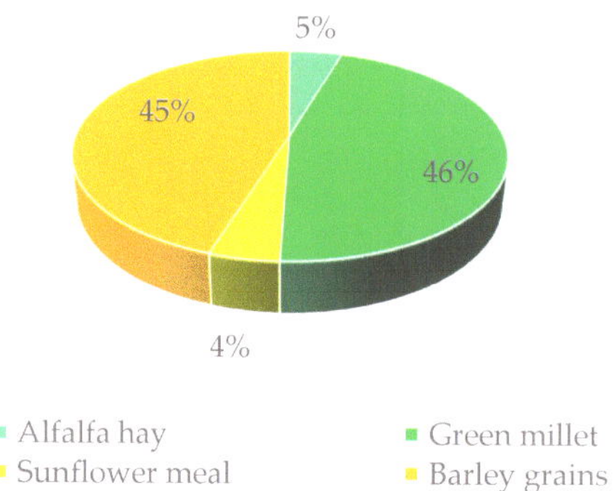

Figure 2. The summer ration structure.

The rations are supplemented with alfalfa hay in both seasons, as well as concentrated fodder, such as corn, barley, and sunflower meal (Table 4).

Table 4. Millet silage winter ratio/head.

Forages	kg/Head/Day	UNC	PDI (g)	Quantity/Head /Period	Price, RON/kg	Value/Head/Period
Alfalfa hay	1.20	0.59	90.00	220	0.90	198
Millet silage	5.00	1.25	86.65	915	0.40	366
Sunflower meal	0.41	0.29	92.66	75	1.17	88
Corn grains	4.00	5.48	292.00	732	1.08	791
Total		7.61	561.31			1442
Norm		*7.60*	*562.00*			

UNC—nutritive unit meat; PDI—digestible protein; RON—Romanian national currency. Source: Authors' own elaboration.

The structure of the winter ration includes 47% millet silage, 38% corn grains, 11% alfalfa hay, and 4% sunflower meal (Figure 3).

Within the income and expenditures budget, variable costs represent 83%, the rest being fixed costs. The cost of forages occupies 66.5% of the variable expenses, constituting the main input. Per kg of live weight, the cost of feed is 6.14 RON. Total cost per kg is 10.52 RON, and the delivery market price is, on average, 11.45 RON, which leads to a rate of return of 8.9%. The rate of net income with subsidies increases to 14.4% (Table 5).

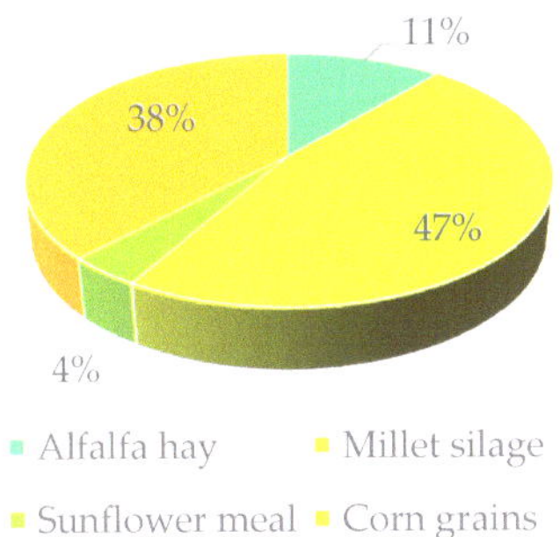

Figure 3. The winter ration structure.

Table 5. Budget of income and expenditures on farm.

Indicators	Average Daily Gain 1000 g, Life Weight at Sale 450 kg/Head		
	RON/Head	RON/kg	Value/Farm
Value of production	5422.5	12.05	542,250
Value of main production	5152.5	11.45	515,250
Subsidies	303.4	0.67	30,340
Gross product	5725.9	12.72	572,590
Total costs	5003.6	11.12	500,357
Costs for main production	4733.6	10.52	473,357
Variable costs	4151.9	9.23	415,191
Forages	2762.0	6.14	276,196
Biologic material	1200.0	2.67	120,000
Energy cost	30.0	0.07	3000
Medicines	75.0	0.17	7500
Other material costs	10.0	0.02	1000
Supply	60.6	0.13	6055
Insurance cost	14.4	0.03	1440
Fixed costs	851.7	1.89	85,165
Labor cost	715.2	1.59	71,520
General expenses	61.2	0.14	6115
Interest rates	42.0	0.09	4200
Amortization costs	33.3	0.07	3330
Taxable income	418.9	0.93	41,893
Taxes and fees	41.9	0.09	4189
Net income + subsidies	680.4	1.51	68,044
Rate of return%	8.9	8.9	8.9
Net income rate%	14.4	14.4	14.4
Production cost	4733.6	10.52	473,357
Price	5152.5	11.45	515,250

Source: Authors' own elaboration.

Comparing the economic and risk indicators of the activity in the two ruminant farms, it is found that the production value per kg of live weight is higher in sheep than in cattle by 32.7% (16 RON compared to 12 RON), but, at the same time, the production cost is higher in sheep compared to cattle, by 18.3%. The margin on variable costs is 87% higher in sheep than in cattle. Break-even point in physical units is reached for sheep at 12.93 kg/head and for cattle at 317.43 kg/head (Table 6).

As illustrated in Figure 4, the value for main production is 39.7% higher in the sheep farm than in the cattle one.

In Figure 5, different categories of expenses per kg of live weight from the two farms are comparatively highlighted, and it can be seen that, apart from fixed expenses, respectively, those with labor force, the other categories of unit costs are lower for cattle than in sheep farms. This could be explained by the differences regarding the period of exploitation of the two species, the animals weight at delivery, as well as the specific allocations of inputs (Figure 5).

Table 6. Comparison between farm economic and risk indicators.

Indicators	Measure Unit	Sheep	Cattle	Sheep/Cattle
Value of total production	RON/kg	16.00	12.05	1.33
Value of main production	RON/kg	16.00	11.45	1.40
Total costs	RON/kg	12.44	11.12	1.12
Costs for main production	RON/kg	12.44	10.52	1.18
Variable costs	RON/kg	10.74	9.23	1.16
Material costs	RON/kg	10.18	8.99	1.13
Fixed costs	RON/kg	1.70	1.89	0.90
Labor cost	RON/kg	1.00	1.59	0.63
Production cost	RON/kg	12.44	10.52	1.18
Unitary price	RON/kg	16.00	11.45	1.40
Profit/kg	RON	3.56	0.93	3.83
Rate of return	%	28.60	8.85	3.23
Margin on variable costs	RON	5.26	2.82	1.87
Margin on variable costs %	%	32.87	23.43	1.40
Break-even point in value units	RON	206.94	3634.62	
Break-even point in physical units	kg/head	12.93	317.43	
Operating risk rate	%	32.33	70.54	0.46
Security index		0.68	0.29	2.34

RON—Romanian national currency. Source: Authors' own elaboration.

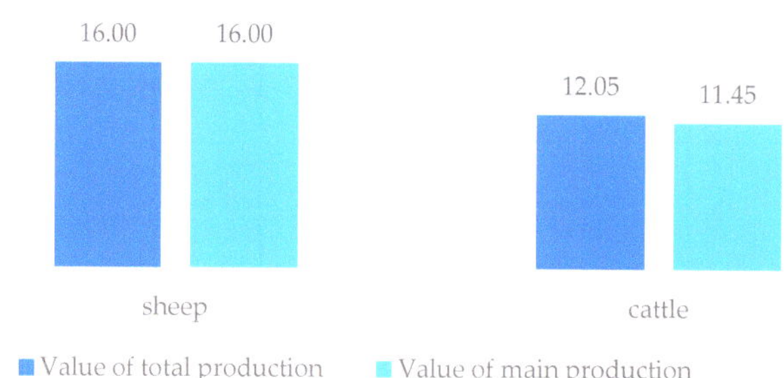

Figure 4. Farms comparative unitary values of production.

From the perspective of economic sustainability, the calculations indicated that in both activities, the costs are covered by the revenues obtained and are profit-producing, which ensures the resumption of production cycles, as the issue of financial stability is a component of economic sustainability [32]. In the case of the sheep farm, the profit is higher than that of cattle (8.85% for cattle and 28.6% for sheep), and this is also due to the higher delivery price (Figure 6). The two categories have different delivery channels in Romania, as well as different consumer preferences. Young, fattened sheep are valued much better on foreign markets, especially the Arab ones, and young bull meat, although more often delivered internally, is still not at the top of consumers' preferences, but poultry and pork.

Any economic activity is accompanied by a certain level of risk, due to a complex of factors taking place and which can influence it, and the calculation of the risk indicators in this case (operating risk rate and security index) indicates a higher risk in the case of the cattle farm, the indicator being the result of the ratio between the break-even point in value units and the value of the main production, which is more advantageous in the case of the sheep farm.

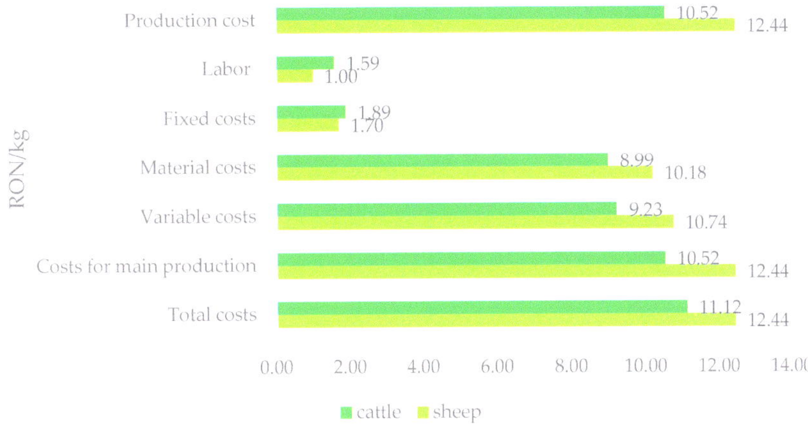

Figure 5. Farms comparative unitary expenditures.

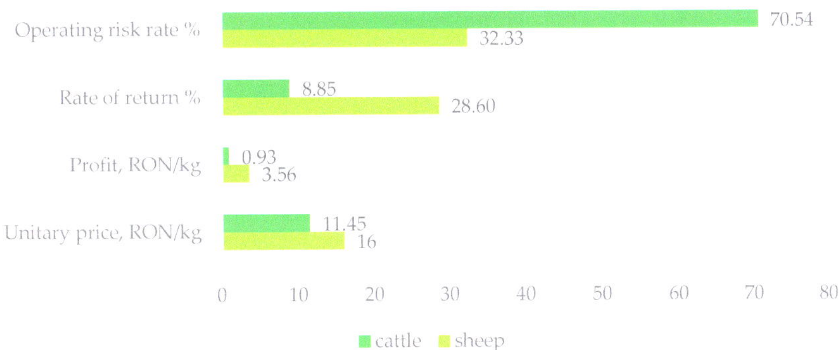

Figure 6. Farms comparative profitability and risk values.

4. Discussion

In sustainable systems, limited natural resources are used over time and subject to conditions of uncertainty [32]. Economic sustainability is a key concern for most farmers, given that it can influence the business continuity decision. Many variables have been found to be significant in ensuring farm profitability, including farm management, quality and availability of agricultural resources, financial management, finding alternatives in conditions of economic vulnerability, etc. [33,34]. The profit ratio is used to compare profitability between farms, and in this case, the value being higher for the sheep farm than the cattle farm is due both to the technological specificities, the consumption of different inputs, the different duration of the productive cycle, which is higher for cattle compared to sheep, as well as and due to the level of sale prices for animals; for sheep, they are higher [34].

The paper presents scenarios regarding the use of alternative fodder in the new environmental and climate conditions, highlighting the obtaining of profit, which can represent a measure that ensures economic sustainability, by calculating the technical–economic indicators that farmers can approach.

Increasing farm production, by this understanding the level of average production, has a significant role in increasing profitability for different farm sizes, but it is even more important for small ones, which also have a higher economic risk [35]. Previous research has shown that different expenditure categories are negatively related to revenue and profitability [36].

Due to increasing consumer demand for food safety, direct-to-consumer marketing channels have been promoted to shorten the supply chains and reduce the number of intermediaries, thus reducing the number of manipulations, especially for fresh products. However, previous research shows that the number of farms using direct-to-consumer marketing channels is still low, and although direct marketing is promoted, wholesale markets are the most profitable marketing channels [37]. In Romania, live cattle are generally delivered to slaughterhouses, respectively, meat processing units or as live animals for export (to a lesser extent), and sheep are generally sold live to customers in Arab countries, or some from Western Europe. This means that animals are sold more as raw materials without added value, which affects the level of profitability.

Resilience and efficiency are components of the sustainability of ruminant productions, and therefore, studying and finding solutions regarding the ability of their production systems to survive and adapt to current climate trends and future challenges are of great importance [38]. In general, sheep are raised on pasture in relatively large flocks and have a higher thermal tolerance compared to large ruminants—cattle [39].

Nutritional management represents one of the pillars of the production and sustainability of livestock in terms of intensification productions. Feeding carries the major costs and greatly influences the quality of products obtained and the impact on the environment [40]. Climate change will intensify, and farmers will be affected, as they rely on climate-sensitive livelihoods. Appropriate adaptation strategies for farmers depend on a clear description of the effects of climate change. In the research conducted by Chingala et al. [41], it was found that high-income farmers have the capital to acquire resources to improve pasture productivity, so income level was one of the major economic factors influencing farmers' perceptions of the impact of climate changes on beef production. This indicates the importance of considering socio-economic factors when developing strategies to adapt to climate change and reduce vulnerability for beef producers.

How can feed alternatives contribute to a reduction in meat production decline? Estimates indicate that replacing or supplementing fodder rations with alternative forages in conditions where classic fodder, such as corn, can be affected and reduced quantitatively and qualitatively can contribute to ensuring the nutritional principles necessary to obtain the planned production level and to avoid the decrease in production of meat.

Many livestock production systems consider increasing efficiency and environmental sustainability features. Production trends in different animal husbandry systems have been associated with the development of science and technology as well as the increase in animal numbers. In the future, production is expected to be increasingly influenced by competition for natural resources and food and forages. The demand for animal products in the future will be determined by socio-economic factors, to which climatic factors will be added [42]. Despite the many challenges facing livestock producers, including economic pressure and feed availability, livestock production is projected to continue to support growth in world meat production. Under heat stress, production can be improved by changes in diet composition [43].

The contribution of ruminants to food security and other dimensions of sustainability will be affected by climate change, although the magnitude of the impact remains unknown. Livestock and climate change studies often focus on describing adaptation practices, like the present study. Climate-related impact risks result from the interaction of climate-related hazards with the exposure and vulnerability of natural systems [44,45].

Livestock rearing is a major factor in increasing the overall contribution to household livelihoods, and on the other hand, livestock can provide immediate income that helps cover important household expenses. In general, the demand for meat has increased among the population, which is increasingly urbanized [44].

Beef cattle are raised globally, following different management practices, and those in colder regions are vulnerable to high environmental temperatures. The use of natural resources such as feed in different production systems is different, and finding solutions to prevent their depletion will have to be a constant challenge [46].

By focusing on the use of drought-resistant crops, such as millet and sorghum, this study involved analysis of resource efficiency and resilience in the production system—two very important facets of the bioeconomy. A bioeconomy system minimizes waste, restores resources, and regenerates land. Replacing water-demanding crops, which include maize, with sorghum and millet could contribute to that since it optimizes the use of natural resources, such as water, at equal or higher levels of productivity in fragile climates. Indeed, sorghum has proved that it may afford high water use efficiency—even outcompeting other crops like corn in dry areas—while still offering excellent nutritional value for livestock [47]. Besides that, the sorghum plant succeeds in semi-arid conditions, eventually reducing overdependence on other water-consuming crops, adding to a more sustainable and regenerative agricultural system in a major way [13,14,47].

Millet was also found to be very adaptable to marginal lands with high levels of salinity, waterlogging, and pest infestations [48]. Therefore, it thrives under conditions that would hardly support other cereal crops. The crop's relatively short growing season, usually three to four months, and a resistant nature against most diseases reduce the level of input required for its cultivation, hence positioning millet as more viable where means are limited [47–49]. Another critical reason for its increased importance is the crop's high value as a fodder source for livestock, considering that the crops are rich in protein; this effectively reduces feed costs and contributes to attaining food security through sustainable livestock production [50]. Therefore, millet makes very important contributions to bioeconomy applications beyond being a food crop. When the grains are harvested, the straw becomes valuable forage for animals, hence embracing the no-waste model, which is an important aspect of the bioeconomy approach [51,52]. It is resilient to adverse abiotic environmental stresses like salinity and drought, hence requiring minimal chemical use, consequently reducing environmental pollution while optimizing yield [53]. The contribution to the bioeconomy in this case is through optimized resource use and circularity in resource flow.

Aside from this, ensuring that the economic calculations of fattening farms for sheep and cattle are optimized will enhance profitability and contribute to the circular economy by efficiently using inputs like feed. This, in turn, makes more effective use of the available resources in a very strategic manner—a core principle of the circular economy closing the resource loops. It has been observed that diversification with sustainable crop alternatives like sorghum reduces the application of water and enriches the soil health, hence making sorghum a very feasible crop within the perspective of the circular economy [54].

5. Conclusions

Following the recommendations of animal nutrition specialists to introduce substitute feed, resilient to the new environmental and climate conditions, into the forage ration of ruminants for meat is one of the solutions to adapt to the effects of drought. The use of sorghum and millet, as green fodder in summer rations and as silage in winter rations, to completely or partially replace some drought-sensitive forages, such as green maize, can provide the necessary nutrients to achieve the planned average daily weight gain.

The calculations of the technical–economic and risk indicators of the activity indicated the fact that, both in the case of young sheep for fattening and of cattle, the production activity can be completed with obtaining a certain level of profit. The economic impact is given by the financial results that farms obtain under the action of different factors. In conditions of adaptation to climate change, the rate of return indicator, which had positive values of 8.85% for cattle and 28.6% for sheep, indicated that adaptation measures can have economically valid responses. Results underlined that these crops, included in rations, reduce not only the vulnerability of farming to climate-induced shocks but also contribute to the circular economy through the optimization of resources and minimization of waste. Therefore, the practices of sustainable feeding guarantee the continuity of activities on the farm with increased productivity and at reduced cost.

There are, however, some limitations that have to be considered for further research, including the geographical specificity of the present study, limited crop diversity, and the addition of several more economic parameters that describe market fluctuations. Nevertheless, this study provides a good understanding of how resilient agricultural practices, coupled with circular economy principles of reducing material inputs such as water and energy, are in agreement with promoting economic and environmental sustainability in ruminant meat production. These are the kinds of strategies that, as climate change continues to threaten agricultural systems worldwide, will prove necessary for farm profitability and food security over the coming decades.

Author Contributions: Conceptualization, R.C.; methodology, R.C. and S.R.; software, R.C., S.R. and V.D.; formal analysis, R.C., S.R., V.D., P.A.T.-R. and A.M.M.; investigation, R.C., S.R. and V.D., data curation, V.D., S.R. and A.M.M.; writing—original draft preparation, R.C. and S.R.; writing—review and editing, R.C. and S.R.; project administration, R.C. All authors have read and agreed to the published version of the manuscript.

Funding: This research was partially funded by Ministry of Agriculture and Rural Development, the ADER 22.1.2. and ADER 22.1.4 projects. Number: 22.1.2; 22.1.4.

Institutional Review Board Statement: This study did not require ethical approval.

Informed Consent Statement: Not applicable.

Data Availability Statement: The data presented in this study are available on request from the corresponding author. The data are not publicly available due to privacy.

Conflicts of Interest: The authors declare no conflicts of interest.

References

1. Sarıçiçek, Z. The effects of climate change on animal nutrition, production and product quality and solution suggestions. *Black Sea J. Agric.* **2022**, *5*, 491–509. [CrossRef]
2. Agovino, M.; Casaccia, M.; Ciommi, M.; Ferrara, M.; Marchesano, K. Agriculture, climate change and sustainability: The case of EU-28. *Ecol. Indic.* **2019**, *105*, 525–543. [CrossRef]
3. Joy, A.; Dunshea, F.R.; Leury, B.J.; Clarke, I.J.; DiGiacomo, K.; Chauhan, S.S. Resilience of Small Ruminants to Climate Change and Increased Environmental Temperature: A Review. *Animals* **2020**, *10*, 867. [CrossRef] [PubMed]
4. Rusu, T.; Moraru, P.I. Impact of climate change on crop land and technological recommendations for the main crops in Transylvanian Plain, Romania. *Rom. Agric. Res.* **2015**, *32*, 103–111.
5. Dincă, G.; Netcu, I.-C.; El-Naser, A. Analyzing EU's Agricultural Sector and Public Spending under Climate Change. *Sustainability* **2024**, *16*, 72. [CrossRef]
6. Henry, B.K.; Eckard, R.J.; Beauchemin, K.A. Review: Adaptation of ruminant livestock production systems to climate changes. *Animal* **2018**, *12* (Suppl. 2), s445–s456. [CrossRef]
7. Geletu, U.S.; Usmael, M.A.; Mummed, Y.Y.; Ibrahim, A.M. Quality of cattle meat and its compositional constituents. *Vet. Med. Int.* **2021**, *1*, 7340495. [CrossRef]
8. Theodoridis, A.; Vouraki, S.; Morin, E.; Rupérez, L.R.; Davis, C.; Arsenos, G. Efficiency analysis as a tool for revealing best practices and innovations: The case of the sheep meat sector in Europe. *Animals* **2021**, *11*, 3242. [CrossRef]
9. Gambelli, D.; Solfanelli, F.; Orsini, S.; Zanoli, R. Measuring the Economic Performance of Small Ruminant Farms Using Balanced Scorecard and Importance-Performance Analysis: A European Case Study. *Sustainability* **2021**, *13*, 3321. [CrossRef]
10. Cheng, M.; McCarl, B.; Fei, C. Climate Change and Livestock Production: A Literature Review. *Atmosphere* **2022**, *13*, 140. [CrossRef]
11. McCuistion, K.C.; Selle, P.H.; Liu, S.Y.; Goodband, R.D. Chapter 12 Sorghum as a Feed Grain for Animal Production. In *Sorghum and Millets*; Elsevier: Amsterdam, The Netherlands, 2019. [CrossRef]
12. Popa, M.; Schitea, M.; Petcu, E.; Petrescu, E.; Dobre, Ș.C.; Petcu, V. Evaluation of New Alfalfa Genotypes for Forage, Quality and Seed Yield Potential under Different Field Trials. *Rom. Agric. Res.* **2024**, *41*, 477–488. [CrossRef]
13. Manole, D.; Giumba, A.M.; Ganea, L. Sorghum, an alternative in complementarity with corn, adapted to climate changes. Amzacea Village, Constanta County, Romania. *Sci. Pap. Ser. Manag. Econ. Eng. Agric. Rural. Dev.* **2023**, *23*, 501–512.

14. Ronda, V.; Aruna, C.; Visarada, K.B.R.S.; Venkatesh Bhat, B. Chapter 14—Sorghum for Animal Feed. In *Wood-Head Publishing Series in Food Science, Technology and Nutrition, Breeding Sorghum for Diverse End Uses*; Aruna, C., Visarada, K.B.R.S., Bhat, B.V., Vilas Tonapi, A., Eds.; Woodhead Publishing: Sawston, UK, 2019; pp. 229–238. ISBN 9780081018798. [CrossRef]
15. Ran, T.; Fang, Y.; Wang, Y.T.; Yang, W.Z.; Niu, Y.D.; Sun, X.Z.; Zhong, R.Z. Effects of grain type and conditioning temperature during pelleting on growth performance, ruminal fer-mentation, meat quality and blood metabolites of fattening lambs. *Animal* **2021**, *15*, 100146. [CrossRef] [PubMed]
16. Arsenopoulos, K.V.; Katsarou, E.I.; Mendoza Roldan, J.A.; Fthenakis, G.C.; Papadopoulos, E. Haemonchus contortus parasitism in intensively managed cross-limousin beef calves: Effects on feed conversion and carcass characteristics and potential associations with climatic conditions. *Pathogens* **2022**, *11*, 955. [CrossRef]
17. Sun, H.X.; Gao, T.S.; Zhong, R.Z.; Fang, Y.; Di, G.L.; Zhou, D.W. Effects of corn replacement by sorghum in diets on performance, nutrient utilization, blood parameters, antioxidant status, and meat colour stability in lambs. *Can. J. Anim. Sci.* **2018**, *98*, 723–731. [CrossRef]
18. Soldado, D.; Bessa, R.J.B.; Jerónimo, E. Condensed Tannins as Antioxidants in Ruminants—Effectiveness and Action Mechanisms to Improve Animal Antioxidant Status and Oxidative Stability of Products. *Animals* **2021**, *11*, 3243. [CrossRef]
19. Lin, L.; Lu, Y.; Wang, W.; Luo, W.; Li, T.; Cao, G.; Du, C.; Wei, C.; Yin, F.; Gan, S.; et al. The Influence of High-Concentrate Diet Supplemented with Tannin on Growth Performance, Rumen Fermentation, and Antioxidant Ability of Fattening Lambs. *Animals* **2024**, *14*, 2471. [CrossRef]
20. Proietti, I.; Frazzoli, C.; Mantovani, A. Exploiting Nutritional Value of Staple Foods in the World's Semi-Arid Areas: Risks, Benefits, Challenges and Opportunities of Sorghum. *Healthcare* **2015**, *3*, 172–193. [CrossRef] [PubMed]
21. Wang, M.; McCarl, B.A. Impacts of Climate Change on Livestock Location in the US: A Statistical Analysis. *Land* **2021**, *10*, 1260. [CrossRef]
22. Ravikesavan, R.; Sivamurugan, A.P.; Iyanar, K.; Pramitha, J.L.; Nirmalakumari, A. Millet cultivation: An overview. In *Handbook of Millets-Processing, Quality, and Nutrition Status*; Springer: Singapore, 2022; pp. 23–47.
23. Kurbanbayev, A.; Zargar, M.; Yancheva, H.; Stybayev, G.; Serekpayev, N.; Baitelenova, A.; Mukhanov, N.; Nogayev, A.; Akhylbekova, B.; Abdelkader, M. Ameliorating Forage Crop Resilience in Dry Steppe Zone Using Millet Growth Dynamics. *Agronomy* **2023**, *13*, 3053. [CrossRef]
24. Hassan, Z.; Sebola, N.; Mabelebele, M. The Nutritional Use of Millet Grain for Food and Feed: A Review. *Agric. Food Secur.* **2021**, *10*, 1–14. [CrossRef] [PubMed]
25. Renganathan, V.G.; Vanniarajan, C.; Karthikeyan, A.; Ramalingam, J. Barnyard millet for food and nutritional security: Current status and future research direction. *Front. Genet.* **2020**, *11*, 500. [CrossRef]
26. Daduwal, H.S.; Bhardwaj, R.; Srivastava, R.K. Pearl millet a promising fodder crop for changing climate: A review. *Theor. Appl. Genetics.* **2024**, *137*, 169. [CrossRef]
27. Nasrullah; Khoso, A.N.; Soomro, J.; Marghazani, I.B.; Kakar, M.-U.-H.; Baloch, A.H.; Brohi, S.A.; Arain, M.A. Comparative investigation of feeding habits and apparent digestibility of maize, millet and sorghum fodders in sheep and goat. *Pak. J. Agric. Res.* **2020**, *33*, 433–439.
28. de Assis, R.L.; de Freitas, R.S.; Mason, S.C. Pearl Millet Production Practices in Brazil: A Review. *Exp. Agric.* **2018**, *54*, 699–718. [CrossRef]
29. Ganapathy, K.N.; Hariprasanna, K.; Tonapi, V. Breeding for Enhanced Productivity in Millets. In *Millets and Pseudo Cereals*; Woodhead Publishing: Sawston, UK, 2021; pp. 39–63.
30. Vagnoni, E.; Franca, A.; Breedveld, L.; Porqueddu, C.; Ferrara, R.; Duce, P. Environmental performances of Sardinian dairy sheep production systems at different input levels. *Sci. Total Environ.* **2015**, *502*, 354–361. [CrossRef]
31. Chetroiu, R.; Dragomir, V. Research on the breakeven point in milk and meat production at ruminants. *Agrar. Econ. Rural. Dev. Trends Chall.* **2022**, *13*, 133–141.
32. Sulewski, P.; Kłoczko-Gajewska, A.; Sroka, W. Relations between Agri-Environmental, Economic and Social Dimensions of Farms' Sustainability. *Sustainability* **2018**, *10*, 4629. [CrossRef]
33. Chetroiu, R.; Cișmileanu, A.E.; Cofas, E.; Petre, I.L.; Rodino, S.; Dragomir, V.; Marin, A.; Turek-Rahoveanu, P.A. Assessment of the Relations for Determining the Profitability of Dairy Farms, A Premise of Their Economic Sustainability. *Sustainability* **2022**, *14*, 7466. [CrossRef]
34. Ferrazza, R.D.A.; Lopes, M.A.; Prado, D.G.D.O.; Lima, R.R.D.; Bruhn, F.R.P. Association between technical and economic performance indexes and dairy farm profitability. *Rev. Bras. Zootec.* **2020**, *49*, e20180116. [CrossRef]
35. Kryszak, Ł.; Guth, M.; Czyżewski, B. Determinants of farm profitability in the EU regions. Does farm size matter? *Agric. Econ./Zemědělská Ekon.* **2021**, *67*, 90–100. [CrossRef]
36. Tey, Y.S.; Brindal, M. Factors influencing farm profitability. *Sustain. Agric. Rev.* **2015**, *15*, 235–255.
37. Lee, B.; Liu, J.-Y.; Chang, H.-H. The choice of marketing channel and farm profitability: Empirical evidence from small farmers. *Agribusiness* **2020**, *36*, 402–421. [CrossRef]
38. Tüfekci, H.; Çelik, H.T. Effects of climate change on sheep and goat breeding. *Black Sea J. Agric.* **2021**, *4*, 137–145. [CrossRef]
39. Koluman-Darcan, N.; Silanikove, N. The advantages of goats for future adaptation to climate change: A conceptual overview. *Small Rumin. Res.* **2018**, *163*, 34–38. [CrossRef]

40. Simões, J.; Abecia, J.A.; Cannas, A.; Delgadillo, J.A.; Lacasta, D.; Voigt, K.; Chemineau, P. Managing sheep and goats for sustainable high yield production. *Animal* **2021**, *15*, 100293. [CrossRef]
41. Chingala, G.; Mapiye, C.; Raffrenato, E.; Hoffman, L.; Dzama, K. Determinants of smallholder farmers' perceptions of impact of climate change on beef production in Malawi. *Clim. Chang.* **2017**, *142*, 129–141. [CrossRef]
42. Yuan, X.; Li, S.; Chen, J.; Yu, H.; Yang, T.; Wang, C.; Huang, S.; Chen, H.; Ao, X. Impacts of Global Climate Change on Agricultural Production: A Comprehensive Review. *Agronomy* **2024**, *14*, 1360. [CrossRef]
43. Togoe, D.; Mincă, N.A. The Impact of Heat Stress on the Physiological, Productive, and Reproductive Status of Dairy Cows. *Agriculture* **2024**, *14*, 1241. [CrossRef]
44. Godde, C.M.; Mason-D'Croz, D.; Mayberry, D.E.; Thornton, P.K.; Herrero, M. Impacts of climate change on the livestock food supply chain; A review of the evidence. *Glob. Food Secur.* **2021**, *28*, 100488. [CrossRef]
45. Sterie, C.M.; Dragomir, V. Global trends on research towards agriculture adaptation to climate change. *Sci. Pap. Ser. Manag. Econ. Eng. Agric. Rural. Dev. DOAJ Dir. Open Access J.* **2023**, *23*, 759–766.
46. Gonzalez-Rivas, P.A.; Chauhan, S.S.; Ha, M.; Fegan, N.; Dunshea, F.R.; Warner, R.D. Effects of heat stress on animal physiology, metabolism, and meat quality: A review. *Meat Sci.* **2020**, *162*, 108025. [CrossRef]
47. Bhattarai, B.; Singh, S.; West, C.P.; Ritchie, G.L.; Trostle, C.L. Water depletion pattern and water use efficiency of forage sorghum, pearl millet, and corn under water limiting condition. *Agric. Water Manag.* **2020**, *238*, 106206. [CrossRef]
48. Mirza, N.; Marla, S.S. Finger millet (*Eleusine coracana* L. Gartn.) breeding. *Adv. Plant Breed. Strateg. Cereals* **2019**, *5*, 83–132.
49. Crookston, B.; Blaser, B.; Darapuneni, M.; Rhoades, M. Pearl millet forage water use efficiency. *Agronomy* **2020**, *10*, 1672. [CrossRef]
50. Meena, R.P.; Joshi, D.; Bisht, J.K.; Kant, L. Global scenario of millets cultivation. In *Millets and Millet Technology*; Springer: Singapore, 2021; pp. 33–50.
51. Chaparro, M.L.; Sanabria, P.J.; Jiménez, A.M.; Gómez, M.I.; Bautista, E.J.; Mesa, L. A circular economy approach for producing a fungal-based biopesticide employing pearl millet as a substrate and its economic evaluation. *Bioresour. Technol. Rep.* **2021**, *16*, 100869. [CrossRef]
52. Tagade, A.; Sawarkar, A.N. Valorization of millet agro-residues for bioenergy production through pyrolysis: Recent inroads, technological bottlenecks, possible remedies, and future directions. *Bioresour. Technol.* **2023**, *384*, 129335. [CrossRef]
53. Harish, M.S.; Axay, B.; Bhagirath, S. Millet production, challenges, and opportunities in the Asia-pacific region: A comprehensive review. *Front. Sustain. Food Syst.* **2024**, *8*, 1386469. [CrossRef]
54. Visarada, K.B.R.S.; Aruna, C. Sorghum: A bundle of opportunities in the 21st century. In *Breeding Sorghum for Diverse end Uses*; Woodhead Publishing: Sawston, UK, 2019; pp. 1–14.

 *sustainability*

Article

From Conventional to Organic Agriculture: Influencing Factors and Reasons for Tea Farmers' Adoption of Organic Farming in Pu'er City

Hao Li [1,2], Shuqi Yang [1], Juping Yan [1], Wangsheng Gao [1], Jixiao Cui [2,*] and Yuanquan Chen [1,*]

1   College of Agronomy and Biotechnology, China Agricultural University, Beijing 100107, China; wsdlh@cau.edu.cn (H.L.); yshuqi@cau.edu.cn (S.Y.); yanjvping@163.com (J.Y.); gaows@cau.edu.cn (W.G.)
2   Institute of Environment and Sustainable Development in Agriculture, Chinese Academy of Agricultural Sciences, Beijing 100875, China
*   Correspondence: cuijixiao@caas.cn (J.C.); chenyq@cau.edu.cn (Y.C.)

Abstract: As the global pursuit of sustainable agricultural practices continues, organic farming is gaining increasing attention. In Pu'er, one of China's major tea-producing regions, the factors influencing tea farmers' willingness to adopt organic agriculture have not yet been fully studied. This study integrates the diffusion of innovations theory and the theory of planned behavior, using field surveys to thoroughly analyze the key factors and reasons affecting tea farmers in Pu'er in adopting organic farming practices. The findings indicate that perceptions of the economic benefits of organic farming are the primary drivers of farmers' willingness to adopt. Experience with organic agriculture training and positive views on environmental and health benefits also significantly enhance the willingness to adopt organic farming. Contrary to common assumptions, education level, age, and household income have minimal influence on adoption willingness. However, low-income families that rely on tea cultivation are more inclined to adopt organic farming. Policymakers should prioritize economic incentives, strengthen training support, and enhance the promotion of the benefits of organic agriculture, while simplifying certification processes and expanding market channels to facilitate the transition of tea farmers to organic agriculture. This study offers insights into the sustainable tea industry and organic farming promotion.

Keywords: organic agriculture; tea farming; planting intentions; diffusion of innovations theory; theory of planned behavior; Pu'er city

Citation: Li, H.; Yang, S.; Yan, J.; Gao, W.; Cui, J.; Chen, Y. From Conventional to Organic Agriculture: Influencing Factors and Reasons for Tea Farmers' Adoption of Organic Farming in Pu'er City. *Sustainability* **2024**, *16*, 10035. https://doi.org/10.3390/su162210035

Academic Editors: Fotios Chatzitheodoridis, Efstratios Loizou and Achilleas Kontogeorgos

Received: 12 September 2024
Revised: 28 October 2024
Accepted: 14 November 2024
Published: 18 November 2024

1. Introduction

Amid the persistent challenges of environmental crises, population growth, and the limitations of traditional food systems, it has become increasingly vital to identify sustainable pathways for economic development, particularly within the agricultural sector. In this context, alternative farming practices and changes in traditional food consumption patterns have attracted growing scholarly attention [1,2]. Organic agriculture, in particular, is regarded as a viable strategy, as it addresses both environmental sustainability and consumer health concerns [3,4]. However, to support the continued expansion of the organic sector, especially in the global tea market, there is a critical need for more in-depth studies that examine the factors influencing farmers' adoption of organic practices. This research seeks to contribute to this field of study by examining the adoption of organic farming practices among tea farmers in Pu'er City, Yunnan Province, a region central to China's tea industry.

Many studies have explored the environmental and social impacts of organic versus conventional production methods [5–7]. Compared with traditional agriculture, organic agriculture has shown significant advantages in reducing agricultural greenhouse gas emissions, improving soil fertility, and ensuring public food safety [8]. However, organic agriculture is often accompanied by higher labor input requirements, and yields are on average 25% lower than conventional agriculture [9]. Increases in labor inputs and declining yields can pose significant barriers to farmers' adoption of organic production methods. Nonetheless, studies have shown that organic farming yields are relatively more stable, and organic products often command a higher market premium, meaning that organic farmers may enjoy higher incomes. While these advantages and challenges have been widely discussed, there is still a lack of systematic research influencing farmers' willingness to adopt organic production methods. In developing countries, socio-economic and institutional factors determine whether farmers adopt organic farming practices, with economic constraints often cited as the main obstacle. The transition to organic farming often requires a significant upfront investment, both in terms of capital and time, which can be overwhelming for many smallholder farmers [10,11]. In addition, these concerns are further exacerbated by potential yield reductions within organic farming converters and uncertain economic returns to future markets for organic products [12]. Therefore, financial incentives such as government subsidies will play an important role in encouraging farmers to adopt organic farming [13]. The level of education of farmers and the ease of access to effective information are also key factors influencing the adoption of organic farming by farmers. Some studies have shown that farmers with higher levels of education are more likely to adopt organic farming methods because they are better equipped to understand the complexities of organic farming practices and the long-term benefits that organic agriculture offers [14]. In addition, access to information, e.g., formal training programs, extension services, or peer access to information, can have a significant impact on a farmer's decision making regarding their willingness to adopt organic farming practices [15]. Cultural and psychological factors are equally important in influencing farmers' adoption behavior. In areas where conventional farming practices are deeply entrenched, farmers may be reluctant to adopt new methods due to potential crop failures and concerns about unknown risks [16,17].

Pu'er, located in Yunnan Province in southwestern China, is one of the country's most significant tea-producing regions, accounting for a substantial share of China's high-quality tea exports [18]. According to the Pu'er Tea and Coffee Industry Development Center, as of 2021, tea plantations in Pu'er cover an area of 1.75 million mu, generating an output value of approximately CNY 33.8 billion, with the primary tea industry contributing about CNY 6.8 billion [19]. The tea sector constitutes 32% of Pu'er's GDP and directly benefits more than 1.2 million residents. As such, the promotion of organic tea farming in Pu'er has the potential to profoundly impact both the local economy and the global tea market. However, the intense competition in the tea industry, coupled with the deep-rooted traditional business model of tea farming, makes it challenging for farmers to perceive the short-term economic benefits of transitioning to organic methods [15]. Additionally, the shift to organic production requires significant investments in time and resources to make adjustments and optimize practices, which adds to the complexity of the transition process [2,20,21]. Socio-economic factors, such as tea farmers' economic returns, market demand for tea, and policy support, play a crucial role in their decision-making processes [22]. Equally important, however, are the psychological factors such as risk perception, traditional attitudes, and trust in new technologies that influence farmers' willingness to adopt organic farming practices [23]. Despite the rapid global expansion of the organic agricultural sector, existing research on the willingness of tea farmers to adopt organic methods tends to focus narrowly on technical or policy factors [24,25], often neglecting the broader socio-economic and psychological drivers. This gap limits a comprehensive understanding of farmers' decision-making processes and poses challenges for designing effective organic agriculture extension policies. This study seeks to bridge these gaps by investigating not only the specific drivers that influence the adoption of organic farming but also the role of external factors, such as government policies, market incentives, and social influences, in shaping farmers' choices. Addressing these challenges is critical for enhancing production methods, optimizing resource utilization, and strengthening the agricultural trading system, all of which are essential for stimulating exports and fostering regional development.

To fill the gaps in existing research, diffusion of innovations theory (DIT) and theory of planned behavior (TPB) provide valuable theoretical frameworks for understanding the complexities of the adoption of organic farming practices by tea farmers. DIT explains the adoption process of innovations through five key attributes: comparative advantage, compatibility, complexity, trialability, and observability [26]. These attributes help to understand the acceptance of organic farming practices by tea farmers. For example, the relative advantages of organic farming in reducing environmental pollution and improving tea quality may be an important incentive to attract tea farmers to make the transition [27]. However, if organic farming is due to high technological complexity, tea farmers may feel intimidated and thus take a wait-and-see attitude towards the transition. In addition, trialability and observability in DIT directly affect the decision-making process of tea farmers to adopt organic farming practices. For example, tea farmers verify the possible risks associated with organic transition through small-scale trials or visualize cases of a successful adoption of organic methods by other farmers to observe farmers' willingness to adopt organic farming [28]. The TPB can further dissect the decision-making process of tea farmers adopting organic farming. TBP explores how three core factors, including attitudes, subjective norms, and perceived behaviors, influence behavioral intentions [29]. This is reflected in the following aspects. First, in the context of the willingness to adopt organic methods of production, tea farmers' attitude of whether they believe that organic farming can bring higher economic benefits, or a better ecological environment, will directly affect their willingness to adopt organic farming [30]. Secondly, tea farmers may be more inclined to switch to organic farming if they feel pressure from their peers or the market that organic farming is a more socially desirable option. This also suggests that subjective normative pressure from the family, community, or society may also have an impact on tea farmers' decision-making process to adopt organic farming [31]. Finally, if tea farmers perceive that they lack the necessary resources, technology, or market support, they may abandon or delay their plans to convert. This reflects the influence of perceived behavioral control on behavioral intention in TPB theory, i.e., tea farmers' confidence in their own ability and awareness of the difficulty of organic farming may be important influences [32]. In the context of tea cultivation in Pu'er, DIT is employed to ascertain which of the advantages of organic agriculture (for example, increased income and reduced environmental impact) are most likely to motivate the adoption of such practices. Concurrently, TPB provides a supplementary viewpoint that emphasizes the influence of attitudes, subjective norms, and perceived behavioral control in shaping adoption decisions. Thus, by combining DIT with TPB, a more comprehensive understanding of the psychological and behavioral dynamics of tea farmers in Pu'er City in the face of organic farming practices can be achieved.

This study aims to explore the multiple influencing factors affecting tea farmers' willingness to adopt organic farming practices in Pu'er City, Yunnan Province. To this end, the diffusion of innovations theory and the theory of planned behavior are employed to conduct a multi-dimensional systematic analysis. The primary objectives of this study are as follows: (1) to analyze the complex relationship between tea farmers' socio-economic backgrounds and their attitudes and behavioral willingness to adopt organic farming practices through field surveys, and to explore the key factors affecting the adoption of organic production methods by farmers with different characteristics; (2) to study how tea farmers' perceptions of organic agriculture affect their willingness to adopt organic farming from the cognitive and psychological perspectives, and reveal the role of these perceptions in decision-making processes. The findings of this study will provide a theoretical basis and practical reference for the development of sustainable agriculture policies in Pu'er and other tea-producing regions.

2. Materials and Methods

2.1. Study Area

Pu'er is located in the southwest of Yunnan Province (Figure 1) and is one of the major tea-producing areas in China. In 2021, Pu'er had a tea plantation area of 116,700 hectares, with an output of 136,700 tons, and the city's organic tea plantation area reached 28,027 hectares, making the tea plantation area the largest in the country among prefecture-level cities. The area's varied elevations, ranging from lowlands to high mountains, create microclimates that are conducive to growing high-quality tea. These conditions make Pu'er an important site for examining the factors that influence tea farmers' willingness to adopt organic farming practices.

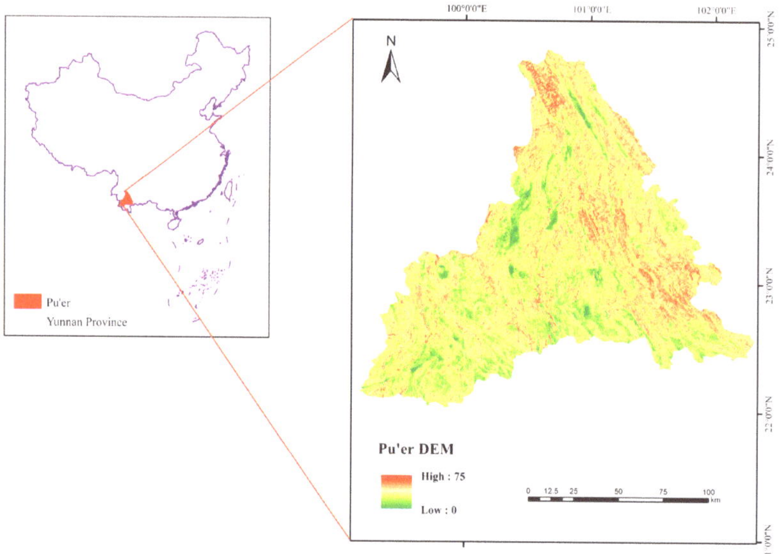

Figure 1. Geographic location of surveyed areas in China.

2.2. Data Collection

In this study, data were collected through a survey of 162 tea farmers in Pu'er City. The tea farmers surveyed represented a diverse range of age groups, educational backgrounds, income levels, and years of agricultural experience. The survey instrument consisted of structured questions distributed in collaboration with local agricultural offices and tea associations to ensure a representative sample of tea farmers. The use of questionnaires facilitated the establishment of a rapport with the tea farmers, which was essential for acquiring reliable and candid responses.

The questionnaire comprised three principal sections. The initial section encompassed inquiries pertaining to the respondents' age, gender, educational background, household size, number of agricultural laborers, annual household income, percentage of agricultural income, duration of tea cultivation, and size of the tea plantation. The second section pertains to awareness of organic farming, encompassing farmers' familiarity with organic standards, knowledge of organic certification systems, and perceptions of potential income, environmental impacts, and health-related benefits associated with adopting organic farming practices. Additionally, the objective conditions of tea farmers' willingness to adopt organic practices are examined, including their awareness of government subsidies for organic agriculture and the influence of successful examples in their vicinity on their decision to adopt organic farming.

2.3. Theoretical Framework

In this study, the diffusion of innovations theory and the theory of planned behavior were combined to understand the decision-making process of tea farmers adopting organic farming practices in Pu'er. The theoretical framework (see Figure 2) demonstrates how external and internal factors interact with each other to influence tea farmers' decisions. DIT explains the process of the diffusion of organic agriculture as an innovation in communities by emphasizing five key attributes: comparative advantage, compatibility, complexity, testability, and observability. In this study, these five factors were captured in the questionnaire through the design of specific questions. For example, the questionnaire measures the impact of 'comparative advantage' by assessing farmers' perceptions of the economic benefits and environmental advantages of organic agriculture; 'compatibility' by assessing farmers' perceptions of organic agriculture in relation to existing farming methods; and 'complexity' by assessing farmers' perceptions of the technical and technological advantages they may encounter during the conversion process. "Complexity" was measured by investigating the technical and economic barriers that farmers may encounter during the conversion process. In addition, the questionnaire included a survey on the feasibility of organic agriculture in small-scale trials and the visibility of success stories to assess 'testability' and 'observability'. At the same time, the TPB was used to analyze farmers' decision-making factors at the psychological level. In the design of the questionnaire, questions were developed to address three key elements of the TPB: attitude, subjective norms, and perceived behavioral control. Some of the questions in the questionnaire were designed to understand farmers' perceptions of the economic benefits of organic farming and environmental protection to assess their attitudes; other questions focused on community expectations and social pressures to measure the impact of subjective norms on farmers; and we also investigated whether farmers had received training and technical support to understand their perceptions of the difficulty of transitioning to organic farming to assess perceived behavioral control. To better understand the relationship between external and internal factors, the theoretical frameworks of DIT and TPB were tightly integrated in the study. The five attributes of DIT are driven by government support and market incentives; e.g., through government-funded pilot projects and demonstration cases, farmers can more easily observe and experiment with the effects of organic agriculture, thus enhancing "observability". The three psychological components of TPB are driven by market incentives and social influences; e.g., market incentives and the demonstration of success stories enhance farmers' positive attitudes, while positive social expectations from the community reinforce subjective norms. In addition, government support and market conditions boosted farmers' confidence in overcoming barriers, thus strengthening their perceived behavioral control. By integrating DIT and TPB, we designed a theoretical framework to capture the multiple considerations of tea farmers in Pu'er City when adopting organic agriculture. This approach not only helped us to clarify the influence of economic and technical factors, but also revealed the critical role of social and psychological factors in farmers' decision-making process.

2.4. Analytical Methods

2.4.1. Descriptive Analysis

In order to more accurately describe the basic situation of tea farmers in Pu'er City, this study summarized and analyzed the socio-demographic and socio-economic characteristics of the respondents using frequency distribution statistics. The use of frequency distribution statistics enabled the description of the various characteristics of the respondents, including gender, age, literacy, income level, and land ownership, in order to demonstrate the distribution of different characteristics within the tea farmers' group. This included the presentation of the gender ratio, the distribution of tea farmers across different age groups, the structure of the literacy levels of the tea farmers, and the distribution of their incomes. Furthermore, frequency distribution can elucidate the extent of the concentration of respondents on specific characteristics. For instance, it can reveal whether a particular income level or literacy level is dominant among tea farmers, which in turn reflects certain commonalities or differences in the socio-economic conditions of tea farmers.

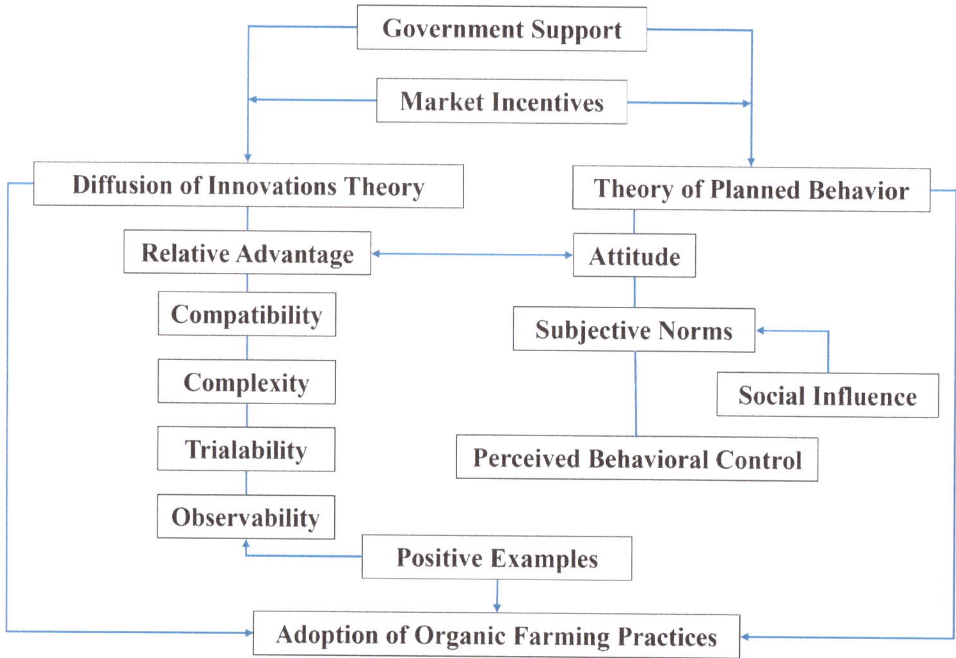

Figure 2. Theoretical framework.

2.4.2. Correlation Analysis

To gain a more comprehensive understanding of the relationship between different variables, this study employed a correlation matrix to examine the correlation between socio-demographic factors and respondents' perceptions of organic farming, as well as their willingness to adopt organic farming practices. The correlation matrix not only demonstrated the linear relationship between the variables but also facilitated the identification of potentially significant associations. The analysis enabled the researchers to elucidate the extent to which different socio-demographic characteristics influence tea farmers' willingness to adopt organic farming practices. This included investigating whether farmers with higher levels of education are more inclined to adopt organic farming or whether there are significant differences in perceptions of organic farming among different age groups. To illustrate the strength and direction of the correlation coefficients, heat maps were employed for visualization purposes. The findings of this analysis can provide robust data support for the development of more precise policies and interventions, which can more effectively promote the promotion and application of organic agriculture among tea farmers in Pu'er City.

2.4.3. Regression Analysis

In order to explore in more depth which factors significantly influence tea farmers' willingness to adopt organic farming practices, this study constructed a multiple regression analysis model. Through the use of a regression analysis, the study was able to determine the extent to which each of the independent variables influenced tea farmers' willingness to adopt. This analysis helped to identify key factors that should be prioritized when promoting organic agriculture. Regression modeling allows policymakers to implement more precise interventions based on data-driven results, thus promoting the effective promotion of organic agriculture. The specific steps of the regression analysis are as follows:

(1) Variable definition: Several independent variables were selected for the purpose of predicting tea farmers' willingness to adopt organic agriculture, which was defined as the dependent variable. The independent variables included the respondents' level of education, age, annual income, experience with tea planting, training in organic agriculture, and perceptions regarding the potential for organic agriculture to increase income and improve the environment. The dependent variable was the willingness of tea farmers to adopt organic agriculture.

(2) Data standardization: Prior to conducting the regression analysis, all independent variables were standardized to eliminate discrepancies in magnitude across variables of varying scales. This guarantees that the regression coefficients accurately represent the degree of influence exerted by each factor on the dependent variable.

(3) Regression model: A multiple linear regression model was employed, formulated as follows:

$$Y = \beta_0 + \beta_1 X_1 + \beta_2 X_2 + \cdots + \beta_n X_n + \varepsilon \tag{1}$$

where Y represents the willingness to adopt organic farming; β_0 is the intercept of the regression model; β_1, β_2, ..., β_n are the coefficients for each independent variable X_1, X_2, ..., X_n; and ε is the error term.

(4) Estimation of coefficients: In this study, the least squares method was used to estimate the regression coefficients. The least squares method determines the best-fit line by minimizing the squared error between the predicted and actual values to calculate the regression coefficients for each independent variable.

(5) Interpretation of regression results: By analyzing the positivity, negativity, and magnitude of the regression coefficients, we derived the direction and strength of the influence of each independent variable on the tea farmers' willingness to adopt organic farming.

3. Results

3.1. Statistical Characteristics of Tea Farmers in Pu'er City

The demographic characteristics of the tea farmers in Pu'er City reveal that the majority of respondents are middle-aged, with 59.3% of them aged between 40 and 59 years. Specifically, 30.4% are aged 50–59 years, and 28.9% are aged 40–49 years. Gender distribution shows that 59.9% of the respondents are male, while 40.1% are female. In terms of education level, more than half of the tea farmers (53.5%) have only completed elementary education or less, with 30.6% having completed junior high school, and only 7.0% possessing a college degree or higher. Regarding land ownership, the majority of tea farmers (42.0%) manage tea plantations ranging from 0.33 to 0.67 ha, while 19.1% have less than 0.33 ha, and 17.3% operate plantations larger than 1.33 ha. The annual household income of most respondents (46.3%) falls within the USD 2600–6700 range, with 23.8% earning less than USD 2600, and only 4.3% earning more than USD 20,200. Additionally, 37.0% of the tea farmers have 11–20 years of experience in tea cultivation, indicating a relatively experienced farming community, while 25.3% have 5–10 years of experience, and 21.0% have 21–30 years of experience. These results indicate that the tea farming community in Pu'er City is predominantly composed of middle-aged, male farmers with relatively low levels of formal education. Most farmers manage small- to medium-sized tea plantations and have moderate income levels, with a significant portion having substantial experience in tea cultivation (Figure 3).

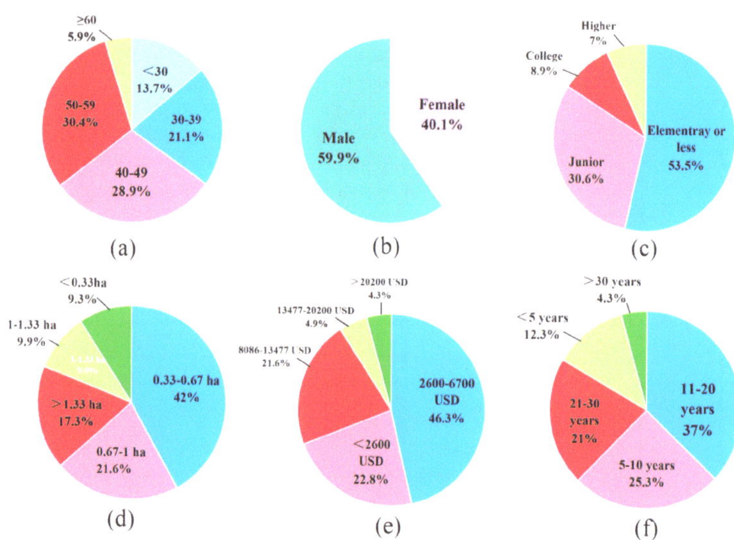

Figure 3. Demographic characteristics of tea farmers in Pu'er City, Yunnan Province. (**a**) Age distribution; (**b**) gender distribution; (**c**) education level; (**d**) area of tea plantations; (**e**) distribution of annual household income; and (**f**) number of years of tea cultivation.

3.2. Willingness to Adopt Organic Farming Practices Among Tea Farmers

Figure 4 shows the willingness of tea farmers in Pu'er to adopt organic farming practices and analyzes different demographic and socio-economic factors. The data show that all age groups below 60 years old are more willing to adopt organic farming, while older farmers (>60 years old) are less willing to adopt organic farming (Figure 4a). In total, 76.76% of farmers with university education or higher education expressed willingness to adopt organic farming, while 72.94% of farmers with only primary education were also willing to adopt organic farming (Figure 4b). In terms of tea garden size, farmers managing medium-sized plots (0.33–0.67 ha) expressed the greatest willingness, accounting for about 64.71% of this group (Figure 4c). In terms of experience in tea cultivation, farmers with 11–20 years of cultivation experience showed the most neutral willingness to adopt organic production, while farmers with less than 5 years and 5–10 years of cultivation experience were more willing to adopt organic agriculture (Figure 4d). Farmers with an annual income of less than USD 2600 had the highest willingness to adopt organic farming, which was about 83.78% (Figure 4e). In addition, the share of income from tea cultivation in household income also had a strong impact on willingness; farmers who relied on tea cultivation for a larger share of their income were more likely to adopt organic farming practices (Figure 4f). Training in organic farming was a strong determinant, with 84.27% of trained farmers willing to adopt organic farming practices, while only about 58.90% of untrained farmers were willing to adopt organic farming practices (Figure 4g). In addition, farmers who perceived organic tea production to be beneficial to the environment and health were more likely to adopt organic production practices, with more than 70% of farmers indicating a willingness to adopt these practices (Figure 4h). Finally, management style also influences adoption; independent operations or those with a more rigid management structure are more likely to adopt organic production practices than those individual farmers with more flexible management (Figure 4i).

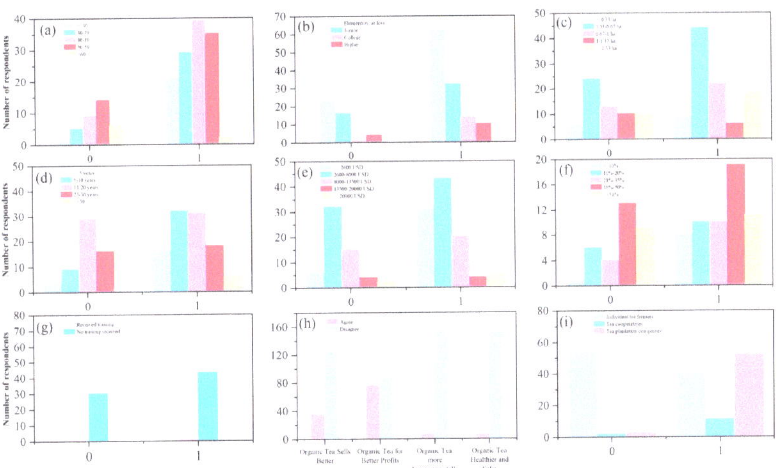

Figure 4. The willingness of tea farmers to adopt organic farming practices. (**a**) Age distribution; (**b**) education level; (**c**) size of tea plantation; (**d**) tea-growing experience; (**e**) annual household income; (**f**) proportion of income from tea cultivation; (**g**) training in organic agriculture; (**h**) perception of organic tea production; (**i**) management practices. Willingness (1 = Yes, 0 = No).

3.3. Income and Education and Willingness to Adopt Organic Agriculture

The distribution of income by education level demonstrates that farmers with higher education levels possess higher income levels in comparison to those with lower education levels (Figure 5). Additionally, approximately 76.76% of farmers in this group are inclined to adopt organic production methods, in contrast to 72.94% of low-income farmers (Figure 4e). This indicates that farmers with higher levels of education may possess a greater capacity to comprehend the long-term advantages of organic agriculture. Additionally, they often have more resources (financial and informational) at their disposal, which can facilitate the adoption of organic practices. Consequently, the rate of adoption is higher among this group. Conversely, farmers with low levels of education and lower incomes may be more inclined to adopt organic production methods as a means of enhancing their income. These findings indicate that to enhance the overall adoption rate, it is essential to customize policy support or policy guidance to encompass diverse income and education levels. For instance, targeted assistance, such as subsidies and training programs, is crucial for farmers with lower incomes or lower education levels. Conversely, for those with higher incomes or more advanced education, it is vital to disseminate information about the long-term benefits of organic tea to promote the active adoption and implementation of organic tea production methods within this demographic.

3.4. Analysis of Regression Coefficients for Willingness to Adopt Organic Farming

Figure 6 illustrates the coefficients of a regression analysis of the various predictors that affect the willingness of tea farmers to adopt organic farming practices. The regression coefficients indicate the direction and magnitude of the effect of each predictor variable on the willingness to adopt organic farming. Positive values indicate an increase in willingness, while negative values indicate a decrease in willingness. The results demonstrated that the belief held by tea farmers that organic farming can increase their income exerts the greatest influence on their willingness to adopt, with a regression coefficient of 1.63. This indicates that economic gain is the most significant factor driving the adoption of organic farming by tea farmers. Tea farmers who had received organic agriculture training also exhibited a stronger willingness to adopt, with a regression coefficient of 0.99. This suggests that training in organic farming methods can increase tea farmers' awareness of organic agriculture and play an important role in increasing their willingness to adopt organic agriculture. Furthermore, tea farmers who believe that organic farming can improve the environment are also more likely to adopt the method, with a regression coefficient of 0.88. However, the regression coefficient of the education level is −0.32, which suggests that tea farmers with higher education levels may be more risk-averse with regard to the economic viability of organic farming or the challenges associated with its implementation. This may potentially lead to a reduced inclination towards adopting this approach. The regression coefficients for age and annual household income were 0.01 and 0.00, respectively, which were close to zero, indicating that these factors did not have a significant effect on the intention to adopt. The regression coefficient for tea-growing experience was −0.06, indicating that experienced tea farmers may be more inclined to adhere to their established conventional growing methods rather than transitioning to organic farming practices. These findings elucidate the multifaceted factors influencing tea farmers' adoption of organic agriculture, with financial gain and training support identified as the primary drivers, while education level and traditional experience may act as inhibitors to some extent.

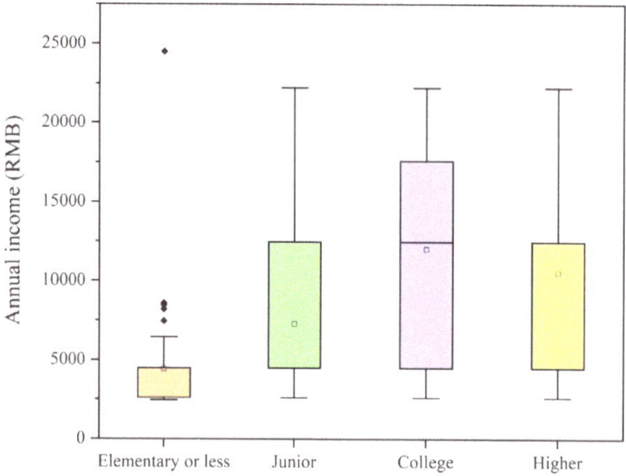

Figure 5. Distribution of annual income across different education levels.

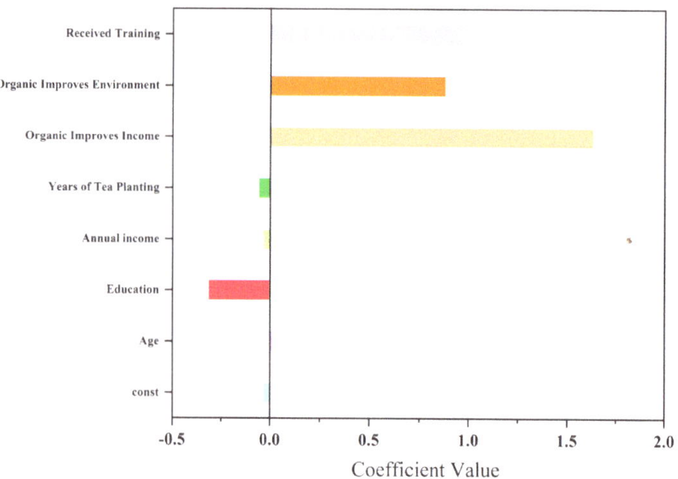

Figure 6. Regression coefficients for willingness to adopt organic farming.

3.5. Analysis of Relevance of Willingness to Adopt Organic Agriculture

Figure 7 presents a correlation analysis of the demographic factors of tea farmers in Pu'er City, their awareness of organic agriculture, and their willingness to adopt organic farming. This analysis provides further insight into the key factors influencing tea farmers' adoption of organic agriculture. The results indicate that knowledge of organic standards shows a strong positive correlation with the willingness to adopt organic farming (0.47), suggesting that the deeper the farmers' understanding of organic standards, the stronger their willingness to adopt organic practices. This finding highlights the importance of access to information and knowledge in promoting adoption decisions. Additionally, the correlation between receiving training and the willingness to adopt organic farming practices is 0.28, further demonstrating the role of training in enhancing farmers' awareness and willingness to adopt organic agriculture. In contrast, tea farmers' perceptions of organic agriculture's environmental benefits (0.42) and the belief that organic tea is healthier (0.29) also exhibit positive correlations with their willingness to adopt organic agriculture, indicating that awareness of environmental protection and health factors influences farmers' decision making. Traditional socio-economic factors such as age (0.01) and annual income (0.0014) show very weak correlations, consistent with the findings in Figure 6, further suggesting that these factors have little relationship with the willingness to adopt organic farming. Overall, the awareness of organic agriculture, training, and perceptions of environmental and health benefits are positively associated with the adoption of organic farming practices among tea farmers. Therefore, increasing the promotion of organic agriculture and enhancing training efforts may be key approaches to encouraging the widespread adoption of organic farming among tea farmers in Pu'er City.

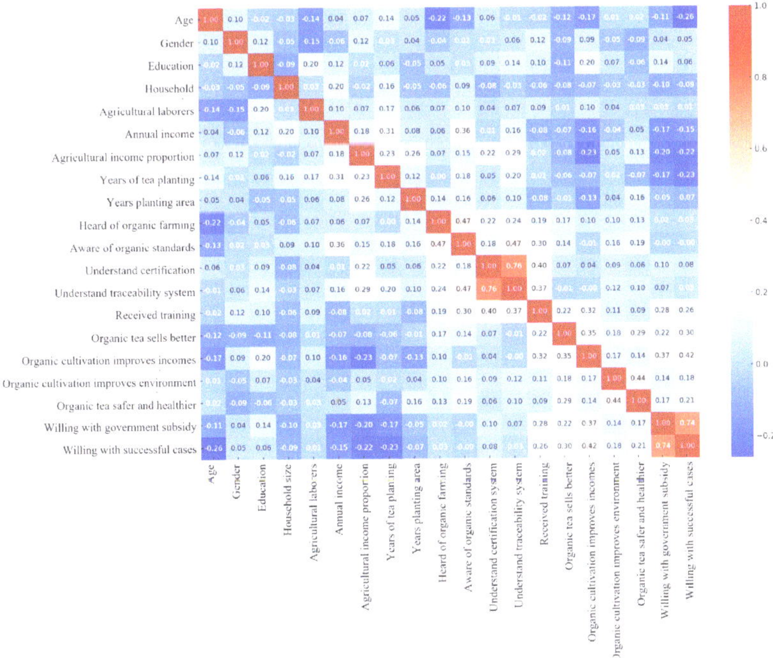

Figure 7. Correlation analysis of tea farmers' willingness to adopt organic farming.

4. Discussion

4.1. Comparison with Previous Studies

In this study, the factors influencing tea farmers' willingness to adopt organic farming share similarities with, but also differ from, existing research on farmers' adoption of organic farming practices. Previous studies have consistently emphasized that education and income are significant factors influencing the adoption of organic agriculture. The regression analysis in this study revealed a negative relationship between education level and the willingness to adopt organic farming, a finding that initially appears to be counterintuitive [16,21,30]. This trend may be attributed to the fact that more highly educated tea farmers tend to have a heightened awareness of the potential risks and uncertainties associated with organic farming. Prior research indicates that individuals with higher education levels often engage in more comprehensive cost–benefit analyses, which can lead them to adopt a more cautious stance toward economically volatile practices [33,34]. In the specific context of Pu'er's tea farmers, those with higher educational backgrounds may be more cognizant of the challenges related to organic certification standards and the substantial initial investments required, thereby reducing their inclination to transition to organic methods. This observation contrasts with the findings of numerous studies that suggest farmers with higher levels of education are generally more likely to adopt organic farming, as they are better equipped to navigate the complexities of organic certification [35,36]. In Pu'er City, while tea farmers with higher education backgrounds showed a greater willingness to adopt organic practices (Figure 4), the overall educational level in the region remains relatively low (Figure 3). Consequently, the frequency analysis indicates that the presence of highly educated farmers does not necessarily establish education as a decisive factor for adoption. Instead, our regression analysis suggests that the challenges perceived by well-educated farmers in terms of meeting stringent certification criteria may deter them from fully embracing organic agriculture (Figure 6). The relationship between income and the adoption of organic farming is similarly complex. Traditional perspectives posit that higher-income farmers are more inclined to adopt organic practices, as they possess the financial resilience to absorb potential short-term losses [37]. However, this study found no significant correlation between household annual income and tea farmers' willingness to adopt organic farming (Figure 6). Interestingly, farmers with an annual household income below USD 2600, who rely heavily on tea farming, comprising 36–50% of their total household income, were 24.81% more likely to transition to organic farming. This suggests that low-income farmers might perceive organic agriculture as an economic strategy to alleviate financial pressures associated with conventional farming. These findings are consistent with other studies in developing regions, where small-scale farmers often view organic production as an opportunity to differentiate their products and tap into premium markets [26,37,38]. Overall, these findings provide a deeper understanding of the socio-economic factors influencing the adoption of organic farming, particularly within the distinct cultural and market context of Pu'er's tea industry. The results suggest that variables such as education and income do not function independently; rather, they are shaped by local challenges and perceptions of economic feasibility. This study offers a novel perspective by integrating DIT and TPB, allowing for an analysis that encompasses both socio-economic and psychological influences on tea farmers' decision-making processes. Unlike earlier studies that tend to examine broader determinants, this research employs a dual-theoretical framework combining DIT and TPB to offer a comprehensive perspective. This integrated approach enables a more thorough examination of not only the socio-economic factors but also the psychological dynamics influencing the willingness of tea farmers to adopt organic practices. The insights generated from this study, particularly in the unique context of Yunnan's tea sector, provide a valuable contribution to the discourse on sustainable agriculture.

4.2. Policy Implications

A significant correlation was observed between farmers' knowledge of organic agriculture and their awareness of organic certification standards and traceability systems, as well as their willingness to adopt organic practices (the highest correlation was at 47%). Additionally, a 36% correlation was identified between annual income and willingness to adopt organic practices (Figure 7). However, these correlations reflect a more complex reality in the specific context of this study. Due to the generally low education levels and limited financial capital among tea farmers in Pu'er (Figure 3), they face challenges in understanding and implementing organic farming practices. As a result, relying solely on the willingness of highly educated tea farmers in Pu'er to adopt organic production may not lead to significant effects [39–41]. Policymakers should simplify the organic certification process or provide more intuitive, easily implementable technical support. Tailored training programs that combine local context with the needs of farmers with lower education levels are essential to help them overcome knowledge and cognitive barriers, ensuring that training effectively translates into a willingness to adopt organic farming practices [42,43]. Furthermore, the study reveals that households with an annual income below USD 2600 are more likely to adopt organic agriculture, especially when tea income accounts for 36–50% of total household income (Figure 6). This finding contrasts with previous studies, which have suggested that higher-income farmers are better equipped to bear the initial costs of transitioning to organic agriculture. However, in this study, low-income tea farmers in Pu'er may view organic agriculture as an economic strategy to mitigate the pressure of low prices in traditional agricultural markets by producing differentiated, high-value-added products. This implies that policymakers should focus on the needs of these farmers and provide direct economic support, such as subsidies or loans, while creating stable market channels (e.g., ensuring consistent market access or price guarantees for organic products) to offer long-term economic incentives for low-income farmers [44,45]. Such a policy framework should not only offset the economic costs of transitioning to organic farming but also ensure that farmers can achieve sustainable income through the price premium of organic products [46,47]. Finally, the study finds a significant positive correlation between farmers' perceptions of the environmental benefits of organic agriculture and the health benefits of organic tea, and their willingness to adopt organic farming practices. Given Pu'er's tea-centric industrial structure and the increasing consumer awareness of health and environmental issues, policymakers could further promote and publicize the ecological and public health benefits of organic agriculture [44,46]. This would also provide new opportunities for Pu'er to develop a more sustainable organic tea supply chain.

In conclusion, policymakers aiming to encourage the adoption of organic agriculture among tea farmers in Pu'er City must consider a variety of factors, avoiding a one-size-fits-all approach that fails to address the unique circumstances of different farmer groups. First and foremost, policies should account for the economic conditions and educational levels of various groups of tea farmers, as well as their perceptions of the environmental and health benefits of organic agriculture. Training and support programs must be flexible and targeted, effectively addressing the specific barriers faced by tea farmers during the transition to organic production [48–50]. Additionally, guidance on organic farming techniques and organic certification requirements should be actively promoted to assist farmers in connecting with the organic market and to boost their confidence in making the transition [51,52]. Implementing such comprehensive policies and services can more effectively promote the development of organic tea cultivation in Pu'er, establishing a mutually beneficial model for both farmers and the environment. To further enhance the adoption of organic farming among low-income tea farmers, policymakers should consider optimizing the organic certification process by reducing costs and simplifying requirements, making it more accessible to small-scale farmers. Additionally, targeted subsidies could be introduced to cover initial transition costs, such as purchasing organic inputs or compensating for yield reductions during the conversion period. This could involve creating a tiered certification system, where basic organic standards are recognized initially, with incremental incentives for more advanced certification levels as farmers gain experience. By implementing these tailored strategies, the development of organic agriculture in Pu'er City will become more sustainable, offering a successful model for other tea-producing regions.

4.3. Limitations

This study aimed to explore the factors influencing tea farmers' willingness to adopt organic farming from multiple perspectives. However, it is acknowledged that the study was not without limitations. Firstly, the questionnaire utilized in this study solely collected data regarding the current attitudes and behaviors of tea farmers, which may restrict the ability to draw conclusions regarding the causal relationship between the factors influencing the adoption of organic farming and subsequent actual behaviors [53]. Future research could employ longitudinal methods to track farmers' behavioral and psychological changes, thereby gaining a deeper understanding of how farmers' willingness to adopt organic farming practices may change in response to changes in market conditions or government policies [54,55]. Secondly, it is possible that farmers may overstate their willingness to adopt organic farming practices due to social pressures or due to an inaccurate recollection of past experiences associated with organic farming [56]. Future research could address the limitations of this study by increasing the scale and number of studies and introducing third-party observations [57]. Additionally, this study focused on the influence of socio-economic and psychological factors on tea farmers' willingness to adopt organic methods. Future research could further examine the influence of market dynamics on organic adoption intentions by analyzing fluctuations in tea prices, changes in demand for organic products, and competition in the tea industry [58].

5. Conclusions

This study combines the diffusion of innovations theory and the theory of planned behavior to explore the influencing factors and their underlying reasons for tea farmers' adoption of organic agricultural production methods in Pu'er City, Yunnan Province. The results of the study showed that tea farmers' perception of the economic benefits of organic agriculture was the most important factor influencing their willingness to adopt. In the regression analysis, the perception that organic agriculture can increase income was 0.99, indicating that technical training plays an important role in increasing the adoption rate. The regression coefficient of 0.88 for farmers who had a positive perception of organic farming in improving the environment and health showed the significant influence of environmental and health benefits in enhancing adoption intentions. Contrary to the conventional view, the effect of the education level was not significant in this study. The data showed that tea farmers with a higher education level showed more caution in adopting organic farming with a regression coefficient of -0.32. In contrast, low-income households with an annual income of less than USD 2600 and whose main source of income was tea were 24.81% more likely to use organic farming practices than in other scenarios. The novelty of this study lies in the first application of the integration of DIT and TPB theories to the study of tea farmers' adoption of organic agriculture in Pu'er City, revealing the complex relationship among economic, environmental, and psychosocial factors. Through an in-depth analysis of tea farmers' behaviors in this specific context in Pu'er City, we not only provide new insights on organic agriculture adoption, but also suggest specific policy recommendations. In order to promote organic agriculture more effectively, policymakers should pay attention to tea farmers' economic status, educational background, and level of awareness of environmental and health benefits. This study suggests streamlining the organic certification process, providing technical training and economic support, and promoting successful organic farming practices in order to develop flexible and targeted policies. In addition, the government and relevant stakeholders should actively develop stable market channels and provide price guarantees to reduce the economic risks faced by tea farmers during the transition process. Overall, this study not only provides a theoretical basis and policy recommendations for the development of organic agriculture in Pu'er, but also provides useful references for the promotion of organic agriculture in other tea-producing regions.

Author Contributions: Conceptualization, H.L. and S.Y.; methodology, H.L.; validation, H.L., J.C. and Y.C.; formal analysis, H.L.; investigation, H.L. and J.Y.; resources, W.G.; data curation, H.L. and J.Y.; writing—original draft preparation, H.L.; writing—review and editing, J.C. and Y.C.; visualization, H.L.; supervision, J.C. and Y.C.; project administration, W.G.; funding acquisition, J.C. and Y.C. All authors have read and agreed to the published version of the manuscript.

Funding: This research was funded by the Major Science and Technology Project of Yunnan Province, grant number 202202AE090029.

Institutional Review Board Statement: Not applicable.

Informed Consent Statement: Informed consent was obtained from all subjects involved in the study.

Data Availability Statement: The datasets used during the current study are available from the corresponding author on reasonable request.

Conflicts of Interest: The authors declare no conflicts of interest.

References

1. Çakmakçi, R.; Salik, M.A.; Çakmakçi, S. Assessment and Principles of Environmentally Sustainable Food and Agriculture Systems. *Agriculture* **2023**, *13*, 1073. [CrossRef]
2. Arhin, I.; Li, J.J.; Mei, H.L.; Amoah, M.; Chen, X.; Jeyaraj, A.; Li, X.H.; Liu, A.J. Looking into the future of organic tea production and sustainable farming: A systematic review. *Int. J. Agric. Sustain.* **2022**, *20*, 942–954. [CrossRef]
3. Carrié, R.; Smith, H.G.; Ekroos, J. Sensitivity to agricultural inputs and dispersal limitation determine the response of arable plants to time since transition to organic farming. *J. Appl. Ecol.* **2024**, *61*, 1227–1242. [CrossRef]
4. Panday, D.; Bhusal, N.; Das, S.; Ghalehgolabbehbahani, A. Rooted in Nature: The Rise, Challenges, and Potential of Organic Farming and Fertilizers in Agroecosystems. *Sustainability* **2024**, *16*, 1530. [CrossRef]
5. Cobb, D.; Feber, R.; Hopkins, A.; Stockdale, L.; O'Riordan, T.; Clements, B.; Firbank, L.; Goulding, K.; Jarvis, S.; Macdonald, D. Integrating the environmental and economic consequences of converting to organic agriculture: Evidence from a case study. *Land Use Pol.* **1999**, *16*, 207–221. [CrossRef]
6. Meng, F.Q.; Qiao, Y.H.; Wu, W.L.; Smith, P.; Scott, S. Environmental impacts and production performances of organic agriculture in China: A monetary valuation. *J. Environ. Manag.* **2017**, *188*, 49–57. [CrossRef]
7. Shin, E.; Shin, Y.; Lee, S.W.; An, K. Evaluating the Environmental Factors of Organic Farming Areas Using the Analytic Hierarchy Process. *Sustainability* **2024**, *16*, 2395. [CrossRef]
8. Misselbrook, T.H.; Menzi, H.; Cordovil, C. Preface—Recycling of organic residues to agriculture: Agronomic and environmental impacts. *Agric. Ecosyst. Environ.* **2012**, *160*, 1–2. [CrossRef]
9. Wang, P.C.; Liu, F.C.; Lee, D.C.; Lin, M.Y. Environmental Knowledge, Values, and Responsibilities Help to Enhance Organic Farming Intentions: A Case Study of Yunlin County, Taiwan. *Agriculture* **2023**, *13*, 1476. [CrossRef]
10. Feledyn-Szewczyk, B.; Kopinski, J. Productive, Environmental and Economic Effects of Organic and Conventional Farms-Case Study from Poland. *Agronomy* **2024**, *14*, 793. [CrossRef]
11. Heinrichs, J.; Kuhn, T.; Pahmeyer, C.; Britz, W. Economic effects of plot sizes and farm-plot distances in organic and conventional farming systems: A farm-level analysis for Germany. *Agric. Syst.* **2021**, *187*, 102992. [CrossRef]
12. Martín-García, J.; Gómez-Limón, J.A.; Arriaza, M. Conversion to organic farming: Does it change the economic and environmental performance of fruit farms? *Ecol. Econ.* **2024**, *220*, 108178. [CrossRef]
13. Liang, H.Z.; Meng, Y. Impact of direct payments and non-financial support on smallholder income from environmentally friendly agriculture in Tohoku region, Japan. *J. Environ. Manag.* **2024**, *351*, 119698. [CrossRef] [PubMed]
14. Karipidis, P.; Karypidou, S. Factors that Impact Farmers' Organic Conversion Decisions. *Sustainability* **2021**, *13*, 4715. [CrossRef]
15. Rust, N.A.; Stankovics, P.; Jarvis, R.M.; Morris-Trainor, Z.; de Vries, J.R.; Ingram, J.; Mills, J.; Glikman, J.A.; Parkinson, J.; Toth, Z.; et al. Have farmers had enough of experts? *Environ. Manag.* **2022**, *69*, 31–44. [CrossRef]
16. Quoc, T.N.; Tiet, T. The role of peer influence and norms in organic farming adoption: Accounting for farmers' heterogeneity. *J. Environ. Manag.* **2022**, *320*, 115909. [CrossRef]
17. David, L.; Streith, M.; Michaud, A.; Dambrun, M. Organic and Conventional Farmers' Mental Health: A Preliminary Study on the Role of Social Psychological Mediators. *Sustainability* **2024**, *16*, 1926. [CrossRef]
18. Liu, H.C.; Jiang, M.M.; Li, Q.W. Nonstereoselective dissipation of sulfoxaflor in different Puer tea processing. *Food Sci. Nutr.* **2020**, *8*, 4929–4935. [CrossRef]
19. Wang, S.A.; Qiu, Y.; Gan, R.Y.; Zhu, F. Chemical constituents and biological properties of Pu-erh tea. *Food Res. Int.* **2022**, *154*, 110899. [CrossRef]
20. Qiao, Y.H.; Halberg, N.; Vaheesan, S.; Scott, S. Assessing the social and economic benefits of organic and fair trade tea production for small-scale farmers in Asia: A comparative case study of China and Sri Lanka. *Renew. Agric. Food Syst.* **2016**, *31*, 246–257. [CrossRef]

21. Zhen, H.Y.; Qiao, Y.H.; Ju, X.H.; Hashemi, F.; Knudsen, M.T. Organic conversion tea farms can have comparable economic benefits and less environmental impacts than conventional ones-A case study in China. *Sci. Total Environ.* **2023**, *877*, 162698. [CrossRef] [PubMed]
22. Doanh, N.K.; Thuong, N.T.T.; Heo, Y. Impact of Conversion to Organic Tea Cultivation on Household Income in the Mountainous Areas of Northern Vietnam. *Sustainability* **2018**, *10*, 4475. [CrossRef]
23. Zheng, R.R.; Ma, Y.L.; Liu, L.X.; Jiang, B.Y.; Ke, R.M.; Guo, S.S.; He, D.C.; Zhan, J.S. Synergistic Improvement of Production, Economic Return and Sustainability in the Tea Industry through Ecological Pest Management. *Horticulturae* **2022**, *8*, 1155. [CrossRef]
24. Deka, N.; Goswami, K. Economic sustainability of organic cultivation of Assam tea produced by small-scale growers. *Sustain. Prod. Consum.* **2021**, *26*, 111–125. [CrossRef]
25. Karki, L.; Schleenbecker, R.; Hamm, U. Factors influencing a conversion to organic farming in Nepalese tea farms. *J. Agric. Rural Dev. Trop. Subtrop.* **2011**, *112*, 113–123.26.
26. Martens, B.V.D.; Goodrum, A.A. The diffusion of theories: A functional approach. *J. Am. Soc. Inf. Sci. Technol.* **2006**, *57*, 330–341. [CrossRef]
27. Parra-Lopez, C.; De-Haro-Giménez, T.; Calatrava-Requena, J. Diffusion and adoption of organic farming in the southern Spanish olive groves. *J. Sustain. Agric.* **2007**, *30*, 105–151. [CrossRef]
28. Li, J.; Jiang, R.; Tang, X.Y. Assessing psychological factors on farmers' intention to apply organic manure: An application of extended theory of planned behavior. *Environ. Dev. Sustain.* **2024**, *26*, 2467–2491. [CrossRef]
29. Issa, I.; Hamm, U. Adoption of Organic Farming as an Opportunity for Syrian Farmers of Fresh Fruit and Vegetables: An Application of the Theory of Planned Behaviour and Structural Equation Modelling. *Sustainability* **2017**, *9*, 2024. [CrossRef]
30. Jiang, Y.H.; Arafat, Y.; Letuma, P.L.; Ali, L.; Tayyab, M.; Waqas, M.; Li, Y.C.; Lin, W.W.; Lin, S.; Lin, W.X. Restoration of Long-Term Monoculture Degraded Tea Orchard by Green and Goat Manures Applications System. *Sustainability* **2019**, *11*, 1011. [CrossRef]
31. Badgley, C.; Moghtader, J.; Quintero, E.; Zakem, E.; Chappell, M.J.; Avilés-Vázquez, K.; Samulon, A.; Perfecto, I. Organic agriculture and the global food supply. *Renew. Agric. Food Syst.* **2007**, *22*, 86–108. [CrossRef]
32. Wang, Y.C.; Chen, C.T.; Li, R.Y.; Lu, Y.H.; Chiang, L.C. Climate risk analysis of low-altitude tea gardens in central Taiwan using a Bayesian network. *Environ. Monit. Assess.* **2024**, *196*, 809. [CrossRef] [PubMed]
33. Xie, S.W.; Feng, H.X.; Yang, F.; Zhao, Z.D.; Hu, X.D.; Wei, C.Y.; Liang, T.; Li, H.T.; Geng, Y.B. Does dual reduction in chemical fertilizer and pesticides improve nutrient loss and tea yield and quality? A pilot study in a green tea garden in Shaoxing, Zhejiang Province, China. *Environ. Sci. Pollut. Res.* **2019**, *26*, 2464–2476. [CrossRef] [PubMed]
34. Yang, X.K.; Chen, Q.H.; Lin, N.M.; Han, M.Z.; Chen, Q.; Zheng, Q.Q.; Bin, G.; Liu, F.B.; Xu, Z.Y. Chinese consumer preferences for organic labels on Oolong tea: Evidence from a choice experiment. *Int. Food Agribus. Manag. Rev.* **2021**, *24*, 545–561. [CrossRef]
35. Qiao, Y.H.; Martin, F.; Cook, S.; He, X.Q.; Halberg, N.; Scott, S.; Pan, X.H. Certified Organic Agriculture as an Alternative Livelihood Strategy for Small-scale Farmers in China: A Case Study in Wanzai County, Jiangxi Province. *Ecol. Econ.* **2018**, *145*, 301–307. [CrossRef]
36. Phromthep, P.; Torut, B. Comparing the Collaboration of Smallholder Farmers through Participatory Guarantee System Practices in Northeastern Thailand. *Sustainability* **2024**, *16*, 4186. [CrossRef]
37. Tshotsho; Lippert, C.; Zikeli, S.; Krimly, T.; Barissoul, A.; Feuerbacher, A. The role of management and farming practices, yield gaps, nutrient balance, and institutional settings in the context of large-scale organic conversion in Bhutan. *Agric. Syst.* **2024**, *220*, 104057. [CrossRef]
38. Wu, S.; Li, S.W. Collaboration to Address the Challenges Faced by Smallholders in Practicing Organic Agriculture: A Case Study of the Organic Sorghum Industry in Zunyi City, China. *Agriculture* **2024**, *14*, 726. [CrossRef]
39. Bloom, S.M.; Duram, L.A. A framework to assess state support of organic agriculture. *J. Sustain. Agric.* **2007**, *30*, 105–123. [CrossRef]
40. Sano, D.; Prabhakar, S. Some Policy Suggestions for Promoting Organic Agriculture in Asia. *J. Sustain. Agric.* **2010**, *34*, 80–98. [CrossRef]
41. Youngberg, G.; DeMuth, S.P. Organic agriculture in the United States: A 30-year retrospective. *Renew. Agric. Food Syst.* **2013**, *28*, 294–328. [CrossRef]
42. Barbosa, M.W. Government Support Mechanisms for Sustainable Agriculture: A Systematic Literature Review and Future Research Agenda. *Sustainability* **2024**, *16*, 2185. [CrossRef]
43. Martínez-Moreno, M.M.; Buitrago, E.M.; Yñiquez, R.; Puig-Cabrera, M. Circular economy and agriculture: Mapping circular practices, drivers, and barriers for traditional table-olive groves. *Sustain. Prod. Consum.* **2024**, *46*, 430–441. [CrossRef]
44. Aznar-Sánchez, J.A.; Velasco-Muñoz, J.F.; López-Felices, B.; del Moral-Torres, F. Barriers and Facilitators for Adopting Sustainable Soil Management Practices in Mediterranean Olive Groves. *Agronomy* **2020**, *10*, 506. [CrossRef]
45. Miah, M.A.M.; Bell, R.W.; Haque, E.; Rahman, M.W.; Sarkar, M.A.R.; Rashid, M.A. Conservation agriculture practices improve crop productivity and farm profitability when adopted by Bangladeshi smallholders in the Eastern Gangetic Plain. *Outlook Agric.* **2023**, *52*, 11–21. [CrossRef]
46. Simin, M.T.; Glavas-Trbic, D.; Petrovic, M.; Komaromi, B.; Vukelic, N.; Radojevic, V. Can organic agriculture be competitive? *Custos Agron. Line* **2020**, *16*, 429–444.

47. Zander, K.; Nieberg, H.; Offermann, F. Financial relevance of organic farming payments for Western and Eastern European organic farms. *Renew. Agric. Food Syst.* **2008**, *23*, 53–61. [CrossRef]

48. Breustedt, G.; Latacz-Lohmann, U.; Tiedemann, T. Organic or conventional? Optimal dairy farming technology under the EU milk quota system and organic subsidies. *Food Policy* **2011**, *36*, 223–229. [CrossRef]

49. Zhang, Y.N.; Gao, X.P.; Sun, B.W.; Liu, X.B. Hydrodynamics, Diagenesis and Hypoxia Variably Drive Benthic Oxygen Flux in a River-Reservoir System. *Water Resour. Res.* **2024**, *60*, e2023WR035449. [CrossRef]

50. Pawlewicz, A. Change of Price Premiums Trend for Organic Food Products: The Example of the Polish Egg Market. *Agriculture* **2020**, *10*, 35. [CrossRef]

51. Emeana, E.M.; Trenchard, L.; Dehnen-Schmutz, K.; Shaikh, S. Evaluating the role of public agricultural extension and advisory services in promoting agro-ecology transition in Southeast Nigeria. *Agroecol. Sustain. Food Syst.* **2019**, *43*, 123–144. [CrossRef]

52. Jin, G.Q.; Chen, H.X.; Zhang, Z.T.; Jiang, Q.H.; Liu, Z.Y.; Tang, H.W. Transport of Phosphorus in the Hyporheic Zone. *Water Resour. Res.* **2022**, *58*, e2021WR031292. [CrossRef]

53. McCann, E.; Sullivan, S.; Erickson, D.; DeYoung, R. Environmental awareness, economic orientation, and farming practices: A comparison of organic and conventional farmers. *Environ. Manag.* **1997**, *21*, 747–758. [CrossRef] [PubMed]

54. Sooriyakumar, K.; Sivashankar, S.; Sireeranhan, A. Farmers' willingness to pay for the ecosystem services of organic farming: A locality study in Valikamam area of Sri Lanka. *Appl. Ecol. Environ. Res.* **2019**, *17*, 13803–13815. [CrossRef]

55. Kondylis, F.; Mueller, V.; Zhu, S. Measuring agricultural knowledge and adoption. *Agric. Econ.* **2015**, *46*, 449–462. [CrossRef]

56. Wu, J.P.; Qiu, G.Q.; Wu, C.M.; Jiang, W.W.; Jin, J.H. Federated learning for network attack detection using attention-based graph neural networks. *Sci. Rep.* **2024**, *14*, 19088. [CrossRef]

57. Qiao, D.; Li, N.J.; Cao, L.; Zhang, D.S.; Zheng, Y.A.; Xu, T. How Agricultural Extension Services Improve Farmers' Organic Fertilizer Use in China? The Perspective of Neighborhood Effect and Ecological Cognition. *Sustainability* **2022**, *14*, 7166. [CrossRef]

58. Koppenberg, M. Markups, organic agriculture and downstream concentration at the example of European dairy farmers. *Agric. Econ.* **2023**, *54*, 161–178. [CrossRef]

sustainability

Essay

Empirical Research on Factors Influencing Chinese Farmers' Adoption of Green Production Technologies

Xiaojuan Fan [1,*], Guanghui Meng [2] and Qingming Zhang [3]

[1] College of Public Administration, Shandong Agricultural University, Taian 271018, China
[2] College of Economics and Management, Shandong Agricultural University, Taian 271018, China; mengguanghui@sdau.edu.cn
[3] School of Law, Shandong University of Technology, Taian 271000, China; alanzhangchina@hotmail.com
* Correspondence: fanxiaojuan@sdau.edu.cn

Abstract: During a critical period of structural reform in China's agricultural supply chain, accelerating the promotion and application of green production technologies emerges as a pivotal strategy to ensure the quality and safety of agricultural products while advancing agricultural modernization. This study empirically examines the factors influencing farmers' adoption of green production technologies using an ordered logistic model based on survey data collected from 533 respondents in Shandong Province. The survey targeted regions where major economic crops such as corn and soybeans are cultivated, employing simple random sampling to ensure the data's representativeness and reliability. The findings underscore several critical factors influencing farmers' willingness to adopt green production technologies, including the presence of quality inspections, evaluations of restrictions on prohibited pesticide use, sales performance of green products, availability of government subsidies, and traceability of agricultural products. To foster greater adoption of green production technologies and propel the transformation of China's agriculture, it is recommended to advocate and guide green agricultural practices, enhance green agricultural subsidy policies, and strengthen agricultural product market management systems. These measures are essential for ensuring sustainable agricultural development in China.

Keywords: green production technology; willingness to adopt; quality and safety; affecting factors

Citation: Fan, X.; Meng, G.; Zhang, Q. Empirical Research on Factors Influencing Chinese Farmers' Adoption of Green Production Technologies. *Sustainability* **2024**, *16*, 5657. https://doi.org/10.3390/su16135657

Academic Editors: Fotios Chatzitheodoridis, Efstratios Loizou and Achilleas Kontogeorgos

Received: 29 May 2024
Revised: 25 June 2024
Accepted: 30 June 2024
Published: 2 July 2024

1. Introduction

Since the 18th National Congress of the Communist Party of China, based on the basic national policy of resource conservation and environmental protection, the Chinese government has vigorously promoted the construction of an ecological civilization and has advocated for the upgrading of the agricultural industry structure and the practice of green production development concepts. However, the traditional resource-consuming production methods have long caused increasing concerns regarding non-point source pollution in agriculture and food safety, with ecological and environmental issues becoming increasingly prominent. Currently, China is at a critical period of agricultural supply-side structural reforms, thus necessitating effective measures to improve the agricultural production environment to ensure the safety and quality of agricultural products, thereby achieving a harmonious development of agriculture and rural areas that is environmentally friendly. Promoting green agricultural production technologies is not only an important measure to alleviate environmental pollution and transform agricultural production methods but also a main focus of the national ecological civilization construction and agricultural supply-side structural reforms.

The Ministry of Agriculture and Rural Affairs in China, along with other departments, has issued a series of important policy decisions that provide robust support for the adoption of green production technologies. However, in practice, farmers are the key entities in adopting these technologies, and their behaviors are influenced by various factors.

Among these factors, farmers' awareness of and willingness to adopt green production technologies are critical prerequisites for the successful promotion of these technologies.

Currently, Chinese agriculture primarily relies on traditional, resource-intensive, and extensive management practices. The fundamental transformation towards greener methods hinges on increasing farmers' willingness to adopt green production technologies. Therefore, analyzing the subjective and objective factors influencing farmers' willingness to adopt these technologies as well as understanding the mechanisms behind each influencing factor hold significant theoretical and practical relevance. This analysis is essential for promoting green production technologies and transforming agricultural production methods.

Green agricultural production technologies are crucial tools for the development of modern agriculture in China [1]. However, the adoption of these technologies by farmers has stagnated in practice, severely hindering the progress of modern agriculture. Consequently, many scholars have conducted research on the willingness to adopt green production technologies. Yi Funan's empirical analysis [2] revealed that over 90% of farmers are willing to adopt green planting technologies. Lu Lei [3] used the Heckman two-stage model to explore the impact of information access channels on farmers' adoption of green agricultural technologies, concluding that modern media, technical training, and communication among production and business entities significantly promote the adoption intensity of green agricultural technologies among farmers. Chen Nan [4] found that village regulations and climate awareness levels have a significant positive impact on farmers' adoption of green production technologies, and factors such as gender and age also play important roles. Li Yanzi et al. [5] developed a multi-tiered structural model for the adoption of green industry technologies, discovering that farmers' education level, family nature, and the intensity of government regulation are deep-seated influencing factors in technology adoption. Zhang Huiyi [6] explained the mechanism of farmers' adoption of green pest-control technologies and empirically analyzed the influence of government intervention and market incentives on the adoption level of these technologies by farmers. Xiong Ying et al. [7] analyzed the game between relevant stakeholders in the process of adopting green production technologies and found that the main reason for farmers' lack of motivation to adopt green pest-control technologies is the lack of high-quality but low-priced safe agricultural products under market information asymmetry. Additionally, the occurrence of moral hazard among farmers due to market failures also leads to their low willingness to adopt green pest-control technologies. Yao Xing'an et al. [8] believed that although government subsidies can significantly promote farmers' adoption of green agricultural technologies, the risks associated with these technologies can also hinder their adoption. Zhou Li et al. [9] studied farmers' adoption behavior regarding green agricultural technologies by taking soil testing and formulated fertilization technology as an example. They found that whether farmers adopt formulated fertilization is mainly influenced by technology promotion measures and also discovered that labor-supply factors play an important intermediary role. Fan Tianwei et al. [10], through field surveys, found that land trusteeship services can significantly promote farmers' adoption of green agricultural production technologies and increase the probability of non-green production farmers adopting green production technologies. Wang Ruonan et al. [11] believed that demonstration effects and policy support measures can significantly enhance farmers' willingness and ability to adopt green production technologies and suggested a reasonable layout of agricultural technology promotion institutions to fully realize the income-increasing effects of various green production technologies.

Agricultural pollution is also a major issue faced by many other developed and developing countries, including the United States, the European Union, the Netherlands, and New Zealand [12–14]. Wauters and Mathijs [15] conducted a review of the factors influencing the adoption of green measures in the United States, Canada, and Australia. In EU countries, Dessart et al. [16] provided a policy-oriented review examining the behavioral factors influencing the adoption of various sustainable agricultural practices in Europe.

Baregheh considered that the adoption of technology is a continuous process of innovation [17]. Esser [18] proposed that the adoption process of green technologies by farmers involves a set of acceptance criteria, including technological attributes, implementation costs, estimated benefits, time requirements, and risks. According to Ajzen [19], individuals periodically adjust their attitudes, subjective norms, and perceived behavioral control based on the outcomes of their behavior. Farmers' environmental and economic attitudes as well as their sources of information significantly influence the adoption of green technologies in organic agriculture [20]. Therefore, the adoption of green production technologies by farmers primarily depends on their environmental [21,22] or environmental awareness [23,24] and economic attitudes [25–27].

At the macro level, regulatory policies rarely achieve beneficial outcomes directly [28,29]. Some scholars took a more pragmatic approach by returning to an understanding of the specific factors influencing farmers' conservation behaviors through empirical research and analysis [30].

The present article identifies family farms as a crucial component of new agricultural business entities. As primary adopters of green production technology, family farms are the entry point for analysis. Shandong Province, China's leading agricultural region, has seen rapid development of family farms in recent years. By 2021, the province nurtured 87,000 family farms, making it one of the leading regions in China for promotional efforts and well-developed supporting policies. Thus, using Shandong Province as the research sample helps to deeply and representatively reveal the factors influencing farmers' willingness to adopt green production technologies. The study targets family farms across 16 cities in Shandong Province, using a combination of onsite interviews and random field inspections to gather data. Using the ordered logistic model to analyze factors affecting farmers' willingness to adopt green production technologies, this paper proposes targeted policy recommendations. The aim is to contribute to the research on farmers' green production behaviors and provide empirical evidence for promoting green agricultural production technologies and the modernization of agriculture.

2. Materials and Methods

2.1. Methodology

This study employed an ordered logistic model to analyze survey data. The survey targeted 533 respondents in Shandong Province, focusing on regions cultivating major economic crops like corn and soybeans.

The study also utilized simple random sampling to ensure the representativeness and reliability of the data.

2.2. Specific Model Used

The authors utilized an ordered logistic model to examine factors influencing farmers' adoption of green production technologies. This model is suitable for analyzing relationships between multiple independent variables and an ordered categorical dependent variable (such as different levels of farmers' adoption of green production technologies).

3. Data Selection and Model Construction

This paper selects the Shandong Province Family Farm Survey as the data support, combined with the literature and practical analysis; selects appropriate influencing variables; and establishes an ordered logistic model to analyze the factors influencing farmers' willingness to adopt green production technologies.

3.1. Data Sources

This article utilizes data from a survey conducted on family farms in 16 cities across Shandong Province from July to August 2021. The survey employed simple random sampling, with rigorous execution and verification methods of randomization. A total of 549 questionnaires were distributed to family farms engaged in planting economic crops such as corn and beans, among others. The survey yielded 533 valid responses, achieving a validity rate of 97.1%, indicating high reliability. In analyzing factors influencing farmers' willingness to adopt green production technologies, particular attention was given to the presence of outliers in some variables. In the process of selecting factors influencing farmers' willingness to adopt green production technologies, the presence of outliers in some variables was considered. To avoid interference with the experimental results, 11 invalid samples were excluded, ultimately utilizing 522 valid datasets.

The limitations and deficiencies of the study are as follows:

- Limited coverage of influencing factors due to constraints in sample selection and data collection methods;
- Inadequate consideration of economic factors, costs of technological transformation, and impacts on agricultural ecological systems;
- Potential biases or limitations in the statistical methods used for analysis;
- Insufficient depth in exploring certain variables crucial to understanding the research question;
- Challenges in generalizing findings beyond the specific context of the study area.

3.2. Model Setting

Given that this paper analyzes factors influencing farmers' willingness to adopt green production technologies, the dependent variable is the farmers' willingness to adopt these technologies. The willingness presents the five possibilities of "very unwilling", "unwilling", "neutral", "willing", and "very willing" as non-continuous ordered categorical variables. Therefore, drawing on the model used by Li Jiazhen et al. [31] in discussing the analysis of loan accessibility, an ordered logistic model was established, specifically expressed as follows:

$$\ln \frac{P(y \leq j)}{1 - P(y \leq j)} = \alpha_j + \sum_{i=1}^{n} \beta_i x_i \tag{1}$$

This can be equivalently transformed:

$$P(y \leq j \mid x_i) = \frac{l^{(\alpha_j + \sum_{i=1}^{n} \beta_i x_i)}}{1 + l^{(\alpha_j + \sum_{i=1}^{n} \beta_i x_i)}} \tag{2}$$

Y is the dependent variable representing farmers' willingness to adopt green production technologies; α_j is the constant term; x_i represents the independent variables, which are the factors influencing farmers' willingness to adopt green production technologies; β represents the regression coefficients.

3.3. Variable Selection

3.3.1. Dependent Variable

In this study, the dependent variable is farmers' willingness to adopt green production technologies. This is categorized into five levels of willingness, represented as Y = 1, 2, 3, 4, and 5, corresponding to "very unwilling", "unwilling", "neutral", "willing", and "very willing", respectively.

3.3.2. Explanatory Variables

To explain the factors influencing farmers' willingness to adopt green production technologies, this study considered the existing literature [32–34] and empirical survey data. Consequently, the variables selected for analysis include gender, age, education level, government subsidy amount, land management area, green product sales performance, presence of sales quality inspection, traceability of agricultural products, assessment of the enforcement of banned pesticide use restrictions, and membership in cooperatives. For detailed variable selection and settings, see Table 1.

Table 1. Variables and Definitions of Model Variable.

Type	Variable Name	Variable Definition	Mean	Standard Deviation
Dependent variable	Willingness to adopt green production technologies (Y)	"Very Unwilling" = 1; "Unwilling" = 2; "Neutral" = 3; "Willing" = 4; "Very Willing" = 5	4.251	0.035
Explanatory variables	Sex (X_1)	"Male" = 1; "Female" = 0	0.755	0.019
	Age (X_2)	"40 years and under" = 1; "40 to 50 years" = 2; "50 to 60 years" = 3; "Above 60 years" = 4	2.188	0.035
	Level of education (X_3)	"Junior High School and Below" = 1; "High School or Technical/Vocational School" = 2; "Community College or Technical Institute " = 3; "Bachelor's Degree and Above" = 4	2.056	0.038
	Government subsidy amount (X_4)	"No Subsidy" = 1; "0 to 50,000 yuan" = 2; "50,000 to 100,000 yuan" = 3; "Above 100,000 yuan" = 4	1.851	0.048
	Land management area (X_5)	"0 to 32.94 acres" = 1; "32.94 to 65.88 acres" = 2; "65.88 to 98.82 acres" = 3; "98.82 to 131.76 acres" = 4; "over 131.76 acres" = 5	2.307	0.061
	Sales performance of green products (X_6)	"Very Difficult" = 1; "Quite Difficult" = 2; "Moderate" = 3; "Relatively Easy" = 4; "Very Easy" = 5	2.713	0.042
	Presence of sales quality inspection (X_7)	"Yes" = 1; "No" = 0	0.854	0.015
	Traceability of agricultural products (X_8)	"Yes" = 1; "No" = 0	0.715	0.020
	Assessment of enforcement intensity against the use of banned pesticides (X_9)	"Very Small" = 1; "Somewhat Small" = 2; "Moderate" = 3; "Somewhat Large" = 4; "Very Large" = 5	3.891	0.046
	Membership in a cooperative (X_{10})	"Yes" = 1; "No" = 0	0.291	0.020

4. Empirical Analysis

The analysis began with the descriptive statistical evaluation of the selected variables, followed by testing each explanatory variable to address potential multicollinearity issues. The subsequent empirical analysis examined the mechanisms by which various factors influence farmers' willingness to adopt green production technologies. Lastly, robustness checks were conducted to ensure the authenticity and reliability of the findings.

4.1. Descriptive Statistics of Sample Variables

After excluding 11 samples with outliers, descriptive statistics were performed based on 522 valid sample data. Detailed statistical analysis results are presented in Table 2.

Table 2. Descriptive statistics of variables.

Sample Characteristics	Categorical Indicators	Frequency	Percentage	Sample Characteristics	Categorical Indicators	Frequency	Percentage
Sex (X_1)	Male	394	75.5	Sales Performance of Green Products (X_6)	Very difficult	55	10.5
	Female	128	24.5		Quite difficult	153	29.3
Age (X_2)	40 years and under	100	19.2		Moderate	218	41.8
	40 to 50 years	247	47.3		Relatively easy	79	15.1
	50 to 60 years	152	29.1		Very easy	17	3.3
	Above 60 years	23	4.4	Presence of Sales Quality Inspection (X_7)	Yes	446	85.4
Education Level (X_3)	Junior high school and below	149	28.5		No	76	14.6
	High school or vocational/technical secondary school	226	43.3	Traceability of Agricultural Products (X_8)	Yes	373	71.5
	Junior college or higher vocational education	116	22.2		No	149	28.5
	Bachelor's degree and above	31	5.9	Assessment of Enforcement Intensity against the Use of Banned Pesticides (X_9)	Very small	20	3.8
Government Subsidy Amount (X_4)	No subsidy	279	53.4		Quite small	33	6.3
	0 to 50,000 yuan	118	22.6		Moderate	105	20.1
	50,000 to 100,000 yuan	49	9.4		Relatively large	190	36.4
	Above 100,000 yuan	76	14.6		Very large	174	33.3
Land Management Area (X_5)	0 to 32.94 acres	196	37.5	Membership in a Cooperative (X_{10})	Yes	152	29.1
	32.94 to 65.88 acres	148	28.4		No	370	70.9
	65.88 to 98.82 acres	71	13.6				
	98.82 to 131.76 acres	36	6.9				
	Over 131.76 acres	71	13.6				

4.2. Multicollinearity Test

Before conducting regression analysis on the data, it is essential to perform a multicollinearity test on the variable data. If the variance inflation factor (VIF) is less than 10, it indicates that there is no collinearity issue among the independent variables; if the VIF is 10 or higher, it indicates the presence of collinearity issues. The test results are shown in Table 3, indicating that there are no multicollinearity issues among the independent variables.

Table 3. Multicollinearity Test.

Model	Unstandardized Coefficient		Standardized Coefficients	t	Significance	Collinearity Statistics	
	B	**Standard Error**	**Beta**			**Tolerance**	**VIF**
X_1	−0.031	0.080	−0.017	−0.389	0.698	0.965	1.036
X_2	0.010	0.045	0.010	0.231	0.817	0.928	1.078
X_3	0.029	0.041	0.031	0.714	0.475	0.926	1.080
X_4	0.043	0.032	0.059	1.365	0.173	0.957	1.045
X_5	0.017	0.025	0.030	0.692	0.490	0.949	1.054
X_6	0.101	0.037	0.120	2.770	0.006	0.945	1.058
X_7	0.235	0.099	0.103	2.370	0.018	0.945	1.058
X_8	0.172	0.077	0.096	2.245	0.025	0.957	1.045
X_9	0.138	0.033	0.181	4.205	0.000	0.959	1.043
X_{10}	0.073	0.075	0.041	0.964	0.336	0.985	1.015

4.3. Analysis of Factors Influencing Farmers' Willingness to Adopt Green Production Technologies

Following the successful multicollinearity test of the variable data, models were estimated using SPSS 22. An ordered logistic regression model was employed to conduct an empirical analysis of the factors influencing farmers' willingness to adopt green production technologies. Model 1 incorporates all variables into the regression analysis to assess their collective impact, while model 2 refines the approach by excluding statistically insignificant variables and reassesses the regression with the remaining significant variables. Detailed results of these analyses are presented in Table 4, providing insights into the dynamics of green technology adoption among farmers.

Table 4. Factors Influencing Farmers' Willingness to Adopt Green Production Technologies.

Item	Model 1				Model 2			
	Estimate	**Standard Error**	**Wald**	**Significance**	**Estimate**	**Standard Error**	**Wald**	**Significance**
X_1	0.062	0.202	0.094	0.759				
X_2	0.041	0.112	0.136	0.712				
X_3	0.029	0.103	0.081	0.776				
X_4	0.137	0.081	2.882	0.090 *	0.138	0.079	3.047	0.081 *
X_5	0.008	0.063	0.014	0.905				
X_6	0.242	0.092	6.819	0.009 ***	0.235	0.091	6.587	0.010 ***
X_7	0.542	0.245	4.889	0.027 **	0.559	0.244	5.238	0.022 **
X_8	0.529	0.192	7.611	0.006 ***	0.538	0.191	7.926	0.005 ***
X_9	0.420	0.083	25.524	0.000 ***	0.415	0.083	25.236	0.000 ***
X_{10}	0.177	0.190	0.876	0.349				

Note: ***, **, and *, respectively, indicate that differences are statistically significant at the 1%, 5%, and 10% levels.

Based on the regression results from model 1 and model 2, the sales performance of green products, the traceability of agricultural products, and the evaluation of enforcement intensity against the use of banned pesticides all passed the 1% significance level test. The presence of sales quality inspections passed the 5% significance level, while the amount of government subsidies passed the 10% significance level. Variables such as gender, age, level of education, land management area, and membership in cooperatives did not pass the significance tests, indicating that they do not have a significant correlation with farmers' willingness to adopt green production technologies.

(1) The sales performance of green products has a significant positive impact on farmers' willingness to adopt green production technologies, indicating that each improvement in the sales status of green agricultural products increases the probability of farmers adopting green production technologies by 23.5%. As the quality of life and health awareness improve in China, consumers in first-, second-, and third-tier cities are increasingly focusing on the green, healthy, and safe aspects of agricultural products and are willing to pay a

premium for green and organic agricultural products. As the ultimate recipients and payers for agricultural products, consumers' awareness of green agricultural products and their related actions have a strong positive influence on producers. This awareness of green and safe food fully utilizes the consumer market's role, prompting supply chain stakeholders to adopt green agricultural production technologies and produce high-quality green products.

(2) The traceability of agricultural products passed the 1% significance level test in both model 1 and model 2, with positive regression coefficients. On the one hand, utilizing traceability systems such as blockchain technology for agricultural product quality and safety can present all the relevant information about the product's production and sales to consumers, significantly gaining their trust and effectively increasing the sales ratio of green agricultural products. On the other hand, the continuous improvement and development of agricultural product quality and safety traceability systems also urges farmers towards safe production practices, enhancing the use of green production technologies, ensuring the quality and safety of agricultural products, and fostering a sense of responsibility among producers.

(3) The assessment of enforcement intensity on the use of banned pesticides has a significant positive impact on farmers' willingness to adopt green production technologies. Specifically, farmers perceive that stricter enforcement against the use of banned pesticides correlates with a stronger inclination to adopt green production techniques. The root cause of safety issues in agricultural products is the irrational application of pesticides, leading to residues that not only pose food safety risks but also cause ecological damage due to their externalities. Therefore, legal measures, as a means of controlling the quality and safety of agricultural products at the source, can effectively regulate and adjust pesticide use among farmers. This enforcement also motivates farmers to adopt green production technologies, thereby reducing the safety risks associated with agricultural products.

(4) The presence of sales quality inspections passed the 5% significance level test in both model 1 and model 2, with positive regression coefficients. Sales quality inspection, serving as the final safeguard for the quality and safety of agricultural products, is directly linked to consumer safety and the effective functioning of social order. Establishing and improving the system for quality inspection of agricultural products can urge producers and sellers to strictly adhere to production and sales standards, thereby reducing the likelihood of substandard products circulating in the market. Additionally, the enhancement of the agricultural product quality inspection system continuously encourages farmers to adopt green agricultural production technologies, reducing hazards such as pesticide residues and ensuring the quality and safety of agricultural products.

(5) The amount of government subsidies is significantly positively correlated with farmers' willingness to adopt green production technologies. Specifically, for each unit increase in government subsidy amount, the probability of farmers adopting green production technologies increases by 13.8%. The government's green agricultural subsidy system, as an effective mechanism for incentivizing the development of green agriculture, not only facilitates the implementation of green agricultural production standards and regulates the farming practices of agricultural producers but also enhances the quality and safety levels of agricultural products. Moreover, it encourages producers to actively adopt green technologies and methods of production, effectively controlling and eliminating endogenous pollution in agriculture.

4.4. Robustness Checks

The primary analysis in the preceding sections utilized an ordered logistic regression model to examine the factors influencing farmers' willingness to adopt green production technologies. To further substantiate the reliability of the model estimates, an ordered probit model was initially employed. The statistical significance of the variables remained unchanged, which lends some degree of robustness to the empirical results. Additionally, to further assess the reliability of the original model estimates, this paper employed a method of removing certain extreme sample values. Specifically, a 1% winsorization was applied to both the government subsidy amount and land management area variables to eliminate outliers. Subsequently, the model was re-applied to perform a secondary test. The results, as presented in Table 5, show that the estimated coefficients of the variables are fairly consistent, and their significance did not change substantially. Therefore, the conclusions drawn in the previous sections are robust and valid. This consistency in findings supports the robustness of the conclusions and confirms the reliability of the model estimates, ensuring their applicability in scholarly discussions on sustainability.

Table 5. Robustness test.

	Ordered Logistic	Ordered Probit	Applying a 1% Winsorization to the Government Subsidy Amount	Applying a 1% Winsorization to the Land Management Area
X_1	0.062	0.003	0.039	0.048
X_2	0.041	0.022	0.042	0.064
X_3	0.029	0.031	0.044	0.038
X_4	0.137 *	0.075 *	0.140 *	0.134 *
X_5	0.008	0.012	0.016	0.014
X_6	0.242 ***	0.154 ***	0.245 ***	0.237 **
X_7	0.542 **	0.317 **	0.531 **	0.495 **
X_8	0.529 ***	0.273 **	0.511 ***	0.415 ***
X_9	0.420 ***	0.225 ***	0.419 ***	0.526 ***
X_{10}	0.177	0.105	0.164	0.167

Note: ***, **, and *, respectively, indicate that differences are statistically significant at the 1%, 5%, and 10% levels.

5. Research Results

This study provides an analysis of the key factors influencing farmers' adoption of green production technologies in China's agricultural sector. The significant determinants we identified include the presence of quality inspections, evaluations regarding restrictions on prohibited pesticide use, the market performance of green products, availability of government subsidies, and the traceability of agricultural products.

The findings highlight the crucial role of these factors in influencing farmers' decisions towards adopting sustainable agricultural practices. Specifically, it is recommended to promote and provide guidance on green agricultural practices, strengthen subsidy policies for supporting green technologies, and enhance agricultural product market management systems. These measures are essential for driving the transformation and sustainable development of agriculture in China.

6. Policy Recommendations for Promoting Green Agricultural Production

The Chinese government actively promotes the use of green production technologies in agriculture through various regulatory mechanisms. However, as previously discussed, there are several prominent weaknesses that, unless effectively addressed, could hinder the true green transformation of Chinese agriculture. Addressing the quality and safety issues of agricultural products not only requires farmers to control their own production behaviors but also necessitates dual constraints from both external government and market forces. Only through a complementary approach of production and management can farmers be motivated to adopt green agricultural production technologies and ensure the quality and safety of agricultural products. The specific recommendations are as follows.

6.1. Improving Green Certification Standards and Systems

The premium price of green products can motivate farmers to adopt green production technologies and safer pesticide practices. However, "free-riding" often results in a lemon market, which undermines consumer confidence in green agricultural products. To address this issue, it is crucial to strengthen quality certification management, improve certification systems, and ensure the quality of certified products. These measures will effectively enhance the authority and credibility of product certifications, thereby boosting consumer confidence. Additionally, developing a comprehensive green standards system will ensure that Chinese agricultural products are recognized as green and sustainable in the international market. Furthermore, optimizing market environments and standardizing trading mechanisms can expand the market for green agricultural products. By leveraging consumer demand, differentiated pricing strategies can incentivize farmers to adopt green production methods [33] and supply high-quality, safe green agricultural products to the market.

6.2. Enhancing the Traceability System for Agricultural Products and Increasing Transparency of Production Information

To improve the construction of databases that manage information throughout the production and processing stages of agricultural products, it is essential to record and manage quality and safety data for agricultural products before they enter the market. This effort aims to create a traceable data chain within the agricultural product supply chain, promoting the interoperability and sharing of traceability data. Additionally, efforts should be intensified to educate and promote awareness of agricultural product quality and safety traceability among grassroots rural farmers, effectively enhancing the public's awareness and trust in the traceability system. This will guide consumers to strengthen their understanding and consciousness of food quality and safety, ultimately enhancing consumer confidence and food security.

6.3. Synchronously Enhancing Regulation of Banned Pesticides and Promoting Green Production

Through guidance and publicity by relevant governmental departments, the awareness and attitudes of farmers towards green agricultural production are enhanced [35], thereby increasing their inclination towards adopting green production methods. Simultaneously, government agencies intensify enforcement efforts, strengthening the regulation of pesticide production and sales. This includes further specifying the applicable scopes of various pesticides to maximize the safety of their use. Additionally, by publicizing the environmental benefits of green agricultural practices and highlighting the adverse effects of excessive use of fertilizers and pesticides, the role of green agricultural production bases as models is reinforced. This helps farmers tangibly perceive the economic and environmental benefits of green agricultural practices, reducing their risk perception associated with green production and increasing their willingness to adopt such methods.

6.4. Innovating Subsidy Approaches through Enhanced Transparency and Traceability

Government subsidies support farmers adopting sustainable and environmentally friendly production methods. This initiative encourages agricultural professionals to collaborate in promoting green production practices, aiming to elevate the industry's overall standards for green production. A traceable subsidy system was established to ensure that all allocations and utilizations of subsidies are clearly documented, aligning with the WTO's requirements for subsidy transparency and preventing international trade disputes due to unclear subsidy policies. Enhancing the transparency of subsidy policies includes the public disclosure of information such as subsidy amounts, beneficiaries, and procedural details, ensuring that all subsidy actions comply with the provisions of the SCM Agreement. Concurrently, regular audits by independent third-party entities ensure that the distribution and use of subsidies adhere to established policies specifically targeting green production, further affirming the commitment to sustainability and regulatory compliance.

6.5. Innovating Subsidy Methods through Enhanced Transparency and Traceability

Government subsidies [36,37] as well as green production risk compensation mechanisms [38] are provided to support farmers adopting sustainable and environmentally friendly production methods, encouraging agricultural practitioners to unite and collectively promote green production models, thus elevating the industry's overall green production standards. Establishing a traceable subsidy system ensures that all subsidy flows and utilizations are clearly recorded, meeting the WTO's requirements for subsidy transparency and averting international trade disputes due to ambiguous subsidy policies. Enhancing the transparency of subsidy policies, including the public disclosure of subsidy amounts, beneficiaries, and procedural details, ensures that all subsidy measures comply with the stipulations of the SCM Agreement. Concurrently, regular audits conducted by independent third-party institutions verify that the disbursement and utilization of subsidies adhere to the established policies for green production. This approach not only fosters transparency but also reinforces the credibility and efficacy of the subsidy framework in promoting environmental sustainability.

7. Conclusions

This study provides a comprehensive analysis of the actual attitudes of Chinese farmers, represented by major agricultural provinces, towards adopting green production technologies, along with the diverse influencing factors. It offers empirical evidence and policy recommendations for advancing China's agricultural modernization and aligning with global agricultural sustainability goals. Through rigorous data analysis and valuable conclusions, the research serves as a crucial reference for optimizing strategies to promote green agricultural technologies in China, enhancing food safety standards, and fostering sustainable agricultural development. It contributes significantly to practical applications and advocacy for green policies.

Author Contributions: Conceptualization, X.F. and G.M.; formal analysis, investigation, and methodology, G.M. and X.F.; supervision, Q.Z.; writing—original draft preparation, G.M. and X.F.; writing—review and editing, X.F. and Q.Z. All authors have read and agreed to the published version of the manuscript.

Funding: This research did not receive any external funding.

Institutional Review Board Statement: Approval for the study was not required in accordance with China's legislation, The Personal Information Protection Law, and citing relevant legislation (Accessed on 1 November 2021), Interim Measures for Scientific and Technological Ethics Review (Accessed on 1 December 2023).

Informed Consent Statement: Informed consent was obtained from all subjects involved in the study.

Data Availability Statement: The data presented in this study are available upon request from the authors.

Conflicts of Interest: The authors declare no conflicts of interest.

References

1. Huang, Y.; Luo, X.; Li, R.; Zhang, J. Farmers' cognition, external environment, and willingness to produce green agriculture—Based on survey data from 632 farmers in Hubei Province. *Resour. Environ. Yangtze Basin* **2018**, *27*, 680–687.
2. Yi, F.; Wang, F.; Ke, Y.; Bai, Z.; Chen, S. Research on Farmers' Cognition, Economic Incentives, and Adoption Behavior of Green Pest Control Technologies: Based on Survey Data from 347 Betel Nut Plantation Households in Wanning City, Hainan Province, China. *For. Econ.* **2023**, *7*, 52–53.
3. Hou, X.; Liu, T.; Huang, T. Farmers' Adoption of Green Agricultural Technologies and Income Effects. *J. Northwest AF Univ.* **2019**, *3*, 121–122.
4. Chen, N. Factors influencing farmers' green production behavior from the perspective of rural revitalization. *Agric. Technol.* **2021**, *41*, 144–149.
5. Li, Y.; Bai, J.; Wang, J.; Liu, F. Analysis of factors influencing the advancement of high-quality wheat green industry technologies—A case study of high-quality wheat production areas in Hebei Province. *Sci. Technol. Manag. Res.* **2020**, *40*, 240–248.
6. Zhang, H. Analysis of the impact of government intervention and market incentives on farmers' adoption of green pest control technologies. *Fujian Tea* **2020**, *42*, 55–56.
7. Xiong, Y.; Guo, Y. Game analysis of relevant stakeholders in the adoption of green control technologies. *Agric. Prod. Qual. Saf.* **2019**, *3*, 87–92.

8. Yao, X.; Nie, Z. Study on the behavior of empty-nest farmers adopting green agricultural technologies from the perspective of green agricultural subsidies. *J. Xinyang Agric. For. Univ.* **2017**, *3*, 20–24.

9. Zhou, L.; Feng, J.; Cao, G. Study on the Adoption Behavior of Green Agricultural Technology by Farmers—A Survey of Farmers in Hunan, Jiangxi, and Jiangsu. *Rural. Econ.* **2020**, *3*, 93–101.

10. Fan, T.; Yu, Y. The Impact of Land Trusteeship on Farmers' Adoption of Green Agricultural Production Technologies—A Case Study of M Village in Shandong Province. *Anhui Agric. Sci. Bull.* **2021**, *3*, 13–16.

11. Wang, R.; Han, X.; Cui, M.; Zheng, F. The Income-Enhancing Effect of Farmers' Adoption of Green Production Technologies: A Perspective from Quality Economics. *Res. Agric. Mod.* **2021**, *3*, 462–473.

12. Heggie, K.; Savage, C. Nitrogen yields from New Zealand coastal catchments to receiving estuaries. *N. Z. J. Mar. Freshw. Res.* **2009**, *43*, 1039–1052. [CrossRef]

13. Ongley, E.D. *Control of Water Pollution from Agriculture*; FAO Irrigation and Drainage Paper 55; Food and Agriculture Organization of the United Nations: Rome, Italy; GEMS/Water Collaboration Centre Canada Centre for Inland Waters: Burlington, ON, Canada, 1996.

14. Ongley, E.D.; Xiaolan, Z.; Tao, Y. Current status of agricultural and rural non-point source pollution assessment in China. *Environ. Pollut.* **2010**, *158*, 1159–1168. [CrossRef]

15. Wauters, E.; Mathijs, E. The adoption of farm level soil conservation practices in developed countries: A meta-analytic review. *Int. J. Agric. Resour. Gov. Ecol.* **2014**, *10*, 78–102. [CrossRef]

16. Dessart, F.G.; Barreiro-Hurlé, J.; van Bavel, R. Behavioural factors affecting the adoption of sustainable farming practices: A policy-oriented review. *Eur. Rev. Agric. Econ.* **2019**, *46*, 417–471. [CrossRef]

17. Baregheh, A.; Rowley, J.; Sambrook, S. Towards a multidisciplinary definition of innovation. *Manag. Decis.* **2009**, *47*, 1323–1339. [CrossRef]

18. Esser, K. Factors influencing the adoption of green technologies by farmers: A review. *Sustainability* **2018**, *10*, 2774.

19. Ajzen, I. The theory of planned behavior. Organ. *Behav. Hum. Decis. Process.* **1991**, *50*, 179–211. [CrossRef]

20. Serebrennikov, D.; Thorne, F.; Kallas, Z.; McCarthy, S.N. Factors Influencing Adoption of Sustainable Farming Practices in Europe: A Systemic Review of Empirical Literature. *Sustainability* **2020**, *12*, 9719. [CrossRef]

21. Läpple, D.; Van Rensburg, T. Adoption of organic farming: Are there differences between early and late adoption. *Ecol. Econ.* **2011**, *70*, 1406–1414. [CrossRef]

22. Läpple, D.; Kelley, H. Spatial dependence in the adoption of organic drystock farming in Ireland. *Eur. Rev. Agric. Econ.* **2015**, *42*, 315–337. [CrossRef]

23. Dabiah, A.T.; Alotibi, Y.S.; Herab, A.H. Attitudes of Agricultural Extension Workers toward the use of Electronic Extension Methods in Agricultural Extension in the Kingdom of Saudi Arabia. *Int. J. Agric. Biosci.* **2023**, *12*, 104–109.

24. Mzoughi, N. Farmers adoption of integrated crop protection and organic farming: Do moral and social concerns matter. *Ecol. Econ.* **2011**, *70*, 1536–1545. [CrossRef]

25. USDA NRCS. USDA NIFA Conservation Effects Assessment Project (CEAP) Fact Sheets. USDA NRCS NIFA. 2011. Available online: http://www.soil.ncsu.edu/publications/NIFACEAP (accessed on 2 February 2018).

26. Pannell, D.J.; Marshall, G.R.; Barr, N.; Curtis, A.; Vanclay, F.; Wilkinson, R. Understanding and promoting adoption of conservation practices by rural landholders. *Aust. J. Exp. Agric.* **2006**, *46*, 1407–1424. [CrossRef]

27. Greiner, R.; Gregg, D. Farmers' intrinsic motivations, barriers to the adoption of conservation practices and effectiveness of policy instruments: Empirical evidence from northern Australia. *Land Use Policy* **2011**, *28*, 257–265. [CrossRef]

28. Chouinard, H.H.; Wandschneider, P.R.; Paterson, T. Inferences from sparse data: An integrated, meta-utility approach to conservation research. *Ecol. Econ.* **2016**, *122*, 71–78. [CrossRef]

29. Savage, J.; Ribaudo, M. Improving the Efficiency of Voluntary Water Quality Conservation Programs. *Land Econ.* **2016**, *92*, 148–166. [CrossRef]

30. Liu, T.; Bruins, R.J.F.; Heberling, M.T. Factors Influencing Farmers' Adoption of Best Management Practices: A Review and Synthesis. *Sustainability* **2018**, *10*, 432. [CrossRef] [PubMed]

31. Li, J.; Li, M. The Impact of Farmers' Professional Cooperatives on the Accessibility of Financing for Farmers. *J. Jishou Univ. (Nat. Sci. Ed.)* **2019**, *40*, 83–90.

32. Peng, C. The Basic Framework of China's Agricultural Subsidies, Policy Performance, and Directions for Momentum Transformation. *Theor. Explor.* **2017**, *3*, 18–25.

33. Chen, W. Institutional Constraints and Policy Recommendations for the Green Transformation of Farmers' Production Under the Strategy of Rural Revitalization—Based on In-depth Interviews with 47 Conventional Production Farmers. *Exploration* **2018**, *3*, 136–145.

34. Yang, C.; Qi, Z.; Huang, W.; Chen, X. The Impact of Benefit Perception on Farmers' Adoption Behavior of Green Production Technologies—A Heterogeneity Analysis Based on Different Production Stages. *Resour. Environ. Yangtze Basin* **2021**, *30*, 448–458.

35. Nie, W.; Zuo, T.; Chen, J. Analysis of Factors Influencing Farmers' Perception of Agricultural Green Development and Adoption of Green Production Behaviors. *J. Northeast. Agric. Univ. (Soc. Sci. Ed.)* **2020**, *18*, 9.

36. Min, J.; Kong, X. Research Progress on Agricultural Non-Point Source Pollution in China. *J. Huazhong Agric. Univ. (Soc. Sci. Ed.)* **2016**, *2*, 59–66.

37. Yu, W.; Luo, X.; Li, R.; Xue, L.; Huang, L. Study on the Discrepancy between Willingness and Behavior of Farmers to Adopt Green Technologies from the Perspective of Green Cognition. *Resour. Sci.* **2017**, *8*, 1573–1583.
38. Zhang, M.; Zhang, C.; Li, F.; Liu, Z. Green Finance as an Institutional Mechanism to Direct the Belt and Road Initiative towards Sustainability: The Case of China. *Sustainability* **2024**, *16*, 6164. [CrossRef]

MDPI AG
Grosspeteranlage 5
4052 Basel
Switzerland
Tel.: +41 61 683 77 34

Sustainability Editorial Office
E-mail: sustainability@mdpi.com
www.mdpi.com/journal/sustainability